C0-AKZ-583

R. M. Kamp D. Kyriakidis Th. Choli-Papadopoulou (Eds.)

Proteome and Protein Analysis

Springer
Berlin
Heidelberg
New York
Barcelona
Hong Kong
London
Milan
Paris
Singapore
Tokyo

R. M. Kamp D. Kyriakidis
Th. Choli-Papadopoulou (Eds.)

Proteome and Protein Analysis

With 147 Figures

Prof. Dr. Roza Maria Kamp
Technische Fachhochschule Berlin, FBV (Biotechnologie)
Seestraße 65
D-13347 Berlin
Germany
E-mail: kamp@tfh-berlin.de

Prof. Dr. Dimitris Kyriakidis
Aristotelian University Thessaloniki
School of Chemistry, Laboratory of Biochemistry
Thessaloniki, 54006
Greece
E-mail: kyr@chem.auth.gr

Ass. Prof. Dr. Theodora Choli-Papadopoulou
Aristotelian University Thessaloniki
School of Chemistry, Laboratory of Biochemistry
Thessaloniki, 54006
Greece
E-mail: tcholi@chem.auth.gr

Selected papers presented at XIIth International Conference on Methods in Protein Structure Analysis (MPSA) September 1998, Halkidiki, Greece

ISBN 3-540-65891-2 Springer-Verlag Berlin Heidelberg New York

Library of Congress Cataloging-in-Publication Data
Proteome and protein analysis / R.M. Kamp, D. Kyriakidis, Th. Choli – Papadopoulou (eds.). p.cm. Includes bibliographical references and index.
ISBN 3-540-65891-2 (alk. paper)
1. Proteins-Analysis. I. Kamp, R. M. (Roza Maria), 1951– . II. Kyriakidis, D. (Dimitris), 1948– . III. Choli-Papadopoulou, Th. (Theodora), 1956–
QP551.P7553 1999 572'.633-dc21

Cover design: d & p, D-69121 Heidelberg
Typesetting: Mitterweger & Partner Kommunikationsgesellschaft mbH
Production: PRO EDIT GmbH, D-69126 Heidelberg
SPIN 10689115 31/3136 5 4 3 2 1 0 – Printed on acid free paper

To Don Sheer Memory

Dr. Donald Sheer and his wife, Diane, were aboard Swissair Flight 111 when it crashed off the coast of Nova Scotia on September 2, 1998. There were no survivors. The news of their untimely deaths came as a shock to all of those who knew the couple.

Don was Consulting Scientist in the Analytical Division of Millipore Corp. where he was active in the development of new separation products and applications for the research, biotech and pharmaceutical industries. Don had worked in the Millipore/ Amicon organizations for over 5 years and had contributed greatly to setting the product development targets for innovative products. Don worked closely with scientists in many fields of technology and his ability to establish collaborative projects was legendary.

Don's most recent effort was developing applications for a new Millipore product named ZipTips™, which are novel sample preparation devices containing absorptive resins that allow the

cleanup and concentration of microliter volumes of solution. Don was traveling to Halkidiki, Greece to present a paper at the XIIth International Conference on Methods in Protein Structure Analysis on this new technology.

Before Millipore, Don worked as Associate Director of Protein Chemistry at Ares Technology (Serono Labs) and as a Senior Scientist at Genesis Pharmaceuticals in the Boston area. Prior to that he worked in the Protein Applications Group at Applied Biosystems Inc. in Foster City, CA. He also spent five years as a Biotechnology Training Fellow at the National Cancer Institute in Bethesda, MD.

Don received his doctorate degree in Biochemistry from the University of Arizona (1983) and an undergraduate degree in Chemistry from Ohio State (1976).

Don was an unique character dedicated to his work and was known to many researchers around the world. He will be missed by all.

May 1999

Rick Garretson
Millipore Corporation

Preface

This book includes selected papers presented at the XIIth International Conference on Methods in Protein Structure Analysis (MPSA), which took place in Halkidiki, Greece, in September 1998, under a scientific atmosphere, good weather and the spirit of the greek philosopher Aristotle, who was born in that area of Greece. To insure the highest quality and the most rapid publication, the MPSA Book Advisory subcommittee has chosen for the book the most interesting and useful presentations from the meeting.This book includes papers covering new, sensitive and rapid methodologies, as well as various aspects of the Protein Chemistry field, reflecting the multidisciplinary nature of the problems concerning the analysis of the total cell proteins, namely proteome, a term which was first used in 1994 at the Siena 2D-Electrophoresis meeting. So, we finally decided on the title of "Proteome and Protein Analysis", which indeed corresponds to our thoughts and to the wishes of most scientists, as we believe, in the protein area.

Several methodologies from different fields, like Edman degradation, DNA sequencing, mass spectrometry have to be combined for an elaborate study of proteins' function in biological processes. For example the combination of mass spectrometry techniques with proteome analysis answers a large number of scientific questions.

Different mass spectrometry methods such as MALDI, ESI- or Nano Electrospray are described and discussed for their use in proteome and protein analysis. The importance of the mass spectrometry in proteome and protein analysis was acknowledged with the presentation of the Edman Award 1998 to three mass spectrometry experts: Matthias Mann, Matthias Wilm and John Yates.

We believe that this book contains new and interesting contributions to bioscience, describing new methodologies, new developments on structural studies, functional implication of proteins, NMR analysis, database searches of multi-alignments and mass spectrometry analysis of proteins.

The chapters have been written by scientists with extensive experience in these fields. The contributors of this volume are all active researchers in various aspects of protein study and they have attempted to guide the reader by introducing briefly their subject and giving details for experimental work, which in their own experience give good results.

We express our great sorrow at the loss of both Dr. Don Sheer, coworker of Millipore, and his wife, whose tragic death, after an airplane crash on the way from USA to Greece, was announced before the opening ceremony of the XIIth International MPSA.

We would like to acknowledge all contributors of this volume, the participants of the XIIth MPSA for their active contribution as well as Springer-Verlag for their kind offer to publish this work. We are grateful to the conference Book Advisory subcommittee Drs C.W. Anderson, E. Appella and B. Wittmann-Liebold for all their efforts to select the most important topics.

We hope that this volume will stimulate further research in the field of Protein Chemistry and be beneficial because of the new and interesting ideas it introduces.

May 1999 The Editors

Contents

Part I
Proteomics and Mass Spectrometry

CHAPTER 1

Strategies and Methods for Proteome Analysis

D. R. GOODLETT[1] A. TIMPERMAN[1], S. P. GYGI[1], J. WATTS[1], G. CORTHALS[1], D. FIGEYS[1] and R. AEBERSOLD[1]

A proteome has been defined as the protein complement expressed by the genome of an organism (Wilkins, et al. 1996). In multicellular organisms the proteome is the protein complement expressed by a tissue or differentiated cell. The most common approach to proteome analysis involves separation of proteins by one- or two-dimensional gel electrophoresis (IEF/SDS-PAGE), enzymatic cleavage of selected proteins, tandem mass spectrometry (MS/MS) of peptides, and finally data interpretation by computer routines which also search databases (Fig. 1.1). In most cases the availability of protein and DNA sequences in public databases eliminates the need for complete protein sequence analysis. Protein sequences are more rapidly identified by partial sequence analysis using tandem mass spectrometry which allows rapid, complete gene identification. As reliable as protein identification by mass spectrometry has become there are still many obstacles that prevent proteome analysis from becoming as automated and routine as genome analysis.

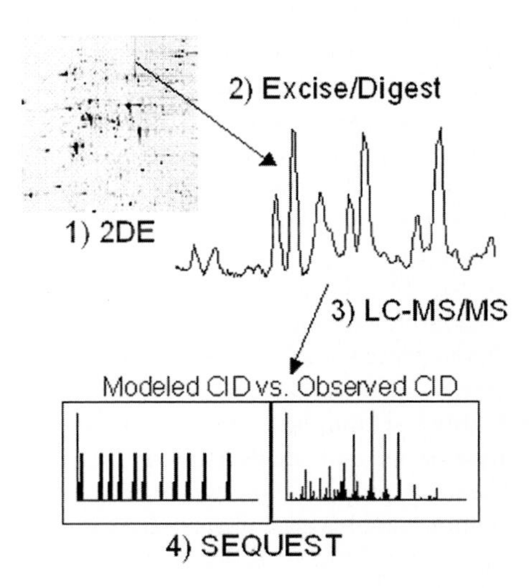

Fig. 1.1. Schematic of Proteomic Analytical Process. Proteins are 1) separated by 2DE and visualized by silver staining, 2) spots of interest are excised and digested with trypsin, 3) peptides are analyzed by automated μLC-MS/MS and 4) data screened for hits against protein or DNA databases using the database search routine SEQUEST

[1] Department of Molecular Biotechnology, University of Washington, Seattle, Washington 98195 USA. E-mail: ruedi@u.washington.edu.

Table 1.1. Comprehensive Proteome Projects

Organisms	References
Saccharomyces cerevisiae	Hodges, P.E., et al. Nucl. Acids Res. 1998, 26, 68–72.
Salmonella enterica	O'Connor, C. D. et al. Electrophoresis 1997, 18, 1483–1490.
Spiroplasma melliferum	Cordwell, S.J., et al. Electrophoresis 1997, 18, 1335–1346.
Mycobacterium tuberculosis	Urquhart, B.L., et al. Electrophoresis 1997, 18, 1384–1392.
Ochrobactrum anthropi	Wasinger, V.C., et al. Electrophoresis 1997, 18, 1373–1383.
Haemophilus influenza	Link, A. J:, et al. Electrophoresis 1997, 18, 1314–1334.
Synechocystis spp.	Sazukam T. & O. Ohara Electrophoresis, 1997, 18, 1252–1258.
Escherichia coli	van Bogelen, R. A., et al. Electrophoresis, 1997, 18, 1243–1251.
Rhizobium leguminosarum	Guerreiro, N. et al., Mol. Plant Microbe Interact., 1997, 10, 506–516.
Dictyostelium discoideum	Yan, J. X., et al. Electrophoresis 1997, 18, 491–497.
Tissues	**References**
Human bladder squamous call carcinoma, Human keratinocytes, Human fubroblasts, Mouse kidney	Celis, J. et al., FEBS Lett. 1996, 398, 129–134.
Human liver & plasma	Appel, R.D., et al. Electrophoresis 1993.
Rat serum	Haynes, P., et al., Electrophoresis 1998, 19, 1484–1492.

As listed in Table 1.1 numerous proteome projects are underway for both whole organisms and specific cell types. In contrast to genome sequencing where numerous prokaryotic and two eukaryotic genomes have been described (Goffeau, et al. 1996; Fleischmann, et al. 1995; Fraser, et al. 1997) none of the listed proteome projects is approaching completion. Difficulties which have prevented completion of proteome projects include:

1) an inability to amplify protein sequences as can be done with nucleic acid sequences,
2) post-translational modifications that in the case of regulatory proteins may be non-stoichiometric and
3) the presence of isoforms as in the case of glycoproteins.

Additional complications in proteome analysis that are more difficult to address arise because of the dynamic nature of proteomes and the variation in physicochemical properties between proteins such as pI, solubility and size. A single genome can, because of variations in cell cycle, state of differentiation or nutrient supply, give rise to different proteomes. In order to prepare reproducibly the same proteome for characterization all parameters affecting cell growth and the purification process must be controlled.

In this chapter we will present various aspects of basic proteome analysis and describe methods employed for protein characterization on a large scale as well

as for single, isolated proteins. We will discuss current limitations in the technology, tips for avoiding pitfalls and circumventing problems. Specifically we hope to provide the reader with an overview and helpful hints on the following subject matter:
- Global Protein Separation
- Sample Preparation for Mass Spectrometry
- Sample Introduction for Mass Spectrometry
- Protein Identification by Tandem Mass Spectrometry
- Challenges of Low Abundance Proteins
- Phosphopeptide Analysis.

1 Global Protein Separation

The method of choice for separation of complex protein mixtures is polyacrylamide gel electrophoresis (PAGE). Separation by PAGE not only provides resolving power but also facilitates subsequent proteolytic digestion because PAGE most frequently is done in a denaturing environment. Partially purified samples of limited complexity can be efficiently separated by one-dimensional PAGE. For complex mixtures such as whole cell lysates a two-dimensional (IEF/SDS-PAGE or 2DE) separation is usually more suitable. The 2DE method has the potential to separate several thousand proteins (Gorg, et al. 1988; Klose and Kobalz, 1995) in a single experiment and therefore provides the opportunity to detect differences in protein expression between two or more samples by comparative analysis of global protein patterns. 2DE is currently the protein separation method on which proteome analysis is based.

Visualization of proteins in gels is usually by Coomassie blue or silver staining or if the protein is radiolabeled by autoradiography. It is difficult to predict a priori which stained protein will be identifiable by MS/MS, but some generic guidelines can be stated. In general a protein visible by Coomassie blue can be easily identified by the methods outlined in this manuscript. Whereas a protein visible as a very faint silver stained spot (<5 ng of protein) may or may not be identifiable unless multiple, identical spots from separate gels are pooled. Regulatory proteins or other proteins present in cells at low abundance are not visible by silver staining of total cell lysates and are therefore impossible to identify without selective enrichment prior to gel electrophoresis.

2 Sample Preparation for Mass Spectrometry

Due to the sensitivities of current mass spectrometric techniques sample contamination is a concern. Careless sample handling or use of buffers and reagents not optimized for high sensitivity applications (Zhang et al 1998) will inevitably lead to contamination with proteins such as kerratin. It is therefore important to avoid contamination by maintaining a clean work environment, wearing gloves and using quality-assured reagents. To avoid other sources of contamination a dedicated gel dryer and gel equipment are recommended for high sensitivity protein analysis. The signal to noise ratio and therefore working sensitivity of the mass spectrometer is dependent on the purity of the reagents used. Organic contaminants such as polymers extracted from plasticware by organics such as acetonitrile and detergents left over from incompletely rinsed glassware are other common contaminants detected in the mass spectrometer. As each laboratory environment is unique, it is difficult to suggest generic rules for optimum performance. However, a laboratory wishing to apply this technology can work out by elimination which grade of chemicals, producer of labware and source of water are best suited.

3 Sample Introduction for Mass Spectrometry

Good sample handling techniques are critical for the success of any analytical project. This is especially true if only a few microliters of total volume exist and the concentration of the working sample is in the nano- to picomolar range (femto- to attomoles/microliter). The major problems are sample contamination and sample loss due to adsorption to wetted surfaces. To avoid nonspecific sample loss the number of sample manipulations between digestion and mass spectrometric analysis should be kept to a minimum. For high sensitivity applications proteolytic digests from electrophoretically separated proteins must be desalted prior to analysis. This minimizes matrix effects from salts which compete with peptides during the ionization process in both MALDI and ESI. For MALDI-TOF and nano-ESI application (Shevchenko, et al. 1996; Wilm, et al. 1996) samples are purified off-line by the use of small extraction devices such as ZIP-Tips (MilliPore). When micro-separation systems such capillary LC or CE are coupled online to ESI-MS instruments the peptide separation system serves to both, desalt/purify and separate the analytes. For CE-MS/MS the sample is pressure injected on a C18 cartridge (1 mm × 50 μm) placed at the head of the CE capillary. The sample is desalted and then eluted with a small plug of organic solvent in to the CE capillary for separation (Figeys et al. 1996; Figeys and Aebersold 1997). For μLC we use 50–100 μm i.d. polyimide coated capillary columns packed with C18 support to separate and desalt peptide digests (Lee et al. 1998).

To achieve high quality collision induced dissociation (CID) spectra from very small amounts of peptides we have modified the SPE-CE-MS/MS method (Figeys et al. 1999) by applying a variable CE voltage to decrease flow rate and thus increase the time available for MS/MS. This peak parking method with SPE-CE (Fig. 1.2) is achieved by an automated (instrument control language) decrease in the applied CE voltage which is initiated whenever the mass spectrometer detects peptide ions present above a preset signal/noise ratio (Figeys et al 1997). Simultaneously to the voltage drop the mass spectrometer is switched from scanning to CID mode. The concentration dependent nature of ESI (Goodlett, et al. 1993) allows one to fragment peptides of low abundance by reducing the flow of liquid without sacrificing sensitivity. Peak parking is advantageous because ion selection by signal/noise works well for abundant peptides in a digest but fails, due to a lack of time, to select lower abundance peptides co-eluting with abundant peptides. Therefore, if complex peptide mixtures are analyzed the lower abundance species, frequently the most interesting biologically, go undetected even though

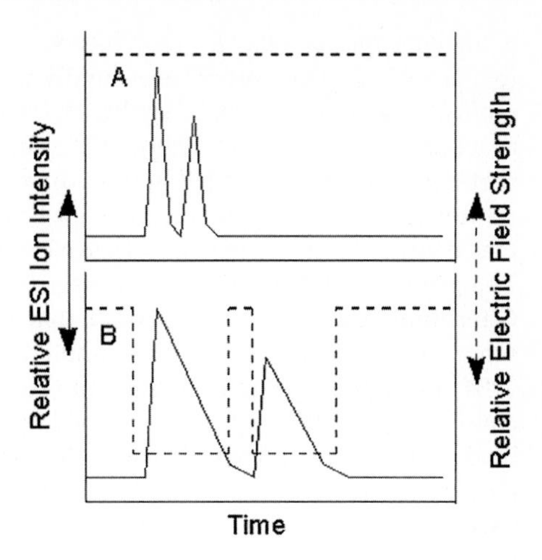

Fig. 1.2. Schematic of Voltage Drop Experiment. Capillary electrophoresis-electrospray ionization-mass spectrometry with constant (A) and variable (B) electrophoretic field strength. A computer routine automatically varies the field strength according to detected ion intensity and switches the mass spectrometer between scanning and tandem MS mode

they are within the range of detection by the mass spectrometer. As a consequence of this, a phosphopeptide generated from a phosphoprotein phosphorylated to low stoichiometry may not be chosen for CID because the non-phosphorylated peptide is present at a much higher concentration. Peak parking allows more time for fragmentation of peptides across a given chromatographic area without sacrificing sample. During μLC introduction to ESI a decrease in pressure at the pump (Davis and Lee 1997) provides the same advantage as electrophoretic peak parking.

One approach to decrease sample loss is the use of microfabricated devices (Figeys et al, 1997). For sample introduction into the mass spectrometer samples are applied to reservoirs and electroosmotically pumped to the MS for identification. The advantage of low flow rates and thus increased analysis time is similar to the Nano-Spray technique application (Shevchenko, et al. 1996; Wilm, et al. 1996) but the chip is more convenient. Several different samples can be loaded concurrently on one chip and analyzed in turn without sample-to-sample cross contamination. We are working to further reduce sample loss by carrying out the entire sample processing stage from digestion/extraction to analysis on a chip.

4
Protein Identification by Tandem Mass Spectrometry

Traditionally, gel separated proteins were identified by methods such as 1) stepwise chemical, N-terminal sequencing by Edman degradation (Matsudaira, 1987; Aebersold, et al. 1986), 2) internal sequencing (Rosenfeld, et al. 1992; Aebersold, et al. 1987) or 3) immunoaffinity techniques comparing behavior of known to unknown protein (Honore, et al. 1993). With the availability of large or even complete DNA and protein sequence databases proteins no longer need to be completely sequenced (Mann and Wilm 1994; Eng, et al. 1994; Yates, et al, 1995). Rather proteins can be identified by comparing the information generated by MS or MS/MS analysis of proteolytic protein digests with the sequence databases using appropriate database search tools. A particularly powerful database search tool is SEQUEST because it uses highly informative peptide CID spectra to automatically search sequence databases (Ducret et al. 1998). In principle and in practice (Santos, et al. 1999), an unknown protein which has never been sequenced can be identified using a single good quality CID spectrum.

Most proteins contain at least a few basic residues such as lysine and arginine that trypsin will recognize and cleave on the C-terminal side of the amide bond. The products of this specific digestion are peptides with protonation sites at the C-terminii (i.e. lysine or arginine) that can, in the absence of interference by other amino acids, generate a complete "ladder" series of y-ions (Roepstorff, 1984) in the tandem mass spectrum (Hunt et al. 1986). For this reason manual confirmation of the sequence provided by available routines like SEQUEST is also made easier when trypsin is used. Note that it is always important to manually check the answers provided by computer search routines because scores are based on a set of rigid rules rather than knowledge of protein chemistry. Other proteolytic enzymes cannot be assumed to produce such a beneficial effect, but are still useful when trypsin fails to produce peptides in the 1000–2000 Da range.

5
Challenges of Low Abundance Proteins

In yeast a total of approximately 6000 genes are present (Goffeau et al. 1996). To determine what fraction of these genes are visualized by 2DE when a total cell lysate is separated by 2DE, a recent study in our laboratory examined the codon bias value for proteins identified after separation by 2DE. The codon bias value is a calculated measure of the degree of redundancy of triplet DNA codons used to produce each amino acid in a given gene sequence. It is a useful indicator of the level of gene products in a cell (Garrels, et al. 1997). It is assumed that codon bias correlates with the protein copy number per cell, with high codon bias indicating a high level of protein expression. Codon bias values calculated for the entire yeast genome showed that most proteins have a codon bias of <0.2 (Fig. 1.3A) and are therefore of low abundance or low copy number per cell. It was shown, however, that the majority of yeast proteins that could be readily identified from silver stained 2DE gels (Gygi et al. 1999) had a codon bias value >0.2 (Fig. 1.3B). Therefore the proteins visible by our most sensitive, generic staining technique, are the most abundant proteins in yeast cells. The proteins not visible by silver staining are likely to be the very proteins of interest as targets for small molecule therapeutics. To make these proteins available for analysis will require new methods to be devised to enrich cell lysates for these proteins.

As mentioned earlier the best 2DE gels can separate in excess of several thousand proteins including regulatory proteins such as kinases, phosphatases, and growth factors. Very specific human cell types are thought to express around 50,000 to 100,000 proteins. Therefore, only a small fraction of the total protein present can be visualized by 2DE. Limitations in loading capacity and resolution of 2DE can be circumvented by fractionation prior to loading. This sort of purification could be as simple as an ammonium sulfate fractionation or bulk cation exchange prior to loading. The problem raised by an examination of codon bias

Fig. 1.3. Codon Bias Values in Yeast. Codon bias value (A) calculated for the entire yeast genome and (B) measured for all yeast proteins visualized on a silver stained 2DE gel

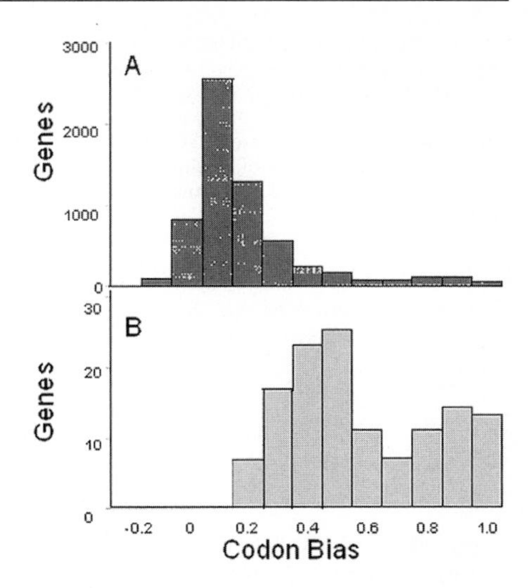

for the yeast genome is more difficult to address because it is often impossible to scale up. If antibodies or other specific reagents are available, then more specific sample enrichment can be achieved and even low abundance proteins can be isolated and identified (Fu et al. 1992; Veals et al. 1992; Carter et al. 1992; Carter et al. 1993).

6 Phosphopeptide Analysis

Identification of phosphopeptides in a mixture of predominantly non-phosphorylated peptides can be difficult because the stoichiometry of phosphorylation is often low. In these cases we prefer to work with ^{32}P labeled proteins because the radiolabel provides a convenient method for observing low abundance phosphopeptides. With a half-life of two weeks initial experiments requiring the radiolabel can be performed immediately after labeling and mass spectrometric analysis delayed until sufficient decay has occurred to avoid contaminating the mass spectrometer. Prior to mass spectrometric characterization of a phosphoprotein it is generally useful to determine the complexity of the phosphrylation events on the protein. A suitable technique for doing so is two-dimensional phosphopeptide (2DPP) mapping (Watts et al. 1994). The technique separates peptides electrophoretically on a cellulose plate and then by hydrophobicity in a second dimension using thin layer chromatography. Spots visualized by autoradiography (Fig. 1.4) can be scraped off, eluted from the cellulose and analyzed by tandem mass spectrometry (Affolter et al 1994). The number of spots visualized by autoradiography provides an estimate of the maximum number of phosphorylated sites and spot intensity provides an indication of the relative stoichiometry of phosphorylation between peptides.

To illustrate the sensitivity that can be achieved after 2DPP, the protein CD3-ζ was in vitro labeled with γ-^{32}P-ATP using the protein kinase p56lck (Watts et al

Fig. 1.4. Two Dimensional Phosphopeptide Map of CD3-ζ peptides. After in vitro labeling CD3-z was digested with trypsin. Recovery of ^{32}P-peptides from the plate ranged from 55–95 %. As assessed by radioactive decay from the date of labeling, spots 2 and 3 contained 8.0 and 1.5 picomoles of peptide respectively

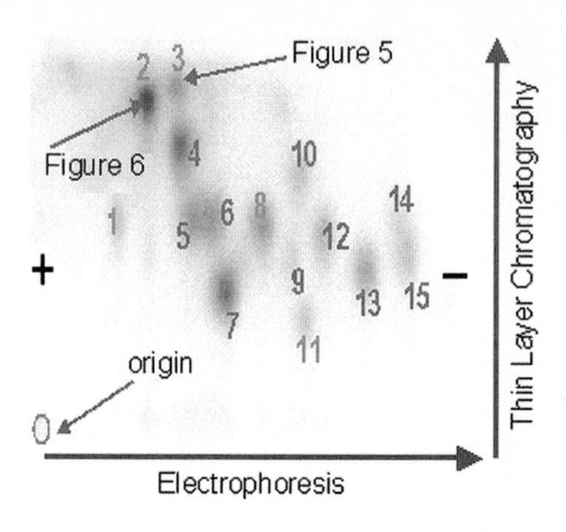

1994). The protein was digested with trypsin, peptides were separated by 2DPP mapping, extracted and quantitated by radioactive decay. Spot 3 (Fig. 1.4) was analyzed by μLC/MS/MS. Injecting 10 % of the extract from spot 1 (150 femtomoles of peptide) produced the spectrum shown in Fig. 1.5. Injecting more than 10 % of a spot often resulted in increased chemical noise originating from the cellulose slurry and failure to observe any peptide signal. The quality of the data was good enough to enable the phosphorylation site to be unambiguously identified within the known amino acid sequence of the protein. Spot 2 was analyzed twice so that a comparison could be made between in-source CID (Fig. 1.6.A) and tandem CID (Fig. 1.6.B). These two different types of fragmentation produce spectra that, when compared, can aid in locating phosphorylated residues. As can be seen in Fig. 1.6A in-source CID produces a cleaner looking spectrum from dissociation of high charge states rather than from a single selected charge state. If suf-

Fig. 1.5. Tandem MS of $[M + 3H]^{3+}$ CD3-z peptide (DTpYDALHMQTLAPR) taken from spot 3 in Fig. 1.4. Data acquired using an auto-MS/MS routine that initiated tandem MS based on preset signal intensity. To ensure acquisition of interpretable data in automated mode the instrument control language routine changes $V_{COFF} = [M + 2H]^{2+}/\text{-}25.5$ by about–1.5V per tandem MS scan. Ten % of spot 3 or 150 femtomoles analyzed by μLC-MS/MS

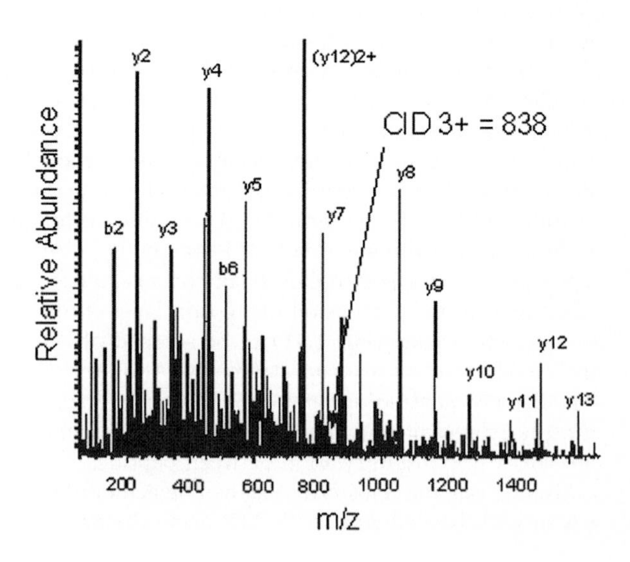

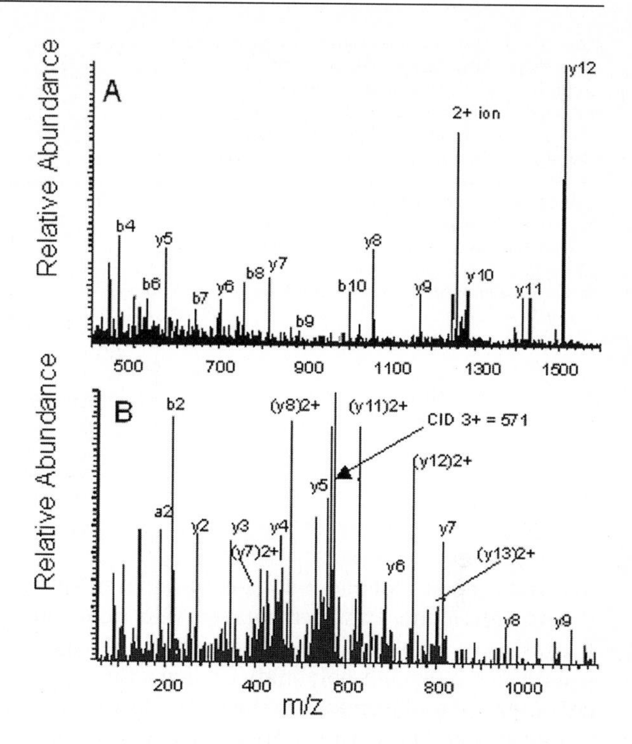

Fig. 1.6. CD3-ζ peptide (SAETAANLQDPNQLpY-NELNLGR) taken from spot 3 in Fig. 1.4 was subjected to μLC A) in source CID where V_{APICID} = -40V and B) tandem MS as described in Fig. 1.5. In both cases 800 fmoles or 10 % of spot 3 was used

ficient sample is available in-source CID is a route for MS/MS/MS experiments on a triple quadrupole mass spectrometer. This would be necessary if the CID spectrum did not lead to unambiguous assignment of the phosphorylation site. A given fragment ion produced by in-source CID is selected by quadrupole one for CID in quadrupole 2. A MS/MS/MS spectrum is produced of ions that formed during in-source CID, were selected by Q1 and fragmented in Q2.

In the absence of a radiolabel diagnostic ions generated by in-source CID can be monitored to identify phosphopeptides eluting from HPLC columns (Carr et al. 1996; Huddelston et al. 1993; Hunter and Games 1994). For phosphopeptides these diagnostic ions are $H_2PO_4^-$ (97u) which results from beta-elimination, $H_2PO_3^-$ (79u) and PO_3^- (63u). When in-source fragmentation (CID) is combined with on-line HPLC a chromatographic trace identifying where a phosphopeptide eluted is produced.

An example of how a diagnostic scan can be useful to identify phosphopeptides in peptide mixtures is shown in Fig. 1.7. After SDS-PAGE of whole yeast lysate a 32 kDa silver stained band was excised and digested with trypsin. To this was added one picomole of phosphorylated angiotensin II (DRApYIHPF). The mixture was pressure injected on a 50 μm id capillary column packed with C18 beads and eluted with a linear gradient of acetonitrile. The triple quadrupole mass spectrometer was set up in negative ion mode and peptides fragmented by in-source CID (Hunter and Games 1994). The mass spectrometer was programmed to automatically cycle between three scan types, which collectively detected the phosphopeptides eluting from the HPLC column and determined

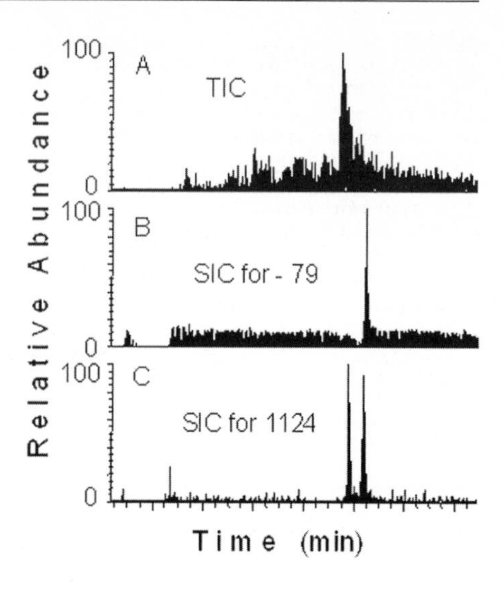

Fig. 1.7. Phosphopeptide diagnostic ion experiment showing A) total ion chromatogram, B) single ion chromatogram for 79 m/z and C) single ion chromatogram for 1124 m/z. One picomole of phosphorylated angiotensin II ($[M-H]^{1-}$ = 1124) was added to the tryptic digest of a 32 kDa silver stained band from a gel where whole yeast lysate was analyzed. The entire mixture was analyzed in negative ion mode by a phosphate diagnostic scan as described in the body of the text. For in source CID V_{APICID} = +70V

their m/z values. There are no special requirements for operating the system in negative ion mode. The microsprayer is the same one used for positive ion work. It consists of a Valco union into which the separation capillary terminates and from which a tapered polyimide capillary serves as the ESI emitter. Voltage for ESI is applied to the union. On the first of three scan types the mass spectrometer fragmented peptides by in-source CID and performed multiple ion monitoring to observe loss of -63, -79 and -97 which served as chromatographic markers for phosphopeptides. During the second scan type the in-source CID continued but now one of the quadrupoles was scanned thus acquiring in negative ion mode the masses of all ions passing through quadrupole 1 at that time. This scan distinguishes between phosphopeptide and non-phosphopepitde because phosphopeptides exhibit, in addition to the $[M-H]^{1-}$ ion, some combination of the following ions: $[M-H_2PO_4]^{1-}$, $[M-H_2PO_3]^{1-}$ and $[M-PO_3]^{1-}$. Finally, the source CID was turned off and a final scan acquired which confirms assignment of the $[M-H]^{1-}$ ion. The chromatographic results of such an experiment can be seen in Fig. 1.4. Panel A shows the total ion chromatogram which contained a number of non-phosphopeptides. Panel B shows the single ion chromatogram for loss of phosphate by in-source CID. This scan is used to identify the elution time of the phosphopeptide. In panel C two peptides were identified with the same nominal m/z of 1124. Only the one that lines up with the phosphate loss in panel B is a phosphopeptide. With the molecular weight of the phosphopeptide known, a second injection is made and tandem MS in positive ion mode performed for the purpose of locating the phosphorylation site.

7 Conclusions

While it is apparent that current 2DE technology has limitations for proteome analysis it is nevertheless extremely useful for large scale protein analysis and for the analysis of specific proteins in a sample. As noted by an examination of codon bias values (Gygi et al. 1998) coverage of an entire proteome is currently out of reach because of the low copy number of most proteins in any given cell. However, those using the technology to ask direct biological questions can be successful (Santos et al. 1998). To provide the best chance of success research questions involving large scale protein analysis should be addressed using a suitable combination of the available biochemical and immunological tools for isolation and purification of proteins. As for the future of mass spectrometers in proteomics it should be possible to identify a protein based on accurate mass measurement of a single peptide using FT-ICR mass spectrometers. Additionally, the wider use of MALDI on instruments such as FT-ICR and ion trap mass spectrometers may facilitate a more judicious use of valuable samples fractionated on-line during ESI-MS/MS at nanoliter/minute flow rates and then re-analyzed on MALDI instruments.

8 Acknowledgments

We would like to acknowledge funding from the National Science Foundation Science and Technology Center for Molecular Biotechnology. DRG acknowledges funding from Kinetek Pharmaceuticals and NIH. Further details of our work including other references can be found at http://weber.u.washington.edu/~ruedilab.

References

Aebersold R H, Leavitt J, Saavadra R A, Hood L E and Kent S B (1987) Proc Natl Acad Sci USA 84:6970–6974
Aebersold R H, Teplow D B, Hood L E and Kent S B (1986) J Biol Chem 261:4229–4238
Affolter M, Watts JD, Krebs DL, Aebersold R (1994) Anal Biochem 15:74–81.
Carr S A, Huddleston M J and Annan R S (1996) Anal Biochem 239:180–192
Carter J R, Franden M A, Aebersold R and McHenry C S (1992) J Bacteriol 174:7013–7025
Carter J R, Franden M A, Aebersold R and McHenry C S (1993) J Bacteriol 175:5604–5610
Davis M T and Lee T D (1997) J Am Soc Mass Spectrom 9:194–201
Ducret A, van Oostveen I, Eng J K, Yates III J R and Aebersold R (1998) *Protein Sci* 7:706–719
Eng J, McCormack A L and Yates III J R (1994) J Amer Soc Mass Spectrom 5:976–989
Figeys D, Ducret A, Yates III J R and Aebersold R (1996) Nature Biotech 14:1579–1583
Figeys D and Aebersold R (1997) Electrophoresis 18:360–368
Figeys D, Ning Y and Aebersold R (1997) Anal Chem 69:3153–3160
Figeys D, Corthals G L, Gallis B, Goodlett D R, Ducret A, Corson M A and Aebersold R (1999) Anal Chem in press
Fleischmann R D, Adams M D, White O, Clayton R A, Kirkness E F, Kerlavage A R, Bult C J, Tomb J -F, Dougherty B A, Merrick J M, McKenney K, Sutton G, FitzHugh W, Fields C, Gocayne J D, Scott J, Shirley R, L. Liu -I, Glodek A, Kelley J M, Weideman J F, Phillips C A, Spriggs T, Hedblom E, Cotton M D, Utterback T R, Hanna N C, Nguyen D T, Saudek D M, Brandon R C, Fine L D, Fritchman J L, Fuhrmann J L, Geoghagen N S M, Gnehm C L, McDonald L A, Small K V, Fraser C M, Smith C O and Venter J C (1995) Science 269:496–512

Fraser C M, Casjens S, Huang W M, Sutton G G, Clayton R, Lathigra R, White O, Ketchum K A, Dodson R, Hickey E K, Gwinn M, Dougherty B, Tomb J F, Fleischmann R D, Richardson D, Peterson J, Kerlavage A R, Quackenbush J, Salzberg S, Hanson M, van Vugt R, Palmer N, Adams M D, Gocayne J, Weidman J, Utterback T, Watthey T, McDonald L, Artiach P, Bowman C, Garland S, Fujii C, Cotton M D, Horst K, Roberts K, Hatch B, Smith H O and Venter J C (1997) Nature 390:580–586
Fu X Y, Schindler C, Importa T, Aebersold R and Darnell Jr J E (1992) Proc Natl Acad Sci USA 89:7840–7843
Garrels J L, McLaughlin C S, Warner J R, Futcher B, Latter G I, Kobayashi R, Schwender B, Volpe T, Anderson D S, Mesquita-Fuentes R and Payne W E (1997) Electrohporesis 18:1347–1360
Goffeau A, Barrell B G, Bussey H, Davis R W, Dujon B, Feldmann H, Gailbert F, Hoheisel J D, Jacq C, Johnston M, Louis E J, Mewes H W, Murakami Y, Philippsin P, Tettelin H and Oliver S G (1996) Science 274:546–549
Goodlett D R, Wahl J H, Udseth H R and Smith R D (1993) J of Microcolumn Separations 5:57–62
Gorg A, Postel W and Gunther S (1988) Electrophoresis 9:531–546
Gygi S P, Rochon Y, Franza B R and Aebersold R (1999) Mol Cell Biol accepted
Hieter B, Vogelstein B and Kinzler K W (1997) Cell 88:243–251
Honore B, Leffers H, Madsen P and Celis J E (1993) Eur J Biochem 218:421–430
Huddleston M J, Annan R S, Bean M F and Carr S A (1993) J Am Soc Mass Spectrom 4:710–717
Hunt D F, Yates III J R, Shabanowitz J, Winston S and Hauer C R (1986) Proc Natl Acad Sci USA 83:6233–6236
Hunter A P and Games D E (1994) Rapid Commun Mass Spectrom 8:559–570
Klose J and Kobalz U (1995) Electrophoresis 16:1034–1059
Lee N, Goodlett D R, Ishitani A, Marquardt H and Geraghty D (1998) J of Immunology 160:4951–4960
Mann M and Wilm M (1994) Anal Chem 66:4390–4399
Matsudaria P (1987) J Biol Chem 262:10035–10038
Patterson S D and Aebersold R H (1995) Electrohporesis 16:1791–1814
Roepstorff P and Fohlman J (1984) Biomed Mass Spectrom11:601
Rosenfeld J, Capdevielle J, Guillemot J C and Ferrara P (1992) Anal Biochem 203:173–179
Shevchenko A, Wilm M, Vorm O and Mann M (1996) Anal Chem 68:850–858
Susin S A, Lorenzo H K, Zamzami N, Marzo I, Brothers G, Snow B, Jacotot E, Costantini P, Larochette N, Goodlett D R, Aebersold R, Pietu G, Prevost M-C, Siderovski D, Penninger J and Kroemer G (1999) Nature 397:441
Veals S A, Schindler C, Leonard D, Fu X Y, Aebersold R, Darnell Jr J E and Levy D E (1992) Mol Cell Biol 12:3315–3324
Watts J D, Affolter M, Krebs D L, Wange R L, Samelson L E and Aebersold R (1994) *J Biol Chem* 269:29520–29529
Wilkins M R, Pasquali C, Appel R D, Ou K, Golaz O, Sanchez J-C, Yan J X, Gooley A A, Hughes G, Humphrey-Smith I, Williams K L and Hochstrasser D (1996) Biotechnology 14:61–65
Wilm M, Shevchenko A, Houthaeve T, Breit S, Schweigerer L, Fotsis T and Mann M (1996) Nature 379:466–469
Yates III J R, Eng J K, McCormack A L and Schieltz D (1995) Anal Chem 67:1426–1436
Zhang Y, Figeys D and Aebersold R (1998) Anal Biochem 261:124–127

CHAPTER 2

Protein Sequencing or Genome Sequencing. Where Does Mass Spectrometry Fit into the Picture?

P. Roepstorff[1]

1 Introduction

Advanced technology for the determination of DNA sequences has become widely available in the last decade and has been used for sequencing cDNAs and entire genomes from a variety of organisms ranging from viruses to man. As a consequence, the amount of DNA sequence information entered in publicly accessible databases has increased exponentially in the last decade (Fig. 2.1). This growth has far exceeded the growth of sequences entered in databases based on

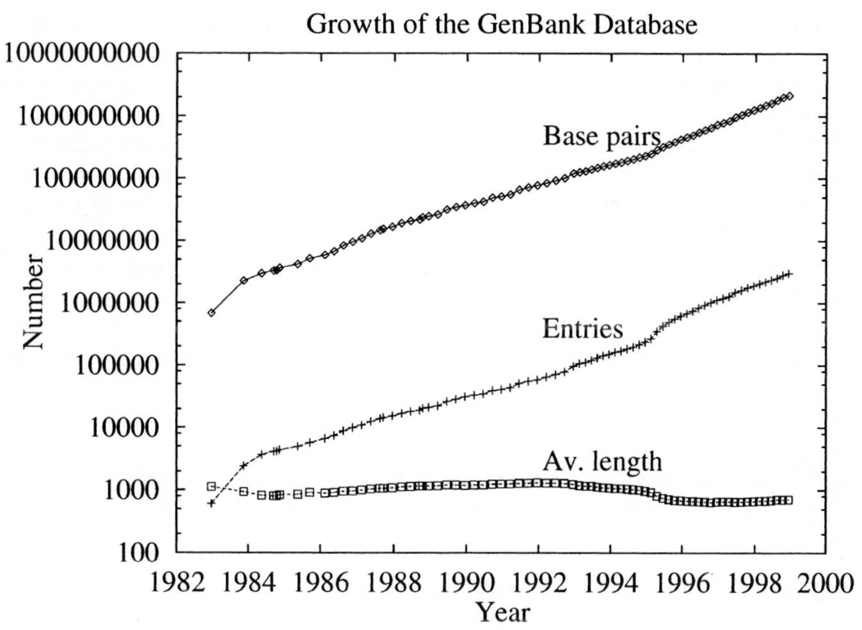

Fig. 2.1. The growth of the GeneBank database in the years 1982–1999 (Courtesy of Dr. S Brunak, Center for Biological Sequence Analysis, Denmark).

[1] Department of Molecular Biology, Odense University, DK 5230 Odense M, Denmark.

Table 1. Mass accuracy and sensitivity routinely achieved by MS of peptides and proteins

mass range		0.5–5 kDA	5–20 kDA	> 20 kDa
Mass Accuracy	MADLI DE-RF-TOF	30 ppm	50–100 ppm	–
	MALDI DE-lin-TOF	100 ppm	100–200 ppm	0.02–0.1 %
	nanoESI all modes	0.1 Da	1 Da	0.01 %
Sensitivity	MALDI all modes	0.1–1 fmoles	1–10 fmoles	0.1–1 pmoles
	nanoESI-quadrupole	10–50 fmoles	50–100 fmoles	1–10 pmoles
	nano-ESI-QTOF	1–10 fmoles	1–10 fmoles	0.5–5 pmoles

de novo protein sequencing. In addition to the genomic sequencing, large-scale partial cDNA sequencing has resulted in another set of data, the so-called expressed sequence tags (ESTs), containing stretches of sequence from a large number of genes from a variety of organisms. The obvious question is therefore: does *de novo* protein sequencing have a role to play in the future? There is no simple answer to this question. The genomic sequences only provide information about the potential of the selected micro-organisms and cell types but do not reflect the actual situation at any given moment, i.e. which proteins are expressed and how they are modified. cDNA sequences or the incomplete ESTs give information on proteins actually expressed, but no information on processing and secondary modification. Therefore, the study of the protein will never be obsolete, but the questions to address will be different.

Independently, but concurrently, mass spectrometric analysis has undergone an equally dramatic development. From being an analytical tool for the analysis of small volatile molecules, new ionization methods, especially electrospray ionization (ESI) (Fenn *et al.* 1989) and matrix assisted laser desorption/ionization (MALDI) (Karas and Hillenkamp 1988), have increased the accessible mass range to include nearly all proteins. Mass accuracy and sensitivity have been improved to allow routine molecular mass determination on the 100 ppm level of peptides and proteins which are available in only mid to low femtomole amounts (Jensen *et al.* 1996). Even better mass accuracy and sensitivity can currently be obtained under optimal conditions (Table 2.1). Mass spectrometry (MS) has been proven ideal for the analysis of peptide and protein mixtures and partial or complete sequence information can be generated from the single components in such mixtures by the so-called MS/MS techniques. In addition, MS is the ideal technique for analysis of post-translational modifications in proteins, thus being the perfect complement to DNA sequencing (Roepstorff 1997).

Below, selected applications of mass spectrometry will be described using recent examples from studies of proteins and peptides in our research group.

2
Proteome – the Next Step After the Genome

Once a genome has been sequenced the next natural step is the analysis of the proteome which, as defined by Wilkins *et al.* 1996, represents: the total protein complement expressed by an organism, a cell or a tissue type. Proteome analysis involves two essential steps: first, the separation and visualization of the proteins,

Fig. 2.2. General strategy for identification of gel-separated proteins. Adapted from Shevchenko *et al.* 1996

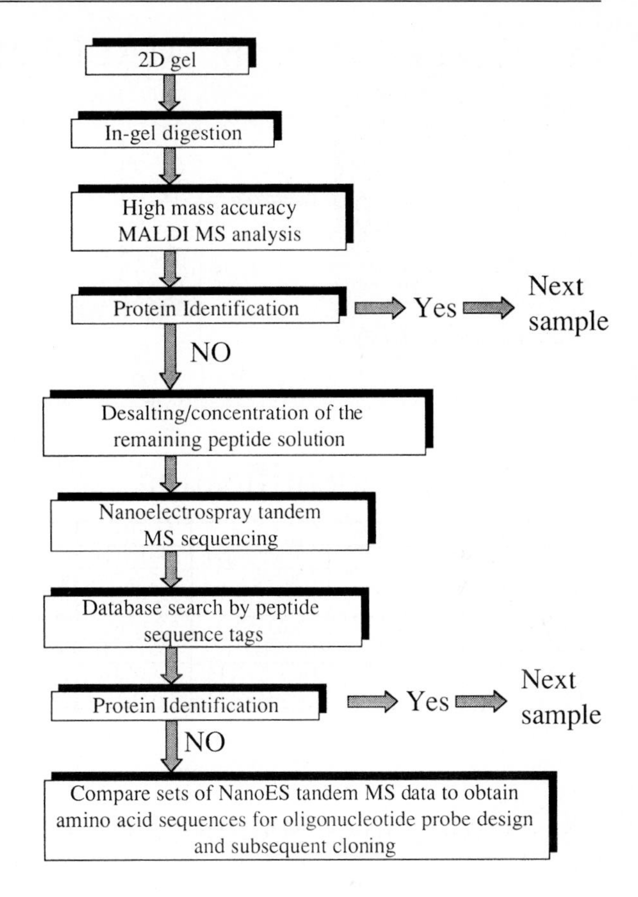

and second, the identification of the proteins relative to the available genomic sequence information. 2-dimensional polyacrylamide gel electrophoresis (2D-PAGE) is the only technique currently available for the separation of all or the majority of the proteins from a given cell type. Identification of proteins is now routinely carried out by mass spectrometry after proteolytic digestion of the proteins in a gel. This can be done either based on peptide maps produced by MALDI MS or partial sequences produced by ESI MS. The general strategy used in our laboratory is shown in Fig. 2.2 (modified after Shevchenko *et al.* 1996). Of these two techniques peptide mapping by MALDI MS is the simplest and most sensitive whereas the sequence-based techniques are more specific. Partial sequences also often allow identification in cases where only partial protein sequence information is available, e.g. in EST databases (Mann 1996). Upon positive identification, the corresponding cDNA can be ordered and sequenced.

The MALDI peptide mapping concept is illustrated in Fig. 2.3 (Jensen *et al.* 1998). An extract of proteins from the β-cells of the islets of Langerhans from rat pancreas was separated by 2D-PAGE (Fig. 3A). The spot labelled with an arrow was excised, partially dried and re-swollen in buffer (50 mM NH_4HCO_3) contain-

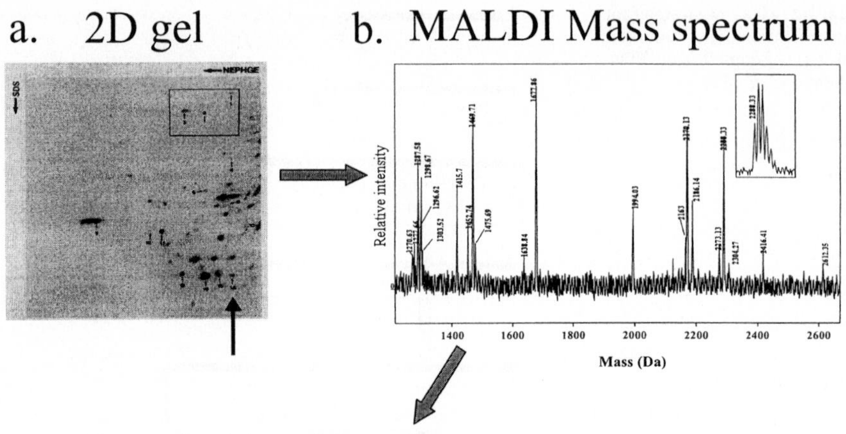

c. Protein identification

	Index	Pept. matched	Acc. number	Mw (Da)	Protein name
1	23532	9	swiss/P08699	27070.44	LEG3_RAT GALECTIN-3
2	143771	9	T/J02962	27229.70	RNBPIGE 1 Rat IgE binding protein
3	144409	9	T/M13697	15670.16	RNIGEBP_1 Rat IgE binding protein
4	125590	5	T/X60200	80330.02	ECTN1000_1 E. coli transposon
5	61878	4	Tnew/U33115	95593.73	SC33115 1 gene: "PSP1"; product
6	65552	4	Tnew/L77216	90710.31	BBLONNA 1 gene: "Ion"; product
7	67081	4	Tnew/Z49988	75189.10	SPMMSAGEN 2 gene:"mmsA"; product

d. Galectin-3 Sequence

MADGFSLNDALAGSGNPNPQGWPGAWGNQPGAGGYPGASYPGAYPGQAPPGGYPGQAPPSAYPGPTGPSA

YPGPTAPGAYPGPTAPGAFPGQPGGPGAYPSAPGAYPSAPGAYPATGPFGAPTGPLTVPYDMPLPGGVMP

RMLITIIGTVKPNANSITLNFKKGNDIAFHFNPRFNENNRRVIVCNTKQDNNWGREERQSAFPFESGKPF

KIQVLVEADHFKVAVNDVHLLQYNHRMKNLREISQLGIIGDITLTSASHAMI

Fig. 2.3. Example of protein identification in proteome studies. (a) NEPHGE gel of proteins from β-cells from rat Langerhans Islets. (b) MALDI spectrum of the peptide mixture derived by in-gel tryptic digestion of the spot indicated with an arrow in the gel. (c) top ranking candidates after a search in the EBI non-redundant database using the Peptide Search program. The three top ranking sequences are the same protein. (d) The sequence of the identified protein rat Galectin-3. The sequence covered by the peptide map is underlined

ing 12 ng/ml trypsin. After digestion the peptides were extracted from the gel, vacuum dried and re-dissolved in 10–20 μl 5 % formic acid. 0.5 μl of this solution was applied to a mass spectrometric target precoated with a matrix (a-cyano-4-hydroxycinnamic acid) nitrocellulose mixture (Kussmann *et al.* 1997) and dried. The surface was washed with 3–10 μl of 0.1 % TFA, dried and the mixture analyzed on a MALDI-TOF instrument in reflector and delayed extraction mode. The resulting mass spectrum shown in Fig. 2.3B was first externally calibrated and then re-calibrated based on a few peaks identified as autodigestion products of trypsin. The spectra are isotopically resolved and with this calibration mode a mass accuracy better than 50 ppm is obtained for the monoisotopic mass. The mass list, in this case containing 20 peptide masses, is then exported into the peptide search program (Mann *et al.* 1994) and the entire EBI non-redundant database searched for matching proteins. The top ranking candidate was Rat Galectin-3 with 9 matching peptides (Fig. 2.3C). Since the species (rat) was correct and the calculated protein mass (27 kDa) was consistent with the molecular mass observed in the gel (32 kDa), the protein is considered to be unambiguously identified. Upon closer examination of the spectrum two additional peaks could be identified as a result of oxidized Methionine residues in the corresponding peptides. The sequence coverage was 39 %. This is rather low. However, examination of the sequence (Fig. 2.3D) revealed that the first 140 N-terminal amino acid residues did not contain any Lys or Arg residues at which tryptic cleavage could take place. The peak corresponding to this large peptide was not observed. The sequence coverage for the remaining part of the protein was 84 %, which is a typical sequence coverage obtained from good quality spectra. Some of the remaining peaks were assigned to keratin-related peptides, keratin being a frequent contaminant when working at low protein levels, and a few to the tryptic autodigestion products used for calibration. It is a precondition for positive identification based on MALDI peptide mapping that the sequence of the protein or the sequence of a highly homologous protein from another species is present in the database and that a sufficient number of peptide signals (> 4–5) are present. Sometimes when working on very low protein levels too few peaks are present or the peptide signals may be suppressed by contaminant-derived signals. In such cases a concentration/purification step might improve the spectra (see micro-purification below).

If the protein cannot be identified based on the peptide map then partial sequence information is derived for selected peaks by performing nanoESI MS/MS on an aliquot of the remaining part of the peptide mixture extracted from the gel. This may be relevant when very few peaks are present, when identification is made based on search in EST databases, or by homology to proteins from other species. Recording of spectra by nanoESI always requires micro-purification to remove salts and other contaminants.

Micro-purification is performed by applying the protein extract to a custom-made miniature column with a bed volume of a few hundred nanoliters made in an Eppendorph gel loader tip (Gobom *et al.* 1999). The extracted peptide mixture dissolved in 5 % formic acid is loaded on the column, washed with 5–10 μl of 5 % formic acid and eluted directly into the nanospray needle with 50 % aqueous methanol, 2 % formic acid. As mentioned above, micro-purification can also be applied with advantage prior to MALDI peptide mapping when the spectra are of poor quality. Inclusion of the micro-purification procedure usually leads to improved sequence coverage. For MALDI the elution is performed with 1 μl of matrix solution containing 70 % acetonitrile directly onto the mass spectrometric target (Gobom *et al.* 1999).

Identification based on peptide sequence tags is illustrated in Fig. 2.4. A human protein complex was isolated by immune precipitation with a mouse antibody directed against one of the components and the components separated by 2D

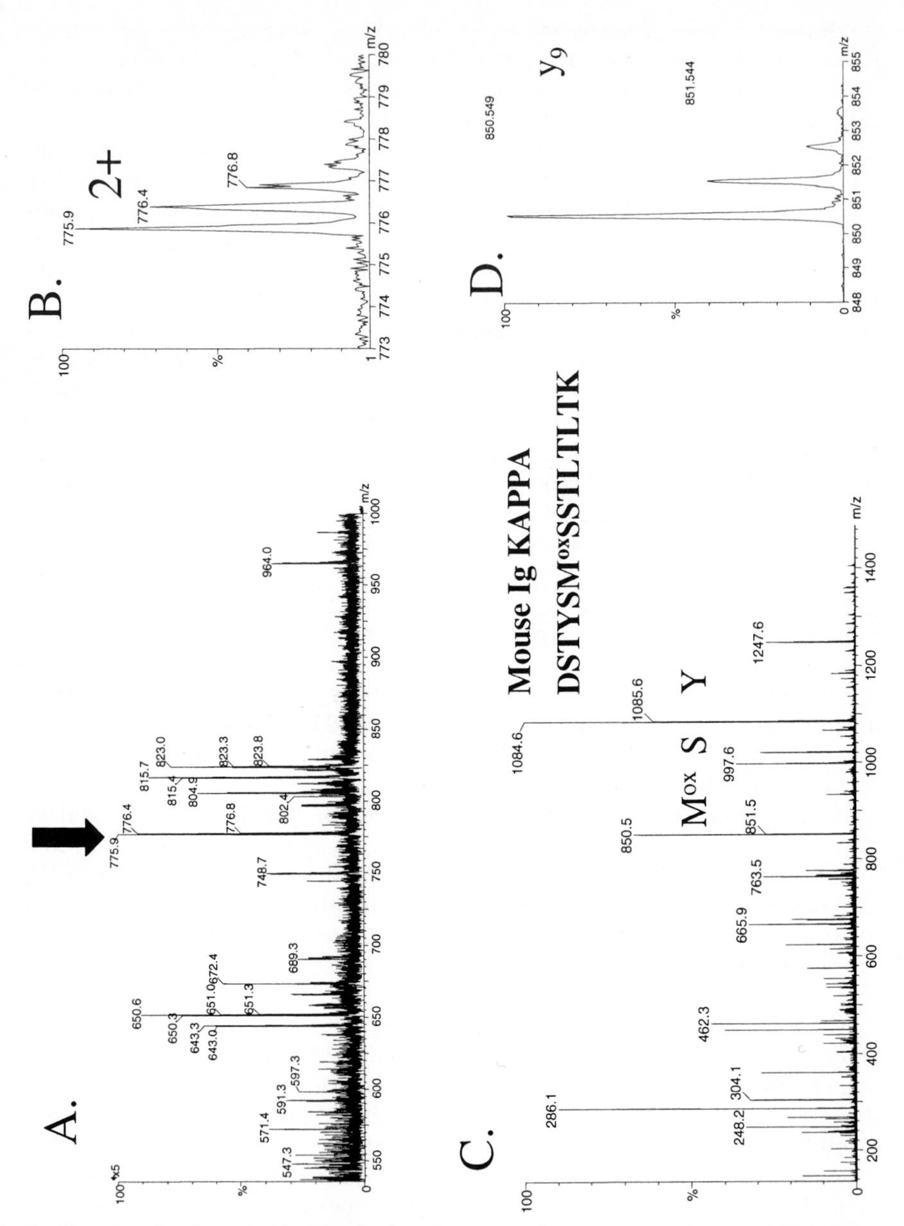

Fig. 2.4. Example of protein identification based on a peptide sequence tag. (A) nanoESI-spectrum obtained on a QTOF mass spectrometer of the peptide mixture obtained after in-gel digestion and micropurification of a protein spot from a 2D-gel of an immune precipitated human protein complex. (B) The resolution is sufficient to identify the selected peak as doubly charged based on the spacing between the isotope peaks. (C) MS/MS spectrum obtained upon selection and collision induced dissociation of the peaks in the isotope envelope at m/z 775–777. Based on singly charged fragment ions (D) a sequence tag "850.5 M^{ox}SY1247.6" can be generated. Upon search in the EBI non-redundant database the peptide is identified as a tryptic peptide from the constant region of a mouse antibody kappa light chain. Thus, the protein in the gel-spot is the antibody used for immune precipitation

PAGE digestion. The extract obtained by *in situ* digestion of one of the spots with trypsin was micropurified, loaded direct in the nanospray needle and analyzed with nanoESI-MS on a micromass Q-TOF mass spectrometer (Fig. 2.4A). A peak derived from a doubly-charged ion was selected for MS/MS in the peptide map. The Q-TOF instrument has sufficient resolution to determine the charged state of an ion based on the mass differences between the isotope signals (Fig. 2.4B). In the MS/MS spectrum of the selected ion (Fig. 2.4C) singly-charged ions are identified (Fig. 2.4D) and a peptide sequence tag is generated (Mann and Wilm 1994). A search in the EBI non-redundant database for this sequence tag identified the peptide as being a tryptic peptide derived from the constant region of a mouse Ig kappa light chain. Peptide sequence tags derived by MS/MS of several of the other peaks present in the spectrum (Fig. 2.4A) resulted in identification of the same protein. Thus, the unknown spot represented the antibody used for immune precipitation.

3
Characterization of Secondary Modifications in Proteins

Once a protein is identified, the next obvious questions are: Is the identified protein post-translationally modified and if so, how? If the purified protein is available then the strategy is to compare its molecular mass determined by MS with that calculated from its DNA sequence. If these masses are different, then the modified sites and the types of modification are identified by mass spectrometric peptide mapping and, when relevant, supplemented with MS/MS of selected peptide ions or degradation with appropriate enzymes, e.g. glycosidases or phosphatases (Burlingame 1996, Bean *et al.* 1995). If the proteins are available only as spots or bands in electrophoretic gels, then it is often not possible to determine the molecular mass of the intact protein, and characterization of post-translational modifications must rely on peptide mapping before and after enzymatic treatments and, when appropriate, MS/MS. In such cases it is essential to obtain complete or very high sequence coverage in the peptide maps (Moertz *et al.* 1996, Wilm *et al.* 1996a).

The most frequent secondary modifications are: proteolytic processing, acylation of the N-terminus and Lys ε-amino groups, phosphorylation of Ser, Thr and Tyr, and N-glycosylation of Asn or O-glycosylation of Ser and Thr. The two first types of modifications are normally easily recognized in the peptide maps whenever the corresponding signals are present. However, suppression of some of the peptide signals in the spectra of the mixtures is a frequent phenomenon. This is especially true for peptides modified with acetic groups, e.g. phosphate groups or sialic acid containing glycans. This can be overcome if sufficient protein is available for separation of the peptides. Glycosylated peptides can easily be recognized by mass spectrometry due to the characteristic peak pattern caused by the glycan heterogeneity (Fig. 2.5) and phosphorylated peptides can be identified by the loss of the phosphate group either due to post source decay in MALDI (Fig. 2.6) or by collision-induced dissociation in ESI.

The peak pattern from the glycopeptides derived by enzymatic digestion of glycoproteins can be used to generate a site specific glycoprofile for a protein

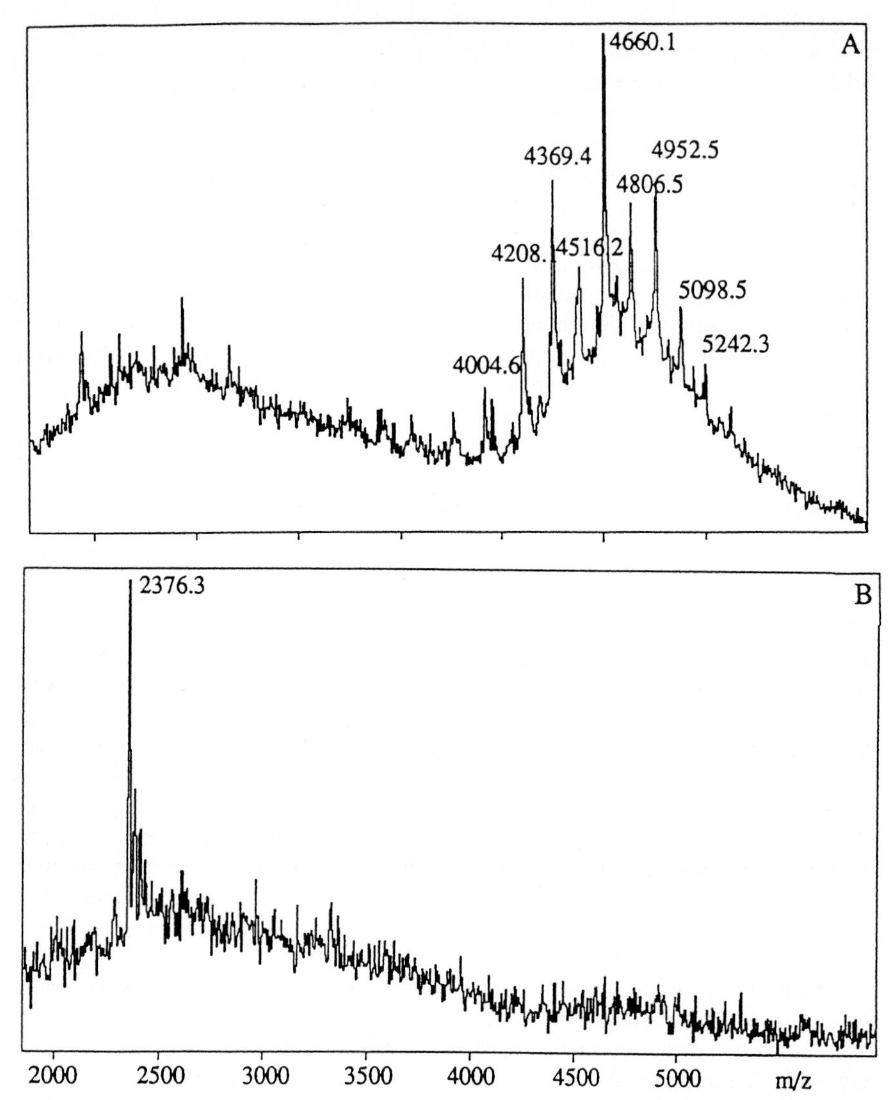

Fig. 2.5 .(A) MALDI spectrum obtained in linear mode of a glycopeptide isolated by HPLC after tryptic digestion of the major cat allergen Fel d1 (Kristensen et al. 1997). The mass differences are indicative of heterogeneity caused by a varying number of sialic acid, galactose and fucose residues. (B) The MALDI spectrum of the same peptide after deglycosylation with PNGase F is used to identify the peptide relative to the sequence

(Rahbek-Nielsen *et al.* 1997, Ploug *et al.* 1998, Olsen *et al.* 1998). Such a glycoprofile is based on the assumption that the glycan structures are of the most common types, e.g. high mannose, complex and hybrid structures. No information on linkage types or alternative branching is obtained. More detailed investigation of the glycan structures is possible by mass spectrometric analysis of the prod-

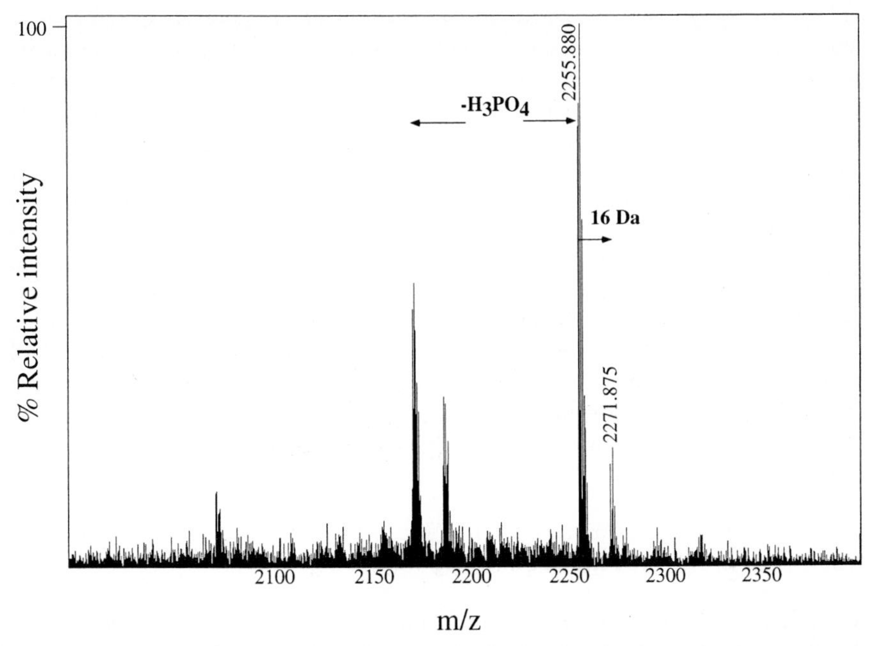

Fig. 2.6 .MALDI spectrum obtained in reflector mode of a phosphorylated peptide. The peptide contains a partially oxidized Met-residue indicated by a peak 16 Da above the molecular ion. The presence of a phosphate group is indicated by the metastable loss of the phosphate group. The loss of phosphate appears as a loss of 84 Da and not the expected 98 Da loss because the instrument is calibrated for MS-mode and not for PSD mode. This observed mass difference is instrument dependent

ucts generated by sequential digestion with glycosidases specific for the residue and linkage types. In our laboratory, this strategy has been successfully applied to the determination of the site specific glycan structures on the major cat allergen (Kristensen *et al.* 1997), a fish rhabdovirus glycoprotein (Einer-Jensen *et al.* 1998) and from gel bands representing differently glycosylated forms of human interferon γ (Moertz *et al.* 1996). In the latter study, one of the glycosylated peptides could be identified directly in the peptide map after removal of the sialic acid residues by treatment of the peptide mixture with neuramidase. Another glycosylated peptide was only observed after separation of the components using narrow bore RP-HPLC. However, isolation of the peptides is not a realistic solution when only minute protein amounts are present in gel spots or bands, and the site specific characterization of glycans from gel-separated glycoproteins still represents a major challenge.

Phosphorylated peptides derived from gel-separated proteins have been identified by differential peptide mapping before and after treatment with alkaline phosphatase. The specific phosphorylated sites in the peptides have been localized by comparing the results obtained by in-gel digestion with different proteolytic enzymes. The phosphopeptides were identified based on either a mass decrease of 80 Da upon phosphatase treatment of the corresponding peaks or if the phosphopeptide signals were suppressed in the spectra by the appearance of new peaks upon phosphatase treatment (Jensen *et al.* 1999, Kussmann *et al.* 1999). Phosphorylated peptides can also be selectively detected by monitoring precursor ions that generate a diagnostic fragment ion at m/z 79 due to loss of the phosphate group in negative ion mode ESI-MS/MS (Carr *et al.* 1996). This method has recently been applied successfully to identify phosphorylated peptides directly in the mixtures derived from gel-separated proteins (Neubauer and Mann 1999). Purification of the phosphorylated peptides from the mixture by affinity chromatography is an alternative to overcome the suppression phenomenon. The phosphopeptides can be isolated directly using a column containing immobilized Fe^{3+} or indirectly using avidin or streptavidin columns after conversion of the phosphopeptides to biotinylated peptides by β-elimination, followed by ethanethiol addition to form S-ethylcystine and biotinylation of the free thiol group (Oda and Chait 1998). Both methods have been demonstrated to enable the identification of phosphopeptides in peptide mixtures generated by in-gel digestion of proteins separated by 2D-PAGE (O.N. Jensen and M.R. Larsen personal communications).

4
When is de novo Protein Sequencing Relevant?

If a protein under identification in proteome studies comes out as 'unknown' in the database search, then sufficient *de novo* sequence information must be generated to allow synthesis of nucleotide probes for subsequent isolation and sequencing of the corresponding cDNA. This has been achieved from silver-stained 2D-gels by nanoESI MS/MS (Wilm *et al.* 1996a, Wilm *et al.* 1996b, Shevchenko *et al.* 1997). However, if no genomic or cDNA information is available for the organism studied, then complete sequencing on the protein level might be indicated as for example in our studies of insect and crustacean cuticle proteins (Andersen *et al.* 1995). A number of mass spectrometric approaches have been suggested for *de novo* protein sequencing (listed in Fig. 2.7). In our experience, however, sequencing of proteins with mass spectrometry as the only sequencing method nearly always fails to give the complete sequence. Some residues always remain unidentified and frequently even longer stretches of the sequence. In addition, the isomeric amino acid residues Ile and Leu cannot be distinguished. At the 1992 MPSA meeting in Berlin we proposed a strategy for protein sequencing based on a combination of Edman degradation and mass spectrometry (Roepstorff and Højrup 1993). Although the different mass spectrometric sequencing procedures play an increasing role in our present version of the strategy, we find that this combined approach is still the best strategy for complete *de novo* protein sequencing when sufficient amounts of protein are available, i.e. in the low to mid picomole range. Due to the inclusion of mass spectrometry this combined approach can be used to sequence the individual proteins in mixtures of proteins as demonstrated by the simultaneous sequencing of three isoforms of a meal worm cuticle protein (Haebel *et al.* 1995). Subsequent isolation of the proteins from single animals by 2D-PAGE and mass spectrometric determination of their molecular mass after electroelution as well as mass spectrometric peptide mapping demonstrated that the three forms represented allelic differences.

Fig. 2.7. Different approaches for mass spectrometric *de novo* protein sequencing

Protein sequencing by mass spectrometry

MS-only based techniques

Collision induced dissociation (CID)

Post source decay (PSD)
(Spengler *et al.* 1992)

Combined methods with MS detection

Ladder sequencing using Edman degradation
(Chait *et al.* 1993)

Ladder sequencing using carboxypeptidases
(Patterson *et al.* 1995)

Partial acid hydrolysis
(Vorm and Roepstorff 1994)

MS supported strategy

Edman degradation combined with MS
(Roepstorff and Højrup 1993)

5 Conclusion

There is no doubt that mass spectrometry will play an increasingly important role as a tool for protein analysis in the post genome era. The information obtained from the mass spectra is perfectly complementary to the information derived from genome or cDNA sequencing. The combination of sensitivity and specificity is unsurpassed by any other technique for protein identification and characterization of post-translational modifications. Albeit not a fully reliable sequencing tool, the sensitivity with which partial or full sequence information can be generated from peptides is extremely useful in many applications. The next major fields in protein studies where mass spectrometry can be expected to gain acceptance are those of protein surface topology and protein interaction studies. The first applications in these fields have already appeared (Ens *et al.* 1998) and the future potential is very high (Roepstorff 1997).

6 Acknowledgments

Drs. O.N. Jensen and M.R. Larsen are acknowledged for supplying data and help with preparing figures and Mrs H.M. Mortensen for critical reading of the manuscript. The Danish Biotechnology Programme and the Danish National Research Foundation are acknowledged for financial support.

References

Andersen SO, Højrup P, Roepstorff P (1995) Insect cuticular proteins. Insect Biochem Molec Biol 25: 153–176.

Burlingame AL (1996) Characterization of protein glycosylation by mass spectrometry. Current Opin Biotechnol 7: 4–10.

Bean MF, Annan RS, Hemling ME, Mentzer M, Huddleston MJ, Carr SA (1995) LC-MS methods for selective detection of post translational modifications in proteins: Glycosylation, phosphorylation, sulfation and acylation. In Techniques in Protein Chemistry VI. (Crabbe J, ed.) Academic Press, San Diego pp. 107–116.

Carr SA, Huddleston MJ, Annan RS (1996) Selective detection and sequencing of phosphopeptides in the femtomole level by mass spectrometry. Anal Biochem 239: 180–192.

Chait BT, Wang R, Beavis RC, Kent SBH (1993) Protein ladder sequencing. Science 262:89–92.

Einer-Jensen K, Krogh TN, Roepstorff P, Lorenzen N (1998) Characterization of Intramolecular Disulphide Bonds and Secondary Modifications of the Glycoprotein from Viral Hemorrhagic Septicemia Virus, a Fish Rhabdovirus. J Virology 72: 10189–10196.

Ens W, Standing KG, Chernushevich IV (eds) (1998) New Methods for the Study of Biomolecular Complexes. NATO ASI Series C Vol. 510, Kluwer Academic Publishers, Dordrecht.

Fenn JB, Mann M, Meng CK, Wong SF, Whitehouse CM (1989) Electrospray ionization for the mass spectrometry of large biomolecules Science 246: 64–71.

Gobom J, Nordhoff E, Mirgorodskaya E, Ekman R, Roepstorff P (1999): A sample purification and preparation technique based on nano-scale RP-columns for the sensitive analysis of complex peptide mixtures by MALDI-MS. J Mass Spectrom 34: 105–116.

Jensen ON, Potelejnikov A, Mann M. (1996) Delayed extraction improves specificity in database searches by MALDI peptide maps. Rapid Comm Mass Spectrom 10: 1371–1378.

Jensen ON, Larsen MR, Roepstorff P (1998) Mass spectrometric identification and microcharacterization of proteins from electrophoretic gels: strategies and applications. PROTEINS: Structure, Function and Genetics Suppl.2, 74–98.

Haebel S, Jensen C, Andersen SO, Roepstorff P (1995) Isoforms of a cuticular protein from larvae of the meal beetle, *Tenebrio moletor*, studied by mass spectrometry in combination with Edman degradation and 2D-PAGE. Protein Science 4: 394–404.

Karas M, Hillenkamp F (1988) Laser desorption ionization of proteins with molecular masses exceeding 10000 daltons. Anal Chem 60: 2299–2301.

Kristensen AK, Schou C, Roepstorff P (1997) Determination of Isoforms, N-Linked Glycan Structure and Disulfide Bond Linkages of the Major Cat Allergen Fel d 1 by a Mass Spectrometric Approach. Biol Chem 378: 899–908.

Kussmann M, Nordhoff E, Rahbek-Nielsen H, Haebel S, Larsen MR, Jakobsen L, Gobom J, Mirgorodskaya E, Kristensen AK, Palm L, Roepstorff P (1997) MALDI-MS Sample Preparation Techniques Designed for Various Peptide and Protein Analytes. J Mass Spectrom 32: 593–601.

Kussmann M, Hauser K, Kissmehl R, Breed J, Plattner H, Roepstorff P (1999) Comparison of *in vivo* and *in vitro* phosphorylation of the exocytosis-sensitive phosphoprotein PP63/Para Fusin by differential MALDI mass spectrometric peptide mapping. Biochemistry in press.

Mann M, Højrup P, Roepstorff P (1994) Use of mass spectrometric molecular weight information to identify proteins in sequence databases. Biol Mass Spectrom 22: 338–345.

Mann M, Wilm M (1994) Error tolerant identification of peptides in sequence databases by peptide sequence tags. Anal Chem 66: 4390–4399.

Mann, M (1996) A shortcut to interesting human genes: Peptide sequence tags, ESTs and computers. Trends Biol. Sci. 21, 494–495.

Moertz E, Sareneva T, Haebel S, Julkunen I, Roepstorff P (1996) Mass spectrometric characterization of glycosylated interferon-g variants separated by gel electrophoresis. Electrophoresis 17: 925–931.

Neubauer G and Mann M (1999) Mapping phosphorylation sites of gel-isolated proteins by nanoelectrospray: Potentials and limitations. Anal Chem. 71: 235–242.

Oda Y, Chait BT (1998) Purification and identification of phosphopeptides. Proc. 46th ASMS Conference on Mass Spectrometry and Allied Topics Orlando, Florida, May 31-June 4

Olsen EHN, Rahbek-Nielsen H, Thøgersen IB, Roepstorff P, Enghild JJ (1998) Post-translational Modifications of Human Inter-a-Inhibitor: Identification of Glycans and Disulfide Bridges in Heavy Chains 1 and 2. Biochemistry 37: 408–416.

Patterson DH, Tarr GE, Regnier FE, Martin SA (1995) C-terminal ladder sequencing via matrix assisted laser desorption/ionization mass spectrometry coupled with carboxypeptidase Y time-dependent and concentration-dependent digestions. Anal Chem 67:3971–3978.

Ploug M, Rahbek-Nielsen H, Nielsen PF, Roepstorff P, Danø K (1998) Glycosylation Profile of a Recombinant Urokinase-type Plasminogen Activator Receptor Expressed in Chinese Hamster Ovary Cells. J Biol Chem 273: 13933–13943.

Rahbek-Nielsen H, Roepstorff P, Reischl H, Wozny M, Koll H, Haselbeck A (1997) Glycopeptide Profiling of Human Urinary Erythropoietin by Matrix-assisted Laser Desorption Ionization Mass Spectrometry. J Mass Spectrom 32: 948–958.

Roepstorff P, Hoejrup P (1993) Methods in Protein Sequence Analysis. K. Imahori & F. Sakiyama (eds.) Plenum Press, New York, pp 149–156.

Roepstorff P (1997) Mass Spectrometry in protein studies from genome to function. Current Opin. Biotechnology 8: 6–13.

Shevchenko A, Jensen ON, Potelejnikov AV, Sagliocco F, Wilm M, Vorm O, Mortensen P, Schevchenko A, Boucherie H, Mann M (1996) Linking genome and proteome by mass spectrometry: Large scale identification of yeast proteins from two dimensional gels. Proc. Nat. Acad. Sci. (USA) 93: 14440–14445.

Shevchenko A, Wilm M, Mann M (1997) Peptide sequencing by mass spectrometry for homology searches and cloning of genes. J. Prot. Chem. 92:481–490.

Spengler B, Kirsch D, Kaufmann R Jaeger E (1992. Peptide sequencing by matrix assisted laser desorption/ionization mass spectrometry. Rapid Commun Mass Spectrom 6: 105–108.

Vorm O, Roepstorff P (1994) Peptide sequence informationderived by partial acid hydrolysis and matrix assisted laser desorption/ionization mass spectrometry. Biol Mass Spectrom 23: 734–740.

Wilkins MR, Pasquali C, Appel RD, Ou K, Golaz O, Sanchez J-C, Jan JX, Gouley AA, Humphrey-Smith I, Williams KL, Hochstrasser EF (1996) From proteins to proteomes: Large scale protein identification by two-dimensional electrophoresis and amino acid analysis. Bio/Technology 14: 61–65.

Wilm M, Neubauer G, Mann M (1996a) Parent ion scan of unseparated peptide mixtures. Anal. Chem. 68: 527–533.

Wilm M, Shevchenko A, Houthaeve T, Breit S, Schweigerer L, Fotis T, Mann M (1996b) Femtomole sequencing of proteins from polyacrylamide gels by nano electrospray mass spectrometry. Nature 379: 466–469.

CHAPTER 3

Reversible Negative Staining of Protein on Electrophoresis Gels by Imidazole-Zinc Salts: Micropreparative Applications to Proteome Analysis by Mass Spectrometry

L. R. Castellanos-Serra[1], E. Hardy[1], W. Proenza[1], V. Huerta[1], L. J. Gonzalez[1], J. P. Le-Caer[2], R. L. Moritz[3] and R. J. Simpson[3]

1 Introduction

Negative staining of proteins on sodium dodecyl sulfate (SDS)-polyacrylamide gels with salts of transition metals such as Cu^{2+} (Lee 1987), Ni^{2+} (Dzandu, 1988) and Zn^{2+} (Dzandu, 1988) was first introduced about 10 years ago. This ingenious type of staining is based on the selective precipitation of metal salts on the gel surface except where proteins are located. Consequently, proteins appear in the gel as transparent, colorless bands (or spots) against a slightly stained background. Unfortunately, reproducibility and sensitivity are highly influenced by the electrophoresis – associated reagents remaining in the gel after completion of the electrophoretic run. Such variability in the gel composition, that leads to differences in sensitivity and reproducibility, is particularly significant when scarce amounts of proteins are present.

In 1990 the group working at C.I.G.B in Havana improved the reproducibility and sensitivity of the metal negative staining procedure by exchanging soluble gel components (buffers, salts etc.,) with imidazole (ImH) prior to metal staining (Fernandez-Patron and Castellanos-Serra 1990). Imidazole is well known in protein chemistry for its strong interaction with transition metals, a property that has been largely exploited for the design of poly-histidine tail fused proteins for immobilized metal affinity chromatography. Imidazole-zinc staining combines several features that are highly appreciated for a detection procedure:

i) speed (15 mins or less) and simplicity (two steps),
ii) highly reproducible sensitivity in the femtomole range, closely approaching that of conventional silver staining,
iii) it does not cause modifications of proteins,
iv) proteins are reversibly fixed in the gel through zinc-mediated complexes and
v) the staining reagents are neither toxic nor environmentally dangerous. This detection procedure is based on the generation of an insoluble salt, zinc imidazolate (ZnIm2), in the gel matrix. The mechanism of negative staining has been investigated, and the influence of common electrophoresis reagents including SDS on gel staining, was determined. (Fernandez-Patron et al. 1998).

[1] Center for Genetic Engineering and Biotechnology (CIGB) P.O. BOX 6162, Havana 10600, Cuba.
[2] Ecole Superieure de Chimie et Physique de la Ville de Paris, 10 rue Vauquelin, Paris 75024 France.
[3] Joint Protein Structure Laboratory, Ludwig Institute for Cancer Research (Melbourne) and the Walter and Eliza Hall of Medical Research, Parkville, Victoria, Australia.

This improved zinc staining has been termed Reverse-staining; the term has a dual connotation, alluding both to the "reverse" character of the stain (background or negative stain) and to the fact that the process can be fully "reversed" to the unstained -original- state. The procedure is being currently used by many research groups for protein detection before MS analysis or in-gel digestion (Cohen and Chait, 1997; Matsui et al, 1997; Gevaert et al, 1998; Scheler et al, 1998) and a staining kit is now commercially available. Recently, it was shown that Reverse-staining is not limited to proteins, but this can be generally applied to the sensitive visualization of other biomolecules – nucleic acids (Hardy et al, 1996) and bacterial lipopolysaccharides (Hardy et al, 1997) – on electrophoresis gels.

Here we present the protocols laying at the interface between gel electrophoresis and microanalysis; all of them based on the advantages derived from protein detection by Reverse-staining (Fig. 3.1). Following each procedure, the reader will find detailed information on practical hints, particular applications and personal experiences from the authors (those 'five ml of experience' which are an essential component of any successful protocol). The usefulness of Reverse-staining and protein elution from microparticles will be exemplified through an optimized procedure for microdigestion of Reverse-stained proteins and its applications to microanalysis of protein digests by mass spectrometry. The efficiency of this procedure is demonstrated by the very high coverage of the protein sequence attained after a single digestion working at the low picomole level. Finally, applications to proteome analysis by capillary chromatography-electrospray mass spectrometry (Moritz et al, 1996b; Zugaro et al, 1998; Reid et al, 1998.) will be discussed.

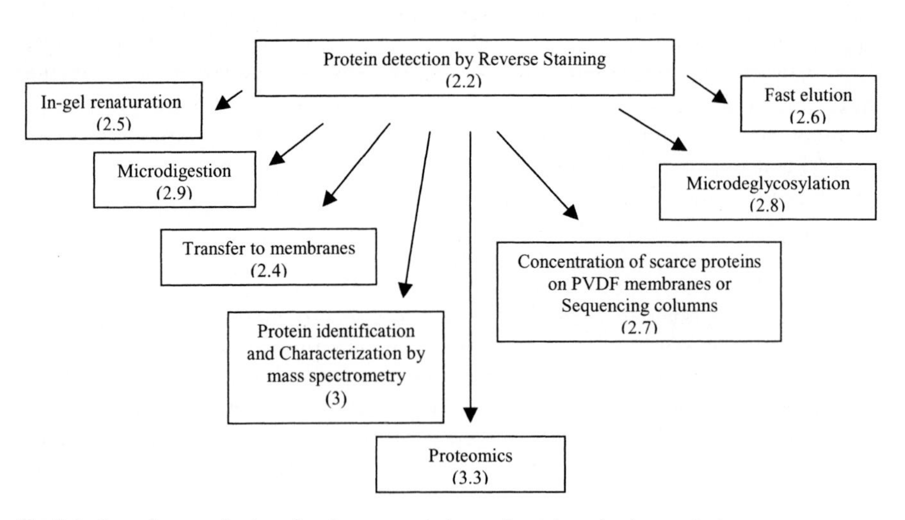

Fig. 3.1. Procedures at the interface between gel electrophoresis and microanalysis

2
Procedures

2.1
Methods

Reagents are analytical or electrophoresis grade from international suppliers (Merck, BioRad, Pharmacia). Solutions are prepared with high quality water (Milli-Q water, MQW). Volumes of staining solutions for mini-gels (about 8 × 10 cm) are ~ 100 mL; for medium-size gels (16 × 18 cm), the volume should be ~ 250 mL.

2.2.
Protein detection on polyacrylamide gels by Reverse-staining

2.2.1
Detection on SDS gels

Procedure *(Fernandez-Patron, and Castellanos-Serra, 1990; Fernandez-Patron et al.1992, Ortiz et al 1992)*
Prepare 10 × stock staining solutions: (a) 2 M ImH, 1 % SDS and (b) 2 M zinc sulfate. For use, dilute each stock solution 1:10 with MQW.

Imidazole step: After electrophoresis, place the gel in a transparent plastic tray, soak for 30 s in MWQ. Incubate the gel for 15 mins in 0.2 M ImH, containing 0.1 % SDS, while shaking continuously.

Zinc step: Discard the imidazole solution. Hold the tray a few centimeters over any dark surface to facilitate the visualization of the staining pattern, then add 0.2 M zinc sulfate while manually shaking. The gel will stain white due to the precipitation of zinc-imidazolate-SDS complex on the gel surface, leaving transparent, colorless (unstained) protein bands. When an adequate staining pattern is attained (after 30–45 s), discard the zinc solution and rapidly rinse the gel 3–4 times in abundant MQW, to avoid overstaining.

Digital Recording of Reverse Stained gels: Place the gel on the scanner screen and cover it with a black cardboard. Alternatively, scan the gel in a dark room while leaving the scanned lid opened. Adjust the image for optimal contrast. Scanning of Reverse-stained gels can also be obtained by using laser-based densitometers as is performed for gels stained with other positive staining methods (Molecular Dynamics, U.S.A.).

Photography: place the gel on a glass plate, held some centimeters above a black surface, while illuminating with fluorescent light.

Comments

i) The stock staining solutions are stable in amber bottles at room temperature for years.
ii) Recommended incubation time in the imidazole solution is 15 mins for 10 %, 12.5 % and 15 % polyacrylamide gels. For 5 %–7.5 % polyacrylamide gels, this

time can be shortened to 10 mins. During this step, soluble components from electrophoresis (glycine, Tris) are substantially replaced by imidazole.

iii) Ready-to-use gels cast on plastic films, as well as Tris-tricine, Tris-acetate and Tris-urea gels can be stained following this protocol.

iv) Gels can be stored in MQW in sealed transparent bags or alternatively, selected bands are excised and conserved at 4 °C in MQW. The N-terminus of proteins conserved in reverse stained gel bands for as long as three years have been successfully sequenced (Castellanos-Serra, unpublished results).

v) Native gels can be stained by this protocol but the incorporation of SDS during staining may be inconvenient for some applications (i.e., determination of the biological activity of the proteins after elution). Alternatively, SDS can be suppressed from the imidazole solution; some proteins detected by imidazole-zinc under native conditions may appear as positively white stained bands.

Troubleshooting

i) Increasing the concentration of imidazole (i.e., 0.3 M or higher) culminates in producing a deeper white background that may obscure less abundant proteins.

ii) The incorporation of SDS into the imidazole solution facilitates the observation of the unstained bands due to the generation of an homogeneous white stained background,. Although small variations in the concentration of SDS from the recommended 0.1 % value can be tolerated without drastic effects on the resulting staining pattern, at higher concentrations (about 0.5 %) the precipitation of zinc imidazolate is inhibited, provoking no staining of the gel.

iii) Prolonging the incubation in zinc 2+ causes an indiscriminate staining of the gel surface, and no detection of the protein bands.

iv) In cases where these problems have occurred, they can be easily remedied by initially destaining the gel and repeating the staining process: reverse stained gels are destained by incubation in 50 mM Tris-0.3 M Glycine pH 8.0 for 10 minutes. The solution is discarded, the gel is washed with water to remove glycine and then, it can be restained.

Advantages

This procedure has a high sensitivity, close to that of silver staining with the advantages of being highly reproducible, simple, rapid, fully reversible and non-destructive.

2.2.2 Double Staining of Coomassie-blue (CB) Stained Gels

Procedure *(Fernandez-Patron et al. 1995)*

CB-stained gels are washed in MQW (two changes of water, 25 mins each, or alternatively, overnight) to remove methanol and acetic acid. Then, the gel is reverse stained as described in **2.2.1**.

Comments

i) After Reverse-staining, the gel will show
   (a) typical CB-stained bands, which appear superimposed on larger transparent bands, originated by the highly sensitive negative staining of the gel and
   (b) reverse stained (negative) bands, which were previously undetected.

ii) The sensitivity is equivalent to that attained after imidazole-SDS-zinc staining on unstained gels.

iii) CB-stained gels can be Reverse-stained after prolonged storage of the gels in sealed polyethylene bags.

iv) Sensitivity after Reverse-staining does not depend on the degree of destaining achieved on the CB stained gels. Both the extensively destained gels (until obtaining an uncolored gel matrix) as well as the briefly destained (until detection of the band) gels give a very high sensitivity upon Reverse-staining (detection limit for streptokinase is 1.6 ng).

Advantages

Re-staining CB-stained gels with imidazole-SDS-zinc can successfully detect proteins that display markedly low affinity for CB as well as low abundant components that fall below the detection limit of CB staining. This procedure complements CB staining, giving a more realistic picture of the complexity of the sample under analysis. This is due to
(a) its higher sensitivity and
(b) the generality of Reverse-staining, which is independent of the nature of the proteins.

2.3 Protein Mobilization and Alkylation in Reverse-Stained Gels

2.3.1 Protein Mobilization Procedure *(Fernandez-Patron et al. 1995)*

Excise the protein bands of interest, incubate (2 × 5 mins) the gel slice in any one of the following solutions to complex zinc ions: (a) phosphate-buffered saline (PBS), pH 7.4 containing 100 mM EDTA di-sodium salt or 0.3 M glycine, (b) 25 mM Tris, 0.3 M glycine, pH 8.0 (c) 1 % citric acid. (d) 25 mM Tris, 50 mM glycine pH 8.0 containing 0.01 M DTT. After 1–2 minutes of incubation (with gentle hand shaking), the Reverse-stained band becomes completely transparent. At the end of the first 5 mins, renew the chelating buffer for additional 5 mins, and then briefly wash with water.

2.3.2 Protein in-Gel Alkylation Procedure *(Moritz et al. 1996b)*

To perform *in-gel* alkylation of the protein band, first complex the zinc ions with solution (d) 25 mM Tris, 50 mM glycine pH 8.0 containing 0.01 M DTT. After 1–2 minutes of incubation (with gentle shaking), the Reverse-stained band becomes completely transparent. Renew the chelating reduction buffer and continue incubation for an additional 30 mins. To *S*-pyridylethylate, add 4-vinylpyridine (2 % v/v) in the reduction buffer containing the gel and incubate in the dark, at 25 °C for 60 mins. After incubation, halt the alkylation by adding β-Mercaptoethanol (2 % v/v) directly to the solution, then discard the solution and wash the gel band extensively in water (3 × 1ml) for 10 mins

Comments

i) Protein fixation through zinc-mediated aggregates should be fully reverted before (or during) any procedure that requires proteins to be removed from the gel or accessed by enzymes or chemical reagents. This step is called 'protein mobilization'; during this step, protein losses from the gel band due to diffusion with the chelating solution (a,b,d) are only 10 % or less (determined with radioactive MW markers) (Castellanos-Serra, unpublished results) while losses due to diffusion in citric acid are significantly lower.

ii) The chelating agent should be chosen to best suit the next analytical step: (a) or (b) for fast passive elution or electrotransfer; (a) for micro-deglycosylation and *in-gel* renaturation; (b) for *in-gel* digestion and electroelution; (d) for protein reduction and alkylation during zinc chelation.

iii) Incubation in citric acid usually gives very clean first cycles upon automatic sequencing, nevertheless, some proteins tend to give lower transfer yields to membranes in comparison with those treated at neutral to basic pH.

iv) Incubation in EDTA or glycine, in general, gives very high yield on transfer and elution. Nevertheless, in one case, we found disappointingly low yields on transfer when a protein was mobilized in EDTA. This particular protein was a Ca (II)-binding venom toxin, that probably self-aggregated on the gel after removal of the metal.

v) Thiol reagents (2ME, DTT, DTE) are very effective zinc chelators; proteins can be reduced during protein mobilization by using conditions described under (d). For an efficient reduction, increase, if necessary, the incubation time and the temperature (i.e., 56 °C, 2 × 10 mins), next follow the procedures for *in-gel* alkylation with 4-vinyl pyridine or other alkylating agents (Moritz et al, 1996b).

vi) After mobilization, proteins can be then electroeluted or digested following conventional procedures. Nevertheless, standard electroelution devices with dialysis membranes may lead to variable, generally low recoveries when working at the low picomole level, as losses associated to adsorption on membranes can be significantly high. In cases such as these, the fast passive elution procedure (see 2.6) is recommended, yielding almost quantitative recoveries at the low picomole range.

2.4 Transfer to Membranes

2.4.1 Gel Blotting

Procedure

1. Apply the procedure for protein mobilization (2.3) to the entire gel.
2. Wash the gel with abundant MQW (2 × 1 min).
3. Follow the established procedures for semi-dry blotting to PVDF or nitrocellulose membranes (0.8 mA/cm^2, for 40–60 mins).
4. After blotting, wash the membrane with abundant MQW and detect as usual.

2.4.2 Single Band Transfer *(Fernandez-Patron et al 1995)*

Procedure

1. Follow the procedure described under 2.4.1 for gel transfer with the following modifications: Excise the band of interest, mobilize the protein (see 2.3) and wash with water (2 × 1 mL × 1 min).
2. Cut the blotting paper and the transfer membrane to fix the size of the excised gel band.
3. Mount and place the individual mini-sandwiches at the center of the semi-dry graphite electrode. Teflon spacers can be placed on the lower electrode at both sides of the mini-sandwiches to afford an adequate stability of the upper electrode (the height of the spacers should be the same or slightly lower than that of the mini-sandwiches).
4. Transfer at 0.8 mA/cm^2 for 40–60 mins.

Comments

i) Ready-to-use membrane disks furnished with the sequencers can be used for mounting the transfer.
ii) When working with small amounts of proteins, several (2–3) tiny bands excised from different lanes can be placed on a single membrane disk, to increase the amount of protein on the blot.

Advantages

Single band transfer allows for optimized transfer conditions (blotting buffers and transfer time) for each protein of interest in the gel, with respect to its electrophoretic mobility. Sensitivity of protein detection on the gel by Reverse-staining is higher than that attained with amido black on the membrane after transfer; this allows for sequence information to be obtained from protein amounts that were previously detectable on the gel but insufficient for detection on the membrane after blotting. In summary, once the conditions for transfer have been established, staining of the membrane can be obviated and the membrane can be directly analyzed, thus, avoiding artifacts from the staining reagent.

2.5
In-Gel Renaturation

Procedure *(Hardy et al. 1996)*
Stain the gel as described under 2.2.1, excise the bands of interest and incubate in Tris-EDTA buffer (see 2.3) to chelate zinc 2+.

In-gel renaturation: Wash the gel slices twice in PBS (1 mL) for 10 min to remove excess EDTA. Soak the gel slice (3 × 1 mL × 10 mins) in PBS containing 0.1 % Triton X-100, and next, wash twice in PBS alone (1 mL), to remove excess detergent.

Protein elution: Proceed as described under 2.6.

Comments

i) The present approach shares some similarities with protein renaturation accomplished in gel-filtration matrices, through buffer exchange from denaturing to non-denaturing conditions. As in such cases, there is not a unique protocol for protein renaturation, therefore, it may be necessary to explore a number of renaturing conditions. This protocol can serve as a guide for further optimization, as it has proved useful for a number of proteins both for determination of specific activity after recovery of the renatured protein in solution and for zymogram detection on gel. For optimization, the following factors can be modified: detergent exchanged with SDS (NP-40, sodium deoxycholate, Triton X-100, octyl glucoside), incubation time (from minutes to h) and temperature (4 °C, 20 °C, 37 °C) during the renaturing step. Alternatively, in cases where the protein is sensitive to detergents, the gel slice can be incubated in 30 % acetonitrile in PBS (2 × 10 mins) to remove SDS, followed by incubation in PBS.

ii) Protein renaturation can be accomplished in the entire gel, which can then be analyzed by zymogram detection.

iii) Proteins should not be reduced before electrophoresis, to facilitate the recovery of biological activity.

iv) For determining specific activity, several bands may be processed in order to get enough material for both protein quantitation and biological assay. Each determination should be done at least in triplicate.

v) Whenever a biological assay is attempted, carry out a control experiment in parallel, by processing an equal number of protein-free gel bands. This permits to evaluate any potential toxicity or interference derived from gel and electrophoresis reagents.

vi) This procedure obviously does not work for proteins formed by several subunits of different molecular weight which are separated on the gel, if the biological activity requires the assembly of a heteromeric structure. In this case, protein elution (see 2.6) can be directly performed after separation and detection under native conditions.

2.6 Fast Passive Elution of Proteins

Procedure *(Castellanos-Serra et al. 1996, 1997)*

Preparation of the crushing syringe (Heukeshoven and Dernick 1991): (Materials: 100 μm and 32 μm metal sieves can be purchased from F. Carl Schroetter, Hamburg or any equivalent distributor. Cut two metal sieve disks (one of 100-μm and one of 32-μm) to fit the internal diameter of a 1 mL (tuberculin) polypropylene syringe. Place the 32-μm disk at the bottom, followed by the 100 μm disk. Press them tightly at the bottom with the aid of the syringe plunger. Flush the syringe with 10X Laemmli running buffer several times, and next with MQW.

Protein elution:

1. Detect proteins by imidazole-SDS-zinc (2.2.3).
2. Cut out selected protein bands, chelate residual zinc and mobilize the protein as described under 2.3.
3. Crush the gel band through the metal sieves by placing the gel band(s) in the crushing syringe and forcing the gel through the sieves to obtain 32-μm gel particles. Collect the gel slurry in a 1.5 mL microtube.
4. Flush the syringe with the selected elution buffer through the sieves (see below) and collect the solution in the tube containing the gel slurry.
5. Incubate in Tris-Gly or ammonium bicarbonate 30–50 mM, pH 8.0 (it is not necessary to add detergents) with moderate vortexing during 2×10 mins and next, 1×1 min in water.
6. Collect the solution at each step by centrifuging for a few seconds. The elution volume should be about twice the volume of the crushed gel at each elution step. For highly hydrophobic proteins, substitute the second incubation in Tris- glycine by an incubation in an aqueous solution containing 20 % formic acid, 25 % isopropanol, 15 % acetonitrile.

Comments

i) Working with total radio-iodinated *E. coli* proteins (radioactivity evenly distributed across all MW ranges) this procedure gave the following yield per MW range: 95 % (21–31 kDa), 94 % (31–45 kDa) 92 % (45–97 kDa), 91 % (higher than 97 kDa). These values are for proteins separated on a 12.5 % PA gel, loaded without previous reduction, detected in 15 mins with ImH-SDS-Zinc and immediately eluted after the run.

ii) Same experiment as in (i) but proteins loaded in Laemmli buffer with mercaptoethanol give a significant drop in the yield for the higher MW range (only 65 % recovered). When this gel was left for two days on the bench and then eluted, protein re-aggregation was observed and the recovery dropped to ~27 % for the high MW range. Therefore, to achieve high yields, proteins (fully alkylated in 8M Urea, before electrophoresis, or loaded without previous reduction) should be eluted from negatively stained or unstained gels immediately after the run.

iii) Generating microparticles of the gel is crucial for achieving a very rapid diffusion. For the same elution times, yields from 100-μm gel particles were found lower than those from 32-μm particles.

iv) To collect the solution, centrifuge only for about 30 s–1 min.
v) To minimize losses due to adsorption onto surfaces, proteins analysed with this protocol are only in contact with polypropylene (two 1.5-mL tubes and the syringe); in our experience when working at the low picomole level, significant losses can arise when using centrifugal concentration. If necessary, concentrate by vacuum-centrifugation (Savant) or, alternatively, for protein sequencing of highly diluted samples, apply the solution to PVDF by the "chromatographic loading" procedure (see 2.7).
vi) For the extraction of highly hydrophobic proteins: prepare the formic acid-isopropanol-acetonitrile solution just before use and do not store it in polypropylene tubes, in order to avoid polymer leakage from the tube to the solution that will generate artifacts detectable by mass spectrometry.

Advantages
This is a simple procedure, which does not require dedicated expensive equipment for achieving a rapid elution of proteins with very high yields even at the low picomole level. The high elution yields obtained in the absence of detergents in the elution buffer make this procedure particularly useful for protein elution for further microanalysis. When combined with the chromatographic loading onto PVDF (see 2.7), it allows the recovery and concentration of low abundant proteins from several lanes/spots.

2.7 Chromatographic Concentration on PVDF Disks or Hewlett-Packard Sequencing Columns of Low-Abundant Proteins Eluted From Gel Bands

Procedure *(Castellanos-Serra et al. 1997)*
PVDF-disks (Materials: Sequencing membrane (Problot, or Immobilon PSQ), zero dead volume on-line filter cartridge (Knauer, Supelco), Teflon O-ring).

1. Cut a disk of sequencing membrane to fit the internal diameter of the zero-dead volume cartridge (for the Knauer cartridge, 6-mm diameter).
2. Wet the membrane with methanol and water. Insert the membrane disk into the cartridge and retain it in place with the aid of the Teflon O-ring.
3. Connect the cartridge to the HPLC injector provided with a 5 mL loop (filled with 0.1 % aqueous TFA) and to the detector set at 215 nm.
4. Pump 0.1 % TFA through the cartridge to obtain a stable baseline.
5. Load the protein eluate into the loop and inject it at a flow of 0.1–0.2 mL/min.
6. After the absorbance value has returned to baseline, pump 0.1 % TFA for about 10 mins.
7. Remove the membrane from the cartridge and store at 4 °C until sequencing.

Hewlett-Packard sequencing column (Materials: Hewlett-Packard reversed-phase sequencing column (top half of the biphasic column), Hewlett-Packard sequencer sample loading funnel, or a Hewlett-Packard sequencing column zero dead volume adapter).

1. Load the sample directly onto the reversed-phase half of the biphasic column or,
2. assemble and connect the cartridge to the HPLC injector provided with a 5 mL loop (filled with 0.1 % aqueous TFA) and to the detector set at 215 nm.
3. Pump 0.1 % TFA through the cartridge to obtain a stable baseline.
4. Load the protein eluate into the loop and inject it at a flow of 0.1–0.2 mL/min. After the absorbance value has returned to baseline, pump 0.1 % TFA at 0.2ml/min for 10 mins.
5. Remove the column from the cartridge and assemble it in the sample loading device.
6. Dry the column with pressurised nitrogen to store until sequencing.

Comments

i) The minor amount of SDS that co-elutes with the protein does not interfere with an efficient retention of the protein to PVDF or the Hewlett-Packard sequencing column, making additional steps for removal of SDS unnecessary.
ii) Retention on Immobilon PSQ membrane at different loading flows was determined with four radioiodinated proteins (BSA, streptokinase, cytochrome c and insulin, 100 picomole each) eluted from SDS-PA gels. For loading flows between 25 and 500 μl/min, more than 90 % of the eluted protein was retained on the PVDF disk. At 1mL/min or higher flows, significant amounts of proteins passed through the membrane.
iii) The capacity of the membrane disk (6 mm^2) under the loading conditions here described was about 400 picomole.
iv) For the Hewlett-Packard sequencing columns, flow rate has little effect on protein retention, however care should be taken not to overpressurize the column and create leaks.

Advantages

This procedure permits the concentration of low abundant proteins eluted from several gel bands by filtering the solution through the sequencing membrane or applying directly to protein sequencing columns. This is accomplished by using standard HPLC equipment. It is an alternative to other established procedures (vacuum loading, centrifugal filtration). The advantages of the present technique are:

(a) a precise control of the loading flow on the membrane (a factor affecting protein retention on the membrane),
(b) it is an unattended operation (particularly useful when loading large volumes of very diluted samples),
(c) it allows to record the time-course for loading and membrane clean-up through UV detection, and
(d) allows for the rapid and efficient desalting of samples with minimal sample loss.

2.8
Micro-Deglycosylation

Procedure

Band conditioning: Immediately after electrophoresis, cut the band of interest. Incubate the band in 50 mM sodium phosphate, 0.3 M glycine, pH 7.5, 2 × 5 mins; and next in water, 2 × 5 mins. *Deglycosylation and analysis:* Crush the gel bands to yield 32 μm particles and collect the gel slurry. To 60 μl of the digestion buffer (10 mM sodium phosphate, 20 % glycerol, 1 % NP40) add 7.5 μl of a 1:10 dilution of PNGase F (BioLabs). This solution is added to the crushing syringe and flushed through the sieves to collect any gel particle remaining in the syringe Vortex for 10 mins. and digest for 2–3 h at 37 °C. Elute the protein as described in 2.6. Relevant information can then be obtained by comparing the electrophoretic profile of the glycosylated and deglycosylated protein both on SDS- and IEF-slab gels.

Advantages

This is a very rapid protocol for *in-gel* deglycosylation of proteins. The reaction is generally complete after 2 h; due to the combination of generating gel microparticles after a fully reversible fixative negative staining, and the use of the recombinant PNGAse F (BioLabs).

2.9
Microdigestion of Reverse Stained Proteins *(Castellanos-Serra et al. 1999)*

Immediately after electrophoresis, cut the bands of interest. Perform simple zinc chelation (see Section 2.3.1) by incubating for 2 × 8 mins. in 1 mL 50 mM Tris, 200 mM glycine, pH 8.3 containing 30 % acetonitrile. Wash briefly (2 × 1 min) with 1 mL 50 mM Tris, pH 8.3. Alternatively, perform *in-gel* alkylation (see Section 2.3.2) by incubating for 1 × 5 mins and 1 × 30 mins in 25 mM Tris, 50 mM glycine pH 8.0 containing 0.01 M DTT. To S-pyridylethylate, add 4-vinylpyridine (2 % v/v) in the reduction buffer containing the gel and incubate in the dark, at 25 °C for 60 mins. After incubation, halt the alkylation by adding β-Mercaptoethanol (2 % v/v) then discard the solution and wash the gel band extensively in water.

Digestion: Crush the gel band through a crushing syringe provided with metal sieves of 100 and 32 μm (see 2.6), to yield 32-μm gel particles. Collect the gel slurry. To 30–60 μl of the digestion buffer (see below), add 0.3–0.5 μg of a proteolytic enzyme. This solution is added to the crushing syringe and flushed through the sieves to collect any gel particles remaining in the syringe. The volume of solution should be just enough to cover the gel slurry. Vortex for 10 mins and then incubate at 37 °C for four-five h. Check during the incubation that the solution remains to cover the gel slurry. If necessary add more buffer.

Recommended Digestion Buffers: Trypsin: 50 mM NH_4CO_3; Glu-C endoproteinase: 25 mM NH_4CO_3, pH 7.8; Asp-N endoproteinase: 50 mM phosphate, pH 8.0; Lys-C endoproteinase: 50 mM Tris, pH 9.3.

Recovery of Peptides: (At each step elution volumes are about twice the volume of crushed gel. Do not store solutions containing organic solvents in polypropylene vials as they may release polymers that can interfere with analysis by mass spectrometry.) Centrifuge the gel for a few seconds and collect the supernatant. Add approximately two-fold of the crushed gel volume of digestion buffer, vortex for 10 mins, centrifuge and collect the solution. Add water to the gel slurry, vortex briefly and collect the solution after centrifugation. Then, add 50 % acetonitrile, 0.1 % TFA, vortex for 10 mins. Collect the solution. Repeat this step. If highly hydrophobic peptides are suspected to be present, replace the second extraction with 50 % acetonitrile by an extraction for 10 mins. with an aqueous solution containing 20 % formic acid, 15 % isopropanol, 25 % acetonitrile. Finally, add 90 % acetonitrile and vortex vigurously for 2 mins. centrifuge and collect the solution. Concentrate the solution by centrifugal evaporation.

Approximate Time Schedule for the Protocol (in mins):

Staining	15
Destaining and removal of SDS	16
Gel crushing	10
Digestion	240
Peptide elution	20–30

Starting with a ready to use mini-gel, this time schedule allows separation, detection, digestion and extraction in one working day.

Comments

i) During gel staining with Im-SDS-Zinc, a minor but significant amount of SDS is incorporated onto the proteins in the gel. In the present protocol, SDS is removed before digestion by incubating the slice in 30 % ACN. During this step, Zn II is also removed by incorporating glycine to the ACN solution. If several bands are processed simultaneously, the volume of 30 % ACN should be scaled up proportionally, to assure complete removal of SDS.

ii) Using total radioiodinated *E- coli* proteins separated on a 12.5 % polyacrylamide gel, losses from the gel lane during zinc chelation and removal of SDS in 30 % ACN (step 2) were determined to be only 0.5–1 %.

iii) For microdigestion, proteins are either loaded in Laemmli sample buffer without reducing agent, fully reduced and alkylated in urea solution before loading to the gel as described above or pyridylethylated in the gel band (Moritz et al, 1996). An incomplete reduction/alkylation can lead to *in-gel* reaggregation of proteins which, in turn, affect the efficiency in digestion and may generate peptides difficult to identify, therefore, special care should be taken during this step to assure full derivatization of cysteine residues.

Advantages

The advantages of this protocol in comparison to the established ones for Coomassie blue-detected proteins are derived from the constant pH during separation, detection and digestion, that avoids protein precipitation in the gel and the possibility of protein modification by stain related chemicals, and from the high mass exchange between gel microparticles and the solution. Consequently, it allows the in-solution-like conditions for digestion, with shorter digestion times and a rapid procedure for peptide recovery. The shorter digestion times also have the advantage of reducing non-specific cleavages and protease auto-proteolysis, while increasing the chance for obtaining longer peptides that facilitate their analysis. There are neither artifact peaks from the staining dye in the chromatogram nor CB-protein adducts detected in MS analysis.

2.10 Sample Preparation for Analysis by MALDI MS

The solution containing the eluted peptides is concentrated under vacuum to a volume of about 5–10 μl. Peptides (about 0.5 μl) are analyzed on dihydroxybenzoic acid spotted on the probe tips of the spectrometer. In cases where the salt content of the sample interferes with an adequate ionization of the peptides, microdesalting is accomplished by using a commercial micro reversed phase column (ZipTip; Millipore). Alternatively, the peptide sample can be desalted on the MALDI probe tip, by spotting it on a nitrocellulose/alpha-cyano-4-hydroxy-trans-cinnamic acid matrix, and rinsing several times with water after the analyte solution has dried completely, as described by Jensen et al. (1997)

2.11 Sample Preparation for Analysis by ESI-MS/MS *(Reid et al, 1998; Zugaro et al, 1998)*

The solution containing the eluted peptides is concentrated under vacuum centrifugation to a volume of about 40–50 μl. Peptides are analyzed by directly loading the digest solution onto the reversed-phase capillary HPLC column connected to the electrospray Ion-trap mass spectrometer. This is accomplished by loading the sample (25–50 μl/Inj) at a higher flow rate of 5–10 μl/min onto a 0.2-mm I.D. column in primary Buffer A (0.1 % v/v TFA). Once completed, the flow rate is reset to the operating flow rate of 1–1.5 μl/min. In this case, the salt content of the sample does not interfere with the ionization of the peptides, as microdesalting and buffer exchange is accomplished during loading onto the micro reversed phase columns.

2.12 Preparation of Capillary Reversed-Phase Microcolumns

Procedure *(Moritz & Simpson, 1992; Moritz et al, 1996a)*
The following step by step procedure for constructing fused-silica microcolumns is illustrated using the construction of a 150 × 0.2-mm I.D. column. First, two lengths of fused-silica tubing need to be cut to form the column body (0.20-mm I.D. × 0.32-mm O.D. × 150mm) and the column fluid transfer line (0.05-mm I.D. × 0.19-mm O.D. × 200mm) with the use of a fused-silica sapphire knife. Next, cut out a column frit, using the column body fused-silica tubing as a punch, from a disk of 0.45μm porosity hydrophilic PVDF (Millipore, cat. HVLP04700) and insert it approximately 5 mm into the column body using the fused-silica exiting tubing. The microcolumn body and exiting tubing are then permanently posi-

tioned in place with a drop of preheated epoxy resin at the junction and heated until the epoxy is set using a hot-air gun. The epoxy is preheated to minimize movement of the epoxy into the column by capillary action. Once the epoxy has set, the column is ready to be slurry packed under high pressure. First, the column is connected to a column packer (Shandon, U.K.) via a slurry reservoir (50-mm × 2-mm I.D) by the use of a reducing union with fingertight nuts and graphite ferrules (Moritz et al, 1996a. The column is then pressure tested to 300 Bar using *n*-propanol. A slurry of reversed-phase silica in *n*-propanol (40mg/200µl) is prepared by sonication for 10 mins and then pippeted into the empty slurry reservoir, and rapidly packed into the prefilled fused-silica column at 300 Bar with *n*-propanol as the packing solvent for 20 mins. Once packed, the fused-silica column is conditioned in 50 % methanol/50 % water at 300 Bar for a further 20 mins and then allowed to depressurize slowly. The fused-silica column is then carefully dismantled and installed into the modified HPLC for gradient micro-high performance chromatography.

2.13 Modification of Standard HPLC for Operation as Capillary HPLC

Flow generation. To achieve accurate flow rates (0.4–20µl/min) and reproducible gradient formation for microcolumn operation, a standard HPLC pumping system is used with a preinjection solvent split (1/16-in tee) installed in the flow line that directs ~95–99 % of the flow through a length (~ 300mm) of 0.075-mm I.D. × 0.320-mm I.D. fused-silica tubing to waste with the remainder directed to the microcolumn. Fused-silica capillary columns are connected directly to a Rheodyne Model 8125 injector fitted with a 50µl sample injection loop. The flow rate through the column was measured with a 5µl Hamilton syringe connected to the exiting fused-silica tubing of the microcolumn with a teflon tubing union and accurately timing the advancing meniscus. Once the flow rate was accurately measured, it could be adjusted by either changing the length of the fused-silica tubing on the split tee or by adjusting the primary pump flow rate (100–500µl/min) to achieve low flow rates in the range of 0.4–20µl/min.

UV detection. Column effluent detection is achieved by modifying the original detector by replacing the standard flow cell with low volume (≤ 0.5µl) flow cell with pathlengths of comparable size (5–10mm). This can be achieved with flow cells constructed from either fused-silica tubing (LC Packings, Netherlands) or fused quartz blocks (Hewlett-Packard, Germany). With the latter, these flow cells can be directly installed into diode-array detectors (DAD) with performance levels of signal to noise ratios close to standard flow cells *(Moritz et al, manuscript in preparation)*. For capillary LC/MS, the eluent from the UV detector was connected directly to the electrospray inlet via a 50-cm length of 0.05-mm I.D. × 0.19 O.D. fused-silica tubing. This tubing, which replaced the standard stainless steel electrospray needle, extended to the tip of the electrospray needle assembly.

3 Applications to Proteome Analysis by Mass Spectrometry

Identification and characterization of proteins isolated from natural sources by polyacrylamide gel electrophoresis has become a routine technique. However, many problems with efficient sample proteolysis and subsequent peptide extraction still plague the researcher during the analysis. By combining the high sensitivity in detection, a fully reversible protein fixation, and high elution yields from gel microparticles, the present method for microdigestion has proved to significantly improve the efficiency of digestion and peptide recovery, allowing confident protein identification when followed by peptide analysis by capillary-HPLC-tandem-MS and MALDI-TOF MS.

Fig. 3.2 shows the peptide maps for a 47-kDa protein (Streptokinase) after 4h with four proteases. For comparison, the peptide map obtained in solution for the digestion with endoproteinase Glu-C is presented, showing a remarkable similarity between *in-gel* and in-solution digestion maps. The analysis of a gel blank shows no artifact peaks.

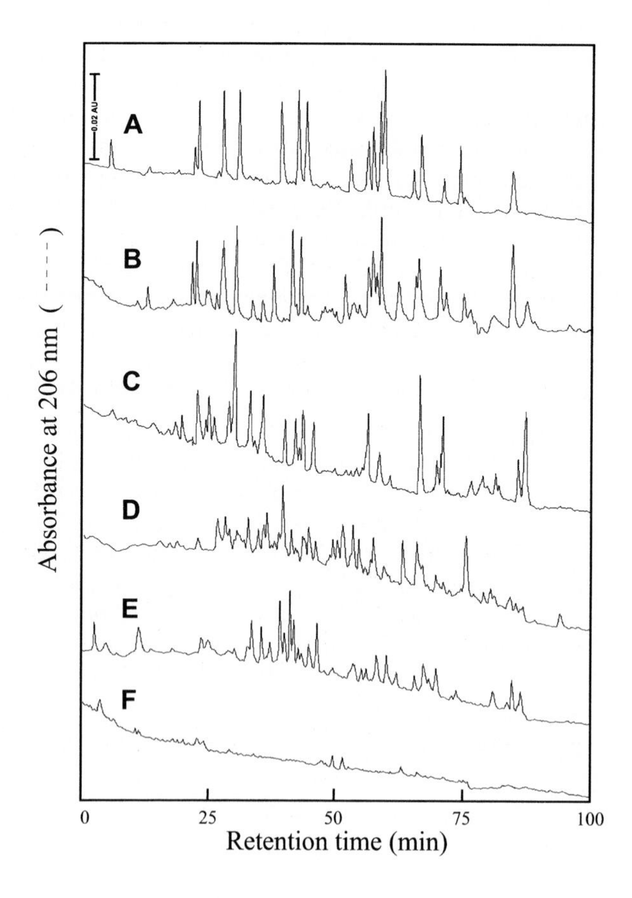

Fig. 3.2. Micro-crushed *In-gel* solution digestion map of recombinant streptokinase. Chromatographic conditions-Column: Vydac C18, 150 × 2.1 mm I.D., flow rate 200 µl/min, detection at 206 nm , Buffer A, 0.1 % TFA (v/v)/water, Buffer B, 0.1 % (v/v)/60 % (v/v) CH3CN, Gradient 0–100 % B in 60 mins *Panel A*: In- solution digestion map of recombinant streptokinase (SK) (50 pmol, 47 kDa). Protease: Glu C, digestion time: 4 h. *Panel B: In-gel* digestion (same conditions as in A). *Panel C: In-gel* digestion (50 pmol SK). Protease: trypsin. Digestion time 4 h. *Panel D: In-gel* digestion (20 pmol SK). Protease: Lys C. Digestion time: 4 h. *Panel E: In-gel* digestion (20 pmol SK). Protease: Asp N. Digestion time: 4 h. *Panel F:* Control map (blank) from digestion of a gel band with Lys C protease. *(Reproduced with permission, Castellanos-Serra, 1999)*

The efficiency in digestion and peptide elution has been evaluated by measuring the recovery of radioiodinated tryptic peptides generated from BSA. Yields after two 10 min incubation of the gel microparticles with either the digestion buffer or in combination with aqueous acetonitrile-0.1 % TFA solutions are 90–96 %. Slightly higher elution yields are obtained if an additional extraction is accomplished with an aqueous solution of formic acid-isopropanol-acetonitrile (98 %).

3.1 MALDI-TOF MS Peptide Mass Fingerprint

Analysis of sequence coverage by MALDI-TOF MS after *in-gel* digestion. Streptokinase (SK), a 47 kDa protein presenting 32 lysine and 18 arginine residues, was selected for evaluating the efficiency of this protocol by analyzing its tryptic map by MALDI-TOF MS. The peptide fraction eluted in 50 % ACN, 0.1 % TFA showed 41 major signals, all identified in the sequence (Fig. 3.3A), and several minor signals corresponding to sodium adducts of the peptides. In addition, two signals corresponded to peptides containing oxidized methionine residues, that accompanied the predominant reduced form and two signals corresponded to trypsin autoproteolysis. The identified peptides on the acetonitrile/TFA eluate covered 90 % of the sequence. A further extraction of the gel pellet under more drastic conditions by using a mixture of aqueous formic acid, isopropanol and acetonitrile showed 12 signals, 11 of them already present in the previously analyzed fraction, and one corresponding to the N-terminal peptide (37 residues) (Fig. 3.3B), for a global coverage of 98.8 % of the sequence. The missing signals corresponded to the tripeptide LLK (185–187) (monoisotopic mass 372.51) and a dipeptide AK (334–335) (monoisotopic mass 217.27), their presence could not be established as they fell below the lower limit for data acquisition (400 Da). The efficiency of the digestion was also evidenced by an analysis of the missed cleavage sites. From the total 42 major signals, 17 arose from total cleavage. The remaining 25 presented one or two missing cleavage sites, 15 of them corresponded to the basic residue followed by proline, and 9 to basic residues linked to aspartic or glutamic acid. These results are equivalent to those we have found for an in-solution tryptic digestion of streptokinase for 4 h (not shown) indicating that the efficiency of gel digestion following the present protocol is very close to that achieved for digestion in solution.

A second example of the application of these procedures is the *in-gel* deglycosylation of human erythropoietin, followed by *in-gel* digestion with trypsin. Erythropoietin (EPO, MW 18.3 kDa) is a relatively highly glycosylated protein, presenting both N- and O-glycosylation, migrating in gel electrophoresis as a 36 kDa protein. It contains 16 tryptic cleavage sites. It was deglycosylated *in-gel* for 2 h, separated by SDS-PAGE and the deglycosylated protein was digested *in-gel* for 4 h with trypsin. According to MALDI-TOF MS analysis, the sequence coverage resulting from a single *in-gel* digestion experiment and considering all possible peptides was 91 % (148 from 165 residues) while coverage considering only the peptides in the range of data acquisition was 95 %. The missing signals corresponded to peptide A^{1}-R^{4} (439.25), and peptide A^{111}-K^{116} (586.34). The dipeptides

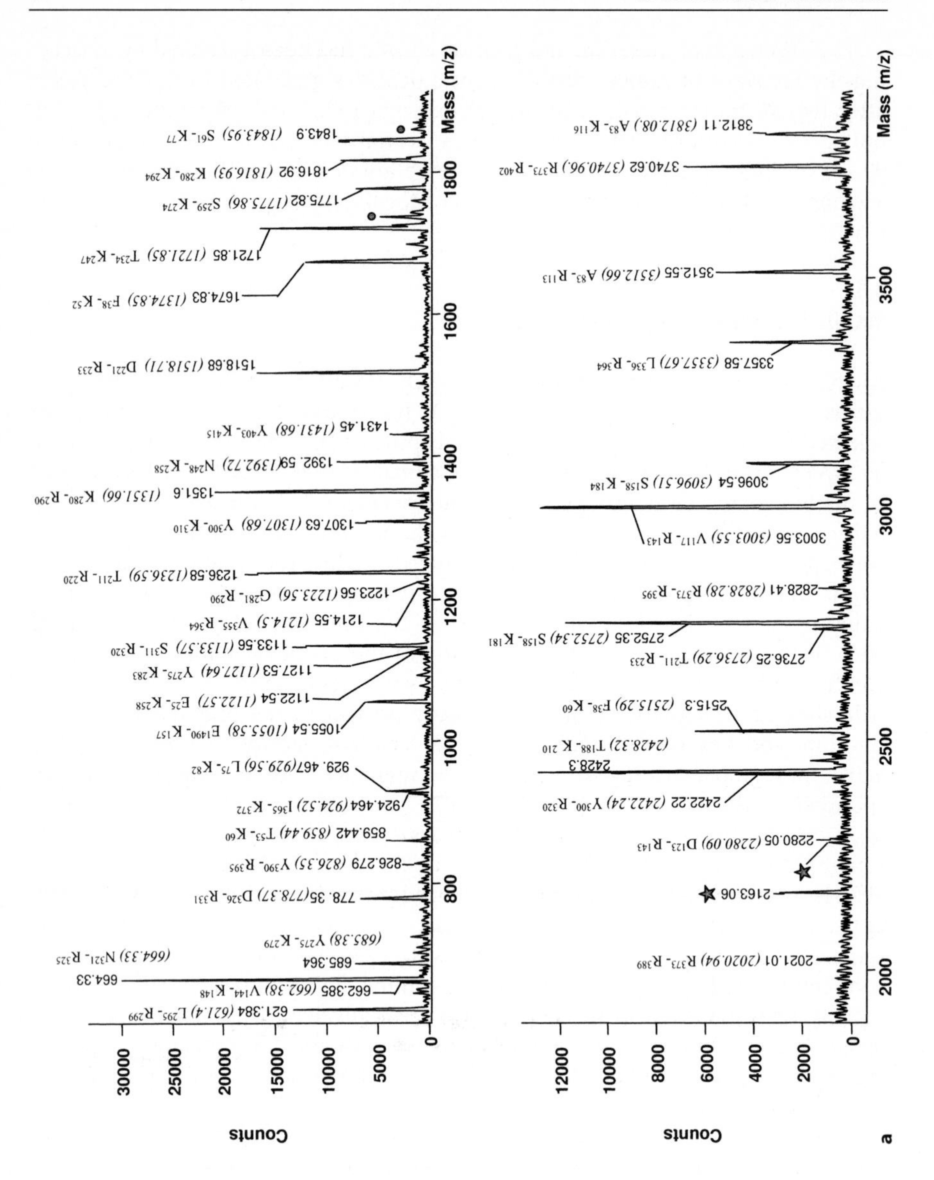

$G^{151}K^{152}$, L^{153}-K^{154} and the C terminal peptide T^{163}-D^{165} (291.10) could not be identified as their mass values fell below the lower limit of data acquisition. All the identified peptides corresponded to total cleavage on the tryptic sites. Starting with the glycosylated protein on a gel band, the entire process was accomplished in one working day: *in-gel* deglycosylation (2 h), re-electrophoresis on a mini-gel (1.8 h), detection (15 mins), gel crushing and digestion with trypsin (4.5 h) and peptide extraction (30 mins).

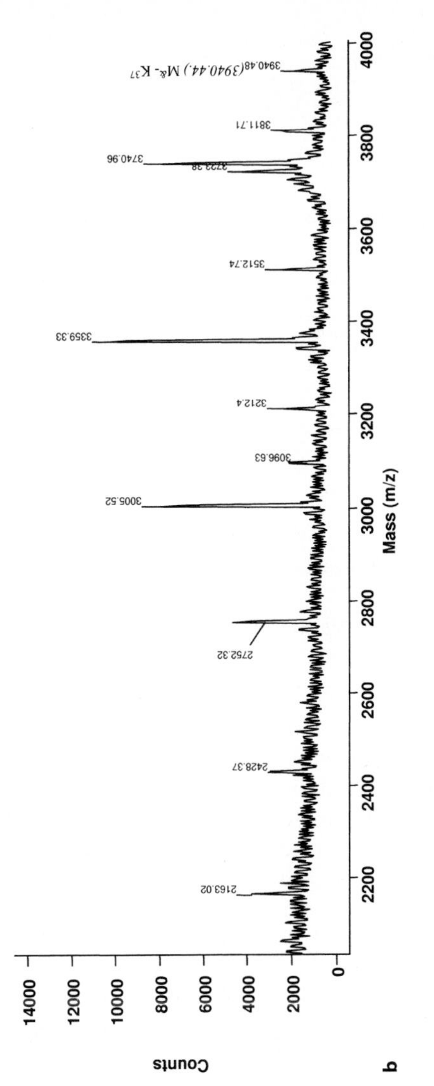

Fig. 3.3. Streptokinase (0.5 µg, 10 pmol) was separated on a 12 % SDS-polyacrylamide gel, detected by Reverse-staining and digested with trypsin in 50 mM ammonium bicarbonate during 4 h. The peptides were extracted by incubating under vortexing for 10 mins the gel microparticles successively once in water, twice in 50 % acetonitrile, 0.1 % TFA and for 1 min once in 80 % ACN. These eluates were concentrated together up to about 50 µl. The gel pellet was then re-extracted with 100 µl of formic acid: isopropanol: acetonitrile: water (20:15:25:40) and the eluate concentrated to about 50 µl. These eluates (0.5 µl) were analyzed on DHB matrix in a PerSeptive Voyager Elite mass spectrometer equipped with a reflectron and delayed extraction. Ion acceleration voltage was 20 kV and 200–256 scans were averaged. The peptide masses were matched with the cloned sequence using the program GPMAW (Lighthouse Data, Denmark). A: Analysis of peptides eluted in aqueous acetonitrile. B: Analysis of peptides eluted in formic acid: isopropanol: acetonitrile: water

3.2 Reversed-Phase Capillary HPLC-MS Analysis

The following procedure was applied for the RP-HPLC separation of peptides recovered from *in-gel* digestion of ZnII Reverse-stained visualized proteins. The combined peptide extracts (see section 2.9) were adjusted to pH 2.0 by the addition of 1 % aqueous TFA (monitored by spotting ~ 0.1µL of the extract onto pH paper) and centrifuged briefly (1–5 secs). For conventional narrowbore (< 2.1 mm I.D.)

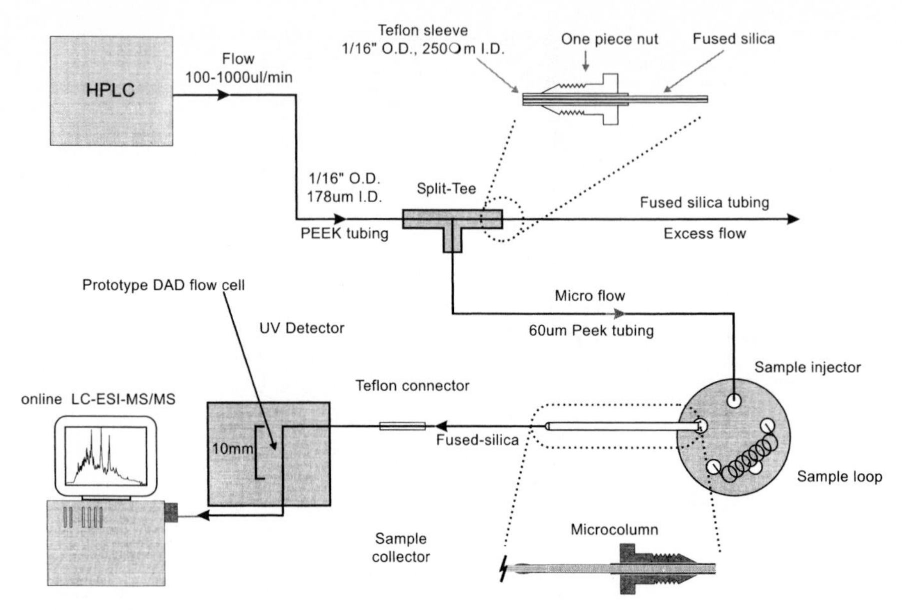

Fig. 3.4. Schematic of the gradient capillary HPLC system. *(Reproduced with permission, Castellanos-Serra et al., 1999)*

chromatography, an in-line filter (2 μm) was placed between the injector and the column. For microcolumn (< 0.3 mm I.D.) chromatography, a standard Hewlett-Packard HPLC was modified (Fig. 3.4), as described in section 2.13, to achieve the low flow rates (~ 1.5μL/min) required for capillary column chromatography. It should be noted, that for the microcolumn HPLC configuration, an in-line filter between the injector and the column was not used and, therefore, care should be taken when filling sample-loading syringes to avoid micro-gel particles.

For the MS analysis of peptides by electrospray (ESI) tandem mass spectrometry, a Finnigan-MAT LCQ ESI-ion trap mass spectrometer was employed (Zugaro et al, 1998). Fig. 3.5 shows a typical high-sensitivity ion-trap capillary LC-MS/MS analysis of a polyacrylamide (PA) gel separated protein, using recombinant interleukin-6 (IL-6) as an example. IL-6 (250ng) was electrophoresed on 4–20 % precast PA gel (Novex), and following visualization by imidazole-zinc Reverse Staining, subjected to *in-gel* digestion with trypsin. Peptides obtained from combined *in-gel* digest extracts were introduced into the ESI-ion trap using a 0.2 mm

→

Fig. 3.5. Micro-crushed *In-gel* digestion map of S-pyridylethylated recombinant human Interleukin-6. Sample: 0.25 μg recombinant IL-6 loaded onto PA gel. Gel Electrophoresis: Novex 4–20 % Pre-cast; Gel Stain: Imidazole-Zinc reverse stain. Sample injection onto RP-HPLC: 20μl/40ul (50 % of total sample digest); Column Column : Brownlee RP-300, 50 mm × 0.2 mm I.D.; Solvent A : 0.1 % aqueous TFA, Solvent B : 0.1 % aqueous TFA / 60 % acetonitrile; Flow Rate : 1.4μl/min, Sheath liquid : 2-methoxyethanol, 3μl/min. *Panel A:xKU> Total-ion-current (TIC) spectrum of a tryptic digest of IL-6; Panel B:* MS/MS TIC spectrum from panel A; *Panel C;* MS/MS spectrum of tryptic peptide T4 from panel A; *Panel D:* SEQUEST output of ranked identified match results. *(Reproduced with permission, Moritz et al, 1996b)*

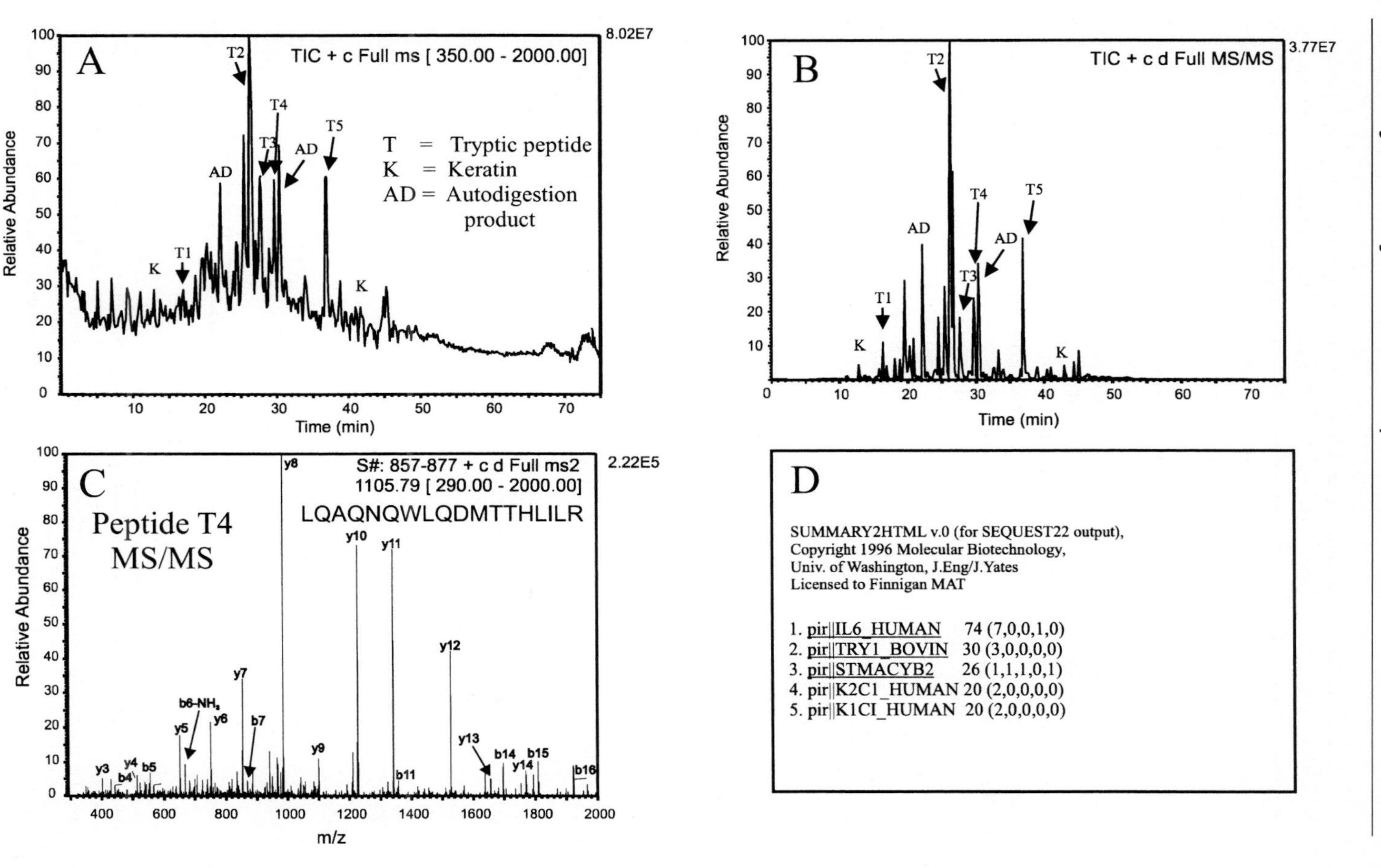
A
TIC + c Full ms [350.00 - 2000.00]
8.02E7
T = Tryptic peptide
K = Keratin
AD = Autodigestion product
Relative Abundance
Time (min)
B
TIC + c d Full MS/MS
3.77E7
Relative Abundance
Time (min)
C
Peptide T4 MS/MS
S#: 857-877 + c d Full ms2
1105.79 [290.00 - 2000.00]
LQAQNQWLQDMTTHLILR
2.22E5
Relative Abundance
m/z
D
SUMMARY2HTML v.0 (for SEQUEST22 output),
Copyright 1996 Molecular Biotechnology,
Univ. of Washington, J.Eng/J.Yates
Licensed to Finnigan MAT
1. pir||IL6_HUMAN 74 (7,0,0,1,0)
2. pir||TRY1_BOVIN 30 (3,0,0,0,0)
3. pir||STMACYB2 26 (1,1,1,0,1)
4. pir||K2C1_HUMAN 20 (2,0,0,0,0)
5. pir||K1CI_HUMAN 20 (2,0,0,0,0)

I.D. column packed with 7-μm particle diameter, 300 Å pore size, C8 packing (Brownlee RP-300) and operated at 1.4 μL /min. The standard Finnigan-ESI source was used with a sheath flow of 2-methoxyethanol introduced at 3 μL/min, with a sheath bath gas of Nitrogen set at 30 (arbitrary unit). The ESI needle was set at +4.5kV and the heated capillary was set at 150 °C. Spectra were recorded in centroid mode with three microscans summed per scan. The spectrometer was operated in automatic-gain-control mode with the trapping limits set at 2×10^8 and 8×10^7 for MS and MS/MS collision activated induced dissociation (CID), respectively, with a maximum injection time set at 200ms for both scan modes. CID experiments were performed as a "triple-play" by selection of the most intense ion observed during MS with a 3-amu isolation window, performing a Zoom-scan™ to determine the charge state, and then performing dissociation using arbitrary 55 % relative collision energy.

For the detection of S-pyridylethylated peptides, in-source-CID was performed by setting the source voltage to 70 % resonance excitation and setting the MS scan to a limited scan range of 104.5–107.5 to detect the characteristic labile 106 Da S-pyridylethyl moiety *(Moritz et al, 1996b)*. The total ion current for MS mode and MS/MS modes obtained from 50 % (~ 5picomol) of the total digest are shown in Fig, 3.5A and 3.5B, respectively

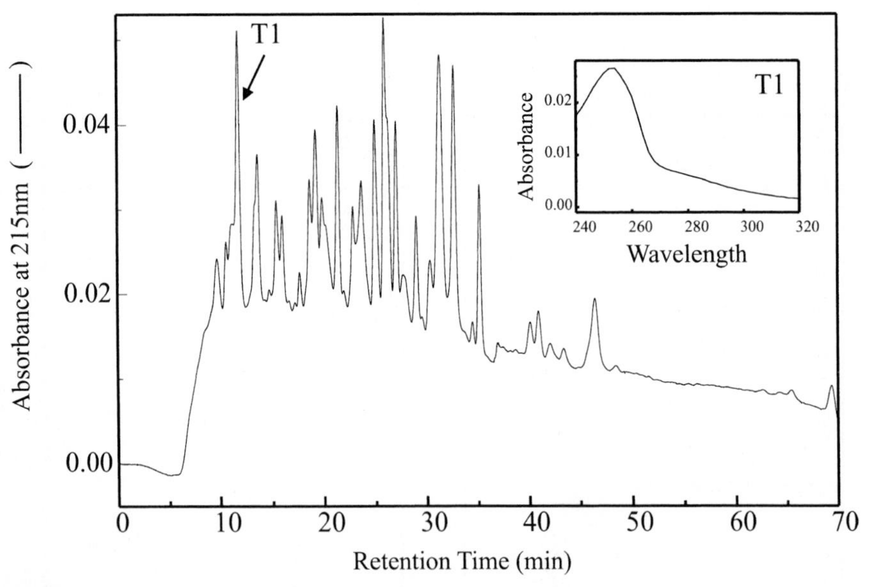

Fig. 3.6. Micro-crushed *In-gel* digestion map of S-pyridylethylated recombinant human Interleukin-6. Sample: 0.5 μg recombinant IL-6 loaded onto PA gel. Gel Electrophoresis: Novex 4–20 % Pre-cast; Gel Stain: Imidazole-Zinc reverse stain. Sample injection onto RP-HPLC: 10μl/40ul (25 % of total sample digest); Column : Brownlee RP-300, 50 mm × 0.2 mm I.D.; Solvent A : 0.1 % aqueous TFA, Solvent B : 0.1 % aqueous TFA/60 % acetonitrile; Flow Rate : 1.4μl/min, Sheath liquid : 2-methoxyethanol, 3μl/min. *Panel A:* Total-ion-current (TIC) spectrum of a tryptic digest of IL-6; *Inset:* UV spectrum from panel A; spectrum of tryptic peptide T4 from panel A

For the high-sensitivity separation of peptides for subsequent analysis, peptides obtained from in-gel digests of ZnII Reversed-stained proteins can also be separated by capillary RP-HPLC and collected manually into 1.5 ml polypropylene tubes (Moritz & Simpson, 1992b). Here, with the aid of DAD and the use of second-derivative spectra, the identification of peptides containing aromatic residues or moieties can be readily identified. Fig. 3.6 shows a typical high-sensitivity UV trace using DAD set at 215 nm of a capillary RP-HPLC separation of a PA gel separated protein, using recombinant interleukin-6 (IL-6) as an example. IL-6 (500ng) was electrophoresed on 4–20 % precast PA gel (Novex), and following visualization by imidazole-zinc Reverse-staining, subjected to *in-gel* alkylation followed by digestion with trypsin. The resultant peptides (25 %, ~ 5picomol) were fractionated by capillary RP-HPLC that was directly coupled to the Hewlett-Packard fused-quartz capillary flow cell. The column effluent was monitored at 5 different wavelengths (200nm, 215nm, 254nm, 280 nm and 290nm, data not shown) simultaneously with UV spectra between 240–320 nm collected "on-the-fly" throughout the separation. The identification of a cysteine-containing peptide was determined by monitoring the UV spectra and peptide T1 was identified with a characteristic absorbance at 254 nm (Fig 3.6, inset).

3.3 Proteomics

In summary, the combined use of the high-sensitivity ZnII Reverse-stain with state-of-the-art mass spectrometry provides a powerful suite of methods for the rapid purification, identification and characterization of low-abundance proteins available. Additionally, these methods provide analysis without the complication of additional contaminants and methods-based protein modification. With the advent of automation, these procedures should be a powerful adjunct to proteome analysis.

References

Castellanos-Serra, L., Fernández-Patrón, C., Hardy, E., Huerta, H. (1996) A procedure for protein elution from reverse stained polyacrylamide gels applicable at the low picomole level: an alternative route to the preparation of low abundance proteins for microanalysis. Electrophoresis 17:1564–1572

Castellanos-Serra, L., Fernández-Patrón, C., Hardy, E., Santana, H., Huerta, V. (1997) High yield elution of proteins from sodium dodecylsulfate-polyacrylamide gels at the low-picomole level. Application to N-terminal sequencing of a scarce protein and to in solution biological activity analysis of on-gel renatured proteins. J. Prot. Chem. 16:415–419

Castellanos-Serra, L., Proenza, W., Huerta, V., Moritz, R. L., Simpson, R. J. (1999) Proteome Analysis of Polyacrylamide Gel Separated Proteins Visualized by Reversible Negative Staining Using Imidazole-Zinc Salts. Electrophoresis 20:732–737

Cohen, S. L., Chait, B. T. (1997) Mass spectrometry of whole proteins eluted from sodium docecylsulfate-polyacrylamide gels. Anal. Biochem. 247:257–267

Dzandu, J., Johnson, F., Wise, G. E. (1988) Sodium dodecylsulfate-gel electrophoresis: Staining of polypeptides using heavy metal salts. Anal. Biochem. 174:157–167

Fernández-Patrón, C., Calero, M., Rodríguez, P. C., García, J. R., Musacchio, A., Soriano, F., Estrada, R., Frank, R., Castellanos-Serra, L., Méndez, E. (1995) Protein Reverse-staining: High-efficiency microanalysis of unmodified proteins detected on electrophoresis gels. Anal. Biochem. 224: 203–211

Fernández-Patrón, C., Castellanos-Serra, L. (1990) Presented at the Eight International Conference on Methods on Protein Sequence Analysis, July 1–6, (Abstract Booklet), Kiruna, Sweden.

Fernández-Patrón, C., Castellanos-Serra, L., Hardy, E., Guerra, M., Estévez, E., Mehl, E., Frank, R.W. (1998) Understanding the mechanism of the zinc-ion stains of biomacromolecules in electrophoresis gels: Generalization of the Reverse-staining technique. Electrophoresis 19:2398–2406

Fernández-Patrón, C., Castellanos-Serra, L., Rodríguez, P. (1992) Reverse-staining of sodium dodecylsulfate polyacrylamide gels by imidazole-zinc salts: Sensitive detection of unmodified proteins. BioTechniques 12: 564–573

Fernández-Patrón, C., Hardy, E., Sosa, A., Seoane, J., Castellanos-Serra, L. (1995) Double staining of Coomassie blue stained gels by imidazole-zinc Reverse-staining: sensitive detection of Coomassie blue undetected proteins. Anal. Biochem. 224:263–269

Gevaert, K., Demos, H., Sklyarova, T., Vandekerckhove, J., Houthaeve, T. (1998) A peptide concentration and purification method for protein characterization in the subpicomole range using MALDI-PSD sequencing. Electrophoresis 19: 909–017

Hardy, E., Pupo, E., Casalvilla, R., Sosa, A., Trujillo, L., López, E., Castellanos-Serra, L. (1996) Negative staining with zinc-imidazole of gel electrophoresis-separated nucleic acids by zinc-imidazole. Electrophoresis 17:1537–1541

Hardy, E., Pupo, E., Castellanos-Serra, L., Reyes, J., Fernández-Patrón, C. (1997) Sensitive Reverse-staining of bacterial lipopolysacharides on polyacrylamide gels by using zinc and imidazole salts. Anal. Biochem. 244:28–32

Hardy, E., Santana, H., Sosa, A., Hernández, L., Fernández-Patrón, C., Castellanos-Serra, L. (1996) Recovery of biologically active proteins detected with imidazole-SDS-zinc (Reverse-staining) on sodium dodecylsulfate gels. Anal. Biochem. 240:150–152

Heukeshoven, J., Dernick, R. (1991) Presented at the Electrophoresis Forum' 91 (Munich), Forum Communication 501–506

Jensen, O.N., Podtelejnikov, A.V., Mann, M. (1997) Identification of the components as simple protein mixtures by high accuracy peptide mass mapping and database searching. Anal. Chem. 69:4741–4750

Lee, C., Levin, A., Branton, D. (1987) Cooper staining: A five-minute protein stain for sodium dodecylsulfate-polyacrylamide gels. Anal. Biochem. 166:308–312

Matsui, N.M., Smith, D.M., Clauser, K.R., Fichmann, J., Andrews, L.E., Sullivan, C.M., Burlingame, A.L., Epstein, L.B. (1997), Immobilized pH gradient two-dimensional gel electrophoresis and mass spectrometric identification of cytokine-regulated proteins in ME-180 cervical carcinoma cells. Electrophoresis, 18:409–417

Moritz, R.L. & Simpson, R.J. (1992a) Application of capillary reversed-phase high-performance liquid chromatography to high-sensitivity protein sequence analysis. J. Chromatogr. 599:119–130

Moritz, R.L. & Simpson, R.J. (1992b) Purification of proteins and peptides for sequence analysis using microcolumn liquid chromatography. J. Microcol. Sep. 4:485–489

Moritz, R.L., Eddes, J.S., Reid, G.E., Simpson R.J. (1996b) *S*-pyridylethylation of intact polyacrylamide gels and *in situ* digestion of electrophoretically separated proteins: A rapid mass spectrometric method for identifying cysteine-containing peptides. Electrophoresis. 17:907–917.

Moritz, R.L., Eddes, J.S., Reid, G.E., Simpson, R.J. (1996a) Capillary HPLC: A method for protein isolation and peptide mapping. Methods. 6:213–226.

Ortiz, M.L., Calero, M., Fernandez-Patron, C., Castellanos, L., Mendez, E. (1992) Imidazole-SDS-Zinc Reverse-staining of proteins in gels containing or not SDS and microsequence of individual unmodified electroblotted proteins. FEBS Letters, 296:300–304

Reid, G.E., Rasmussen, R.K., Dorow, D.S., Simpson, R.J. (1998) Capillary column chromatography improves sample preparation for mass spectrometric analysis: complete characterization of human alpha- enolase from two-dimensional gels following in situ proteolytic digestion. Electrophoresis, 19:946–955

Scheler, C., Lamer, S., Pan, Z., Salnikov, J., Jungblut, P. (1998) Peptide mass fingerprint sequence coverage from differently stained proteins on two-dimensional electrophoresis patterns by matrix-assisted laser desorption/ionization-mass spectrometry. Electrophoresis 19:918–927

Zugaro, L.M., Reid G.E., Hong, Ji., Eddes J.S., Murphy, A.C., Burgess, A.W., Simpson R.J. (1998) Characterization of rat brain stathmin isoforms by two-dimensional gel electrophoresis-matrix assisted laser desorption/ionization and electrospray ionization-ion trap mass spectrometry. Electrophoresis, 19:867–876.

CHAPTER 4

Direct Analysis of Protein Complexes

J. R. Yates[1], A. J. Link[1], D. Schieltz[1], J. K. Eng[1]

1 Introduction

Deciphering the functions of genes discovered by genome sequencing will be a major challenge of the post-genome era. Physiological processes are performed primarily by proteins and these functions are often accomplished with other proteins as components of multi-protein complexes, as part of signal transduction pathways or as ligands for receptors. Identifying the sets of proteins involved in a process will be key to understanding the individual roles of proteins. Approaches to accomplish this goal will encompass at least two types of measurements. First, by measuring expression levels of proteins under different cellular conditions and states, co-regulation of protein expression can be observed. Co-regulation will not necessarily indicate proteins are part of a complex or pathway, but that their functions are regulated as part of the process. A second measurement will encompass dissection of the components of protein complexes. Proteins perform many of their functions in concert with other proteins, by forming stable complexes or through more subtle interactions. A protein's presence in a complex is more reflective of direct involvement in a process then association by co-regulation. Some proteins in complexes may be co-regulated and others may not be. By identifying the interacting proteins as a function of cellular state, insight into the networks of proteins involved in those processes will be obtained. By linking this information to that obtained by expression level measurements, a broader picture will be obtained of those proteins directly involved and those involved in more peripheral aspects of the process. A necessary element to these studies is the ability to rapidly identify the often-complex sets of proteins.

The capability of mass spectrometers to analyze and identify proteins and peptides has improved over the last few years. Refinements to ionization methods have resulted in improved sensitivities and integration with liquid separation methods. In combination with tandem mass spectrometers, these improvements have created a powerful technique to sequence peptides. The fundamental process was described by Hunt et al. in the early 1980's (Hunt et al 1986). Peptide ions are selected in the first mass analyzer and passed into a gas-phase collision cell. Ions are activated to fragment through low-energy gas-phase collisions. The

[1] Department of Molecular Biotechnology, Box 357730, University of Washington, Seattle, WA 98195-7730.

second mass analyzer separates the fragment ions and passes them to a detector. Other types of tandem mass spectrometers have been developed such as ion traps and quadrupole-time-of-flight instruments that are also capable of peptide sequencing. Peptide ions fragment primarily at amide bonds creating a pattern that can be interpreted to reveal the amino acid sequence. A drawback to peptide sequencing using tandem mass spectrometry, however, is that interpretation of the data can be time-intensive. Ionization techniques such as electrospray ionization (ESI) provided better and more robust integration of peptide sequencing methods with liquid chromatography improving the sensitivity of analysis (Covey et al. 1991; Griffin et al. 1991). Sample manipulation is streamlined as protein digests can be directly loaded onto a reversed-phase column, separated, and introduced directly into the tandem mass spectrometer for sequencing. Introduction of chromatographically separated peptides into tandem mass spectrometers is greatly augmented by an under-appreciated technology employing data-dependent instrument control algorithms to control operation of the tandem mass spectrometer (Yates et al. 1995). By using data dependent acquisition, tandem mass spectra can be acquired with higher efficiency than through manual control of conditions. As a result many more tandem mass spectra are acquired, taxing data interpretation.

In 1994 Eng et al discovered the use of protein sequence databases to interpret tandem mass spectra of peptides and identify the protein of origin (Eng et al. 1994). In this process tandem mass spectra are used to search databases using both the molecular weight of the peptide and the fragmentation pattern from the tandem mass spectrum. Each tandem mass spectrum is independently searched through the database, and matched to the amino acid sequence with the best fit. When tandem mass spectra of peptides from the same protein are present and matched to the protein sequence, considerable confidence is added to the identification. Furthermore, Eng et al showed this process capable of identifying proteins present in mixtures (Eng et al. 1994). This protein identification strategy using tandem mass spectra presented a new approach for the analysis of protein mixtures. Subsequently, McCormack et al showed direct identification of proteins in mixtures obtained from several different types of molecular biology experiments (McCormack et al. 1997). Link et al extended this approach to the identification of proteins enriched from subcellular compartments of cells (Link et al. 1997). A large collection of proteins was identified from the periplasmic space of *E. coli*. The strengths of this approach are rapid identification of proteins in mixtures, increased sensitivity, and the potential for comprehensive identification of the proteins present. We have extended this approach to the analysis of protein complexes and developed new technologies to advance the process.

2 Protein Identification in Mixtures

Identification of proteins in mixtures is highly dependent on computer software to process and analyze the data. Our efforts in this area have encompassed the development of computer algorithms to perform subtractive analysis, search databases with increased speeds, and simplification of the data review process.

3
Software for the Analysis of Tandem Mass Spectra

A key element to the identification of proteins in mixtures is the use of computer software to sort through the large amount of tandem mass spectrometry data acquired. To aid in this process, we have developed several computer programs for the analysis of tandem mass spectrometry data. The first program uses concepts of mass spectral library searching to compare tandem mass spectra and can be used for subtractive analyses (Yates et al. 1998).

Tandem mass spectra are compared using correlation analysis techniques. A query CID spectrum is cross-correlated to all the reference spectra in a reference library. The reference spectra can be a set of contaminant spectra or another LC/MS/MS analysis. The cross-correlation scores are normalized against the autocorrelation of the query CID spectrum. A score is then computed using the following relationship:

$$\text{Search score} = 10.0\,\frac{CC_{\text{mod}} \times \text{precurosr}_{\text{ion}}}{AC_{\text{mod}}\ \text{precurosr}_{\text{library}}}$$

Scores near 0 and 10 represent poor and good matches, respectively. As part of the comparison the precursor ion m/z value is used to initially compare spectra. When a small value is used for the comparison, matches occur for spectra with peptides of the same sequence. When the tolerance is decreased, spectra of peptides with small variations in their sequence can still be matched. The algorithm seems to be a robust method for the comparison of peptide tandem mass spectra.

This method provides a good approach to compare spectra for removal of contaminants and to perform subtractive or comparative analysis. Shown in Fig. 4.1A is the database searching results for tandem mass spectra obtained from a set of proteolytically digested Trypanosome brucei proteins. The protein mixture is heavily contaminated with keratin producing good scoring hits of 17 of the top 35 peptides to keratin. To eliminate tandem mass spectra of keratin peptides from the database search process, we have begun accumulating tandem mass spectra of keratin peptides. All tandem mass spectra are compared against a contaminants library and those matching to a contaminant are removed from the database searching queue. The results of the same Trypanosome brucei protein analysis are shown in Fig. 4.1B after subtraction of the tandem mass spectra matching to keratin peptides. All of the keratin tandem mass spectra have been removed by the processing algorithm eliminating any matches to keratin.

This approach can also be used to compare the tandem mass spectra obtained from two different forms of a protein to identify differences. In this manner, a mutated or modified site can be identified. Fig. 4.2 shows a comparison between normal human hemoglobin and a mutated form of hemoglobin. An equal aliquot of each protein was again divided into 4 equal aliquots and then digested with the following proteases; endoproteinase Glu-C, subtilisin, chymotrypsin, and trypsin. After digestion the pools of peptides for each protein were combined and analyzed using LC/MS/MS. By using multiple proteases to digest each protein, sequence coverage greater then 99 % could be achieved. The tandem mass spectra derived from the mutated protein were subtracted from those obtained for the

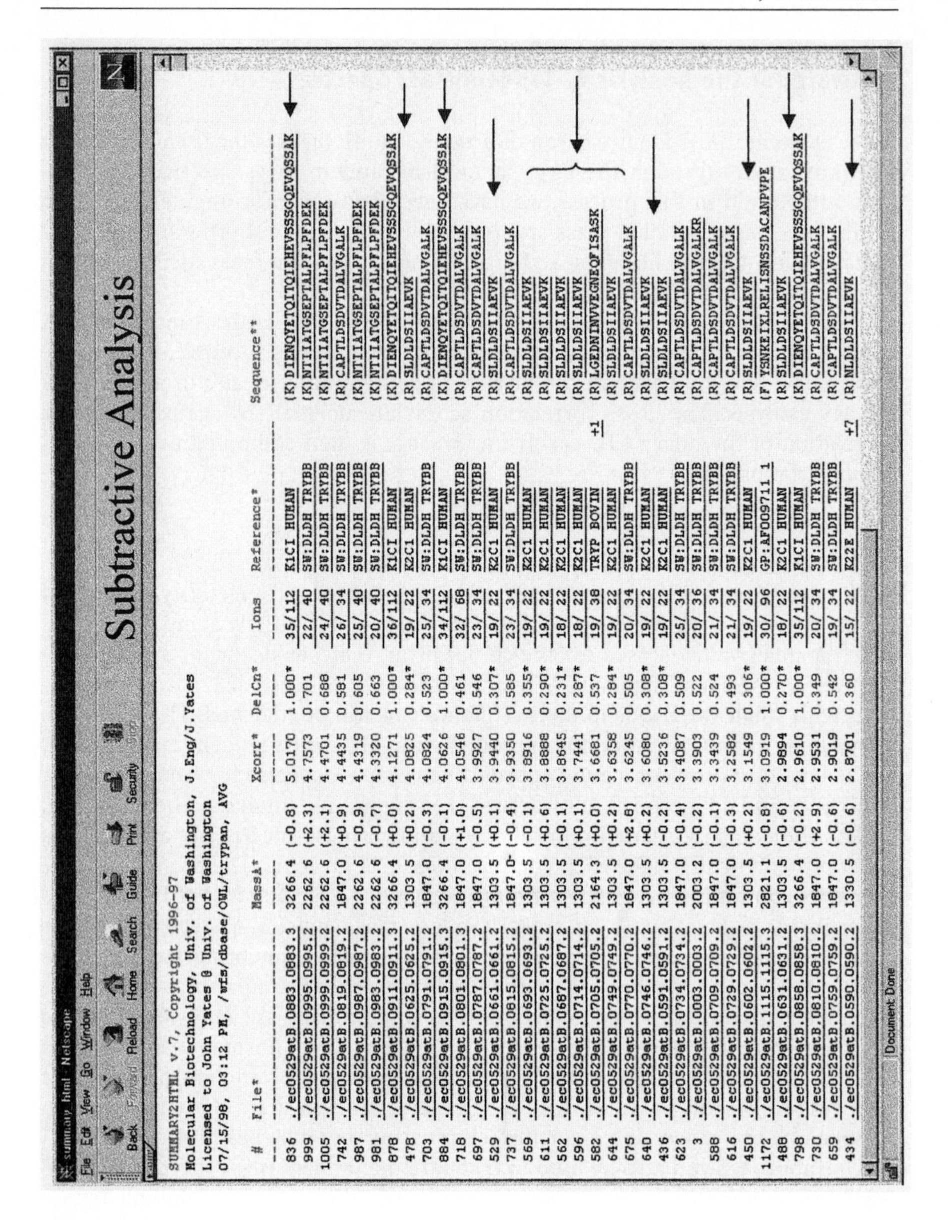

Fig. 4.1A. SEQUEST output for a database search of a collection of tandem mass spectra acquired during the analysis of proteolytically digested Trypanosome brucei protein mixture. Peptides from keratin represent 17 of the top 35 matches. Peptides from keratin are indicated with an arrow

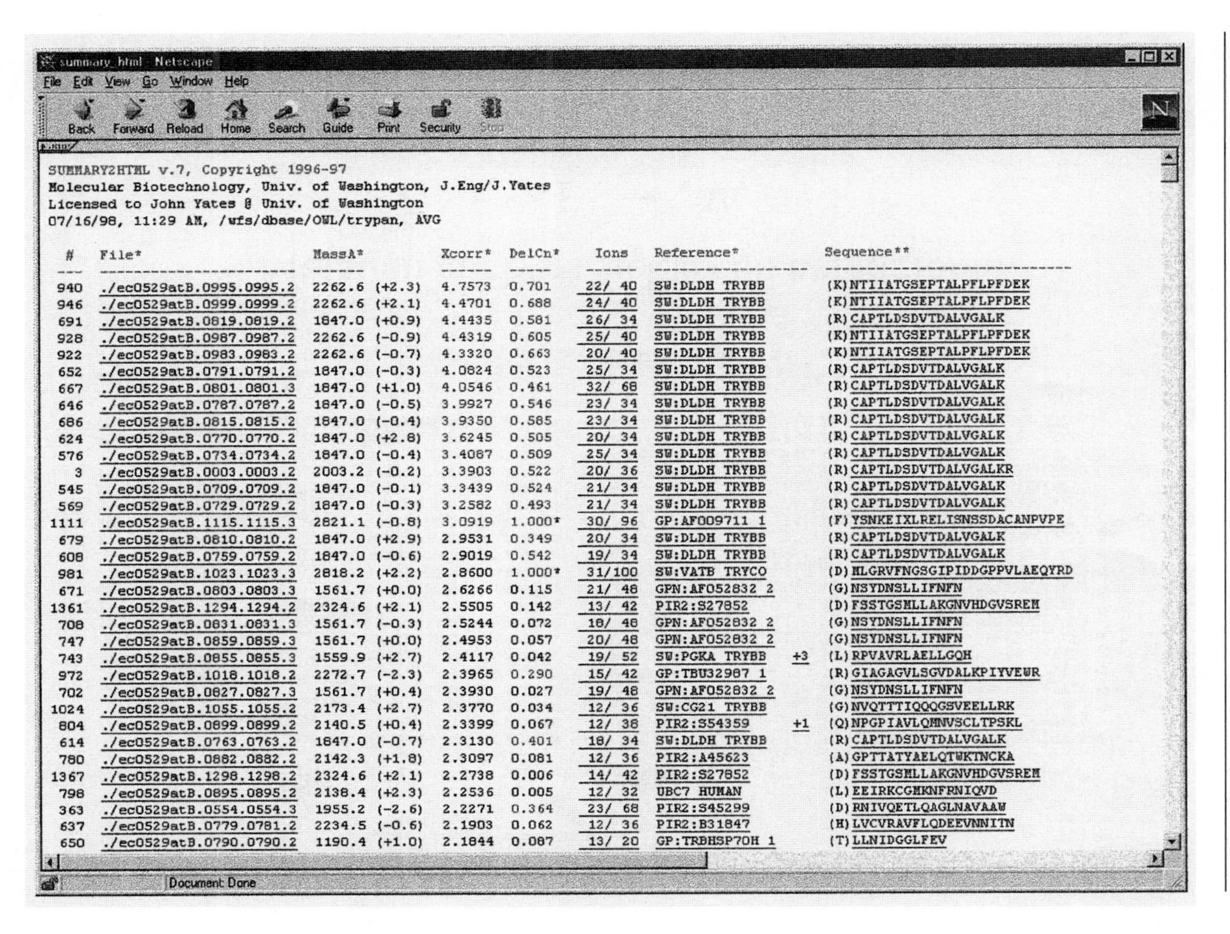

SUMMARY2HTML v.7, Copyright 1996-97
Molecular Biotechnology, Univ. of Washington, J.Eng/J.Yates
Licensed to John Yates @ Univ. of Washington
07/16/98, 11:29 AM, /wfs/dbase/OWL/trypan, AVG

#	File*	MassA*	Xcorr*	DelCn*	Ions	Reference*		Sequence**
940	./ec0529atB.0995.0995.2	2262.6 (+2.3)	4.7573	0.701	22/ 40	SW:DLDH TRYBB		(K)NTIIATGSEPTALPFLPFDEK
946	./ec0529atB.0999.0999.2	2262.6 (+2.1)	4.4701	0.688	24/ 40	SW:DLDH TRYBB		(K)NTIIATGSEPTALPFLPFDEK
691	./ec0529atB.0819.0819.2	1847.0 (+0.9)	4.4435	0.581	26/ 34	SW:DLDH TRYBB		(R)CAPTLDSDVTDALVGALK
928	./ec0529atB.0987.0987.2	2262.6 (-0.9)	4.4319	0.605	25/ 40	SW:DLDH TRYBB		(K)NTIIATGSEPTALPFLPFDEK
922	./ec0529atB.0983.0983.2	2262.6 (-0.7)	4.3320	0.663	20/ 40	SW:DLDH TRYBB		(K)NTIIATGSEPTALPFLPFDEK
652	./ec0529atB.0791.0791.2	1847.0 (-0.3)	4.0824	0.523	25/ 34	SW:DLDH TRYBB		(R)CAPTLDSDVTDALVGALK
667	./ec0529atB.0801.0801.3	1847.0 (+1.0)	4.0546	0.461	32/ 68	SW:DLDH TRYBB		(R)CAPTLDSDVTDALVGALK
646	./ec0529atB.0787.0787.2	1847.0 (-0.5)	3.9927	0.546	23/ 34	SW:DLDH TRYBB		(R)CAPTLDSDVTDALVGALK
686	./ec0529atB.0815.0815.2	1847.0 (-0.4)	3.9350	0.585	23/ 34	SW:DLDH TRYBB		(R)CAPTLDSDVTDALVGALK
624	./ec0529atB.0770.0770.2	1847.0 (+2.8)	3.6245	0.505	20/ 34	SW:DLDH TRYBB		(R)CAPTLDSDVTDALVGALK
576	./ec0529atB.0734.0734.2	1847.0 (-0.4)	3.4087	0.509	25/ 34	SW:DLDH TRYBB		(R)CAPTLDSDVTDALVGALK
3	./ec0529atB.0003.0003.2	2003.2 (-0.2)	3.3903	0.522	20/ 36	SW:DLDH TRYBB		(R)CAPTLDSDVTDALVGALKR
545	./ec0529atB.0709.0709.2	1847.0 (-0.1)	3.3439	0.524	21/ 34	SW:DLDH TRYBB		(R)CAPTLDSDVTDALVGALK
569	./ec0529atB.0729.0729.2	1847.0 (-0.3)	3.2582	0.493	21/ 34	SW:DLDH TRYBB		(R)CAPTLDSDVTDALVGALK
1111	./ec0529atB.1115.1115.3	2821.1 (-0.8)	3.0919	1.000*	30/ 96	GP:AF009711 1		(F)YSNKEIXLRELISNSSDACANPVPE
679	./ec0529atB.0810.0810.2	1847.0 (+2.9)	2.9531	0.349	20/ 34	SW:DLDH TRYBB		(R)CAPTLDSDVTDALVGALK
608	./ec0529atB.0759.0759.2	1847.0 (-0.6)	2.9019	0.542	19/ 34	SW:DLDH TRYBB		(R)CAPTLDSDVTDALVGALK
981	./ec0529atB.1023.1023.3	2818.2 (+2.2)	2.8600	1.000*	31/100	SW:VATB TRYCO		(D)MLGRVFNGSGIPIDDGPPVLAEQYRD
671	./ec0529atB.0803.0803.3	1561.7 (+0.0)	2.6266	0.115	21/ 48	GPN:AF052832 2		(G)NSYDNSLLIFNFN
1361	./ec0529atB.1294.1294.2	2324.6 (+2.1)	2.5505	0.142	13/ 42	PIR2:S27852		(D)FSSTGSMLLAKGNVHDGVSREM
708	./ec0529atB.0831.0831.3	1561.7 (-0.3)	2.5244	0.072	18/ 48	GPN:AF052832 2		(G)NSYDNSLLIFNFN
747	./ec0529atB.0859.0859.3	1561.7 (+0.0)	2.4953	0.057	20/ 48	GPN:AF052832 2		(G)NSYDNSLLIFNFN
743	./ec0529atB.0855.0855.3	1559.9 (+2.7)	2.4117	0.042	19/ 52	SW:PGKA TRYBB	+3	(L)RPVAVRLAELLGQH
972	./ec0529atB.1018.1018.2	2272.7 (-2.3)	2.3965	0.290	15/ 42	GP:TBU32987 1		(R)GIAGAGVLSGVDALKPIYVEWR
702	./ec0529atB.0827.0827.3	1561.7 (+0.4)	2.3930	0.027	19/ 48	GPN:AF052832 2		(G)NSYDNSLLIFNFN
1024	./ec0529atB.1055.1055.2	2173.4 (+2.7)	2.3770	0.034	12/ 36	SW:CG21 TRYBB		(G)NVQTTTIQQQGSVEELLRK
804	./ec0529atB.0899.0899.2	2140.5 (+0.4)	2.3399	0.067	12/ 38	PIR2:S54359	+1	(Q)NPGPIAVLQMNVSCLTPSKL
614	./ec0529atB.0763.0763.2	1847.0 (-0.7)	2.3130	0.401	18/ 34	SW:DLDH TRYBB		(R)CAPTLDSDVTDALVGALK
780	./ec0529atB.0882.0882.2	2142.3 (+1.8)	2.3097	0.081	12/ 36	PIR2:A45623		(A)GPTTATYAELQTWKTNCKA
1367	./ec0529atB.1298.1298.2	2324.6 (+2.1)	2.2738	0.006	14/ 42	PIR2:S27852		(D)FSSTGSMLLAKGNVHDGVSREM
798	./ec0529atB.0895.0895.2	2138.4 (+2.3)	2.2536	0.005	12/ 32	UBC7 HUMAN		(L)EEIRKCGMKNFRNIQVD
363	./ec0529atB.0554.0554.3	1955.2 (-2.6)	2.2271	0.364	23/ 68	PIR2:S45299		(D)RNIVQETLQAGLNAVAAW
637	./ec0529atB.0779.0781.2	2234.5 (-0.6)	2.1903	0.062	12/ 36	PIR2:B31847		(H)LVCVRAVFLQDEEVNNITN
650	./ec0529atB.0790.0790.2	1190.4 (+1.0)	2.1844	0.087	13/ 20	GP:TRBHSP70H 1		(T)LLNIDGGLFEV

Fig. 4.1B. SEQUEST output after applying a subtractive analysis algorithm to remove all tandem mass spectra of keratin peptides

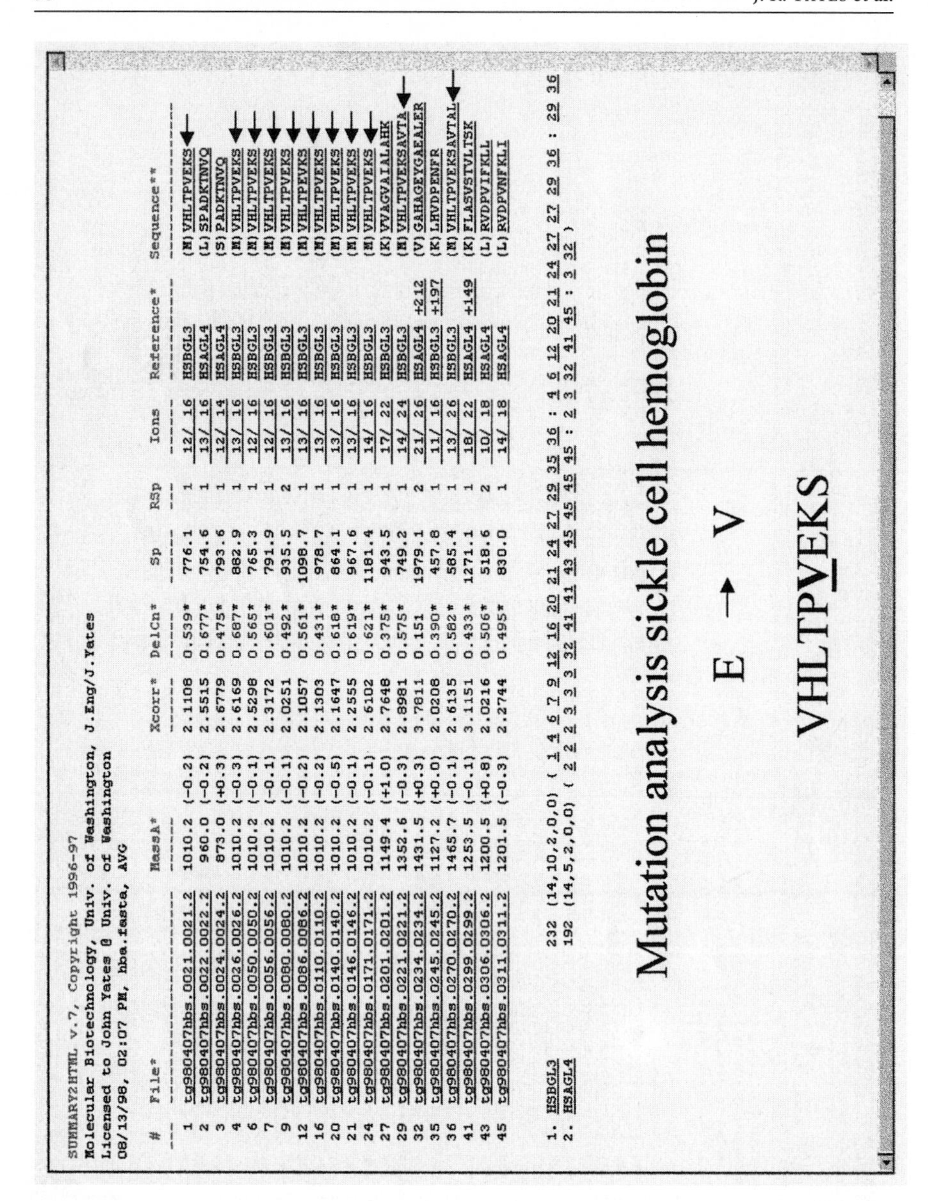

SUMMARY2HTML v.7, Copyright 1996-97
Molecular Biotechnology, Univ. of Washington, J.Eng/J.Yates
Licensed to John Yates @ Univ. of Washington
08/13/98, 02:07 PM, hba.fasta, AVG

#	File*	MassA*		Xcorr*	DelCn*	Sp	RSp	Ions	Reference*		Sequence**
1	tg980407hbs.0021.0021.2	1010.2	(-0.2)	2.1108	0.539*	776.1	1	12/ 16	HSBGL3		(M) VHLTPVEKS
2	tg980407hbs.0022.0022.2	960.0	(-0.2)	2.5515	0.677*	754.6	1	13/ 16	HSAGL4		(L) SPADKTNVQ
3	tg980407hbs.0024.0024.2	873.0	(+0.3)	2.6779	0.475*	793.6	1	12/ 14	HSAGL4		(S) PADKTNVQ
4	tg980407hbs.0026.0026.2	1010.2	(-0.3)	2.6169	0.687*	882.9	1	13/ 16	HSBGL3		(M) VHLTPVEKS
6	tg980407hbs.0050.0050.2	1010.2	(-0.1)	2.5298	0.565*	765.3	1	12/ 16	HSBGL3		(M) VHLTPVEKS
7	tg980407hbs.0056.0056.2	1010.2	(-0.1)	2.3172	0.601*	791.9	1	12/ 16	HSBGL3		(M) VHLTPVEKS
9	tg980407hbs.0080.0080.2	1010.2	(-0.1)	2.0251	0.492*	935.5	2	13/ 16	HSBGL3		(M) VHLTPVEKS
12	tg980407hbs.0086.0086.2	1010.2	(-0.2)	2.1057	0.561*	1098.7	1	13/ 16	HSBGL3		(M) VHLTPEVKS
16	tg980407hbs.0110.0110.2	1010.2	(-0.2)	2.1303	0.431*	978.7	1	13/ 16	HSBGL3		(M) VHLTPVEKS
20	tg980407hbs.0140.0140.2	1010.2	(-0.5)	2.1647	0.618*	864.1	1	13/ 16	HSBGL3		(M) VHLTPVEKS
21	tg980407hbs.0146.0146.2	1010.2	(-0.1)	2.2555	0.619*	967.6	1	13/ 16	HSBGL3		(M) VHLTPVEKS
24	tg980407hbs.0171.0171.2	1010.2	(-0.1)	2.6102	0.621*	1181.4	1	14/ 16	HSBGL3		(M) VHLTPVEKS
27	tg980407hbs.0201.0201.2	1149.4	(+1.0)	2.7648	0.375*	943.5	1	17/ 22	HSBGL3		(K) VVAGVAIALAHK
29	tg980407hbs.0221.0221.2	1352.6	(-0.3)	2.8981	0.575*	749.2	1	14/ 24	HSBGL3		(M) VHLTPVEKSAVTA
32	tg980407hbs.0234.0234.2	1431.5	(+0.3)	3.7811	0.151	1979.1	2	21/ 26	HSAGL4	+212	(V) GAHAGEYGAEALER
35	tg980407hbs.0245.0245.2	1127.2	(+0.0)	2.0208	0.390*	457.8	1	11/ 16	HSBGL3	+197	(K) LHVDPENFR
36	tg980407hbs.0270.0270.2	1465.7	(-0.1)	2.6135	0.582*	585.4	1	13/ 26	HSBGL3		(M) VHLTPVEKSAVTAL
41	tg980407hbs.0299.0299.2	1253.5	(-0.1)	3.1151	0.433*	1271.1	1	18/ 22	HSAGL4	+149	(K) FLASVSTVLTSK
43	tg980407hbs.0306.0306.2	1200.5	(+0.8)	2.0216	0.506*	518.6	2	10/ 18	HSAGL4		(L) RVDPVIFKLL
45	tg980407hbs.0311.0311.2	1201.5	(-0.3)	2.2744	0.495*	830.0	1	14/ 18	HSAGL4		(L) RVDPVNFKLI

1. HSBGL3 232 (14,10,2,0,0) (1 4 6 7 9 12 16 20 21 24 27 29 35 36 : 4 6 12 20 21 24 27 27 29 36 : 29 36
2. HSAGL4 192 (14,5,2,0,0) (2 2 2 3 3 3 32 41 41 43 45 45 45 45 : 2 3 32 41 45 : 3 32)

Fig. 4.2. SEQUEST output from the analysis of the mutant protein responsible for sickle cell anemia. Tandem mass spectra were acquired from both the normal and mutant protein after proteolytic digestion of separate aliquots using the proteases trypsin, endoproteinase Glu-C, subtilisin, and chymotrypsin. Approximately 99 % of the sequence was represented by tandem mass spectra after LC/MS/MS analysis. Tandem mass spectra acquired from the normal protein were subtracted from that of the mutant protein. The remaining spectra were then subjected to a database search using the SEQUEST-SNP program

normal protein. The remaining spectra were searched through a hemoglobin database. As shown in Fig. 4.2 a majority of the remaining tandem mass spectra match to sequences corresponding to the mutated site of E → V, the mutation responsible for sickle cell anemia.

4 Database Searching for Protein Identification

The most important element of protein identification in mixtures is the ability to search sequence databases using tandem mass spectra. Protein identification based on tandem mass spectrometry data was introduced in 1994 by Eng et al. (Eng et al. 1994) The process uses both the molecular weight of the peptide and the fragmentation pattern produced in collision induced dissociation. A peptide's molecular weight is indicative of amino acid composition while the fragmentation pattern is directly related to the amino acid sequence of the peptide. In the CID process peptides fragment primarily at amide bonds thus producing patterns that are indicative of the sequence of amino acids. Eng et al utilizes the molecular weight of the peptide as a first criterion to identify amino acid sequences in the database. Once candidate sequences are found, the peptide's sequence ions are predicted. The ions predicted by the sequence are compared to the ions present in the tandem mass spectrum. The intensities of sequence ions present in the spectrum are summed. A closeness-of-fit measure is calculated for each sequence within the molecular weight tolerance. These scores are ranked and the top 500 sequences are then compared to the tandem mass spectrum using correlation analysis. This analysis is performed by creating a model tandem mass spectrum for each sequence and comparing this model spectrum to the experimental tandem mass spectrum using a cross-correlation function. The better the fit between the two "spectra" the greater the value of the correlation.

A typical LC/MS/MS analysis using data-dependent acquisition will create hundreds or more tandem mass spectra. To increase search speed through a database the analysis can be multiplexed using multi-processor computers. The analysis is well suited for this approach since each spectrum can be sent to a different processor for analysis resulting in an almost linear increase in speed. We have created a version of SEQUEST for multiplexing the analysis of tandem mass spectra across multiple computers. The software uses the Parallel Virtual Machine (PVM) message passing software to create a virtual parallel processing computer. A computer system has been built based on the Beowolf configuration using 12 alpha based 533 MHz processors running the Linux operating system. At the start of the analysis each computer processor is sent a spectrum. As each processor completes a search, the results are transmitted back to the master computer with a request to send another spectrum. The total processing time of 500 spectra through the non-redundant protein sequence databases (313,000 entries) is 1 minute and 37 seconds. The Unigene EST sequence (53,000 entries) database can be searched with a 3-frame translation in 1 minute and 16 seconds. These improved search speeds provide a greater level of efficiency for the analysis of protein mixtures and offer the possibility of real-time data analysis.

5
Post Search Data Analysis

Autoquest is a script that runs a collection of programs. Included in Autoquest is a program to process search results into a summarized and easily readable format. In addition to preprocessing data files, performing searches, and running a summary, the following data review process occurs: The first step filters the summary file to a single entry per tandem mass spectrum. This process entails selecting between the better search result of the +2 and +3 charge state of the same spectrum. The initial criteria for making this selection between charge states is the raw cross correlation score. If the correlation score of one charge state is significantly larger than the other then that result is selected. Otherwise, the determination is based on a combination of the preliminary score rankings, the raw correlation scores, and the peptide sequences and how well they follow the cleavage rules of trypsin. After the first step is completed, another program generates an HTML formatted file containing search statistics and displays the proteins and peptides identified. The program can process the results of more then one search. The result lists each protein and the peptides identified as well as the number of times the peptide was identified and its relative score. The user has control over the scoring criteria (cross correlation scoring ranges and requiring peptides to be tryptic within each scoring range) and can interactively filter and view the results via a web browser. What makes this program particularly useful is the ability to analyze multiple SEQUEST runs together so that identifications and statistics can be tracked across several analyses.

6
Protein Mixture Analysis

Protein mixture analysis is made possible by the capability of tandem mass spectrometers to characterize a single component in a mixture of many. To perform protein mixture analysis, protein mixtures are first digested with a protease such as trypsin. The complicated peptide mixture is then subjected to separation online followed by automated tandem mass spectrometry data acquisition. There are several fundamental reasons for adopting this approach. First, the diverse chemistry inherent to proteins results in a wide range of solubility. Proteins most frequently missing in a 2-D SDS-PAGE analysis are those of limited solubility (Garrels et al. 1997). We expect each protein will produce at least one, if not many, soluble peptides within the scan range of the mass spectrometer. Second, the technology to sequence peptides using tandem mass spectrometry is well established and robust (Hunt et al. 1992; Hunt et al. 1986). The limit of detection for protein identification is more dependent on the mass spectrometer then the separation, visualization, and manipulation techniques necessary to isolate a homogeneous protein. Lastly, the number of steps involved in sample handling is reduced. Manipulation is performed with complex mixtures of peptides or proteins at stages where sample losses would be minimal because the large amount of peptide/protein present acts as a carrier. Most significant sample loss generally occurs during attempts to manipulate small quantities of homogeneous material.

Direct analysis of protein mixtures has the advantage that proteolytic digestion is performed in solution. Proteins can be aggressively denatured, reduced and alkylated and then digested. We denature proteins in 8-M urea and then digest using trypsin after dilution of the solution to 2-M urea. The collection of peptides is then separated using HPLC. To maximize sensitivity microcolumn liquid chromatography has been combined with microelectrospray ionization. Since the low flow rates of the HPLC help to focus the electrospray at the interface of the mass spectrometer and increase the elution concentration of peptides, sensitivity is increased. Gatlin et al. developed a microcolumn microelectrospray interface that eliminates the use of frits and junctions and splits the HPLC flow at the head of the column (Gatlin et al. 1998). This configuration is good for the analysis of reasonably complex protein mixtures and has been applied to the analysis of protein mixtures obtained in molecular biology experiments.

7 Multi-Dimensional Separations

Identification of proteins contained in mixtures presents a particular challenge to separations. Since an experiment can have several goals including component identification and sequence coverage to identify posttranslational modifications, it is important to achieve the best resolution possible to allow acquisition of mass spectral data for as many peptides as possible. To achieve comprehensive component identification we have used multi-dimensional liquid chromatography to improve the resolution of peptide mixtures. We use a system that combines strong cation exchange and reversed-phase chromatography. By combining these two types of separations, peptides are separated by charge and then by hydrophobicity. To maximize the sensitivity of the method, we sought to transfer as much material as possible between the two separation media. The method also had to be automated to use low flow rates. Shown in Fig. 4.3 is the configuration of the system. To transfer material between the ion exchange column and the reversed-phase column separation a step gradient is used. A step gradient of KCl is used to elute peptides which are then retained on the reversed-phase column. The solvent is then switched and salt washed off the column. A linear reversed-phase

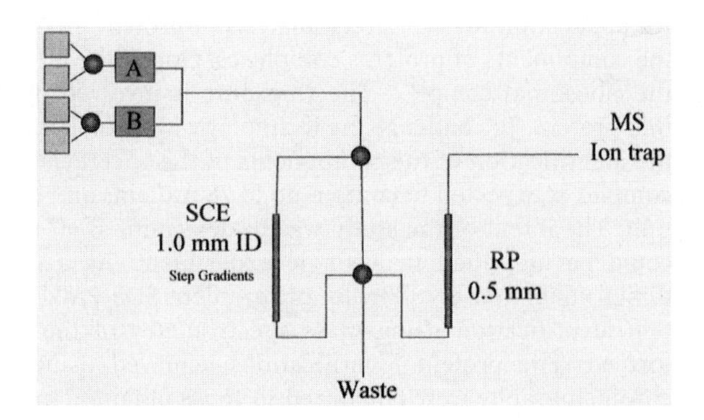

Fig. 4.3. HPLC configuration used to perform multi-dimensional separation of the S. cerevisiae ribosomal complex

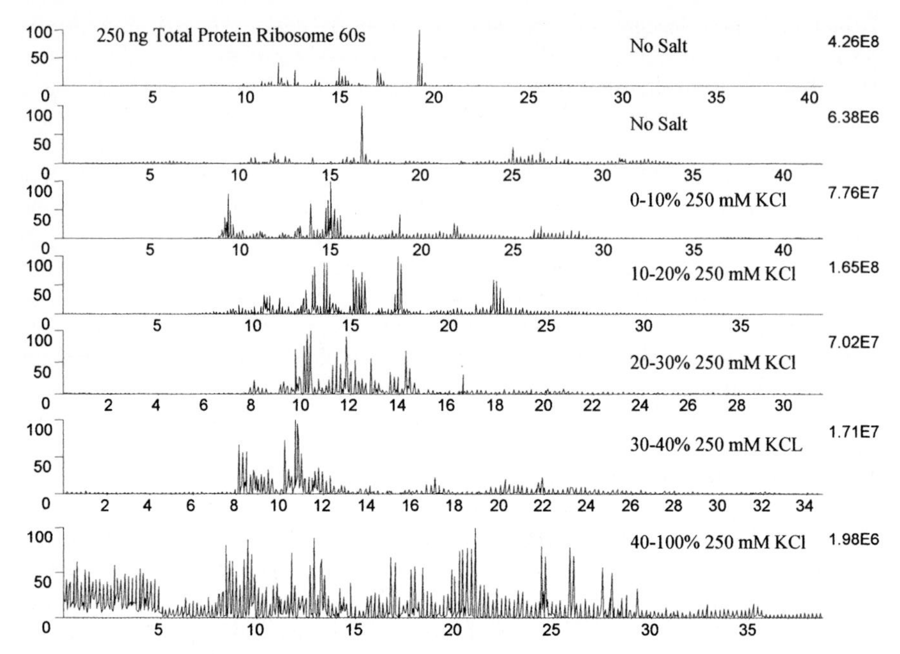

Fig. 4.4. Multidimensional separation of the 60s subunit of the yeast ribosomal complex. 250 ng of protein was injected onto the column. The effluent from the reversed-phase column was fed directly into the tandem mass spectrometer

gradient of 0–60 % acetonitrile is used to separate peptides by hydrophobicity. A second set of peptides is then eluted off the ion exchange column and the reversed-phase separation repeated. The resolution of the separation can be increased by using finer salt steps and/or longer reversed-phase gradient separations. Fig. 4.4 shows several of the salt gradient steps used to separate the peptides generated by digestion of the yeast ribosomal complex.

8 Application to Protein Complexes

An application for the direct analysis of protein mixtures is the identification of the components of proteins complexes. One of the largest complexes of the cell is the ribosomal complex. The ribosome is involved in the translation of mRNA into protein. To challenge the technology for protein identification, we undertook the identification of the components of the S. cerevisiae ribosomal complex. The complex is expected to contain up to 78 proteins and consist of a 40s and 60s subunit. The intent of the study was to determine if all the proteins of the complex could be identified in a single experiment. As a control for the study, a 2-dimensional non-equilibrium pH gradient SDS-PAGE separation coupled to protein identification using mass spectrometry of the protein complex was performed. The protein identifications achieved using multi-dimensional liquid chromatography were compared to those obtained from the 2-D gel experiment.

A total of 75 proteins were identified using direct analysis of the peptide mixture. Three proteins previously shown to be part of the complex were not found. One of the proteins has a molecular weight of 3700 Dalton and pI of 12.76. When digested with trypsin a collection of small peptides are produced which would be difficult to observe with the scan ranges employed in the mass spectrometry experiment. Two other polypeptides with molecular weights of 6600 and 6700 Da, respectively, should have been detected in the experiment. To determine if these polypeptides are present, we used non-data dependent tandem mass spectrometry to select several ions expected to be present from each protein. The expected limit of detection for this experiment is in the mid-attomole range on an ion trap mass spectrometer. Tandem mass spectra were not obtained for the expected peptides suggesting these proteins are not present within the detection limit of the experiment. A total of 12 proteins of the ribosomal complex were identified using liquid chromatography that were not identified in the 2-D gel experiment. There are several potential advantages to the direct analysis approach. First protein solubility is less of a factor in the analysis of proteins. Vigorous denaturation is possible as well as the use of detergents and organic solvents to solubilize the proteins. Second the technique is less sensitive to the molecular weight of a protein as long as peptides are produced that are within the scan range of the mass spectrometer. Third the process is fully automated. Potential pitfalls of the approach encompass the digestion of the protein and the complexity of the peptide mixture produced. Since the method requires the creation of peptides within the scan range of the mass spectrometer, very acidic or very basic proteins can produce peptides too large or too small, respectively. An analysis of the yeast genome shows 98 % of the proteins in the genome will create peptides in the mass range of 800–2800 Dalton after proteolysis with trypsin.

9
Conclusions

Unraveling the mechanisms of biological processes will require fast and sensitive analytical technology. Traditional protein analysis depends on gel electrophoresis to isolate proteins prior to sequence analysis. This step potentially compromises efficient protein recovery by requiring extraction of the protein from the polyacrylamide gel matrix. By digesting proteins in solution and analyzing the resulting complex mixture of peptides using tandem mass spectrometry, less material is lost during sample manipulation. The ability to directly identify the components of protein mixtures allows the analysis of the proteins localized to subcellular spaces, proteins immunoprecipitated, proteins interacting in affinity interaction chromatography, and protein complexes. As technologies and methodologies improve the identification and quantification of proteins in total cell lysates will be possible.

References

Covey TR, Huang EC, Henion JD (1991) Structural characterization of protein tryptic peptides via liquid chromatography/mass spectrometry and collision-induced dissociation of their doubly charged molecular ions. Anal. Chem. 63: 1193–1200
Eng JK, McCormack AL and Yates III JR (1994) An approach to correlate tandem mass spectral data of peptides with amino acid sequences in a protein database. J am Soc Mass Spectrom 5: 976–989
Garrels JI, McLaughlin CS, Warner JR, Futcher B, Latter GI, Kobayashi R, Schwender B, Volpe T, Anderson DS, Mesquita-Fuentes R, Payne WE (1997) Proteome studies of Saccharomyces cerevisiae: identification and characterization of abundant proteins. Electrophoresis 18: 1347–1360
Gatlin C, Kleemann G, Hays L, Yates III, JR (1998) Protein identification at the low femtomole level from silver stained gels using MicroLC/MicroESI/MS/MS, Anal Biochem 263: 93–101
Griffin PR, Coffman JA, Hood LE, Yates III JR (1991) Structural studies of proteins by capillary HPLC electrospray tandem mass sprectrometry. Int J Mass Spectrom Ion Proc 111: 131–149
Hunt DF, Henderson RA, Shabanowitz J, Sakaguchi K, Michel H, Sevilir N, Cox AL, Appella E, Engelhard VH (1992) Characterization of peptides bound to the class I MHC molecule HLA-A2.1 by mass spectrometry. Science 255: 1261–1263
Hutn DF, Yates III JR, Shabanowitz J, Winston S, Hauer CR (1986) Protein sequencing by tandem mass spectrometry. Proc Natl Acad Sci USA 83: 6233–6237
Link AJ, Carmack E, Yates III JR (1997) A strategy for the identification of proteins localized to subcellular spaces: application to E. coli periplasmic proteins. Int J Mass Spectrom Ion Proc 160: 303–316
McCormack AL. Schieltz DM, Goode B, Yang S, Barnes G, Drubin D, Yates I JR (1997) Direct analysis and identification of proteins in mixtures by LC/MS/MS and database searching at the low-femtomole level. Anal Chem 69: 767–776
Yates JR, Morgan SF, Gatlin CL, Griffin PR, Eng JK (1998) Method to compare collision induced dissociation spectra of peptides: potential for lilbrary searching and subtractive analysis. Anal Chem 67: 1426–1436

CHAPTER 5

De novo Sequencing of Proteins With Mass Spectrometry Using the Differential Scanning Technique

M. Wilm[1], G. Neubauer[1], L. Taylor[2], A. Shevchenko[1], A. Bachi[1]

1 Abstract

Protein identification in complete sequence databases or partial sequence databases is done preferentially by mass spectrometric techniques. However, protein de novo sequencing with mass spectrometry is more difficult to achieve. Generally, it is done by C-terminal labeling of the peptide. C-terminal fragment ions are identified in the tandem mass spectrum by the additional mass of the label. This very often allows to assign the correct amino acid sequence to a fragment spectrum of a tryptic peptide. When sub-isotopically resolved tandem mass spectra generated with a triple quadrupole machine are to be interpreted, methylation is the preferred labeling method since the mass shift is at least 14 Da. When working with a quadrupole time of flight machine which generates isotopically resolved fragment spectra, 50 % ^{18}O labeling is the method of choice. In this article a method is presented, called differential scanning, which addresses the major limitations of methylation and ^{18}O labeling preserving at the same time the sensitivity of the analysis. For the triple quadrupole based de novo sequencing there is the need to do two separate tandem MS investigations, one for the unmodified peptide mixture and the second for the methylated one. For the quadrupole time of flight based technique the difficulty is to clearly and unambiguously identify the 50 % ^{18}O labelled fragment ions via their 1:1 $^{16}O/^{18}O$ isotopic cluster in case of overlap with chemical noise ions and other ions mimicking the characteristic isotopic distribution. When the differential scanning technique is used additional information is generated to identify the C- terminal fragment ions. This information can be used to generate a simplified y ions only tandem MS spectrum and it can be exploited to improve computer algorithms to read out the amino acid sequence automatically.

[1] European Molecular Biology Laboratory, Meyerhofstrasse 1, D-69012 Heidelberg, Germany.
[2] PE-Sciex, 71 Four Valley Drive, Concord, Ontario, Canada, L4K 4V8.

2
Introduction

Mass Spectrometry is efficiently used to identify proteins whose sequence had been deposited in databases. For a protein whose complete sequence is known a precise mass measurement of an ensemble of peptides produced by enzymatic digestion of the protein can be sufficient to unambiguously retrieve its sequence in a database (Jensen ON et al. 1996, Jensen ON et al. 1997, Jensen ON et al. 1996). More specific and proven identifications can be obtained when the peptides are subjected to tandem mass spectrometric investigations. Protein identities can be revealed by reading out partial amino acid sequences from peptide fragment spectra (sequence tag approach (Mann M 1996, Mann M and Wilm M 1994)) or by correlating fragment spectra of peptides to predicted spectra from peptides in the database (Eng JK et al. 1994, Yates JR et al. 1995). Tandem MS based protein identification can be done when only partial sequences are known, as for proteins reflected in EST databases (Mann M 1996). On picomole levels of protein it is possible to reach a high throughput of dozens of identified proteins per day (McCormack AL et al. 1997).

De novo protein sequencing with the aim to design oligonucleotide probes for cloning the protein is a more difficult task. Cloning a protein with degenerate oligonucleotide primers can be time consuming even when the peptide sequences are absolutely correct. Therefore, a de novo sequencing technique should not only have a high sensitivity but a very high reliability as well. Two mass spectrometric techniques have been used to this end: MALDI high-energy collision induced dissociation (Medzihradszky KF et al. 1996) and nano electrospray low energy tandem mass spectrometry with C-terminal labeling of the peptides (Hunt DF et al. 1986, Shevchenko A et al. 1997, Wilm M et al. 1996). More than 10 novel proteins could be cloned following peptide sequencing with the latter approach (Bruyns E et al. 1998, Chen R-H et al. 1998, Lingner J et al. 1997, McNagny KM et al. 1997, T. Kawata et al. 1997, Walczak H et al. 1997, Wilm M et al. 1996).

The rational behind C-terminal peptide labeling is that C-terminal fragment ions (y ions) can be unambiguously identified in the spectrum by the additional mass of the label. Since trypsin cleaves after basic amino acids, the C-terminal residue of a tryptic peptide is a charge retention site. Therefore, C-terminal fragments can be extracted by electrical fields from the collision zone of a tandem MS mass spectrometer. As a consequence, fragment spectra of tryptic peptides contain often long y ion series. When they can be identified via the incorporated label long amino acid sequences can be retrieved.

C-terminal peptide labeling can be achieved by esterifying the peptides with methanol (Hunt DF et al. 1986) or by partial incorporation of ^{18}O isotopes when the protein is digested in 50 % ^{18}O water (Shevchenko A et al. 1997). Methylation increases the mass of y ions by 14 Da per free carboxyl group. This method was employed for de novo sequencing when using a triple quadrupole mass spectrometer in our laboratory. Peptides were first analyzed in unmodified and then in esterified form. Two conditions – precise amino acid spacings between adjacent y ions and a 14 Da mass shift per free carboxy group in the fragment of the methylated peptide – allow to read out systematically correct amino acid sequences. The disadvantages of the technique are that the peptides have to be analyzed twice and that the chemical modification tends to increase the chemical noise level in the electrospray spectrum.

The more elegant approach is to label the C-terminus by an ^{18}O isotope during the digestion. A mixture of ^{16}O and ^{18}O isotopes of every cleaved peptide is produced. When they are selected for fragmentation simultaneously all y ions are distinguishable by their unusual 1:1 $^{16}O/^{18}O$ isotopic pattern (Shevchenko A et al. 1997). Since the two isotopes differ only by 2 Da, isotopically resolved tandem mass spectra should be acquired. This is achievable without any loss in sensitivity using a quadrupole – time of flight machine (Morris HR et al. 1996, Shevchenko A et al. 1997). Long amino acid sequences can be read out identifying y ions by their specific isotopic distribution and their precise amino acid spacings.

The identification of y ions is limited by the fidelity with which the 1:1 ratio of the ^{16}O to the ^{18}O isotope is reflected in the spectrum. Some y ions may have been generated with a low efficiency so that the isotope peaks contain only 5–10 ions with a limited statistical representation of the real isotopic distribution. Or, chemical noise ions on the same m/z value disturb the isotopic distribution when fragmenting a peptide of low abundance. Here we present the differential scanning technique which addresses this issue. It facilitates peptide de novo sequencing on a triple quadrupole machine by allowing to use the ^{18}O labeling and it improves the sequence read out from fragment spectra generated from labelled peptides with a quadrupole – time of flight mass spectrometer.

3 Methods

3.1 Sample Preparation

Chemicals used were of HPLC grade.

The peptides analyzed were derived from proteins purified in the course of different biological projects. The examples had been selected to demonstrate general principles of the de novo sequencing technique.

The proteins were generally separated on one dimensional SDS gels and Coomassie Blue stained with quantities in the range between 2 pmol–10 pmol. For digestion they were destained in 1:1 water/methanol, 5 % acetic acid, washed in 1:1 water/acetonitrile and acetonitrile, the proteins were reduced with DTT and alkylated with iodoacetamide (Shevchenko A et al. 1996). Proteins were in-gel digested with trypsin either in 50 % ^{18}O water or 33 % ^{18}O water. Isotopically labelled water was purchased from Cambridge Isotope Laboratories (Andover, MA, USA) and purified by distillation before use (Shevchenko A et al. 1997).

After extraction of the peptide mixture in a 1:1 solution of 25 mM ammonium-bicarbonate and acetonitrile and a 1:1 solution of 10 % formic acid and acetonitrile, the peptide mixture was dried down for storage at 4 °C. For the mass spectrometric analysis, the peptides were taken up in 1 μl 80 % formic acid, rapidly diluted to 8 μl with water and desalted on a 100 nL Poros™ R2 column (PerSeptive Biosystems, Cambridge, USA) assembled in pulled glass capillaries (GC 120, Clark Electromedical Instruments, Pangbourne, England) (Wilm M and Mann M 1996). The peptide mixture was eluted directly into a nano electrospray capillary using two times 0.5 μl of (60 % methanol, 5 % formic acid, 35 % water). The gold covered nano electrospray capillary was mounted in the nano electrospray ion source for tandem mass spectrometric analysis.

3.2 Mass Spectrometry

Peptides were fragmented on a triple quadrupole machine (PE-Sciex, API III, Ontario, Canada) and two different quadrupole time of flight machines (QqTOF-prototype of PE-Sciex and Q-TOF from Micromass, Manchester, UK).

On the triple quadrupole machine spectra were acquired with 0.2 Da step-width. The triple quadrupole instrument was controlled in a semi-automated way using AppleScript™ macros. The fragmentation energy was chosen automatically depending in a linear way on the m/z value of the precursor and the range of the fragment spectrum acquired-one setting for the region below and a second, lower collision energy setting for the region above the precursor m/z value. The width of the Q1 selection window for the precursor was reduced to transmit less chemical noise ions when acquiring the low m/z region of the fragment spectrum.

In order to acquire the fragment spectrum of the ^{18}O isotope the selection in Q1 was increased by 1 unit on the m/z scale from the m/z value of the first isotope of the peptide.

In contrast to the acquisitions on triple quadrupole instruments, the instrument parameters were the same for the complete spectrum on the quadrupole time of flight machines. The $^{16}O/^{18}O$ isotope spectra and the ^{18}O isotope spectra were acquired with identical adjustments. In general the transmission of the selected isotopes was controlled with low collision energy settings to exclude reliably the ^{16}O containing peptide ions before the acquisition of the tandem MS spectrum.

3.3 Data Processing

Data of the tandem mass spectra were processed using the program packages IGOR Pro (Wavemetrics, Lake Oswego, USA) and Bio-MultiView (PE-Sciex). Before processing, the spectra acquired with the quadrupole time of flight machines were baseline subtracted, those obtained with the triple quadrupole mass spectrometer smoothed with Bio-MultiView. The data of the two tandem mass spectra of the different isotopes of a peptide were set to parallel m/z values in steps of 0.02 Da (QqTOF data) and 0.05 Da (Q–TOF data) using the program IGOR Pro. For triple quadrupole data the stepwidth was 0.1 Da. The quadrupole time of flight spectra of different isotopes of the same peptide were scaled to each other in windows of 20 Da and subtracted. For triple quadrupole spectra the scaling window was 10 Da wide. Only positive data values were exported to generate the subtracted spectrum. Subtracted spectra were noise filtered in Bio-MultiView with a minimum width requirement for peak consideration of 0.05 Da in the case of quadrupole time of flight data and 0.5 Da for triple quadrupole data.

4 Results

4.1 Differential Scanning Based de Novo Sequencing of Proteins with a Triple Quadrupole Mass Spectrometer

The advantage of ^{18}O based de novo sequencing in comparison with the methylation method is that only one analytical run needs to be performed instead of two. However, using a triple quadrupole mass spectrometer the fragments are generally not isotopically resolved to preserve the required sensitivity. This can lead to misassignments of masses when 50 % ^{18}O labelled peptides are investigated. The fragment ion mass assignment can be off by up to 2 Da from its real mass when its weight is over 1000 Da and its peak is of low abundance. Fragments with masses over 1000 Da have a remarkable isotopic width due to the natural 13C and 15N contribution. If the peak contains about 20 ions in total, the quantity of the different isotopes may not be correctly represented. In this situation 50 % ^{18}O labeling makes a correct mass assignment to the first isotope considerably more difficult than for unlabelled peptides. To circumvent this problem we label peptides only to 33 % when they are investigated on a triple quadrupole machine. With 33 % labeling, the ^{18}O isotope contributes to the fragment peak only when sufficient ions can be accumulated and for fragment ions up to 1300 Da the first isotope is still the most abundant one. Fig. 5.1 demonstrates that 33 % labeling allows to assign the correct first isotopic mass to an abundant and a less abundant ion in the spectrum in comparison to a 50 % labelled peptide.

The purpose of isotopic labeling was to identify y ions throughout the spectrum. With the reduced labeling it is correspondingly more difficult to recognize the y ions in the lower part of the spectrum. To facilitate the y ion recognition the differential scanning technique is applied. Two tandem MS spectra are acquired, one by transmitting the complete $^{16}O/^{18}O$ isotopic distribution into the collision zone and the second by selecting exclusively the ^{18}O containing ions. This can be done without compromising the sensitivity. The resolution of the precursor selecting quadrupole does not need to be increased to transmit only a single isotope. When the first quadrupole is adjusted to high transmission, all masses within a window of about 3 Da are transmitted into the collision zone. But, the transmission curve has a sharply rising flank starting the transmission at the preset m/z value. Setting this value to a sufficiently higher m/z value it is possible to exclude the smaller ^{16}O containing peptides but transmitting the ^{18}O labelled. This is done without changing other parameters of the quadrupole which could affect the overall ion transmission. Therefore, it is possible to generate a spectrum from the complete $^{16}O/^{18}O$ isotopic set of peptides and a second spectrum exclusively from the ^{18}O isotopes. When the two spectra are acquired they are overlaid to visualize differences between them. Fragments which do not contain the ^{18}O isotope will be at the same place in both spectra, whereas C-terminal y

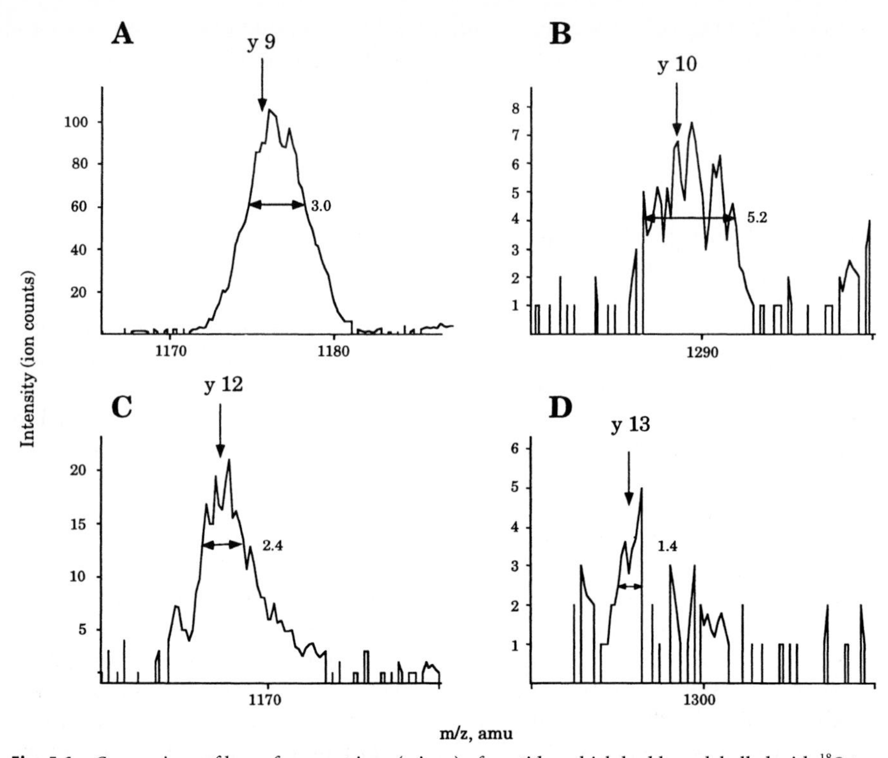

Fig. 5.1. Comparison of large fragment ions (y ions) of peptides which had been labelled with ^{18}O to 50 % (panel A, B) and to 33 % (panels C, D). Panels A and B show details from a fragment spectrum of the peptide QIQEDWELAER, panel C and D of the peptide ADALQAGASQFETSAAK investigated on a triple quadrupole mass spectrometer. The exact position of the first isotope is indicated by the arrows. When sufficient ions had been detected so that the peak shape reflects the relative isotopic abundance it is possible to assign with certainty the location of the first isotope when the peptide is labelled to 50 % (panel A) or to 33 % (panel C). In cases of very low abundant fragment ions misassignments are much more frequent for 50 % labelled than for 33 % labelled peptides (panel B versus panel D)

Fig. 5.2. Fragment spectrum of the peptide NIPGITLLNVSK labelled to 33 % with ^{18}O and investigated with the differential scanning technique on a triple quadrupole machine. Panel A shows the entire tandem MS spectrum, panel B the lower part where y ions are more difficult to identify. b* ions correspond to b ions from the partial peptide PGITLLNVSK (internal ions generated by a double fragmentation process). Panel C shows the same spectrum after the ^{18}O tandem MS spectrum had been subtracted. The comparison between panel B and C shows clearly that all N-terminal ions are suppressed relatively to the C-terminal fragment ions. The subtracted spectrum however is not noise free due to the limited resolution of the original spectrum.
Panel D and E show a subset of the original $^{16}O/^{18}O$ fragment spectrum (panel D) and the ^{18}O fragment spectrum (panel E). The ^{18}O isotope of the y5 ion is in relation to the ^{16}O isotope much more abundant in the second spectrum. This change is the basis of the manual identification of y ions in the overlay of the two spectra

A

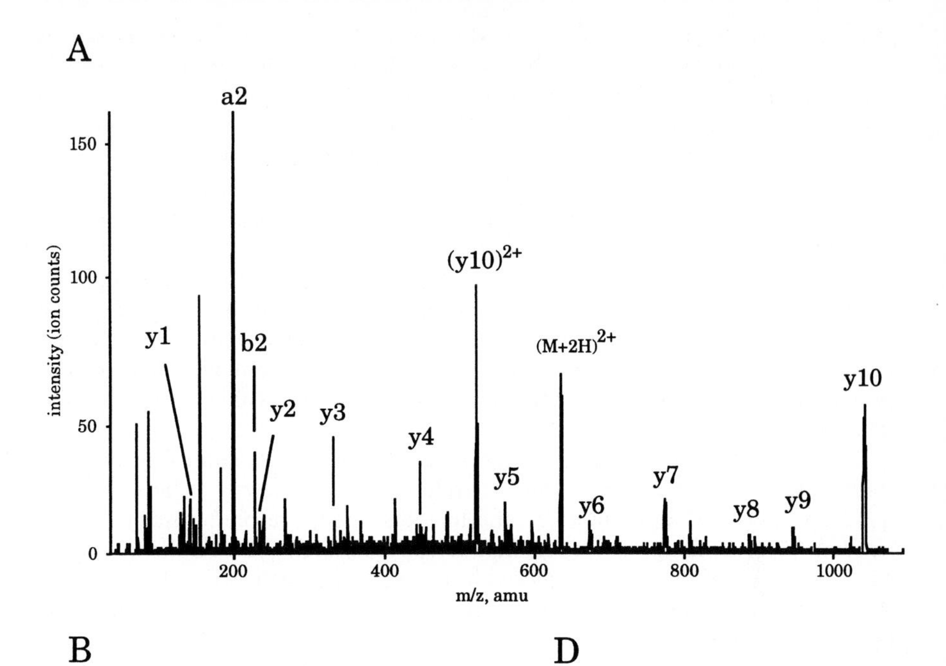
a2
150
100
50
0
intensity (ion counts)
y1
b2
y2
y3
y4
(y10)2+
y5
(M+2H)2+
y6
y7
y8
y9
y10
200
400
600
800
1000
m/z, amu

B

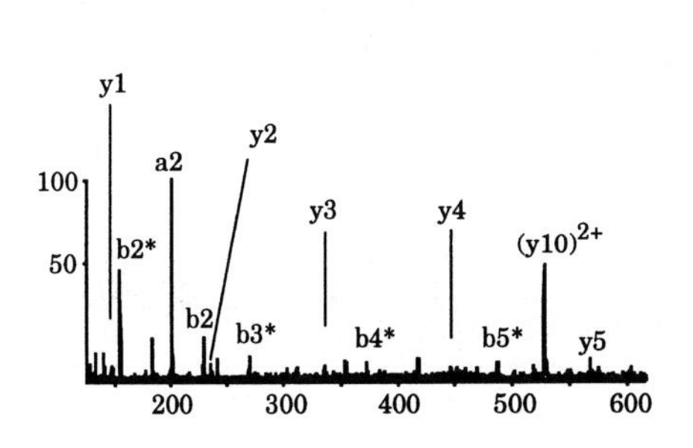
y1
a2
y2
100
50
b2*
b2
b3*
y3
b4*
y4
b5*
(y10)2+
y5
200
300
400
500
600

D

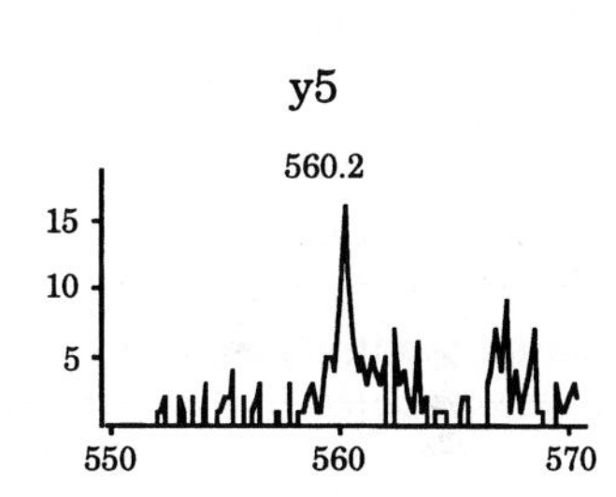
y5
560.2
15
10
5
550
560
570

C

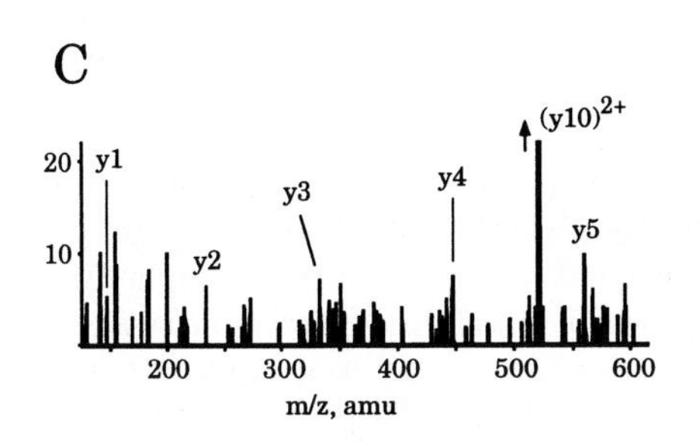
(y10)2+
20
y1
y3
y4
y5
10
y2
200
300
400
500
600
m/z, amu

E

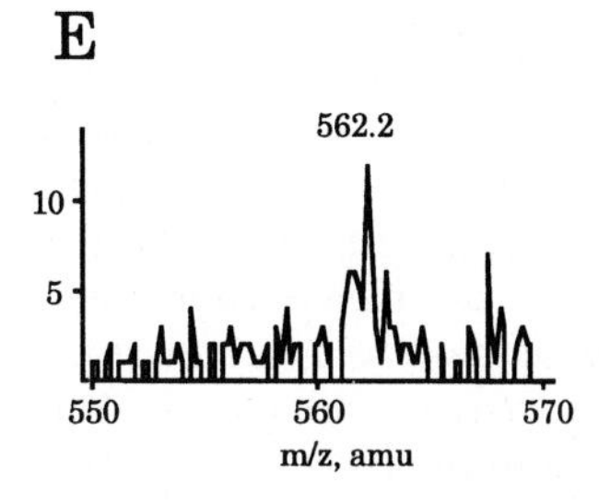
562.2
10
5
550
560
570
m/z, amu

ions will be represented differently–in the first spectrum by both oxygen isotopes and in the second only by the ^{18}O isotope. The second spectrum can be subtracted from the first after having scaled them to the same maxima. This generates a y ion filtered spectrum. Only the ^{16}O-isotope of y ions should be retained in the subtracted spectrum. The limited mass resolution achievable with a triple quadrupole instrument when used in a high transmission mode limits the quality of the filter. We interpret spectra acquired with the differential scanning technique by using the overlaid original data. The subtracted spectrum guides the attention more rapidly to putative y ions. Y ion only spectra are much simpler to interpret than complete tandem MS spectra. They can in principle be generated by an orifice fragmentation process followed by a precursor ion scan for the arginine or lysine y1 ion (Lehmann WD 1998). However, the double fragmentation process limits the sensitivity of this analysis and it is not possible to work on peptide mixtures.

The differential scanning technique relies on the precise control of the ions transmitted into the collision zone. Therefore, it is mandatory to have the first quadrupole Q1 correctly calibrated and to adjust the Q1 resolution such that the window of transmission covers a range of approximately 3 Da on the m/z scale. The ^{16}O isotope of the peptides should be correctly assigned in the Q1 spectrum. We always acquire a well resolved Q1 spectrum of the peptide mixture before the tandem MS analysis begins.

It is more difficult to identify y ions with the 33 % ^{18}O labeling technique using differential scanning than when the peptides are methylated. But since the oxygen labelling method is so much easier to use the majority of proteins sequenced de novo in our laboratory on the triple quadrupole machine are analyzed in this way.

4.2 Differential Scanning Based de Novo Protein Sequencing with a Quadrupole-Time of Flight Mass Spectrometer

The advantage of the differential scanning technique is clearly visible for spectra generated with a quadrupole-time of flight instrument. The high resolution of the spectra allows to use 50 % ^{18}O labeling. The instrument parameters can be adjusted by observing the isotopes transmitted into the collision cell with a low collision energy setting before the tandem MS investigation starts.

Subtraction of the ^{18}O tandem mass spectrum from the ^{16}O/^{18}O spectrum can yield a clear y ions only spectrum (see Fig. 5.3). The subtracted spectrum has a far lower complexity compared to the original one. Because of its simplicity and the high mass accuracy, there is often only one possibility to assemble larger ions into a series of amino acids. The filtered spectrum can serve as a source of information to interpret tandem MS spectra automatically. The original spectra contain more fragment ions like a, b and internal fragment ions but the subtracted spectrum helps in understanding the complete fragmentation pattern due to its relative simplicity. About 50 % of the peptides investigated so far using this technique could be correctly interpreted automatically from the subtracted spectrum alone. When interpreting the original data together with the subtracted spectrum

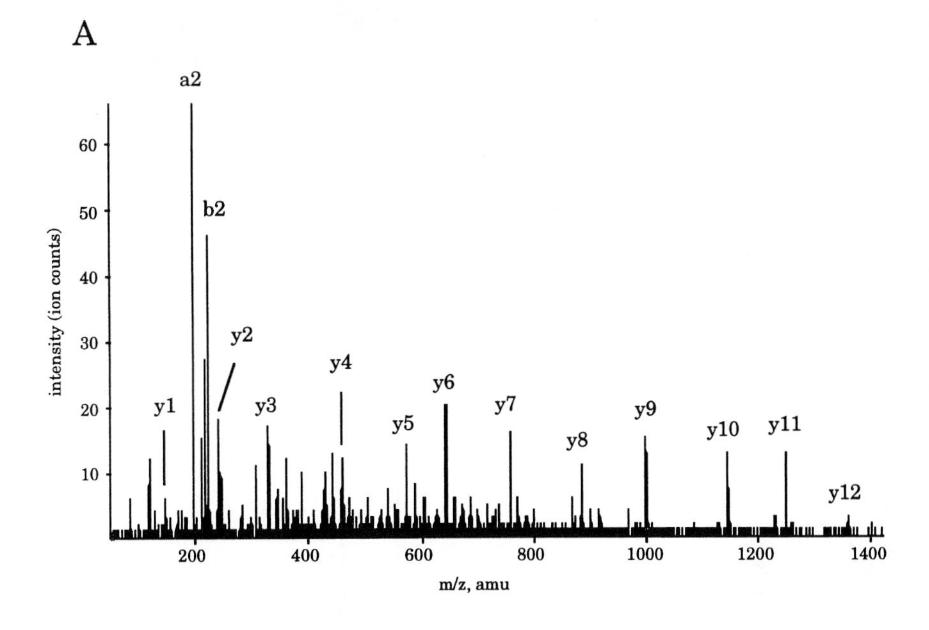

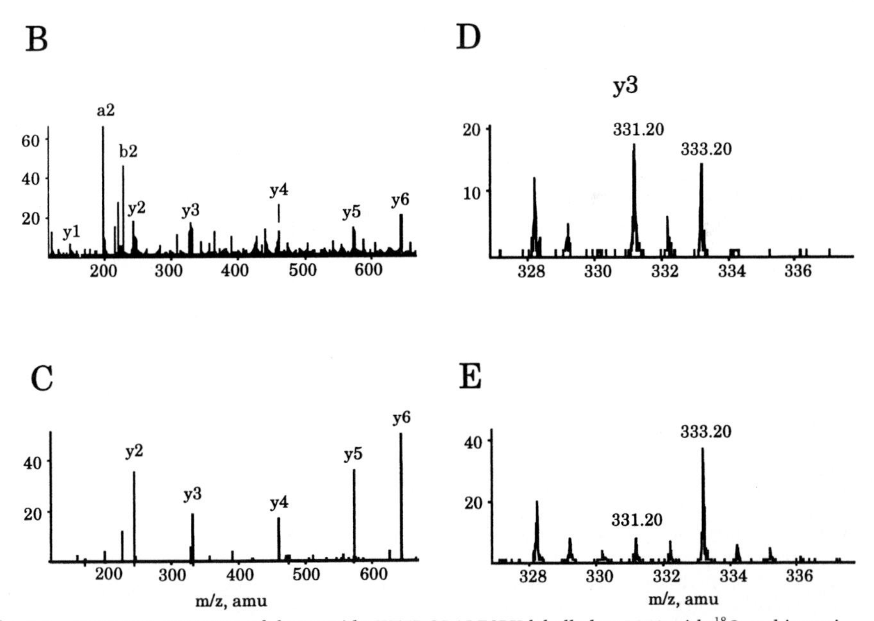

Fig. 5.3. Fragment spectrum of the peptide ILTFDQLALESPK labelled to 50 % with ^{18}O and investigated with the differential scanning technique on a quadrupole–time of flight machine. The panels show the corresponding parts of the fragment spectrum as in figure 5.2 for data acquired with a quadrupole time of flight machine. The high resolution of the original data allows to generate a nearly noise free y ions only spectrum (panel C). The comparison of the original data (panel D and E) shows clearly that only for y ions the ratio of the isotopes change between the two spectra

manually the readout of the amino acid sequence was fast and free of errors. Manual interpretation is superior because it allows to use all the information available, the difference between the $^{16}O/^{18}O$ and the ^{18}O spectrum, the typical isotopic distribution of labelled y ions and additional N-terminal fragment ions in the spectrum.

The differential scanning technique increases the information content of a tandem mass spectrum of a peptide. It allows to filter out y ions from a background of chemical noise ions where the original spectrum does not contain clear analytical information as towhich fragment ion is a y ion. Fig. 5.4 gives an example. The remaining signal at the m/z value of the ^{16}O isotope of the y3 fragment when frag-

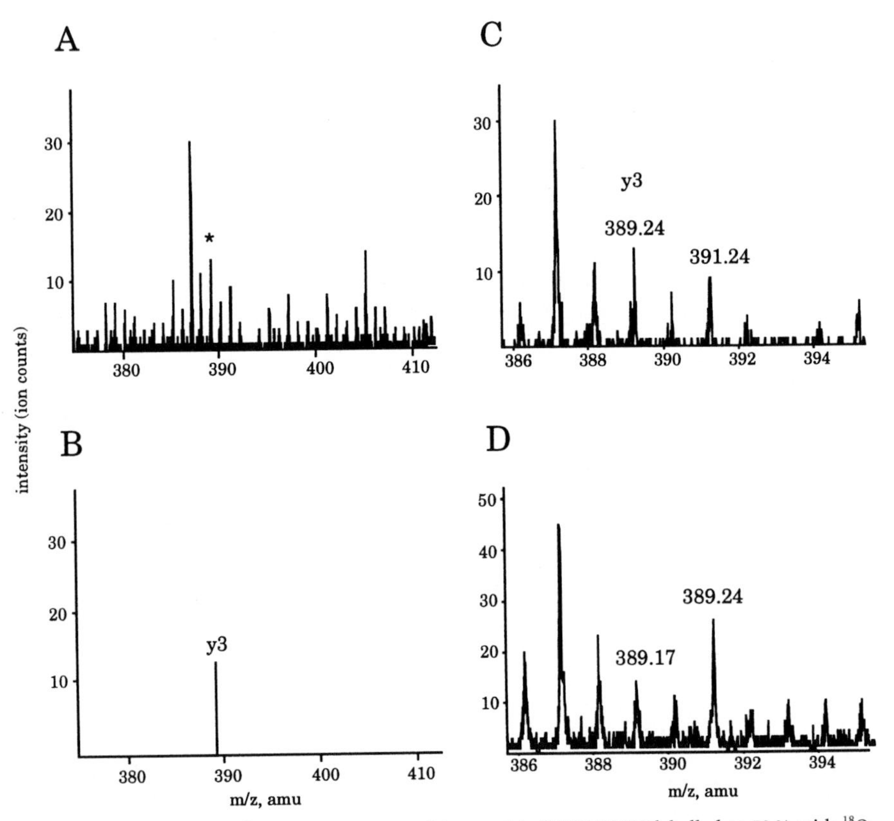

Fig. 5.4. Details from the fragment spectrum of the peptide SNTFVAELK labelled to 50 % with ^{18}O at the C-terminus. Panel A and C are subsets of the tandem MS spectrum generated from the $^{16}O/^{18}O$ precursor, panel D is a subset of the fragment spectrum of the ^{18}O precursor and panel B is one from the subtracted spectrum.
The isotopic ratio between the ^{16}O and the ^{18}O isotope of the y3 ion does not correspond to the 1:1 ratio expected due to contributions from chemical noise ions (panel A, C). This is directly visible in the spectrum generated exclusively from the ^{18}O isotope (panel D). The peak on the m/z value of the ^{16}O isotope is still present but its mass shifted by 0.07 Da indicating that the composition of the ions is different from the one establishing the peak in the $^{16}O/^{18}O$ spectrum. Despite the overlapping noise ions the relative change in intensities between the ^{16}O and the ^{18}O isotope allows to identify the y3 ion (panel C, D). By the relative subtraction of the two spectra it is possible to filter the y3 ion from the spectrum (panel B)

menting the ^{18}O-isotope is not related to the peptide but is part of the chemical background transmitted together with the peptide into the collision cell. The subtracted spectrum emphasizes differences between the two tandem MS spectra and makes the ^{16}O isotope of the y3 ion apparent. This information could not

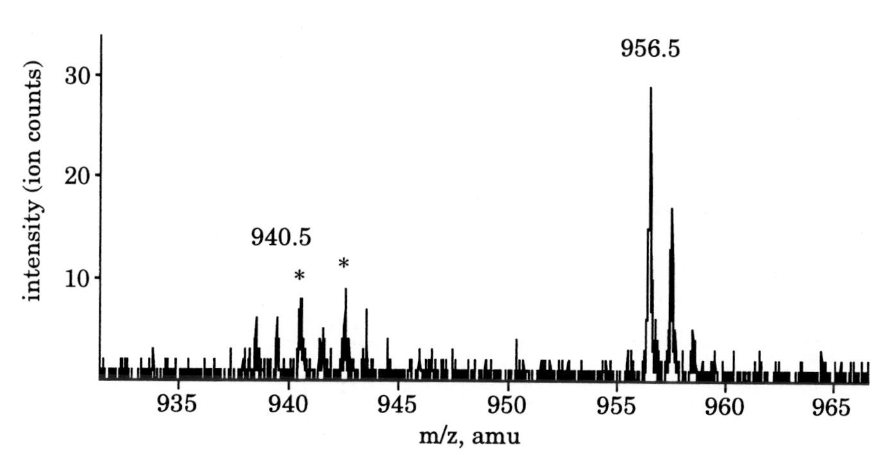

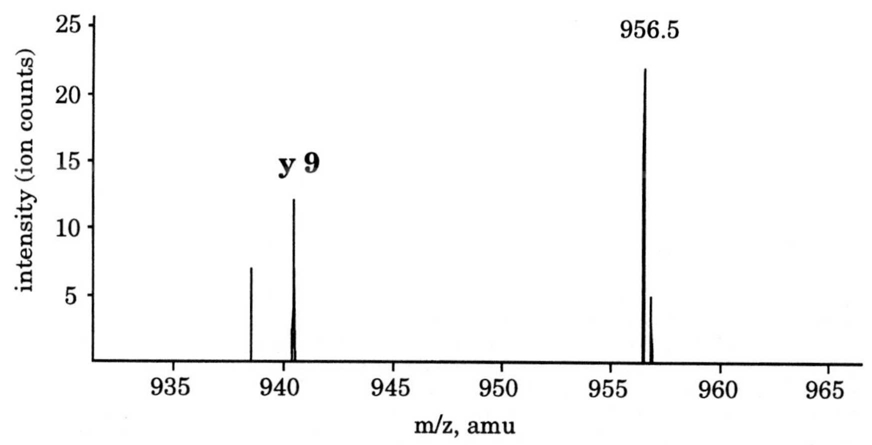

Fig. 5.5. Details of the fragment spectrum of the peptide VAPAPAPAPEVQTK labelled to 50 % with ^{18}O. Panel A shows part of the native spectrum, panel B the corresponding part after subtraction of the spectrum generated from the ^{18}O labelled precursor. The y9 ion is clearly visible in the subtracted spectrum but the 956.5 Da ion was retained as well even though it is obviously not a C-terminal fragment ion (panel A). This effect is caused by the limited perfection in aligning the two original data sets. Therefore, it is important to consider the complete fragments spectrum as well, when reading out the y ion series from the subtracted data set

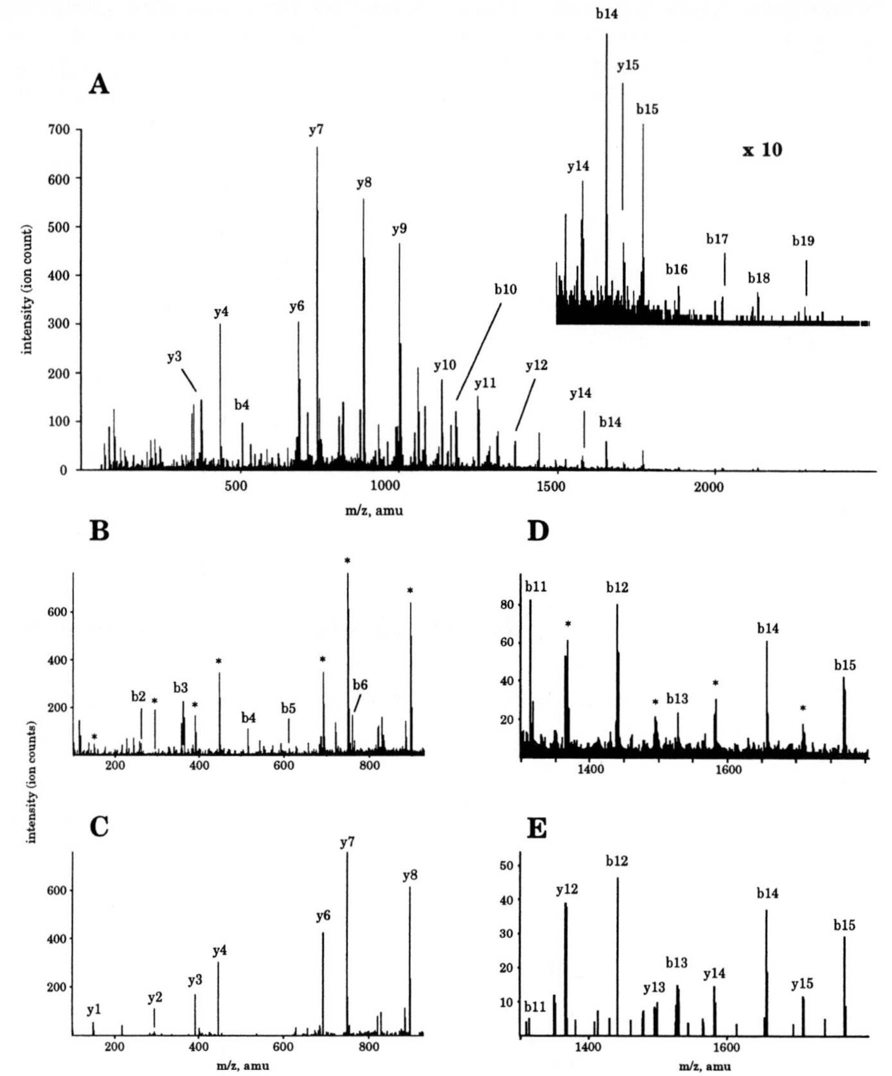

Fig. 5.6. Fragment spectrum of the peptide LFVRPFPLDVQESELNEIFGPFGP(M*)K with M* as an oxidized methionine. The peptide is labelled to 50 % with ^{18}O and was investigated on a triple quadrupole time of flight instrument. Panel A shows the complete spectrum, panel B its low m/z subset, panel D its high m/z region. Panel C and E show the corresponding subtracted data sets. The triply charged precursor has a m/z value of 1008.1. To effectively exclude the ^{16}O contribution the selection had to be moved to m/z 1009.5. This corresponds to a mass shift of 4 Da. By this shift the 12C-only isotope of the ^{18}O labelled peptide is effectively excluded from the second analysis as well. This is visible in the high m/z region of the subtracted spectrum. Non C-terminal ions (like b ions) are effectively suppressed in the low m/z region whereas in the high m/z region the 12C-only isotope of b ions is visible in the subtracted spectrum due to its under representation in the ^{18}O tandem MS spectrum

have been generated from the complete tandem MS spectrum alone because too many other ions mimic the typical 1:1 ^{16}O/^{18}O isotopic ratio.

Y ions can be made visible but, on the other hand, intensive unlabelled fragment ions may still be retained in the subtracted spectrum because of imperfect matching of the two spectra. The character of the ion, however, becomes obvious when considering its normal isotopic distribution in the original spectrum (see Fig. 5.5).

The differential scanning approach is based on the fact that the second tandem MS spectrum is generated exclusively from ions which carry the heavier ^{18}O isotope at the C-terminus. All ions with the ^{16}O isotope should be effectively excluded from the second tandem MS experiment. The natural 13C and 15N isotope contribution of the peptide needs to be taken into account. Therefore, the transmission is often shifted to ions with a higher mass value than (M+2 Da) for the second investigation, with M being the ion mass for the first isotope of the peptide. As a general guideline the selection is shifted by 1 Da on the m/z scale translating into 2 Da for doubly charged and 3 Da for triply charged ions.

The shift to higher than (M+2 Da) mass ions has a side effect. Large unlabelled fragment ions are not eliminated from the subtracted spectrum when a large peptide precursor is fragmented. For a large precursor the ion transmission for the second experiment needs to be shifted considerably higher than to a mass of (M+2 Da) to exclude the 13C and 15N isotopes of the ^{16}O labelled peptide. Therefore, the 12C-only isotope of the ^{18}O labelled peptide is excluded as well from the second experiment. In contrast to the oxygen isotope, which is at a fixed position, the excluded 12C isotope randomly distributes amongst all carbon atoms in the peptide chain. For large fragments, the probabilities add up resulting in an under-representation of the first 12C-only isotope in the fragment spectrum. When subtracting the second spectrum from the first, the 12C-only isotope of a large unlabelled fragment is retained (see Fig. 5.6). Fortunately, the high m/z area of the spectrum is virtually free of chemical noise and y ions can be recognized by their isotopic distribution in the original ^{16}O/^{18}O spectrum.

5 Summary

Differential scanning is a new technique to improve de novo peptide sequencing with tandem mass spectrometry. Y ions are filtered from the fragment spectrum of partially C-terminally labelled peptides. The y ion filter is based on the different isotopic representation of C-terminal ions in a fragment spectrum generated from the complete isotopic ^{16}O/^{18}O distribution and a fragment spectrum of the ^{18}O containing isotopes. Both tandem MS spectra can be generated with the same overall ion transmission of the precursor selecting quadrupole. This technique allows to use ^{18}O isotopic labeling on a triple quadrupole instrument for de novo sequencing. In combination with a quadrupole–time of flight mass spectrometer differential scanning improves the quality and the speed with which spectra can be read out. The technique simplifies tandem MS spectra considerably which can improve automatic interpretation. Currently, it is preferable to determine the amino acid sequences manually. This allows to combine information from the y ion filtered and the original spectrum. The differential scanning technique used on the quadrupole-time of flight mass spectrometer improves mass spectrometric de novo sequencing to a degree that it is absolutely conceivable that mass spectrometry will become the method of choice for sensitive internal protein sequencing.

Using the differential scanning technique on a triple quadrupole machine our group worked on 47 proteins for de novo sequencing in the last eight months, involving two investigators. 32 of them were identified in databases (including EST databases), 10 were identified due to close homology and 3 proteins were cloned from the data.

References

Bruyns E, Marie-Cardine A, Kirchgessner H, Sagolla K, Shevchenko A, Mann M, Autschbach F, Bensussan A, Meuer S and Schraven B (1998) T Cell Receptor (TCR) interacting molecule (TRIM), a novel disulfide linked dimer associated with the TCR/CD3/x- complex, recruits intracellular signalling proteins to the plasma membrane. J. Exp. Med. 188: 561–575

Chen R-H, Shevchenko A, Mann M and Murray AW (1998) Spindle checkpoint protein xmad1 recruits xmad2 to unattached kinetochores. J. Cell Biol. 143: 283–295

Eng JK, McCormack AL and III JRY (1994) An Approach to Correlated Tandem Mass Spectral Data of Peptides with Amino Acid Seequences in a Protien Database. Journal of the American Socienty of Mass Spectrometry 5: 976–989

Hunt DF, Yates JR, Shabanowitz J, Winston S and Hauer CR (1986) Protein sequencing by tandem mass spectrometry. Proc Natl Acad Sci U S A 83 17: 6233–6237

Jensen ON, Podtelejnikov A and Mann M (1996) Delayed Extraction Improves Specificity in Database Searches by Matrix-assisted Laser Desorption/Ionisation Peptide Maps. Rapid Communication in Mass Spectrometry 10: 1371–1378

Jensen ON, Podtelejnikov AV and Mann M (1997) Identification of the components of simple protein mixtures by high accuracy peptide mass mapping and database searching. Anal. Chem. 69: 4741–4750

Jensen ON, Vorm O and Mann M (1996) Sequence Patterns Produced by Incomplete Enzymatic Digestion or one-step Edman Degradation of Peptide Mixtures as Probes for Protein Database Searches. Electrophoreses 17: 938–944

Lehmann WD (1998) Single series peptide fragment ion spectra generated by two-stage collision-induced dissociation in a triple quadrupole. Journal of the American Society of Mass Spectrometry 9 6: 606–611

Lingner J, Hughes TR, Shevchenko A, Mann M, Lundblad V and Cech TR (1997) Reverse Transcriptase Motifs in the Catalytic Subunit of Telomerase. Science 276: 561–567

Mann M (1996) A shortcut to interesting human genes: peptide sequence tags, expressed-sequence tags and computers. Trends in Biological Science 21: 494–495

Mann M and Wilm M (1994) Error-Tolerant Identification of Peptides in Sequence Databases by Peptide Sequence Tags. Analytical Chemistry 66: 4390–4399

McCormack AL, Schieltz DM, Goode B, Yang S, Barnes G, Drubin D and Yates JR (1997) Direct analysis and identification of proteins in mixtures by

LC/MS/MS and database searching at the low-femtomole level. Analytical Chemistry 69: 767–776

McNagny KM, I.Pettersson, Rossi F, A.Shevchenko, M.Mann and Graf T (1997) Thrombomucin, a novel cell surface protein that defines thrombocytes and multipotent hematopoietic progenitors. J Cell Biol 138: 1395–1407

Medzihradszky KF, Adams GW and Burlingame AL (1996) Peptide Sequence Determination by Matrix-Assisted Laser Desorption Ionization Employing a Tandem Doulble Focusing Magnetic-Orthogonal Acceleration Time-of-Flight Mass Spectrometer. Journal of the American Society for Mass Spectrometry 7: 1–10

Morris HR, Paxton T, Dell A, Langhorne J, Berg M, Bordoli RS, Hoyes J and Bateman RH (1996) High Sensitivity Collisionally-activated Decomposition Tandem Mass Spectrometry on a Novel Quadrupole/Orthogonal-acceleration Time-of-flight Mass Spectrometer. Rapid Communication in Mass Spectrometry 10: 889–896

Shevchenko A, Chernushevich I, Ens W, Standing KG, Thomson B, Wilm M and Mann M (1997) Rapid 'de Novo' Peptide Sequencing by a Combination of Nanoelectrospray, Isotopic Labeling and a

Quadrupole/Time-of flight Mass Spectrometer. Rapid Communications in Mass Spectrometry 11: 1015–1024

Shevchenko A, Wilm M, Vorm O and Mann M (1996) Mass Spectrometric Sequencing of Proteins from Silver-Stained Polyacrylamide Gels. Analytical Chemistry 68: 850–858

T.Kawata, Shevchenko A, Fukuzawa M, Jermyn KA, Totty NF, Zhukovskaya NV, A.Sterling, Mann M and Williams JG (1997) SH2 signaling in a lower eukaryote: a STAT protein that regulates stalk cell differentiation in Dictyostelium. Cell 89: 909–916

Walczak H, Degli-Esposti MA, Johnson RS, Smolak PJ, Waugh JY, Boiani N, Timour MS, Gerhart MJ, Schooley KA, Smith CA, Goodwin RG and Rauch CT (1997) TRAIL-R2: a novel apoptosis-mediating receptor for TRAIL. EMBO J. 16: 5386–5397

Wilm M and Mann M (1996) Analytical Properties of the Nanoelectrospray Ion Source. Analytical Chemistry 68: 1–8

Wilm M, Shevchenko A, Houthaeve T, Breit S, Schweigerer L, Fotsis T and Mann M (1996) Femtomole Sequencing of Proteins from Polyacrylamide Gels by Nano-Electrospray Mass Spectrometry. Nature 379: 466–469

Yates JR, Eng JK, McCormack AL and Schieltz D (1995) Method to Correlated Tandem Mass Spectr of Modified Peptides to Amino Acid Sequences in the Protein Database. Anal. Chem. 67: 1426–1436

CHAPTER 6

Characterization of Gel Separated Proteins

A.-CH. BERGMAN[1] M. OPPERMANN[1], U. OPPERMANN[1], H. JÖRNVALL[1]
and T. BERGMAN[1]

Recent developments in preparation and structure determination of proteins have made analysis possible at a sensitivity and with a throughput that was hardly anticipated only a few years ago. In particular, efficient tools for screening and characterization of gene products are important. Protocols and combinations involving sequencer analysis and mass spectrometry have in this regard a high potential (Bergman et al., 1998).

Gel electrophoresis in two dimensions is well suited for preparation of protein material for structural analysis (cf. Patterson 1994) and can replace long purification protocols with column approaches. In addition, SDS/polyacrylamide gel electrophoresis is often the only choice for efficient purification of hydrophobic polypeptides.

For characterization of gel separated proteins regarding amino acid sequence and identity we use a strategy involving in-gel digestion and analysis of proteolytic peptides or electroblotting of the intact protein to a polyvinylidene difluoride (PVDF) membrane for direct sequencer analysis (Fig. 6.1). Both routes are efficient and the choice depends on the amount available but also on time and likelihood of finding a blocked N-terminus (Fig. 6.1).

1 In-Gel Digestion and Further Analysis of Fragments

After electrophoresis, the one-dimensional (1-D) or two-dimensional (2-D) gel is stained, preferrentially with Coomassie blue. Although silver-staining is compatible with the protocol for proteolytic cleavage, the recovery of fragments is low. The protein band or spot is cut and applied to in-gel cleavage with trypsin followed by mass spectrometry (MS) of proteolytic fragments or Microblotter peptide isolation (Fig. 6.1). Peptide mass mapping is powerful for identification (Henzel et al., 1993) and in combination with 2-D gel electrophoresis, hundreds of spots can be analyzed in just one separation (Schevchenko et al., 1996). Mass mapping is frequently carried out by matrix-assisted laser desorption ionization (MALDI) MS but electrospray ionization (ESI) MS is also used to analyze the unseparated peptide mixture. Computer algorithms applied to the mass data obtained, are combined with the cleavage specificity of the enzyme and the esti-

[1] Department of Medical Biochemistry and Biophysics, Karolinska Institutet, SE-171 77 Stockholm, Sweden.

mated size of the protein substrate. The computer screens available protein and nucleic acid databases and the output is generated in the form of a list with possible candidates. If the protein is not in the database, the peptide extract can be further analyzed with tandem mass spectrometry (MS/MS, Fig. 6.1) using the nanoelectrospray ion source (Wilm and Mann, 1996). Via collision-induced dis-

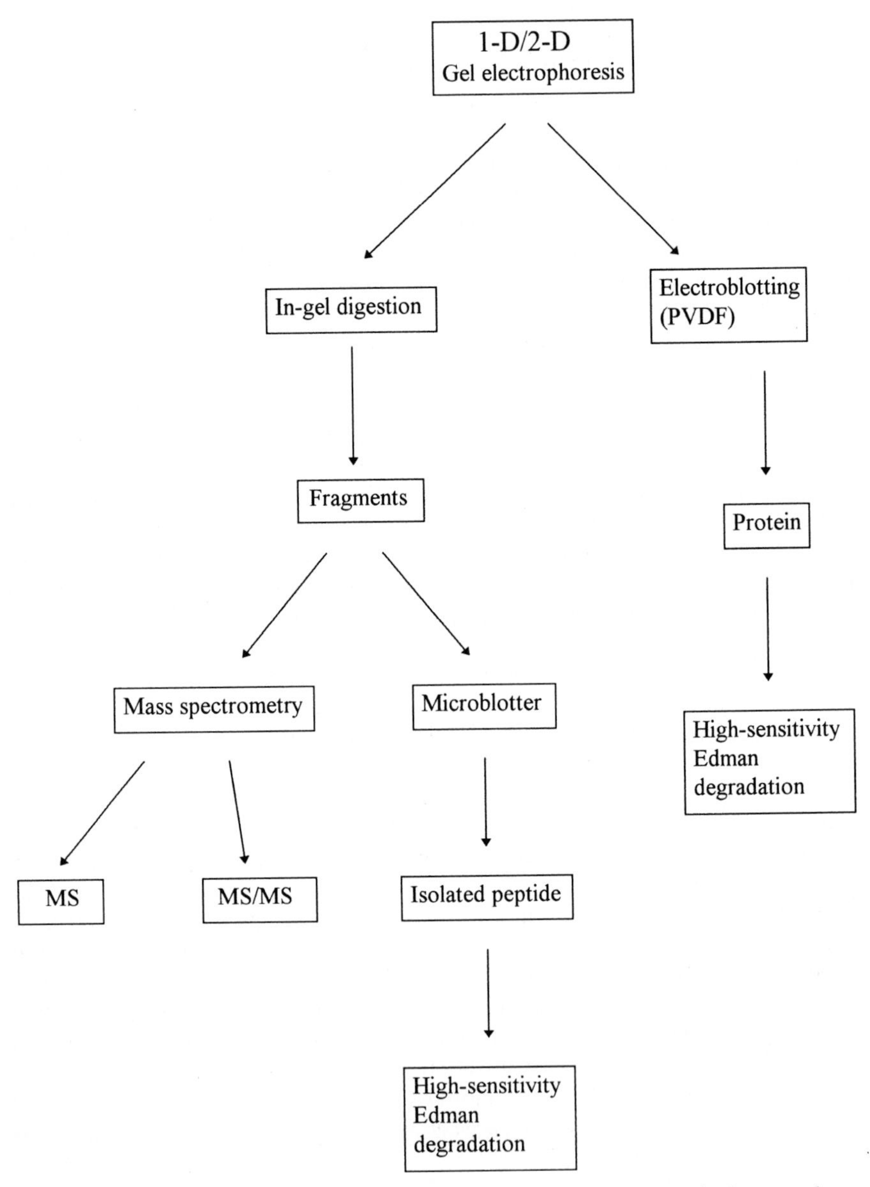

Fig. 6.1. Schematic outline of the recommended strategy for characterization of gel separated proteins (cf. text)

sociation (CID) of proteolytic peptides, sequence stretches of 10–20 residues can be obtained. Even low ng amounts of in-gel digested proteins have provided sequence information helpful for cloning of the corresponding gene (cf. Wilm et al., 1996).

We have applied the in-gel approach to polypeptides of widely different properties and sizes. The technique was tested in studies regarding protein expression in transformed versus normal cells (Bergman et al., 1997; Bergman et al., 1999a).

The amount of a 49 kDa protein separated by 2-D gel electrophoresis was found to be three times higher in c-jun transformed rat fibroblasts (Bergman et al., 1997). To identify this polypeptide the gel spot was cut after staining with Coomassie followed by thorough destaining and washing in 30 % methanol by end-over-end mixing for two days. The latter step is important for removal of as much as possible of the low-molecular-weight contaminants from the gel matrix that will otherwise interfere with subsequent mass spectrometry. This treatment was followed by further washing/equilibration in 0.1 M ammonium bicarbonate, pH 8, mixed 1:1 with acetonitrile. The washed gel pieces were placed in an Eppendorf tube and dried in a Speedvac centrifuge. The dry gel pieces were reconstituted in buffer (0.1 M ammonium bicarbonate, pH 8) containing trypsin (0.2 μg/μl) in a volume just sufficient to cover the gel pieces. Digestion was allowed to proceed for 20 h at 37°C and the liquid was recovered to a new tube. The gel pieces were subsequently extracted with 60 % acetonitrile containing 0.1 % formic acid at 30°C for 30 mins after which the two peptide extracts were combined and dried. The masses of the tryptic fragments were determined via ESI MS. The masses of 13 tryptic fragments were screened in the SwissProt database and found to match the published sequence of rat α-enolase. To further verify the identity, CID MS of fragments was applied to determine parts of the amino acid sequence (Bergman et al., 1997).

A similar strategy was tested to identify genes affected by inhibition of the protein kinase Raf-MEK-ERK signal pathway (Bergman et al., 1999a) which is of major importance for transformation by oncogenes (Cowley et al., 1994). An inhibitor of MEK1 was applied to c-jun transformed rat fibroblasts followed by cell lysis and 2-D gel electrophoresis. Gene products for which expression was altered by MEK1 inhibition were identified both via MALDI MS and ESI MS/MS of fragments from in-gel tryptic digestion. The protocol for in-gel digestion was slightly modified from Bergman et al. (1997). Conditions employed were ammonium bicarbonate at 20 mM for equilibration/digestion, 1 picomole trypsin per μl (or 0.02 μg/μl) for digestion and 0.1 % trifluoroacetic acid (with 60 % acetonitrile) for peptide extraction (Bergman et al., 1999a). The combined data from mass mapping and CID identified several proteins with altered expression in response to MEK1 inhibition. Pirin, a nuclear factor I interacting protein, was down-regulated which was also the case with many other proteins. Truncated proteins were detected and up-regulation after MEK1 inhibition was established for ornithine aminotransferase (Bergman et al., 1999a). The results illustrate the usefulness of the combination 2-D gel electrophoresis, in-gel digestion and MS in studies of signal transduction pathways. The concept was further applied to screening of potential disease-marker proteins in cells from solid tumors of the breast, lung and ovary (Bergman et al., 1999b).

A different approach to identification and sequence analysis of gel separated proteins is to carry out a traditional peptide mapping via reverse-phase HPLC and collect the components for Edman degradation (Fig. 6.1). However, the amounts of individual proteins separated in 2-D gels are seldom above a few picomoles, thus requiring sensitive systems for peptide preparation and sequence analysis. We use an instrument combination involving a capillary HPLC system, the Microblotter, and the Procise cLC sequencer (both from PE-Applied Biosystems) for peptide extracts at the low picomole to sub-picomole level recovered after in-gel tryptic digestion (Fig. 6.1). The C18 chromatography is carried out using a column with inner diameter 0.5 mm and flow rate 5 μl/min. The low flow rate necessitates small-scale fractionation of peptide components for microsequence analysis. The capillary outlet intermittently touches a membrane strip of PVDF while moving. After sample collection, the strip is aligned with the chromatogram for excision of spots corresponding to peptide candidates for structural analysis. The PVDF membrane pieces are mounted in the cartridge of the Procise cLC sequencer which has a matching sensitivity (Fig. 6.1). Peptide amounts in the sub-picomole range can be analyzed for extended degradations (15 cycles or more). The high sensitivity is achieved mainly via miniaturization of vital parts of the sequencer instrument such as the cartridge and the column for phenylthiohydantoin (PTH) identification. The chromatography system can detect below 100 femtomole PTH amino acid.

Trypsin in-gel digestion followed by Microblotter peptide isolation was applied to a 110 kDa insect protein separated by SDS/polyacrylamide gel electrophoresis in the low picomole range. The chromatogram revealed about 20 well separated fragments suitable for sequence analysis. A PVDF-bound peptide fraction corresponding to approximately 1.5 picomole was applied to Procise cLC analysis. The tyrosine residue detected in the first cycle corresponded to 900 femtomole and was followed by a phenylalanine at 870 femtomole in the second cycle. Interpretation of each of the twelve cycles analyzed was straightforward. In the last cycle, a serine residue corresponding to 150 femtomole was clearly assigned which illustrates the sub-picomole capability of the sequencer system. In the PTH analysis, each residue generated a signal that was on average more than 5-fold above the level of the amino acid background. In addition, the carry-over from cycle to cycle was less than 10 %. A database search using the BLAST algorithm (Altschul and Gish, 1996) revealed a strong sequence homology between the protein under investigation and a known Drosophila protein. Of the twelve residues sequenced, ten were identical with the Drosophila sequence and one of the two non-identical residues represented a conservative replacement. In the Drosophila sequence, the first residue is preceded by a lysine residue and the residue at position 13 is arginine. Since these residues both fit trypsin specificity, the analyzed fragment is likely to be thirteen residues in total. Serine detected in position 12 is then penultimate to the C-terminal residue. Still the yield is good, 150 femtomole, showing that wash-out losses are acceptable during degradation.

2 Electroblotting and Direct Sequencer Analysis

The developments in sequencer technology for Edman degradation in recent years now make it interesting to analyze proteins from 2-D gels directly after electroblotting to PVDF membranes (Fig. 6.1). Even proteins that stain very weakly using Coomassie blue, corresponding to 1–2 picomole, can be analyzed. A problem in this approach is when a blocked N-terminus is encountered. Although a new separation followed by in-gel digestion can be performed (provided there is more material), deblocking of the protein directly on the membrane after the initial attempt at Edman degradation would increase the efficiency considerably. Since N-terminal blocking is frequently due to an acetyl group, a protocol for chemical deacetylation (e.g. Gheorghe et al., 1997) applicable at the low picomole level is an important component to make sequence analysis of proteins electroblotted from 2-D separations more useful (Fig. 6.1).

Sequence analysis of 2-D separated proteins electroblotted to PVDF membranes and stained with Coomassie blue was tested using a human liver microsomal protein fraction isolated via standard techniques including ultracentrifugation at 100 000 × g. About 100 μg of total protein was applied to the first dimension, isoelectric focusing, that was carried out using pH 3–10 non-linear gradient IPG strips (Pharmacia) for 25 kVh. The second dimension SDS/polyacrylamide gel electrophoresis employed a 12 % gel and 20 mA to generate good resolution in the size range 10–100 kDa. After separation and staining, several protein containing spots were cut from the PVDF membrane and applied to Procise cLC sequencer analysis. One of the excised spots revealed 70 femtomole of glutamic acid in the first cycle (Table 6.1) with a PTH signal that was roughly 5-fold above

Table 6.1. N-terminal sequences analysis of human 78 kDa glucose regulated protein electroblotted onto PVDF membrane after 2-D separation of a liver microsomal protein fraction

Cycle	Residue	Amount recovered, femtomole (background orrected)
1	Glu	70
2	Glu	100
3	Glu	20
4	Asp	100
5	Lys	120
6	Lys	140
7	Glu	100
8	Asp	50
9	Val	70
10	Gly	50
11	Thr	40
12	Val	10

the level of the background. The sample was analyzed for eleven more cycles and the sequence could be clearly assigned. This was also true for residue 12, valine, that was recovered in only 10 femtomole (Table 6.1). A database search using the BLAST algorithm identified this polypeptide as the 78 kDa glucose regulated protein (GRP) having SwissProt accession number P11021. This protein belongs to the heat shock protein 70 family and is located in the lumen of the endoplasmic reticulum. GRP probably plays a role in the assembly of protein complexes inside the endoplasmic reticulum. Interestingly, an N-terminally truncated form of GRP was found by Rasmussen et al. (1997) when this protein was isolated via 2-D gel electrophoresis applied to a human breast carcinoma cell line. The first three glutamic acid residues were lacking and the sequence started at residue 4, aspartic acid (cf. Table 6.1). The results show that the sub-picomole sequencer capability makes electroblotting to PVDF after 2-D gel separation an attractive alternative for rapid identification via direct sequence analysis of many cellular proteins isolated in amounts down to about 1 picomole.

3 Conclusion

Mass spectrometry and high sensitivity Edman degradation are both important analytical tools for characterization of gel separated proteins (Fig. 6.1). Mass mapping for tentative identification and CID for verification is efficient when the protein is known and in the database. For cloning and when longer sequence stretches are required, high sensitivity Edman degradation is a strong alternative provided amounts are not too low, minimally 3–5 picomole in the gel. However, for intact proteins electroblotted to PVDF, the amount can be substantially smaller (0.5–1 picomole).

4 Acknowledgements

This work was supported by grants from the Swedish Medical Research Council (project 03X-10832), the Swedish Cancer Society (project 1806), the European Commission (BIO4-CT97–2123) and the Emil and Wera Cornell Foundation.

References

Altschul SF, Gish W (1996) Local alignment statistics. Meth Enzymol 266: 460–480

Bergman A-C, Linder C, Sakaguchi K, Sten-Linder M, Alaiya AA, Franzén B, Shoshan MC, Bergman T, Wiman B, Auer G, Appella E, Jörnvall H, Linder S (1997) Increased expression of α-enolase in c-jun transformed rat fibroblasts without increased activation of plasminogen. FEBS Lett 417: 17–20

Bergman A-C, Oppermann M, Jörnvall H, Bergman T (1998) Mass mapping and sensitive Edman degradation of gel separated proteins. J Prot Chem 17: 535–536

Bergman A-C, Alaiya AA, Wendler W, Binetruy B, Shoshan M, Sakaguchi K, Bergman T, Kronenwett U, Auer G, Appella E, Jörnvall H, Linder S (1999a) Protein kinase dependent overexpression of the nuclear protein Pirin in c-JUN and RAS transformed fibroblasts. Cellular and Molecular Life Sciences, in press

Bergman A-C, Benjamin T, Alaiya A, Waltham M, Sakaguchi K, Franzén B, Bergman T, Linder S, Auer G, Appella E, Wirth PJ, Jörnvall H (1999b) Identification of gel-separated tumor marker proteins by mass spectrometry. Electrophoresis, submitted

Cowley S, Paterson H, Kemp P, Marshall CJ (1994) Activation of MAP kinase kinase is necessary and sufficient for PC12 differentiation and for transformation of NIH3T3 cells. Cell 77: 841–852

Gheorghe MT, Jörnvall H, Bergman T (1997) Optimized alcoholytic deacetylation of N-acetyl-blocked polypeptides for subsequent Edman degradation. Anal Biochem 254: 119–125

Henzel WJ, Billeci TM, Stults JT, Wong SC, Grimley C, Watanabe C (1993) Identifying proteins from two-dimensional gels by molecular mass searching of peptide fragments in protein sequence databases. Proc Natl Acad Sci USA 90: 5011–5015

Patterson SD (1994) From electrophoretically separated protein to identification: Strategies for sequence and mass analysis. Anal Biochem 221: 1–15

Rasmussen RK, Ji H, Eddes JS, Moritz RL, Reid GE, Simpson RJ, Dorow DS (1997) Two-dimensional electrophoretic analysis of human breast carcinoma proteins: Mapping of proteins that bind to the SH3 domain of mixed lineage kinase MLK2. Electrophoresis 18: 588–598

Schevchenko A, Jensen ON, Podtelejnikov AV, Sagliocco F, Wilm M, Vorm O, Mortensen P, Shevchenko A, Boucherie H, Mann M (1996) Linking genome and proteome by mass spectrometry: Large-scale identification of yeast proteins from two dimensional gels. Proc Natl Acad Sci USA 93: 14440–14445

Wilm M, Mann, M (1996) Analytical properties of the nanoelectrospray ion source. Anal Chem 68: 1–8

Wilm M, Schevchenko A, Houthaeve T, Breit S, Schweigerer L, Fotsis T, Mann M (1996) Femtomole sequencing of proteins from polyacrylamide gels by nano-electrospray mass spectrometry. Nature 379: 466–469

Part II
Structure Analysis

CHAPTER 7

Protein Structure and Dynamics by NMR in Solution

B. Bersch[1], M. Blackledge[1], B. Brutscher[1], F. Cordier[1], J.-Chr. Hus[1] and D. Marion[1]

1 Introduction

Due to the genome project, a tremendous number of protein sequences have already been obtained, which will continue to increase over the next few years. It is clear from a biological point of view, that this sequential information has to be related to protein function. This can partially be obtained by careful analysis of the sequential data, but unfortunately, functional homology is not always reflected as sequential homology. Another very important connection between sequence and function is the *molecular structure* of a protein. Structural information on the atomic level can currently be obtained from X-ray crystallography, nuclear magnetic resonance (NMR) as well as, under certain circumstances (sequential homology to a protein with known 3D structure), molecular modelling. The contribution of structural biology to the classification and understanding of protein structures is reflected in the ever growing number of molecular coordinates deposited in the Brookhaven Protein Databank (pdb).

Since the introduction of the two-dimensional NMR techniques in the 1970s, many protein structures have been solved using solution NMR spectroscopy. This technique has constantly evolved with the increase of magnetic field strength and computer performance, with the introduction of new, ^{13}N- and ^{13}C edited multidimensional experiments and, very recently, by taking advantage of phenomena well known in solid state NMR.

One very important feature of high resolution NMR is that the molecule of interest is studied in solution, thus allowing the characterization of its dynamic properties under conditions close to those found *in vivo*. In addition, NMR is a sensitive probe for local conformational and/or dynamical variations occuring in a molecule due to ligand-binding, protein-protein interactions or site directed mutagenesis.

In this chapter we will give an informal introduction to modern biomolecular NMR, starting with the necessary information required to initiate an NMR study of a biomolecule: the suitability of a molecule for an NMR investigation and a description of the experimental conditions. The experimental parameters and their possible interpretations will be briefly described before we give some more

[1] Laboratoire de RMN, Institut de Biologie Structurale – Jean Pierre Ebel, CEA-CNRS, 41 rue Jules Horowitz, 38027 Grenoble, France.

concrete examples of recent applications from our laboratory. We hope that this chapter might be a helpful illustration of the various aspects of this technique and an encouragement to take advantage of the possibilities it offers.

2 Sample Requirements

High resolution NMR information on biological macromolecules is usually obtained from solution studies. The most frequently observed nuclei are ^{1}H (100 % natural abundance), ^{13}C (1.1 % natural abundance or isotopic enrichment) and ^{15}N (isotopic enrichment). The obtention of isotopically labelled proteins is presented in section 2.4.

The quality of the experimental data strongly depends on the experimental conditions, so a careful preparation of the NMR sample is primordial.

2.1 Experimental Conditions

The most physiological solvent used for the studies of biological macromolecules is obviously water. However, the use of this solvent has two major inconveniences which can be circumvented by the appropriate choice of experimental parameters. First, the ^{1}H nuclei are generally detected during the experimental aquisition due to their high gyromagnetic ratio (γ), resulting in a much better sensitivity. As the concentration of water protons (110 M) is several orders of magnitude higher than that of the sample (0.5–10 mM), the water signal has to be suppressed, otherwise it would completely mask the signals from the protein and reduce the available dynamic range. Appropriate techniques have been developed: the presaturation of the water resonance, a jump-and-return or a watergate pulse sequence are most commonly used (Hore 1989, Piotto et al. 1992). In addition, several protons capable of forming hydrogen bonds with water have a tendency to exchange with the solvent (exchangeable or labile protons). Most information gained on proteins relies on the observation of the amide protons which do exchange with the solvent (this is the reason why D_2O is rarely used as a solvent for protein studies). Proton-solvent exchange increases with pH above pH ~ 4.0. Therefore, a pH below 7.0 is preferred. Above this value, exchange with the solvent leads to excessive line-broadening and disappearance of several amide resonances.

Buffers used in protein NMR should not contain any non-exchangeable protons. Phosphate is a good choice, but deuterated TRIS and acetic acid are also available. The ionic strength should be kept to a minimum. The temperature is generally set between 15° and 40 °C. Increasing temperature reduces the linewidth, due to faster molecular tumbling, but also reduces the protein stability.

2.2 Choice of the Protein

Two very important criteria for the decision whether a protein can be studied by NMR are the *size* of the protein and its *solubility*.

Concerning the solubility, the protein should be soluble under the above mentioned conditions at a concentration of at least 0.5 mM. For a sample volume between 300 and 500 μl, this means that the quantity of protein required for one sample is at least 15 micromoles.

The maximal size of a protein to be studied by NMR is limited by the characteristics of the technique itself. With increasing molecular size, spectra become more and more crowded and due to increased relaxation effects the sensitivity of the experiment significantly decreases. This has led to the development of heteronuclear techniques (3D, 4D ...), where the frequency of the heteronucleus (^{15}N, ^{13}C or both) to which the proton is linked is added as a supplementary parameter. However, the use of these techniques necessitates isotopic labelling of the protein (see below). Concerning the relaxation effects, partial deuteration of the protein (LeMaster 1990, Sattler et al. 1996) or the constructive use of relaxation interference (TROSY, Pervushin et al. 1997) helps to push the size limit to higher values.

Generally, the following rules apply: a protein which cannot be labelled isotopically should not have more than 100 residues. For proteins up to 150 amino acids, uniform ^{15}N-labelling is sufficient whereas larger proteins should be double- (^{15}N and ^{13}C) or triple- (^{15}N, ^{13}C and ^{2}H) labelled. The actual size limit using the most sophisticated NMR techniques depends on the characteristics of the protein studied. So far, NMR structure determinations have been reported for proteins with molecular weights of up to 40 kD.

2.3 Preparation of the Sample

Obviously, the protein should be of high purity. During the sample preparation, care should be taken not to contaminate the protein with any microorganisms such as bacteria or fungi, which might result in a digestion of the protein. The final solution can be filter sterilized or small quantities of antibiotics or sodium azide might be added to the sample. Sealing the sample under inert athmosphere also helps to avoid damage during storage. Even with the most sophisticated purification procedures, proteases are rarely completely eliminated. Therefore, protease inhibitors are widely used to protect from proteolytic degradation and can also be added to the final sample. Note however, that the aquisition of the experimental data takes several days or even weeks during which the protein should be stable under the experimental conditions given above.

2.4
Obtention of Isotopically Labelled Proteins.

Whereas any source can be used for the obtention of protein samples for homonuclear NMR experiments, the obtention of isotopically labelled proteins necessitates the overexpression of the protein in a convenient host.

Isotopic labelling is achieved by providing completely labelled nitrogen- and/or carbon sources to the culture medium (for a review see McIntosh 1990). To reduce costs, such cultures are usually grown in minimal media supplemented with $^{15}NH_4Cl$ and/or $^{13}C_6$-glucose. As a consequence, the protein should be overexpressed in an organism capable of growing on minimal media such as *E. coli* or *P. pastoris*, the latter is able to grow on methanol as sole carbon source (in this case, the medium should contain ^{13}C-methanol). However, in some cases it might be necessary to increase the expression by the use of rich media which are derived from partially hydrolysed algae protein. In addition, isotopically labelled media are available for insect cells (baculovirus expression system) as well as CHO cells.

Before starting cultures in expensive isotopically labelled media, culture and purification methods should be optimized to a maximum. The use of tags or fusion proteins for an optimized purification is possible but it should be born in mind that the addition of any residues might have consequences for the three-dimensional structure of the protein. Protease mediated cleavage of the fusion protein or the tag is widely used, but does not always give satisfactory results due to side reactions. Very recently, a ubiquitin fusion has been proposed for the expression of small, isotopically labelled peptides which should also work for bigger proteins (Kohno et al. 1998). One advantage of this system is the very specific cleavage reaction by a ubiquitin-specific hydrolase.

If complete deuteration of cells is desired, they must be adapted to deuterated media in a stepwise manner (Katz and Crespi 1966). Note, however, that in contrast to ^{15}N- or ^{13}C-labelling, which have no effect on cell metabolism, deuteration drastically reduces growth and consequently protein production (Haon et al. 1993). In addition, a protocol for the obtention of deuterated proteins which are selectively protonated on the aliphatic sidechains has been described (Rosen et al. 1996).

3
NMR Spectral Parameters

Nuclear magnetic resonance can be used to probe the structure and dynamics of biomolecules, because the spins act as spies of their intimate environment: each nuclear spin interacts with its direct surroundings consisting of other nuclei and electrons. These numerous pair-wise interactions are detected by a smaller number of spectral parameters, such as the chemical shift, the scalar coupling constants or the relaxation rate constants. Structural or dynamical information concerning the protein itself can then be obtained by the conversion of the spectral parameters into structural parameters, such as distances, angles or frequencies of motions. However, one should keep in mind that solution state NMR provides an ensemble average and that measurable spectral parameters represent all conformations interchanging in the ns to ms time range. Therefore, the NMR spectroscopist only gets a blurred picture of the molecule: when several spectral parameters seem mutually inconsistent, this may simply originate from averaging processes during the NMR experiment.

3.1 Chemical Shift

The origin of the chemical shift is that the moving charges of the electron cloud around the nucleus under observation induce a local magnetic field which adds to the applied field, this is called the *shielding* of the nucleus. As a result, nuclei which do not exhibit exactly the same (chemical and structural) environment, have different resonance frequencies or chemical shifts. Although the chemical shift is a small effect ($\approx 10^{-6}$), partially resolved resonances can generally be observed for individual nuclei in a biomolecule. The chemical shift of a resonance (δ) is expressed with respect to a reference line in ppm units: $\delta = (\nu - \nu_{ref})/\nu_o$, where ν_o is the Larmor frequency. The Larmor frequency depends on the magnetic field strength and typically takes values between 500 and 800 MHz for the proton in high resolution NMR spectroscopy. Whereas the separation in Hz between the resonances of two different spins increases with the magnetic field, the chemical shift difference in ppm remains constant. In considering that the NMR line-width is field-independent, it becomes clear that one of the major benefits of higher magnetic fields is the larger separation between resonances (called *spectral resolution*). This accounts for the endless race to higher static magnetic fields which is only limited by production costs and available technology.

A reference is needed for the measurement of the chemical shift; it can be either a small molecule added to the sample (such as tetramethylsilane,TMS or 3-trimethylsilylpropionate, TSP) or a residual solvent line. In the first case, it is assumed that its chemical shift is insensitive to the environmental conditions (pH, ionic strength or temperature) and in the latter case, empirical correction is frequently needed. Several authors have proposed relationships between ^{13}C, ^{15}N and ^{1}H chemical shift scales based on precise knowledge of their gyromagnetic ratio (c.f. Wishart et al. 1995).

The shielding of a nucleus is *anisotropic* (i.e. it varies with the orientation of the molecule in the magnetic field) and can therefore only be described by a rank 2 tensor (σ_{11}, σ_{22},σ_{33}). However, in solution, the molecules generally undergo fast rotation (due to the brownian motion) and do not exhibit any preferential orientation. Because of this averaging, the tensor reduces to its trace: $\sigma = 1/3\ (\sigma_{11} + \sigma_{22} + \sigma_{33})$. In this case, complex relaxation measurements (see below) are required for the detection of *chemical shielding anisotropy* (CSA). On the other hand, in the case of partially aligned molecules, CSA leads to the variation of the observed chemical shift (Ottiger et al. 1997). In this case the chemical shift is not simply given by the trace of the tensor but a weighted average of the individual components.

3.1.1
Chemical Shift and 3D Structure

From the early days of NMR, spectroscopists made a number of attempts at understanding chemical shifts in proteins in terms of 3D structure. In parallel to empirical methods discussed later, theoretical approaches have been developed which steadily increase in speed and accuracy (Oldfield 1995). At present, *ab initio* calculations on a complete protein cannot be carried out in an acceptable amount of time. However, on fragments of two or three amino-acids, excellent correlation for the ϕ, ψ, χ_1 dependence have been found between predicted and experimental shifts. So far, these methods are much better at predicting chemical shift variation than absolute shielding; this is not a severe limitation as experimental data are frequently reported with respect to a reference or a random coil chemical shift. It is expected that in the near future chemical shift theory will be able to directly provide additional structural restraints for structure calculations (Beger and Bolton 1997).

3.1.2
Qualitative Interpretations (Ligand Binding and Chemical Shift Mapping)

Even in absence of detailed understanding of the chemical shift origin, it can be used qualitatively to monitor the interaction of two molecules. The simplest and most popular technique is called "chemical shift mapping", wherein changes in chemical shifts in molecule **A** are recorded during titration with molecule **B**. From the titration curve obtained, a binding constant can be estimated (as for any other spectroscopic technique) but when resonance assignments are available for molecule **A**, the location of resonances that undergo perturbation can be mapped onto its 3D structure. Before interpreting these effects as the putative binding site, it is preferable to use an independent probe (nOe for instance) to confirm that no major conformational transition occurs in molecule **A** due to the interaction. Otherwise, long range effects can be erroneously interpreted as for example multiple binding sites. Numerous examples of this technique are available in the literature: they involve protein-protein interactions (see for instance: Rajagopal et al 1997, McKay et al 1998), RNA-protein interactions (Lee et al 1997) or peptide bound to protein (Qin et al 1995).

3.1.3
Secondary Chemical Shifts

The secondary structure of a protein can often be predicted on the basis of chemical shift information. For the 20 natural amino-acids, the chemical shifts of the backbone and side-chain nuclei have been characterized for random coil conformation. When a protein folds into its native conformation, a deviation from random-coil chemical shift, often referred to as *secondary chemical shift*, is observed. Spera and Bax (1991) and Wishart et al (1992) have reported empirical rules for ^{1}H and ^{13}C spins. For instance, the H^{α} chemical shift in amino acids experiences an upfield shift (with respect to the random coil value) in α-helices and a similar downfield shift in β-extended conformations. This method, later extended to include $^{13}C^{\alpha}$, $^{13}C^{\beta}$ and carbonyl $^{13}C'$ chemical shifts (Wishart and Sykes 1994), is now known as the *chemical shift index* method: outside a central range of shifts (which are interpreted as random-coil) a positive or negative index is assigned to each residue. Larger stretches of positive or negative numbers correspond to regular secondary structures (helices or sheets) and consensus estimates show a predictive accuracy larger than 90 %.

3.1.4 Chemical Exchange

During the NMR measurements, the protein exhibits internal motions, which also influence the chemical shift. Let us consider a spin moving back and forth between two sites A and B: if the motion has a characteristic frequency slower than the chemical shift difference ($\delta_A-\delta_B$), two separate signals are observed. Otherwise, an averaged signal (more or less broadened) is detected. Typical ^1H chemical shift differences are hundreds of Hz and thus only motions in the ms range and slower are not averaged out. For example, rapidly flipping aromatic rings such as Phe and Tyr give rise to two or three 'averaged' ^1H resonances, whereas four or five resonances are observed when they are sterically hindered. This is a valuable structural information, because insights on the compactness of the hydrophobic core of the protein can be obtained (see Wong and Daggett 1998 for example). A weakness of this approach lies in the lack of knowledge of the chemical shifts, when motion is absent. Consequently, only rough estimates of flipping rates can generally be derived, but this approach turns out to be extremely valuable for comparing mutant proteins with wild-type (Gooley and MacKenzie 1990). Amide protons are labile protons that undergo chemical exchange with the water protons. This explains why their chemical shifts are so sensitive to temperature. Unless they are forming a H-bond with a partner, exchangeable protons in side chains (OH of Ser, Thr, or Tyr, $NH_3^+$of Lys ...), which generally exchange more rapidly with the solvent than H^N protons, are extremely broadened by chemical exchange and can no longer be detected. On the other hand, their detection provides valuable evidence of a H-bond, but the acceptor remains unidentified.

3.2 Scalar Coupling Constants

The empirical relation, known as *Karplus relationship*, between three bond J-couplings (^3J) and the intervening dihedral angle, plays a key role in studying protein conformation. The values of the coefficients in the equation $^3J = A \cos^2\theta + B \cos\theta + C$, where θ is the dihedral angle, depend upon the nuclei involved and their direct neighbors. They have been determined for proteins of known conformation or computed by theoretical methods. The Karplus relationship is degenerate, leading to two (and in some cases four) angles for one experimental coupling. One possibility to resolve this ambiguity consists of measuring several ^3J corresponding to different pairs of spins (^1H,^{13}C or ^{15}N) around the same torsion angle; unfortunately, the magnitude of some ^3J between heteronuclei is less than a few Hz and can hardly be measured with precision. On the other hand, the information provided by the coupling constants can directly be used at the structure refinement stage, by defining a penalty function for the coupling constants themselves rather than the inferred dihedral angles (Mierke and Kessler 1992), thus eliminating any explicit conversion.

Two classes of methods have been devised for the experimental determination: the first one is based on the modulation of the intensity of a cross-peak by scalar coupling (quantitative J correlation). In this approach, the intensity of a cross peak (modulated by sin (πJt)) is compared to that of a diagonal peak (modulated by cos (πJt)). The alternative relies on the frequency separation of the two doublet peaks in an additional dimension (E.COSY, see Biamonti et al 1994 for a review) or in separate spectra (S^3E, Meissner et al 1997).

Conformationally flexible molecules give rise to average coupling constants weighted by their probability distribution. When an angle is measured from several couplings, inconsistencies are frequently attributed to internal flexibility, especially for side-chain torsion angle (χ_1, χ_2 ...). Models of various complexity, including either a discrete superposition of staggered conformations or a continuous probability (Dzakula et al. 1992) have been proposed. The improvement brought by the larger number of degrees of freedom should however carefully be questioned to avoid overinterpretation of the data.

Coupling constants have been recently added to other structural restraints for NMR structure determination. Their incorporation offers a reliable means of improving the accuracy of protein structure (Garrett et al. 1994), but, contrary to nOe, they do not contain any long range information which is fundamental for finding the global fold of a protein.

3.3
Relaxation

Once the spins have been perturbed by radio-frequency pulses, the magnetizations (or the coherences) recover to their equilibrium states by an irreversible process called *relaxation*. Relaxation is caused by time-dependent magnetic fields that originate from the random thermal motions present in the sample. Fluctuating magnetic fields at a nucleus may derive from the interaction with other spins (*dipolar relaxation*) or from a modulation of its own chemical shielding (*CSA relaxation*). In a phenomenological description of relaxation, the longitudinal magnetization recovers exponentially with a time constant T_1 (spin lattice relaxation) and the transverse coherences vanish with a time constant T_2 (spin-spin relaxation). Long T_1 values prevent the use of short recycling delays, when multiple scans are acquired, leading to long acquisition times. On the other hand, short T_2 values (in other words, broad resonances) strongly limit the spectral resolution. Unfortunately, large macromolecules with slow overall tumbling rates share these two drawbacks. However, the spin-spin relaxation can be reduced by using complete or partial deuteriation of the molecule (LeMaster 1990).

3.3.1
Nuclear Overhauser Effect

The nuclear Overhauser effect (nOe) is due to the exchange of magnetization between two spins (cross-relaxation) and can be detected as the change in the intensity of an NMR resonance (I) when another spin (S) is irradiated. A requirement for the nOe is that the spin-lattice relaxation of nucleus I is governed by dipole-dipole relaxation with S. Apart from its ability to enhance signals of low sensitivity nuclei (^{15}N, ^{13}C ...), nOe is widely used to obtain information about internuclear distances, essentially between ^{1}H. The nOe has both an interproton distance dependence ($1/r^6$–only short distances < 6 Å are in practice detectable) and an effective correlation time dependence (Neuhaus and Willamson 1989). For the determination of internuclear distances, the latter dependence is often neglected by assuming that it arises from the tumbling of a rigid, approximately spherical molecule. Various complexities in interpretation of nOe data support its rather qualitative use: J-couplings give rise to artefacts, multiple transfers, known as *spin diffusion*, ($I \rightarrow M \rightarrow S$) occur at longer mixing times, conformational flexibility can lead to averaging of nOe and also, a proton which oscillates between two sites, may show incompatible nOe with very distant (> 10 Å) spins. As a result, for protein structure determinations, spectroscopists do not longer try to quantify each nOe precisely, but aim at identifying as many cross-peaks as possible.

On proteins, nOe are measured not by saturation of individual signals but by means of a NOESY 2D experiment: a cross-peak is the evidence that two spins have exchanged magnetization during a mixing period of typically 50–150 ms. For larger ^{15}N/^{13}C labeled proteins, the NOESY scheme can be combined with an ^{1}H-^{15}N/^{13}C correlation experiment to edit the heteronuclear frequency in a third dimension (Marion et al 1989) and thus reduce spectral overlap.

Isotope labeling also turns out to be useful for structure determination of dimers where it is intrinsically impossible to distinguish intermolecular nOe from intramolecular ones. Although distance violations due to incorrect assignment may probably become patent during the structure refinement, it is more convenient to rely on experimental discrimination by using asymmetric labeling (Folkers et al. 1993). On a 1:1 mixture containing unlabelled and ^{15}N/^{13}C-labeled protein, nOe at the dimer interface can be discriminated by heteronuclear filtered or edited experiments (see Handel and Domaille 1996 for an application).

3.3.2
Heteronuclear Relaxation Rate Constants

The function of a protein depends on its ability to explore excited states and is hence intimately coupled to flexibility. A complete description of a structure of a protein will require information on how the structure changes with time. Long before labeled proteins became available, ^{13}C and ^{15}N relaxation experiments were devised to study the dynamics of small molecules. Investigations of protein dynamics most commonly measure the longitudinal (T_1) and transverse (T_2) relaxation time constants as well as the steady state $\{^1H\}X$ nOe (or cross-relaxation rate constant). Three mechanisms contribute to the measurable parameter: the dipolar interaction, the chemical shift anisotropy (see above) and the chemical exchange. Relaxation measurements on proteins have generally focused on nearly ideal systems such as ^{15}N-^{1}H on labeled proteins and $^{13}C^{\alpha}$-^{1}H at natural abundance, in which cases the relaxation of the nucleus of interest is governed by only one other spin. In order to interpret the relaxation data in terms of the motion of the vector connecting the two spins, a few assumptions on the CSA are made which we will not discuss here. It also is important to keep in mind that relaxation data do not contain information about internal motion much slower than rotational diffusion. An exception to this rule are chemical or conformational exchange processes. These effects, which only contribute to transverse relaxation, provide information on μs–ms range motions, but unfortunately, it is rather difficult to assess if the chemical shift for each conformation is not known (see below).

As far as the data analysis is concerned, there are basically two approaches:

1. the *model-free formalism* (proposed by Lipari and Szabo 1982) supposes a separation between overall rotational diffusion and internal motion. There are few cases, such as a protein made of a string of modules, where its validity has been questioned. For each vector, an internal correlation time (τ_i, typically in the picosecond time range) and an order parameter (S^2, varying between 1.0 for completely restricted and 0.0 for completely unrestricted motion) are derived: in regular α-helices or β-sheets, typical S^2 are usually found between 0.8 and 0.9 corresponding to the restricted mobility in the secondary structure elements.
2. the *spectral density mapping* method (Peng and Wagner 1992) analyzes the frequency spectrum of the motions at a number of characteristic frequencies (low frequency, $\approx \omega_x$, $\approx \omega_H$). These values contain contributions from both overall and intramolecular dynamics. This procedure is especially powerful in application to disordered molecules in which the assumption of the Lipari-Szabo method is not valid any more. For example a protein with a flexible end (see van Heijenoort et al. 1998 for an illustration): as compared to the structured core, residues in this part are characterized by larger values of the spectral density function at high frequency and smaller ones at low frequency.

The information obtained from the analysis of relaxation data is manifold. Internal mobility can be analyzed for parts of proteins which look disordered after NMR structure refinement: if ns–ps internal motions are detected for these residues, one can conclude that the lack of a unique conformation is genuine and not just an NMR artifact due to the lack of experimental constraints, spectral overlap or line broadening. From the anisotropy of the overall tumbling (compare 3.5.1. and 4.2), insights can be obtained into the shape of the protein in solution including the solvation layer and possible multimerization. An easy application of relaxation is the comparison of mutant proteins with a wild-type and the study of a macromolecule in presence or absence of a ligand: when used in a relative manner, these measurements become less prone to possible experimental or processing artefacts.

3.3.3 Cross-Correlated Relaxation Rate Constants

There has been recently much excitement about the prospects for using relaxation not only to characterize the dynamical behavior of proteins but also to measure dihedral angles. We have mentioned earlier that the ultimate cause of relaxation is molecular motion: when two relaxation mechanisms are modulated by the same motion (they are said to be *cross-correlated*), spurious effects appear: for instance, the two components of a ^{15}N resonance (split by the H^N coupling) exhibit different line-widths, under slow tumbling conditions. The theoretical background for these effects is outside the scope of this review, but the feature of most interest for a conformational use is their angular dependence in (3 $\cos^2\theta - 1$), where θ defines the relative orientation of the two relaxation mechanisms. This method has been proposed to quantify the dihedral angle underlying two dipole-dipole interactions ($NH-C^{\alpha}H$) or a CSA tensor and a dipole-dipole interaction (C' CSA$-C^{\alpha}H$) (Reif et al 1997, Yang et al 1997). It is expected that these new probes will be widely used in the near future to complement J-couplings, despite their experimental and theoretical complexities (passive coupling, internal motion, etc. ...).

The cross-correlation effect can be directly exploited simply to enhance the spectral resolution: Pervushin et al (1997) have proposed an experiment called TROSY (transverse relaxation optimized spectroscopy) for ^1H-^{15}N correlation. For one of the lines of the correlation peak, a partial cancellation of the transverse relaxation (T_2) arises due to the interference of the CSA and the dipolar relaxation. The authors claim a several-fold increase of the molecular size accessible by NMR with these methods, despite the fact that the predicted narrowing varies substantially from one nucleus to another and reaches its maximum at very high fields (corresponding to ^1H frequencies larger than 1 GHz, actually, the most performant NMR spectrometer operate at 800 MHz).

3.4 Exchange Rate Constants (H–D)

Some hydrogens (such as NH, NH_2, OH) in macromolecules are in continous exchange with hydrogen atoms of the water. Measurements of hydrogen-exchange rates can provide – in a non-perturbing way – information on protein dynamics and stability at an atomic level. The labile hydrogens in a protein are covalently bound and thus a chemical reaction is actually required for the exchange with solvent. In aqueous solution, exchange of peptidic H^N can be catalyzed either by H^+ or OH^- ions, showing a minimal rate between pH 2 and 3. Structural effects can slow down protein hydrogen exchange by large protection factors ($> 10^9$), that always involve hydrogen bonding. Therefore, an exchange reaction requires two steps: a transient opening of the structure and the intrinsic exchange. Two limiting cases can be distinguished, (EX2, EX1) depending upon which step is rate-limiting. Hydrogen exchange is thus related to transient unfolding of the protein, involving either the unfolding of the whole molecule or of secondary structure elements. As compared to other spectroscopic techniques (such as infrared, Raman...), NMR provides the resolution for monitoring each individual site in a protein (see Englander et al 1996 for a review).

Exchange rate constants are measured by dissolving a protonated protein in D_2O and following the disappearance of the ^1H signals by means of any quick correlation experiment. Many experimental procedures are available to alter these rates: denaturant can be added to the solution to partially unfold the protein, pH can be rapidly changed in order to slow exchange and trap some kinetic information (protein folding). All these experiments provide information on the dynamics of the protein on a time scale ranging from seconds to hours.

3.5 Structural Constraints from Anisotropic Rotational Diffusion and Dipolar Coupling Constants

In liquid state NMR it is normally assumed that the molecule of interest undergoes rapid isotropic tumbling in an isotropic environment. In some cases, an anisotropic behavior of the molecule can yield new structural information as some experimental parameters become orientation-dependent. Examples are heteronuclear relaxation data, which can be related to anisotropic rotational diffusion, and residual dipolar couplings, induced by a partial alignment of the molecule in the magnetic field. This information can be used as additional experimental constraints in structure calculations or at least give some insights into the relative orientation of individual chemical bonds or structural elements (such as regions of secondary structure, loops, domains ...).

3.5.1 Determination of the Anisotropic Rotational Diffusion Tensor

In the case of *anisotropic rotational reorientation* of the molecule, the overall reorientation is no longer determined by one global rotational time constant τ_c but by a molecular rotational diffusion tensor D, which is defined by the Euler angles α,β and γ as well as the relative amplitudes of the tensorial components D_{xx}, D_{yy} and D_{zz} (Woessner 1962). In this case, the spin relaxation becomes dependent on the orientation of the interaction vector with respect to this tensor.

It is therefore possible to characterize the rotational diffusion of the molecule by analysing heteronuclear relaxation rates of spins not experiencing significant internal motion contributions to their relaxation. In practice this latter criterion is achieved by analysing the ratio R_2/R_1 which becomes independant of the order parameter S^2 in the fast internal motion limit. Non-linear least-squared fitting has been used to determine the rotational diffusion (i.e. the principle values and the orientation of the rotational diffusion tensor) of a number of molecules in a series of recent studies.

However, in many studies of the global dynamics of molecules of known structure, an axially symmetric tensor was either assumed, or shown, using statistical significance tests, to provide an adequate description of the rotational diffusion. It has nevertheless recently been pointed out that two orthogonal solutions are simultaneously present if the axially symmetric model is used to describe a fully anisotropic system, representing the prolate and oblate approximations to the real tensor (Blackledge et al. 1998).

Knowledge of the rotational diffusion tensor can be related to the size and the shape of the molecule. Interestingly, numerous studies of rotational diffusion using this method have derived values of the diffusion tensor components which, while in general coaxial to the components of the inertia tensor of the protein, imply the presence of a solvation layer of approximately one water molecule thickness around the protein, when compared to hydrodynamic calculations (Garcia de la Torre and Bloomfield 1981).

In addition, the rotational diffusion tensor determined from relatively rigid regions of a molecule (normally secondary structure elements) can be incorporated into a Lipari-Szabo type analysis of local motion to characterize the local mobility. This results in a much greater confidence in the results obtained as a precise knowledge of the component of the auto-correlation function due to the overall tumbling of the molecule is necessary in order for the model-free approach to give realistic information concerning local motions.

3.5.2 Heteronuclear Relaxation Rate Constants as Structural Constraints

The ability to accurately characterize rotational diffusion anisotropy using heteronuclear relaxation rates has led to the observation that the R_2/R_1 ratio can also provide a novel long-range constraint for NMR structure determination (Tjandra et al. 1997). Such constraints no longer aim to describe relative atomic positions of nuclei by means of the scalar coupling or the nOe, which are *local* interactions, but the *absolute* orientation of many internuclear vectors (corresponding to chemical bonds) with respect to a single reference tensor (i.e. a three dimensional reference system, see Fig. 7.1). This orientational information can be used in the form of additional constraints during structure calculation (for a review, see Clore and Gronenborn 1998). In addition, it allows the determination of the *relative orientation* of one protein domain with respect to another, even if they are several tens of Å away. Internuclear nOes do not permit such a determination, as no long-range order information is available and the errors tend to accumulate. Orientational constraints can also be obtained from the analysis of *residual dipolar couplings*, which can be observed if the protein molecules become partially *aligned* within the magnetic field.

3.5.3 Residual Dipolar Couplings

Although theory shows that *scalar* J-couplings do not vary with the magnetic field, several authors have recently reported a field dependence of an apparent J-coupling constant. This unexpected observation arises from the superimposition of a *dipolar* contribution to the true scalar coupling. This effect appears when the protein has a slightly preferred orientation in a magnetic field: consequently, the dipolar interaction – which depends upon the orientation of the internuclear vector in the applied field – does not vanish anymore to zero. Macromolecules may

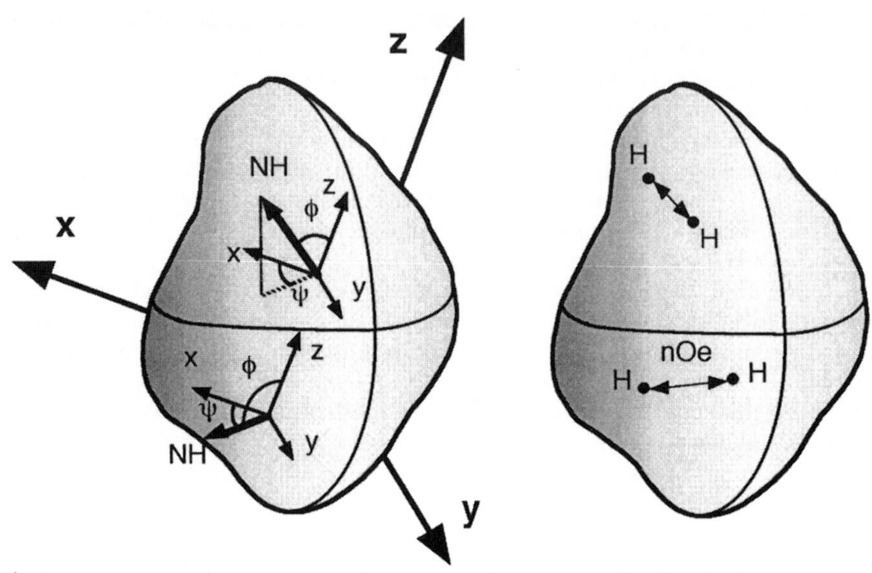

Fig. 7.1. Long range order constraints (left) *versus* nOe distance constraints (right). The experimental parameters R_2/R_1 or the residual dipolar couplings depend on the orientation of the interatomic vector in the external reference frame, defined by the angles ϕ and ψ. Note that *all* interatomic vectors are defined with respect to the same reference system. On the other hand, nOe span a network of short interatomic distances over the whole molecule that does not permit the structural correlation of separated parts of the molecule

tend to orient in a magnetic field for two reasons: either they have some anisotropic magnetic susceptibility, that causes them to naturally align in the magnetic field (Tolman et al 1995), or they are placed in an anisotropic environment and hence adopt some of this order. Such an anisotropic environment can be obtained using bicelles, small discoidal particles of a few hundred Å in diameter formed of a mixture of DMPC (dimyristoylphosphatidylcholine) and DHPC (dihexanoylphosphatidylcholine). These particles orient with their bilayer normal perpendicular to the magnetic field. The induced degree of protein alignment can be controlled by varying the bicelle concentration (Tjandra and Bax 1997, Sanders and Prosser 1998 for a review). Alternatively, filamentous phages (Hansen et al 1998) or mixtures of small organic molecules (Prosser et al 1998) have been used to obtain anisotropic environments for the study of macromolecules in solution. Although an understanding of the mechanism of ordering is not complete at this point, it seems to have a minimal effect on the structure and the molecular tumbling of the protein.

nOe and residual dipolar coupling constants both originate from the dipolar interaction, but the nature of the conformational information provided is in essence very different. nOe give distances between nuclei and residual dipolar coupling constants (when measured on spin pairs with known distance, such as $^1H-^{15}N$) yield angles between the internuclear vector and the preferential alignment axis (compare Fig. 7.1). This long-range information can be very useful for the study of proteins with elongated structure or multiple domain architecture.

4
Selected Applications

In what follows, we would like to illustrate the diversity of information gained from NMR studies on proteins by the presentation of two recent research projects of our laboratory. The first one will give an example of a classical structure determination of a protein module and the use of the chemical shift information for the characterization of ligand-binding. The second one focuses on the obtention of dynamical information on a ^{15}N-labelled protein and the determination of the tensor describing its anisotropic overall tumbling.

4.1
Structure Determination and Ca^{2+} Binding of the EGF-Like Module of the Human Complement Protease C1r

C1, a multimolecular protease, triggers the classical pathway of complement (Arlaud et al. 1987). Its activation and catalytic activity are mediated by two homologous serine proteases, C1r and C1s, which form a Ca^{2+}-dependent tetramer (C1s-C1r-C1r-C1s). These two proteases are *modular* proteins with the following modular composition: CUB-EGF-CUB-CCP-CCP-serine protease. The structural determinants for the tetrameric organization have been shown to be localized in the N-terminal part of these proteins (Thielens et al. 1990). Both C1r and C1s EGF modules contain the typical consensus sequence for Ca^{2+} binding found in many EGF-like modules, and were therefore supposed to play a key role in the intermolecular interaction. In addition, the EGF-like module of C1r (C1r-EGF hereafter) possesses an unusually large loop between the first two cysteines comprising 14 residues as opposed to 2–7 residues in other EGF-like modules. We were interested to determine whether this large loop adopts a well defined structure that could give insight into a possible functional role. Therefore, an NMR investigation was performed in order to determine the molecular structure of C1r-EGF and to study the structural consequences of Ca^{2+} binding. A more detailed description of this study can be found in Bersch et al. 1998.

4.1.1
Structure Determination of *apo* C1r-EGF

C1r-EGF (C1r residues 123–175) has been synthesized chemically (Hernandez et al. 1997). The first step of an NMR structure analysis consists of the assignment of resonances (^{1}H resonances in this case) corresponding to the individual residues. According to a well-established procedure for non-labelled proteins (see Basus 1989 for a detailed description), the intraresidual assignment is achieved by using scalar information by means of COSY or TOCSY-like experiments. Then, the residues are arranged according to the protein sequence using sequential NOESY information. Fig. 7.2 shows the fingerprint region of the NOESY spectrum in which each cross-peak corresponds to a through-space interaction between two protons (H^{N} and H^{α} in F1 and F2 respectively) being less than 6Å apart. From NOESY spectra acquired at different temperatures, 543 nOe could be

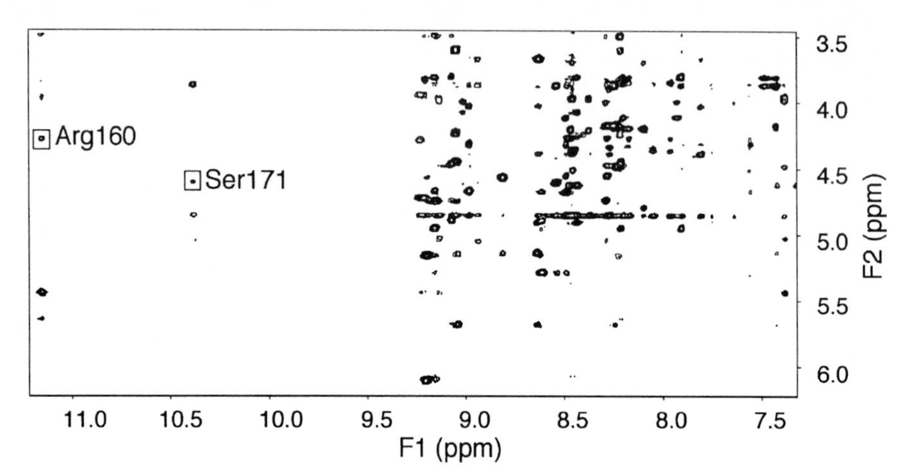

Fig. 7.2. Fingerprint region of the NOESY spectrum aquired on *apo* C1r-EGF (2mM, 15 °C, pH 6.6). The H^N to H^α cross-peaks are indicated for two residues whose H^N resonances are shifted to high frequencies, being an indication for the structured nature of the protein (reprinted with permission from Bersch et al 1998, copyright 1999 American Chemical Society)

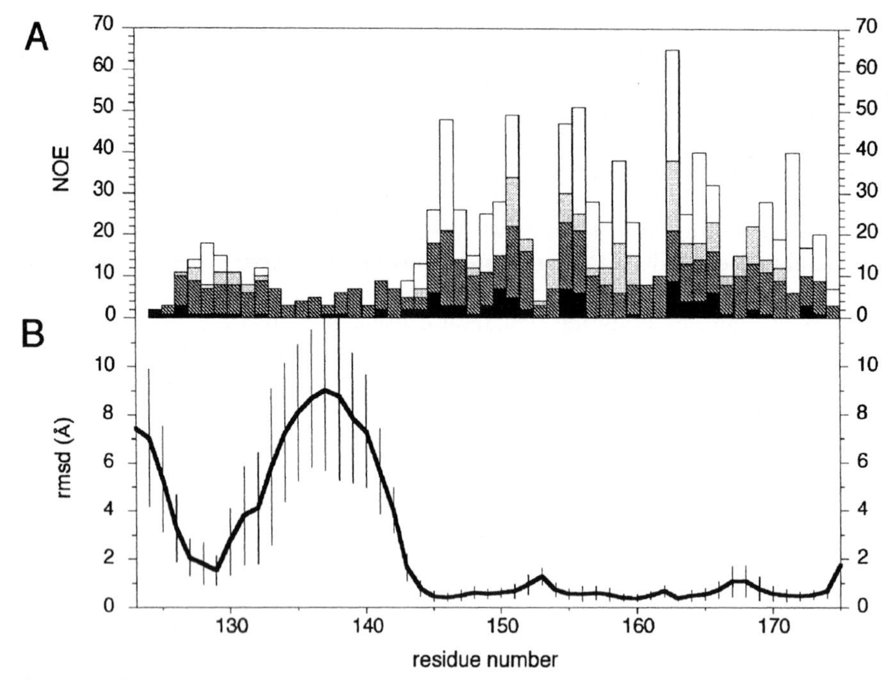

Fig. 7.3. Experimental and structural statistics for *apo* C1r-EGF. A. Number of NOE determined in function of the sequence. Black intraresidual, hatched sequential, light-gray medium-range and white long-range (i > 4) restraints. B. Positional backbone rmsd calculated over the backbone atoms N, C^α, C' with respect to the mean structure, calculated from the superposition of residues 144–174. Error bars indicate the standard deviation for the structural ensemble of 19 structures (reprinted with permission from Bersch et al 1998, copyright 1999 American Chemical Society)

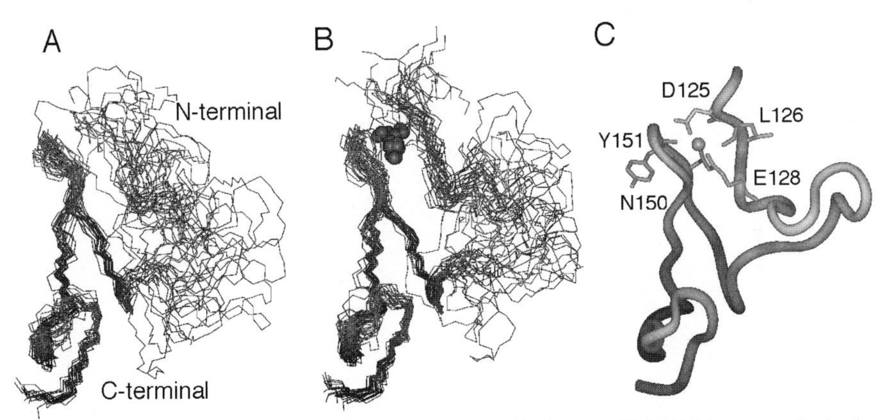

Fig. 7.4. Structures obtained for C1r-EGF. A. NMR ensemble for *apo* C1r-EGF (shown are the backbone atoms N, Ca, C' for 19 structures). The N-terminal part of the protein is highly disordered. B. Structural ensemble resulting of the modelling of the Ca^{2+}-bound C1r-EGF (backbone atoms for 23 structures) shown in the same orientation as in A. Note the slight stabilization of the N-terminal residues. C. Ribbon diagram of an individual structure showing the Ca^{2+} ligands in more detail. The Ca^{2+} ions are represented by spheres

assigned. Their distribution with respect to the protein sequence is shown in Fig. 7.3.A. From this graph it becomes obvious that the N-terminal and C-terminal parts differ significantly with respect to the number of interresidual distances obtained for each residue. This implies that the C-terminal part of the protein is structured whereas the large loop, spanning residues 130–143, is most likely disordered, as no long- or medium-range nOe could be observed.

Structures were then calculated using combined simulated annealing/molecular dynamics with nOe derived distance ranges as input data (Blackledge et al. 1995, Bersch et al. 1998). The resulting structural ensemble, comprising 19 structures, is shown in Fig. 7.4.A and the positional backbone rmsd, calculated with respect to the mean structure is indicated in Fig. 7.3.B. As expected from the distribution of nOe, the C-terminal part is well structured and exhibits the typical EGF-like fold with a major and a minor β-sheet. On the other hand, the N-terminal part, comprising the unusually large loop, is completely disordered. From this structural information and from the fact that this loop contains a high number of charged residues, it may be concluded that it is surface exposed. Its possible role might therefore correspond to the mediation of intra- or intermolecular interactions in C1r or between the different C1 proteins.

4.1.2
Ca^{2+} Binding to C1r-EGF

As has been pointed out in 3.1.1., the chemical shift is a very sensitive probe for the chemical environment of a nucleus. When small aliquots of $CaCl_2$ were added to the C1r-EGF solution, some resonances were shifted with increasing Ca^{2+} concentration. This was direct evidence that the isolated C1r-EGF module was able to bind Ca^{2+}. Fig. 7.5.A shows the changes in chemical shift of Y155-H^{α} as a func-

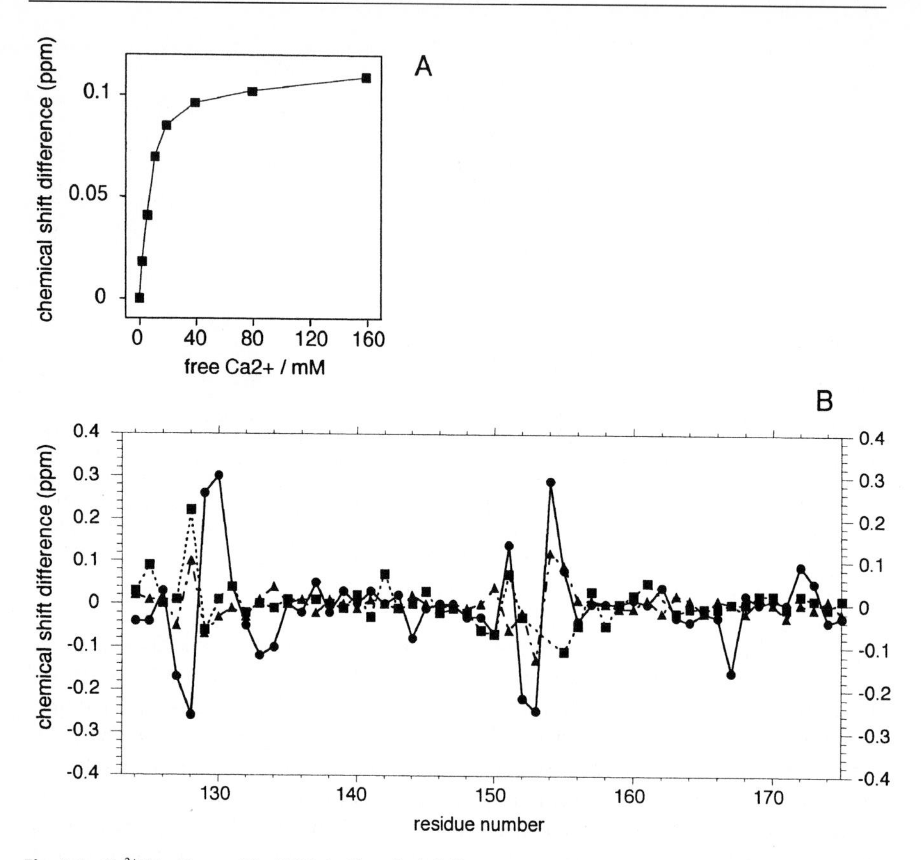

Fig. 7.5. Ca^{2+}-binding to C1r-EGF. A. Chemical shift variation of Y155-H^{α} in function of the free Ca^{2+} concentration (reprinted with permission from Hernandez et al 1997, © 7/1 1999 Munksgaard International Publishers Ltd., Copenhagen, Denmark). B. Chemical shift differences of H^N (●), H^{α} (■) and H^{β} (▲) between apo C1r-EGF and Ca^{2+}-bound C1r-EGF (> 90 % Ca^{2+} saturation) as a function of the protein sequence (reprinted with permission from Bersch et al 1998, copyright 1999 American Chemical Society)

tion of the $CaCl_2$ concentration. The binding could be adequately fitted by a one-site binding equation, leading to an apparent dissociation constant of 10 mM at pH 6.6 (Hernandez et al. 1997). This value is onehundred- to onethousand-fold less than that of the intact C1r protein, which binds Ca^{2+} in the 10–100 μM concentration range. Interestingly, the same behavior has also been observed for the isolated EGF-like modules of coagulation factors IX and X as well as fibrillin (Persson et al. 1998, Valcarce et al. 1993, Handford et al. 1995) and it is now generally admitted that the N-terminal module (the first CUB module in the case of C1r) contributes to the Ca^{2+} binding in a way that has not been determined yet.

A set of NMR experiments was then acquired in the presence of 80 mM Ca^{2+}, corresponding to a > 90% saturation of the Ca^{2+}-binding site, for the assignment of the proton resonances. Chemical shift differences between the *apo*- and the

Ca^{2+}-bound form were plotted as a function of the protein sequence to determine which regions of the protein module were implicated in Ca^{2+} binding (Fig. 7.5.B). Significant variations of the chemical shift occurred in two distinct regions (residues 127–130 and 151–155), which correspond to those residues that have been shown to be Ca^{2+}-ligands in the crystal structures of analogous, Ca^{2+}-binding EGF-like modules (see for example Rao et al. 1995). It can thus be concluded that C1r-EGF binds Ca^{2+} in an analogous manner to other known Ca^{2+}-binding EGF-like modules.

Unfortunately, the quality of the NOESY spectra was not good enough to recalculate the molecular structures from experimental constraints at a sufficient resolution (low signal to noise ratio certainly due to the increased ionic strength and to losses of the peptide during sample preparation). However, those NOESY patterns which could be observed in these spectra were also present in the spectra of the *apo* form and showed identical relative intensities. No new nOe involving the residues close to the Ca^{2+}-binding site could be detected, suggesting that Ca^{2+} binding did not lead to major structural variations. The Ca^{2+}-bound form of C1r-EGF was then modelled in the presence of a Ca^{2+} ion, using the experimental constraints obtained from the *apo* form. Five new distance constraints were added between the Ca^{2+} ion and the carboxyl or carbonyl groups analogous to the Ca^{2+} ligands in the crystal structure of coagulation factor IX (Rao et al. 1995). The resulting structural ensemble and the ligation of Ca^{2+} are shown in Fig. 7.4.B and C respectively. It can be seen, that Ca^{2+} binding induces a significant ordering of the N-terminal part of the peptide. However, this observation is based only on the modelled structures and was not reflected in the experimental nOe data. As it has been proposed that the N-terminal CUB module contributes to Ca^{2+} fixation, it might be that Ca^{2+} binding leads to conformational changes in the module-module interface and/or stabilizes a relative orientation of the two modules. NMR experiments on the CUB-EGF module pair could give further insight into the structural consequences of Ca^{2+} binding. However, such a study may be precluded by the low solubility of this fragment (Thielens et al. 1998).

4.2 Rotational Tumbling and Backbone Dynamics of *Rhodobacter capsulatus* cytochrome c_2

The cytochromes *c* play an important role in the transfer of electrons in a large variety of both eukaryotic and prokaryotic organisms (Pettigrew and Moore 1987). The class I cytochromes *c* have been used as a model for investigations of the relationship between structure and function, in particular to understand the mechanism of electron transfer. The importance of conformational flexibility at the interaction surface for the function of the cytochrome *c* has been recognized and investigated using molecular dynamics simulations and modeling studies (Northrup et al. 1988, Wendolski and Matthew 1989, Banci et al. 1997), but until now the backbone dynamics of class I cytochromes *c* have not been investigated for cytochromes *c* of known structure.

We have determined the solution structure, local backbone dynamics and global hydrodynamics of the cytochrome c_2 from *Rb. capsulatus* (Cordier et al.

1998). Cytochrome c_2 has a broad distribution of helical motifs in angular space permitting the precise determination of the anisotropic diffusion tensor, and thus an accurate parameterization of the local motions of the peptide chain. The precision of the analysis of the anisotropic diffusion tensor has been estimated using Monte-Carlo sampling methods, made possible by an efficient simulated annealing algorithm developed in our laboratory.

4.2.1
Characterization of Local-Motion and Overall Tumbling

Use of a spectral density function assuming an *isotropic* rotational diffusion tensor in the Lipari-Szabo modelfree approach reveals a highly compact protein. However certain irregularities indicate that this model is imperfect – nearly thirty residues require a more complex model than the most rigid one, 10 residues require a second motion on an intermediary timescale and 12 a chemical exchange term. The situation of many of these residues in regions of secondary structure (in which no internal motion or exchange should occur) and the observation that the geometrically orthogonal terminal helices exhibit relaxation on slightly different timescales (Fig. 7.6) persuaded us to analyse the relaxation data using a spectral density function with an anisotropic rotational diffusion tensor.

The orientation and component values of the anisotropic diffusion tensor have been determined using both axially and fully anisotropic models, using selections of residues present in helical regions. In general, we find that the axially symmetric model results in two orthogonal solutions – corresponding to a prolate and an oblate model. The two minima are similarly significant, as shown by confidence limits derived from extensive Monte Carlo simulations, and are both acceptable within these limits. It appears then that the use of the more complex totally anisotropic diffusion tensor is necessary to fully describe the system

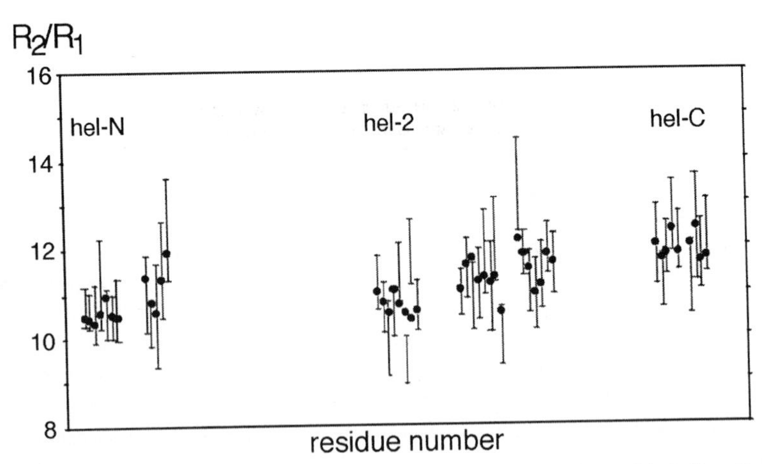

Fig. 7.6. Relaxation of cytochrome c_2. R_2/R_1 is shown for residues situated in helices. Error bars have been calculated from the experimental error. The variations observed within and between the helices reflect the *anisotropic* overall tumbling of the molecule

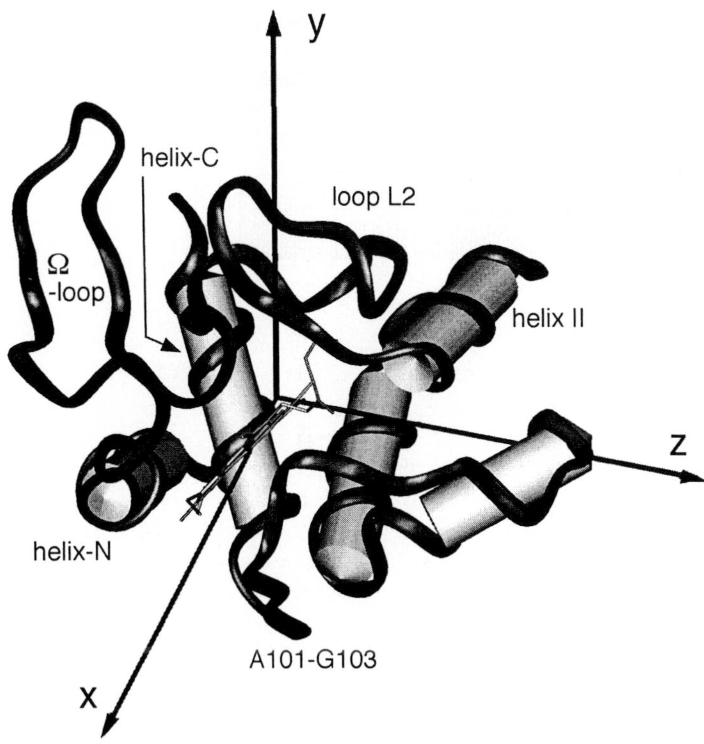

Fig. 7.7. Anisotropic rotational diffusion tensor of cytochrome c_2. The backbone of cytochrome c_2 (NMR structure) is shown in a ribbon representation, helices are given by cylinders. The tensor is characterized by the orientation of its three principal axis as shown. Its component values have been calculated to D_{xx}=1.829, D_{yy} = 1.566 and D_{zz} = 1.405 from the relaxation data

(Fig. 7.7). We have again used robust statistical tests to analyze the significance of the improved fit and shown that the fully anisotropic tensor is justified in this case.

4.2.2 Comparison of Structural and Dynamic Information

The spatial order of the individual NH vectors present in the NMR ensemble has been compared to motional parameters fitted to the model-free approach using the asymmetric anisotropic diffusion tensor. Although the protein is relatively compact, giving rise to high order parameters and low internal flexibility (Fig. 7.8), certain regions of the molecule appear to exhibit coincident dynamic behavior and low structural order. In particular the loop (A101–G103), situated between the heme-methionine binding motif and the C-terminal helix. This tripeptide may be important for electrostatic interactions with electron transfer partners, as K102 makes up part of the positively charged interaction surface common to all cytochrome *c* and c_2 proteins (Tiede et al. 1993), a region which

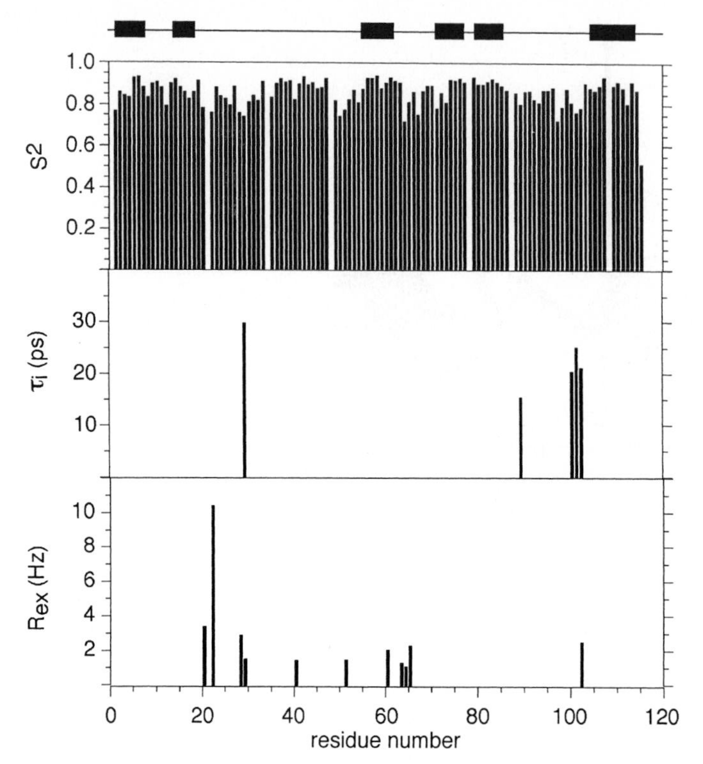

Fig. 7.8. Dynamic parameters for cytochrome c_2 calculated using an anisotropic rotational diffusion tensor. Top, order parameter S^2, middle, internal correlation time constants τ_i and bottom, conformational exchange contribution R_{ex}. The helical segments of the protein are schematically indicated above. The reproduction from Cordier et al 1998 is kindly acknowledged

has been suggested, from chemical shift perturbation measurements (Ubbink and Bendall 1997) to be involved in cytochrome *c* /plastocyanin complex. While this tripeptide is not one of the three regions of the molecule exhibiting higher than average B-factors in both forms of the crystal structure (Benning et al. 1991), the liquid state flexibility may be masked in the crystal state by surface contacts. The residue K91 forms a hinge between the helix IV and the methionine ligand (M96) binding motif and exhibits mobility on the nanosecond timescale, as well as structural disorder in the NMR ensemble. The third region of common disorder contains the residues 29–31 close to the Ω-loop motif behind the histidine heme ligand.

The relaxation data can not only provide confirmation of local structural information, but also global structural information concerning long-range tertiary orientation (see 3.5). For example the relaxation data from helix II, which forms part of the large loop L2 (residues 41–69) covering the bottom of the heme pocket, and provides protection from the solvent for the propionates attached to the porphyrin ring, remains incoherent, exhibiting a periodic variation along the helix which can not be reproduced by the model fit from the fully anisotropic description. In addition, the residue 61 requires a chemical exchange term to fulfill statistical tests even when the fully anisotropic tensor is employed. A possible explanation of the disagreement of the data from helix II, with respect to the rest of the helical data, is that it has a slightly different mean orientation to that determined in the structure calculation. The homologous subdomain of loop L2 has been shown to be the last motif to fold in the cytochrome *c* from mitochondria (Bai et al. 1995), and is absent from certain families of bacterial cytochromes *c* which have otherwise retained very similar secondary and tertiary structural fold without affecting the stability of the protein (Blackledge et al. 1996). The loop may therefore have a certain independence from the rest of the protein fold which allows a different relative orientation of the helix II. Another possible explanation for the comparatively low quality fit for this helix would be the presence of low amplitude slow motions of the subdomain L2, hinged at either residues (41 and 69) or residues (45 and 65), although it is difficult to investigate this type of motion without resorting to unjustifiably complex models.

5
General Conclusions

We have shown that many very different experimental parameters can be obtained from NMR experiments which furnish valuable information both on the molecular structure as well as on the dynamical properties of a protein. The multitude of experimental parameters available clearly represents an advantage of this technique compared for example to radiocrystallography. In fact, very exciting applications of NMR do not only focus on a pure structure determination but aim to investigate features more related to protein function, as for example ligand-binding or protein-protein interactions. Molecular interactions can be monitored by measuring easily obtainable experimental parameters, as for example the chemical shift or heteronuclear relaxation. The possibility of using isotopic filters, by labelling for example only one of the interacting molecules, still increases the performance of such experiments. In addition, the information obtained can usually be assigned to individual residues, offering thus the advantage of a very high resolution. A very interesting approach relying on these features of NMR is now routinely used in the industrial research to screen, select and optimize protein ligands (Shuker et al., 1996). So hopefully, NMR will find an increased entrance into pharmacological and biological research where it should prove to be a very versatile technique for the study of protein properties related to biological function.

6
Acknowledgements

This work has been supported by the Centre National de la Recherche Scientifique, the Commissariat à l'Energie Atomique and Molecular Simulations Inc. We thank Drs GJ Arlaud and JF Hernandez as well as M Cusanovich and M Caffrey for kindly providing us with C1r-EGF and *Rhodobacter capsulatus* cytochrome c_2, respectively, and for the fruitful collaboration that was established between our laboratories. We also want to thank all those who belonged to the Laboratory of Magnetic Resonance at the IBS during the last years and who have contributed to this work by numerous, invaluable discussions. This is publicatoin no 612 of the Institut de Biologie Structurale – Jean-Pierre Ebel.

References

Arlaud GJ, Colomb MG, Gagnon J (1987) A functional model of the human C1 complex. Immunol. Today 8, 106–111.

Bai Y, Sosnick TR, Mayne L, Englander SW (1995) Protein folding intermediates: native state hydrogen exchange. Science 269: 192–197.

Banci L, Gori-Savanalli G, Turano P (1997) A molecular dynamics study in explicit water of the reduced and oxidized form of yeast iso-1-cytochrome c. Eur J Biochem 249: 716–723.

Basus VJ (1989) Proton nuclear magnetic resonance assignments. Meth Enzymol 177: 132–149.

Beger RD, Bolton PH (1997) Protein ϕ and ψ dihedral restraints determined from multidimensional hypersurface correlation of backbone chemical shifts and their use in the determination of protein tertiary structure. J Biomol NMR 10: 129–142.

Benning MM, Wesenberg G, Caffrey MS, Bartsch RG, Meyer TE, Cusanovich MA, Rayment I, Holden MM (1991) Molecular structure of cytochrome c_2 isolated from *Rhodobacter capsulatus* determined at 2.5Å resolution. J Mol Biol 220: 673–685.

Bersch B, Hernandez JF, Marion D, Arlaud GJ (1998) Solution structure of the epidermal growth factor (EGF)-like module of human complement protease C1r, an atypical member of the EGF family. Biochemistry 37: 1204–1214.

Biamonti C, Rios CB, Lyons BA, Montelione GT (1994) Multidimensional NMR experiments and analysis techniques for determining homo- and heteronuclear scalar coupling constants in proteins and nucleic acids. Adv Biophys Chem 4: 51–120

Blackledge MJ, Medvedeva S, Poncin M, Guerlesquin F, Bruschi M, Marion D (1995) Structure and dynamics of ferrocytochrome c_{553} from *Desulfovibrio vulgaris* studied by NMR spectroscopy and restrained molecular dynamics. J Mol Biol 245: 661–681.

Blackledge MJ, Guerlesquin F, Marion D (1996) Comparison of low oxidoreduction potential cytochrome c553 from *Desulfovibrio vulgaris* with the class I cytochrome c family. Proteins: Struct Funct Genet 24: 178–194.

Blackledge M, Cordier F, Dosset P, Marion D (1998) Precision and uncertainty in the characterization of anisotropic rotational diffusion by ^{15}N relaxation. J Am Chem Soc 120: 4538–4539.

Clore GM, Gronenborn AM (1998) New methods of structure refinement for macromolecular structure determination by NMR. Proc Natl Acad Sci 95: 5891–5898.

Cordier F, Caffrey M, Brutscher B, Cusanovich M, Marion D, Blackledge M (1998) Solution structure, rotational diffusion anisotropy and local backbone dynamics of Rhodobacter capsulatus cytochrome c_2. J Mol Biol 281: 341–361.

Dzakula Z, Westler WM, Edison AS, Markley JL (1992) The „CUPID“ method for calculating the continuous probability distribution of rotamers from NMR data. J Am Chem Soc 114: 6195–6199.

Englander SW, Sosnick TR, Englander JJ, Mayne L (1996) Mechanisms and uses of hydrogen exchange. Curr Opin Struct Biol 6: 18–23

Folkers PJM, Folmer RHA, Konigs RNH, Hilbers CW (1993) Overcoming the ambiguity problem encountered in the analysis of nuclear Overhauser magnetic resonance spectra of symmetric dimer proteins. J Am Chem Soc 115: 3798–3799.

Garcia de la Torre JG, Bloomfield VA (1981) Hydrodynamic properties of complex, rigid, biological macromolecules: theory and applications. Q Rev Biophys 14: 81–139.

Garrett DS, Kuszewski J, Hancock T, Lodi PJ, Vuister GW, Gronenborn AM, Clore GM (1994) The impact of direct refinement against three-bond HN-C^α coupling constants on protein structure determinaton by NMR. J Magn Reson 104: 99–103.

Gooley PR, MacKenzie NE (1990) Pro-Ala-35 *Rhodobacter capsulatus* cytochrome c_2 shows dynamic not structural differences. A ^{1}H and ^{15}N NMR study. FEBS Lett 260: 225–288.

Handel TM, Domaille PJ (1996) Heteronuclear (^{1}H, ^{13}C, ^{15}N) NMR assignments and solution structure of the monocyte chemoattractant protein-1 (MCP-1) dimer. Biochemistry 35: 6569–6584.

Handford P, Downing AK, Rao Z, Hewett DR, Sykes BC, Kielty CM (1995). The calcium binding properties and molecular organization of epidermal growth factor-like domains in human fibrillin 1. J Biol Chem 270: 6751–6756.

Hansen MR, Mueller L, Pardi A (1998) Tunable alignment of macromolecules by filamentous phage yields dipolar coupling interactions. Nat Struct Biol 5: 1065–1074.

Haon S, Augé S, Tropis M, Milon A, Lindley ND (1993) Low cost production of perdeuterated biomass using methylotrophic yeast. J Lab Comp Radiopharm 33: 1053–1063.
Hernandez JF, Bersch B, Pétillot Y, Gagnon J, Arlaud GJ (1997) Chemical synthesis and characterization of the epidermal growth factor-like module of human complement protease C1r. J Pept Res 49: 221–231.
Hore PJ (1989) Solvent suppression. Meth Enzymol 176: 64–77.
Katz JJ, Crespi HL (1966) Deuterated organisms: cultivation and uses. Science 151: 1187–1194.
Kohno T, Kusunoki H, Sato K, Wakamatsu K. (1998) A new general approach for the biosynthesis of stable isotope-enriched peptides using a decahistidine-tagged ubiquitin fusion system: an application to the production of mastoparan-X uniformly enriched with ^{15}N and $^{15}N/^{13}C$. J Biomol NMR 12: 109–121.
Lee AL, Volkman BF, Robertson SA, Rudner DZ, Barbash DA, Cline TW, Kanaar R, Rio DC, Wemmer DE (1997) Chemical shift mapping of the RNA-binding interface of the multiple-RBD protein sex-lethal. Biochemistry 36: 14306–14317.
LeMaster DM (1990) Deuterium labelling in NMR structural analysis of larger proteins. Q Rev Biophys 23: 133–74.
Lipari G, Szabo A (1982). Model-free approach to the interpretation of nuclear magnetic-resonance relaxation in macromolecules 1. theory and range of validity. J Am Chem Soc 104: 4546–4559.
Marion D, Kay LE, Sparks SW, Torchia DA, Bax A (1989) 3-Dimensional Heteronuclear NMR of ^{15}N Labeled Proteins. J Am Chem Soc 111: 1515–1517.
McIntosh LP, Dahlquist FW. (1990) Biosynthetic incorporation of ^{15}N and ^{13}C for assignment and interpretation of nuclear magnetic resonance spectra. Q Rev Biophys 23: 1–38.
McKay RT, Pearlstone JR, Corson DC, Gagne SM, Smillie LB, Sykes BD (1998) Structure and interaction site of the regulatory domain of troponin-c when complexed with the 96–148 region of troponin-I. Biochemistry 37: 12419–12430.
Meissner A, Duus JO, Sørensen OW (1997) Spin-state-selective excitation. Application for E. COSY-type measurement of J_{HH} coupling constants. J. Magn. Res. 128: 92–97.
Mierke DF, Kessler H (1992) Combined use of homo- and heteronuclear coupling constants as restraints in molecular dynamics simulations. Biopolymers 32: 1277–1282.
Neuhaus D, Willamson M. (1989) The nuclear Overhauser effect in structural and conformational analysis, VCH publishers, New-York.
Northrup SH, Bowles JO, Reynolds JC (1988) Brownian dynamics of cytochrome c and cytochrome c peroxidase association. Science 241: 67–70.
Oldfield E (1995) Chemical shifts and three-dimensional protein structure. J Biomol NMR 5: 217–225.
Ottiger M, Tjandra N, Bax A (1997) Magnetic field dependent amide ^{15}N chemical shifts in a protein-DNA complex resulting from magnetic ordering in solution. J Am Chem Soc 119: 9825–9830.
Peng JW, Wagner G (1992) Mapping of the spectral densities of N-H bond motions in eglin c using heteronuclear relaxation experiments. Biochemistry 31: 8571–86.
Persson KEM, Astermark J, Björk I, Stenflo J (1998) Calcium binding to the first EGF-like module of human factor IX in a recombinant fragment containing residue 1–85. FEBS Lett. 421: 100–104.
Pervushin K, Riek R, Wider G, Wüthrich K (1997) Attenuated T_2 relaxation by mutual cancellation of dipole-dipole coupling and chemical shift anisotropy indicates an avenue to NMR structures of very large biological macromolecules in solution. Proc Natl Acad Sci USA 94: 12366–71.
Pettigrew GW, Moore GR (1987). Cytochromes c: biological aspects. Springer-Verlag, Berlin, Heidelberg, New-York.
Piotto M, Saudek V, Sklenár V (1992) Gradient-tailored excitation for single-quantum NMR spectroscopy in aqueous solutions. J Biomol NMR 2: 661–665.
Prosser RS, Losonczi JA, Shiyanovskaya IV (1998) Use of a novel aqueous liquid crystalline medium for high-resolution NMR of macromolecules in solution. J Am Chem Soc 120: 11010–11011.
Qin J, Clore GM, Kennedy WM, Huth JR, Gronenborn AM (1995) Solution structure of human thioredoxin in a mixed disulfide intermediate complex with its target peptide from the transcription factor NF-κB. Structure 1995 3: 289–297.
Rajagopal P, Waygood EB, Reizer J, Saier MH Jr, Klevit RE (1997) Demonstration of protein-protein interaction specificity by NMR chemical shift mapping. Protein Sci 6: 2624–2627.
Rao Z, Handford PA, Mayhew M, Knott V, Brownlee GG, Stuart D (1995) The structure of a Ca^{2+}-binding epidermal growth factor-like domain: its role in protein-protein interactions. Cell 82: 131–141.
Reif B, Hennig M, Griesinger C (1997) Direct measurement of angles between bond vectors in high-resolution NMR. Science 276: 1230–1233.
Rosen MK, Gardner KH, Willis RC, Parris WE, Pawson T, Kay LE. (1996) Selective methyl group protonation of perdeuterated proteins. J Mol Biol 263: 627–36.
Sanders CR., Prosser, RS. (1998) Bicelles: a model membrane system for all seasons? Structure 6: 1227–1234.

Sattler M, Fesik SW (1996) Use of deuterium labeling in NMR: overcoming a sizeable problem. Structure 4: 1245–9.

Shuker SB, Hajduk PJ, Meadows RP, Fesik SW (1996) Discovering high-affinity ligands for proteins: SAR by NMR. Science 274: 1531–1534.

Spera S, Bax A (1991) Empirical correlation between protein backbone conformation and C^{α} and C^{β} ^{13}C nuclear magnetic resonance chemical shifts J Am Chem Soc 113: 5490–5492.

Thielens NM, Aude CA, Lacroix MB, Gagnon J, Arlaud GJ (1990) Ca^{2+} binding properties and Ca^{2+}-dependent interactions of the isolated N-terminal alpha fragments of human complement proteases C1r and C1s. J Biol Chem 265, 14469–14475.

Thielens NM, Enrie K, Lacroix M, Jaquinod M, Hernandez JF, Esser AF, Arlaud GF (1998) The N-terminal CUB-EGF module pair of human complement protease C1r binds Ca^{2+} with high affinity and mediates Ca^{2+}-dependent interaction with C1s. J Biol Chem 274: 9149–9159.

Tiede DM, Vashishta AC, Gunner MR (1993) Electron-transfer kinetics and electrostatic properties of the *Rhodobacter sphaeroides* reaction center and soluble c cytochromes. Biochemistry 32: 4514–4531.

Tjandra N, Bax A (1997) Direct measurement of distances and angles in biomolecules by NMR in a dilute liquid crystalline medium. Science 278: 1111–1114.

Tjandra N, Garrett DS, Gronenborn AM, Bax A, Clore GM (1997) Defining long range order in NMR structure determination from the dependence of heteronuclear relaxation times on rotational diffusion anisotropy. Nat Struct Biol 4: 443–9.

Tolman JR, Flanagan JM, Kennedy MA, Prestegard JH (1995) Nuclear magnetic dipole interactions in field-oriented proteins: information for structure determination in solution. Proc Natl Acad Sci USA 92: 9279–83.

Ubbink M, Bendall DS (1997). Complex of plastocyanin and cytochrome c characterized by NMR chemical shift analysis. Biochemistry 36: 6326–6335.

Valcarce C, Selander-Sunnerhagen M, Tämlitz AM, Drakenberg T, Björk I, Stenflo J (1993) Calcium affinity of the NH_2-terminal epidermal growth factor-like module of factor X. Effect of the gamma-carboxyglutamic acid-containing module. J Biol Chem 268: 26673–26678.

van Heijenoort C, Penin F, Guittet E (1998) Dynamics of the DNA binding domain of the fructose repressor from the analysis of linear correlations between the ^{15}N-^{1}H bond spectral densities obtained by nuclear magnetic resonance spectroscopy. Biochemistry 37: 5060–73.

Wendolski JJ, Mathew JB (1989). Molecular dynamic effects on protein electrostatics. Proteins: Struct Funct Genet 5: 313–321.

Wishart DS, Sykes BD, Richards FM (1992) The chemical shift index: a fast and simple method for the assignment of protein secondary structure through NMR spectroscopy. Biochemistry 31: 1647–51.

Wishart DS, Sykes BD (1994) The ^{13}C chemical-shift index: a simple method for the identification of protein secondary structure using ^{13}C chemical-shift data. J Biomol NMR 4: 171–180.

Wishart DS, Bigam CG, Yao J, Abildgaard F, Dyson HJ, Oldfield E, Markley JL, Sykes BD (1995) ^{1}H, ^{13}C and ^{15}N chemical shift referencing in biomolecular NMR. J Biomol NMR 6: 135–140.

Woessner DE (1962) Nuclear spin relaxation in ellipsoids undergoing rotational Brownian motion. J. Chem. Phys. 3: 647–654.

Wong KB, Daggett V (1998) Barstar has a highly dynamic hydrophobic core: evidence from molecular dynamics simulations and nuclear magnetic resonance relaxation data. Biochemistry 37: 11182–92.

Yang D, Konrat R, Kay LE (1997) A Multidimensional NMR Experiment for Measurement of the Protein Dihedral Angle Based on Cross-Correlated Relaxation between ^{1}H-^{13}C Dipolar and ^{13}C' (Carbonyl) Chemical Shift Anisotropy Mechanisms, J Am Chem Soc 119: 11938–11940.

CHAPTER 8

Elucidation of Functionally Significant Structural Modifications by Matrix-Assisted Laser Desorption/Ionization Time-of-Flight Mass Spectrometry with Post-Source Decay Analysis

J. J. Gorman[1], B. L. Ferguson[1], S. Lopaticki[1], J. J. Pitt[1], A. W. Purcell[2] and C. J. Morrow[3]

1 Introduction

Analysis of peptides and proteins by mass spectrometry (MS) became accessible to non-specialist mass spectroscopists with the advent of soft ionization techniques such as plasma desorption (Macfarlane and Thorgerson 1976) and fast-atom bombardment (FAB) (Barber et al. 1981). These techniques enabled ionization and detection of pseudomolecular ions of non-derivatized peptides and small proteins using modest quantities of analytes (Barber et al. 1982; Sundqvist et al. 1984). Sequence data were occasionally evident in FAB ionization spectra as a result of in-source fragmentation (Barber et al. 1982) or post-source collisional activation (Desiderio and Katakuse 1983; Katakuse and Desiderio 1983). Generation of extensive fragmentation was usually achieved using multi analyser instruments with a collision cell between an analyser used to select parent ions for fragmentation and a subsequent analyser used to define daughter ions (Biemann 1990; Carr et al. 1990; Hunt et al. 1990). Consequently, high performance tandem double focussing sector and triple quadrupole instruments with extended performance characteristics were developed.

Development of Matrix-Assisted Laser Desorption/Ionization (MALDI) (Karas and Hillenkamp 1988) and Electrospray Ionization (ESI) (Fenn et al. 1989) led to more general utilization of mass spectrometry by protein chemists. This was largely due to greater sensitivities and extended mass range capabilities of these ionization techniques. However, improved user friendliness of data systems and mass analysers, mainly Time-Of-Flight (TOF) and quadrupoles, has also promoted more universal use. ESI in conjunction with triple quadrupole analysers has produced sequence data on extremely small quantities of peptides (Wilm et al. 1996) as well as molecular weight information on peptides and small proteins (Fenn et al. 1989). Use of Fourier transform ion-cyclotron resonance, ion-trap and hybrid quadrupole TOF or Q-TOF instruments has further enhanced the scope for producing peptide sequence data in conjunction with ESI.

Use of reflecting TOF instruments has revealed that MALDI generated peptide ions undergo fragmentation by post-source decay (PSD) during their relatively

[1] Biomolecular Research Institute and
[2] University of Melbourne, 343 Royal Parade, Parkville, Vic. 3052, Australia and
[3] Ross Breeders, Scotland, UK.

long drift time to the reflector (Kaufmann et al. 1993, 1994; Spengler et al. 1991, 1992). Protocols have been developed to analyse these metastable ions and enable their interpretation in terms of peptide sequence data (Rouse et al. 1995; Spengler 1997). However, analysis of PSD fragmentation is not widely accepted as the method of choice for sequencing peptides and other biopolymers. This is apparently due to a combination of the relative complexity of the experimental protocols used to obtain data, the inherent complexity of the data and the lack of universality of fragmentation with all samples of interest. Despite these reputed limitations, we have found PSD of MALDI generated ions has enabled characterization of structural features of modified peptides. The efficacy of PSD for biopolymer characterization will be exemplified below by elucidation of the structures of modified peptides and an oligosaccharide derivative. These include: a byproduct of peptide synthesis that produced a dominant immunological response; variations in the sites of cleavage activation of the fusion protein precursors of different isolates of Newcastle disease virus (NDV); the disulfide bonding pattern of the attachment protein of human respiratory syncytial virus (RSV); and, a 1-phenyl-3-methyl-5-pyrazolone (PMP) derivative of pentaglucose.

2 Experimental

All data presented herein were acquired using a Bruker Reflex mass spectrometer using experimental protocols described in detail elsewhere (Gorman et al. 1996, 1997; Lopaticki et al. 1998) or as elaborated in the legends to specific figures. Procedures for isolation of peptides for analysis have also been described in detail in previous publications (Gorman et al. 1987, 1988, 1990b; Lopaticki et al. 1998) as has the methodology for preparation and isolation of phenylmethylpyrazolone derivatized oligosaccharides (Pitt and Gorman 1997).

3 Results and Discussion

3.1 Characterization of an Nβ-Butyl Asparagine Modified Peptide with Immunodominance over the Unmodified Sequence

Synthetic peptides are frequently used as tools to study biological structure and function (Kent 1988). However, complications can arise in such studies due to synthetic byproducts in peptide preparations that can behave as superagonists or inhibitors. Such byproducts can dominate the responsiveness of the biological system and give rise to false interpretations. This was recently exemplified in an

→

Fig. 8.1. Parent ion masses of (A) the immunologically active byproduct formed during the synthesis of IMIKFNRL, (B) the Asn6 version of IMIKFNRL, (C) deliberately synthesised the Nβ-butyl-Asn6 version of IMIKFNRL and (D) the α-Asp6 version of IMIKFDRL. These spectra were all collected using delayed extraction and a digitization rate of 1GHz

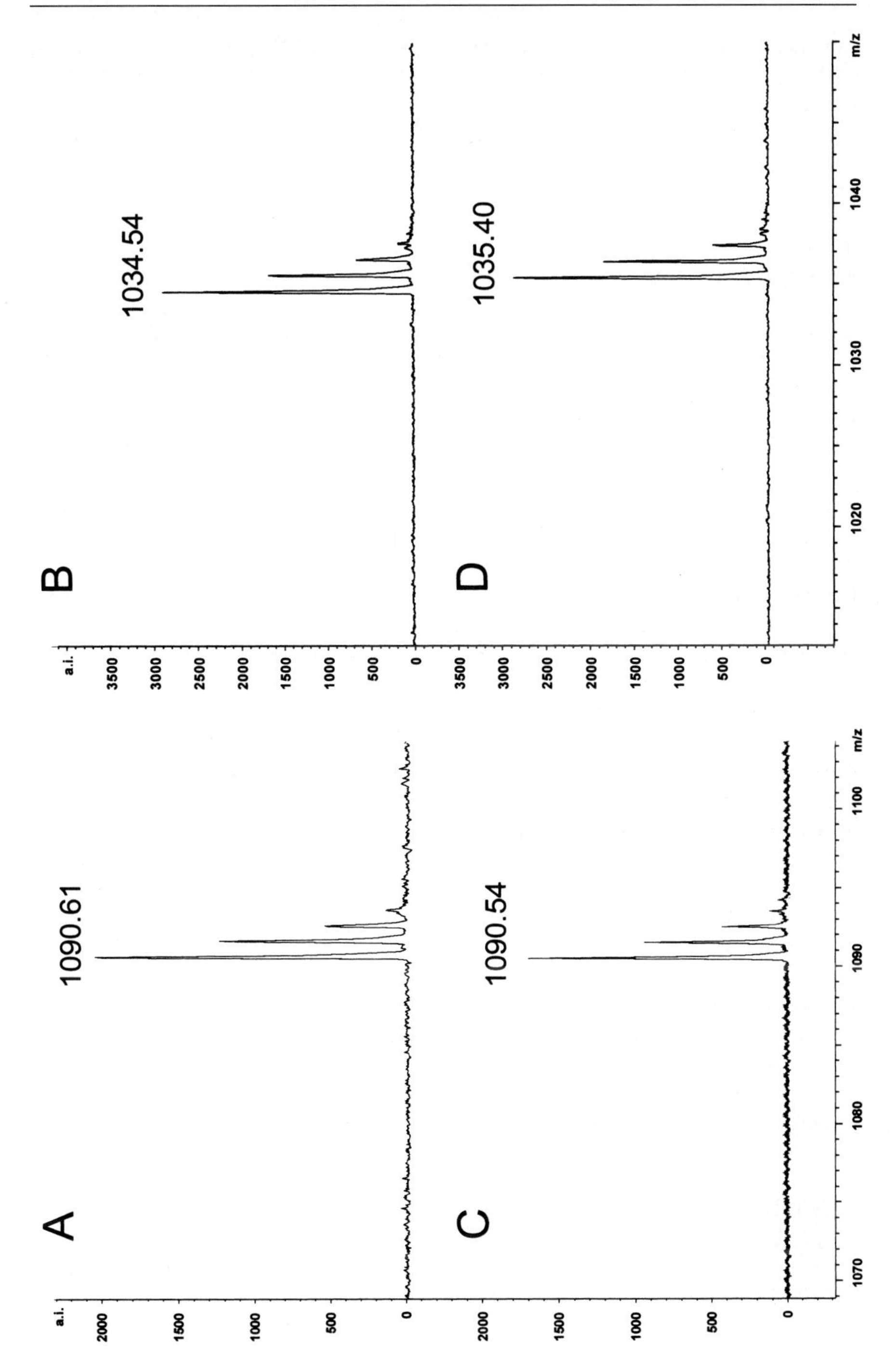
A
1090.61
B
1034.54
C
1090.54
D
1035.40
a.i.
m/z
1070
1080
1090
1100
1020
1030
1040

attempt to obtain cytotoxic T lymphocytes capable of recognition of the peptide, IMIKFNRL (residues 51–58), of the nuclear antigen La(SS-B), bound to major histocompatability antigen molecules (Purcell et al. 1998). A peptide adduct formed in relatively low yield as a byproduct during peptide synthesis appeared to dominate the immunogenic response when the peptide preparation was used as an immunogen.

3.1.1 Molecular Ion Analyses by MALDI-TOF-MS

The active agent involved in binding to major histocompatibility complex molecules such that the complex was recognised by a clonal line of cytotoxic T cells was previously found to be a minor byproduct in the synthetic preparation of residues 51–58 of the nuclear antigen La(SS-B) used for generation of the T cells (Purcell et al. 1998). This byproduct and the target peptide (IMIKFNRL) were isolated from the synthetic peptide mixture by reverse phase HPLC and their masses determined by MALDI-TOF-MS (Fig. 8.1). The immunologically active byproduct had a monoisotopic protonated molecular ion of m/z = 1090.61 (Fig. 8.1A) compared to the theoretical value of 1034.62 for the target sequence. The isolated target peptide produced a protonated molecular ion of m/z = 1034.54 (Fig. 8.1B). This difference between the experimentally observed masses, 56 Da, is consistent with modification by a butyl group as described previously (Purcell et al. 1998). It was previously shown that the byproduct yielded an anomalous PTH-amino acid derivative at position 6 during Edman sequencing indicating that the Asn residue at position 6 was modified (Purcell et al. 1998). Thus, it was postulated that the byproduct was modified due to formation of an Nβ-butyl amide on Asn6 during the synthetic process, which would add 56Da as opposed to a butyl ester which would add 57Da. Consequently, a peptide derivative was deliberately synthesized with this modification and this derivative was found to have the same mass (Fig. 8.1C) as the immunologically active byproduct (Fig. 8.1A). Parenthetically, it should be noted that the deliberately synthesized derivative was found to have the same immunological characteristics as the isolated byproduct (Purcell et al. 1998).

The original synthetic product also contained a stable imide involving Asn6 of the sequence which yielded a mixture of α- and β-linked aspartyl peptides upon treatment with mild base. These aspartyl peptides were isolated and characterized by Edman degradation so as to differentiate the α- and β-linked forms. The α-aspartyl peptide produced a molecular ion 1Da larger than the Asn6 form of the peptide (Fig. 8.1D).

Fig. 8.2. PSD spectra of (A) the immunologically active synthetic byproduct, (B) the Asn6 version of IMIKFNRL, (C) deliberately synthesised the Nβ-butyl-Asn6 version of IMIKFNRL and (D) the Asp6 version of IMIKFNRL. Ions are labelled as belonging to C-terminal (yn and zn) or N-terminal (an, bn and cn) series or as the immonium ion of Nβ-butyl-Asn (indicated by an asterisk) as defined in Fig. 8.3 and Table 8.1. Only ions derived by single peptide backbone bond cleavages are labelled, other ions may have arisen as a consequence of multiple fragmentations and/or side chain cleavages. Positive symbols denote ions 56 Da heavier than fragments predicted for unmodified IMIKFNRL. These spectra were collected using continuous extraction and a digitization rate of 250 MHz

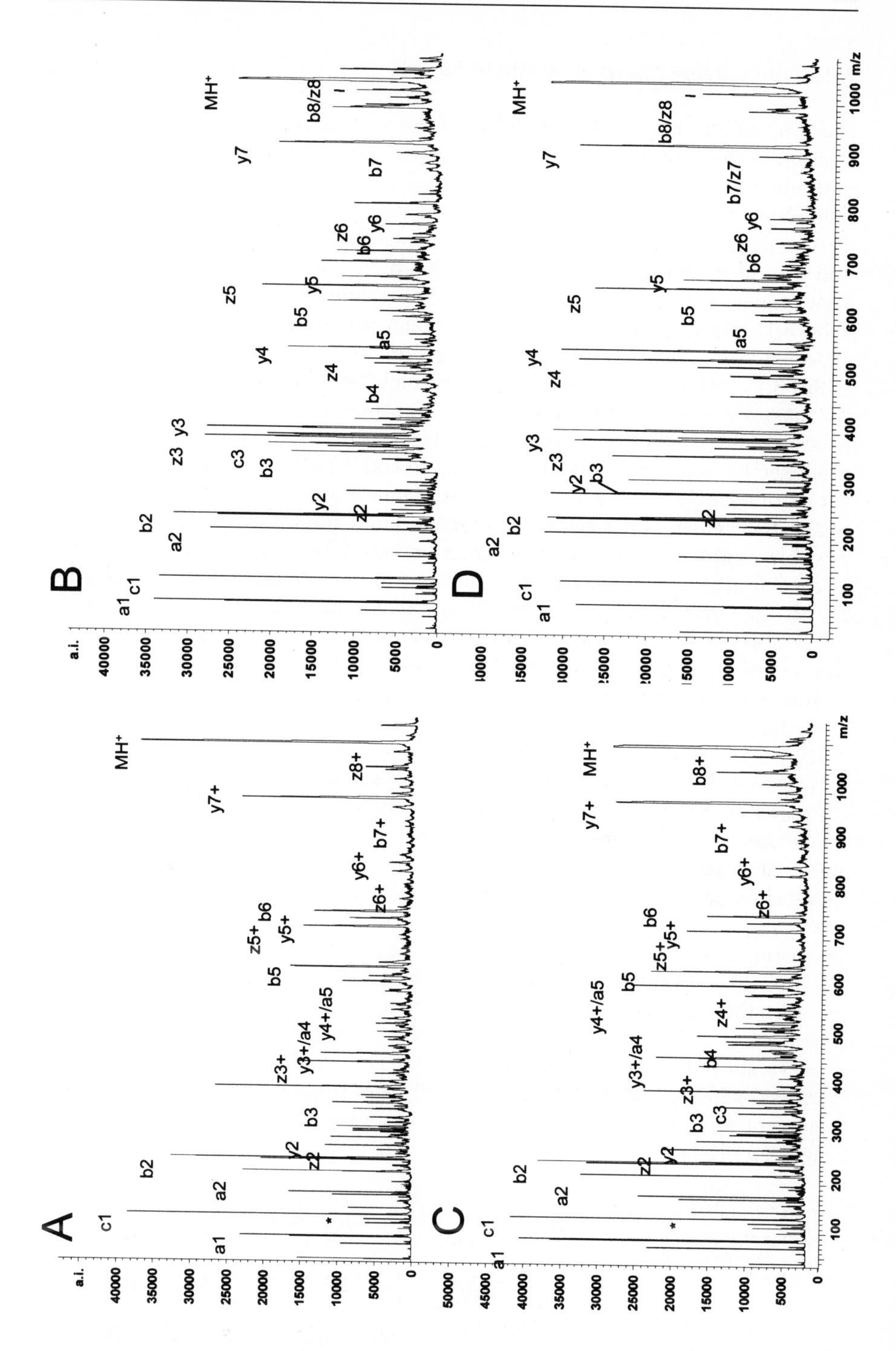
A
B
C
D
a.i.
m/z
MH+
y7+
y7
b8/z8
z8+
b8+
b7/z7
c1
a1
a2
b2

3.1.2 Post-Source Decay Analyses by MALDI-TOF-MS

The availability of these different forms of the peptide allowed a systematic investigation of the previously observed PSD behaviour of the immunologically active byproduct (Purcell et al. 1998). The isolated byproduct was used for analysis of metastable fragments produced by PSD of ions generated by MALDI-TOF-MS. This revealed C- and N-terminal ion series consisting principally of z- and y-type and a- and b-type ions, respectively (Figs. 8.2A and 8.3). The C-terminal ion series appeared to cover all but the last two C-terminal residues (Fig 8.2A and 8.3; Table 8.1). These ions were 56Da heavier than the corresponding fragment ions expected from the unmodified sequence up to and including the z3y3 pair (Table 8.1). There was a subsequent transition between the z3y3 and z2y2 pairs to the masses predicted for the unmodified sequence. This is consistent with modification of Asn6 with a butyl amide. In contrast, the N-terminal ion series were as predicted for the unmodified sequence up to and including the b6 ion. The observation of a single ion that could have been interpreted as an unmodified b6 ion of the N-terminal series indicated that the butyl modification was on Arg7. A possible explanation for this contradiction could be that the butyl group was cleaved from the Asn side chain in addition to peptide bond fission to produce the b6 ion. Further evidence for modification of Asn6 with a butyl moiety was the observation of an ion at m/z = 143.1 which could have represented an immonium ion for Nβ-butyl asparagine (theoretical m/z = 143.21).

Interpretation of the PSD spectrum was ambiguous at some points due to possible coincidences of ions from N- and C-terminal series. Such ambiguities occur for either or both the y3 and a4 (m/z = 458.6 and 458.7, respectively) and y4 and a5 fragment ions (m/z = 605.6 and 605.9, respectively).

Nearly complete C- and N-terminal ion series were observed by PSD analysis of the unmodified peptide (Fig. 8.2B and 8.3; Table 8.1). The only fragmentation not clearly represented was for production of the ion representing the C-terminal Leu residue in the C-terminal ion series. The most important aspect of the fragment ion spectrum of the unmodified peptide was the absence of C-terminal fragment ions shifted by 56Da as seen for the modified sequence (Table 8.1) and no probable Nβ-butyl-Asn immonium ion at m/z = 143.2. These observations support the fragment ion assignments made above for the modified peptide. There was a minor ion representing the a5 fragment, however, an intense y4 ion was evident (Fig. 8.2B) that indicated that the ambiguous y4/a5 assignment in the modified peptide (Fig. 8.2A) was essentially a y4 fragment plus 56 Da. The same argument applies to resolution of the y3/a4 ambiguity in the modified peptide (Fig. 8.2A) in favour of a y3 assignment.

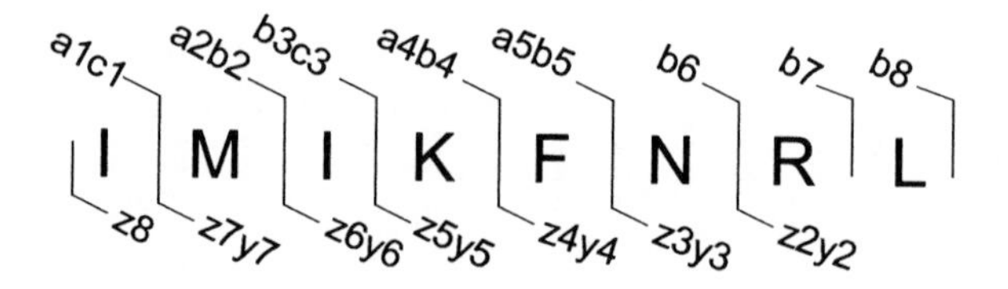

Fig. 8.3. Summary of ions seen in post source decay spectra described in Fig. 8.2 and Table 8.1

Table 8.1. Fragment Ions of Derivatives of the Peptide IMIKFNRL

Ion Type	m/z								
	Derivative								
	Byproduct			Asn6 Sequence		Nβ-Butyl-Asn6		Asp6 Sequence	
	Theory[1]	Obs	Diff	Obs	Diff	Obs	Diff	Obs	Diff
C-terminal									
z2	271.4	271.4	–	271.2	0.2	271.1	0.3	271.4	–
y2	288.4	288.5	0.1	288.3	0.1	288.3	0.1	288.5	0.1
z3	385.5	441.6	56.1	385.4	0.1	441.2	55.7	385.9	0.4
y3	402.5	458.6	56.1	402.3	0.2	458.3	55.8	403.5	1.0
z4	532.6	–	–	532.1	0.5	588.1	55.5	533.3	0.7
y4	549.7	605.9	56.2	549.1	0.6	605.0	55.3	550.2	0.5
z5	660.8	717.3	56.5	660.3	0.5	716.3	55.5	661.8	1.0
y5	677.8	734.3	56.5	677.3	0.5	733.2	55.4	678.8	1.0
z6	774	830.8	56.8	773.6	0.4	829.9	55.9	775.1	1.1
y6	791	848.1	57.1	791.1	0.1	846.9	55.9	792.1	1.1
z7	905.2	–	–	–	–	–	–	905.4	0.2
y7	922.2	978.6	56.4	921.4	0.8	977.6	55.4	922.9	0.7
z8	1018.3	1074.5	56.2	1017.6	0.7	–	–	1018.5	0.2
N-Terminal									
a1	86.2	86.2	–	86.2	–	86.1	0.1	86.2	–
c1	129.2	129.3	0.1	129.2	–	129.1	0.1	129.2	–
a2	217.4	217.5	0.1	217.3	0.1	217.3	0.1	217.5	0.1
b2	245.4	245.4	–	245.1	0.3	245.2	0.2	245.2	0.2
b3	358.5	358.7	0.2	358.3	0.2	358.3	0.2	357.5	0.8
c3	373.5	373.7	0.2	373.4	0.1	373.2	0.3	373.6	0.1
a4	458.7	458.6	0.1	–	–	458.3	0.4	–	–
b4	486.7	–	–	486.5	0.2	486.6	0.1	–	–
a5	605.9	605.9	–	605.6	0.3	605.0	0.9	605.9	–
b5	633.9	634.4	0.5	633.3	0.6	633.7	0.2	633.8	0.1
b6	748	748.1	0.1	747.6	0.4	747.1	0.9	748.5	0.5
b7	904.2	960.6	56.4	903.8	0.4	959.8	55.6	905.4	1.2
b8	1017.3	–	–	1017.3	–	1073.4	56.1	1018.5	1.2

[1] Theoretical fragment ion masses are those calculated for the unmodified Asn6 version of the sequence.

A further check on the validity of the assignments made to ions in the fragment ion spectrum of the immunologically active byproduct was performed by examining the version of the sequence deliberately synthesized with Nβ-butyl-Asn at position 6. The ion series observed for this peptide were essentially the same as those seen for the isolated byproduct except for variable observation of z4, z8 and b8 fragments shifted by 56 Da relative to the unmodified sequence (Fig. 8.2C and 8.3; Table 8.1). The potential Nβ-butyl-Asn immonium ion was observed at m/z = 143.0. An ion corresponding to a b6 fragment was observed at m/z = 747.1, which was a further indication that the corresponding ion seen for the byproduct was produced by a combination of cleavage of the peptide chain and cleavage of the butyl group attached to Asn6.

Analysis of fragments produced by the α-Asp6 form of the peptide (Fig. 8.3D; Table 8.1) provided further support for the assignment of identities to ions in the C-terminal series of the previous spectra. The aspartyl peptide produced a dominant C-terminal ion series, consisting principally of y- and z-type ions, and comparatively less intense ions from the N-terminal series (Fig. 8.3D; Table 8.1). Whereas the C-terminal ions were approximately 1Da heavier than the comparable ions of the unmodified asparaginyl peptide up to and including the z3/y3 pair, the z2/y2 ion pair was equivalent to the corresponding pair of the unmodified sequence. No ions were observed at the position postulated above to correspond with the probable Nβ-butyl-Asn immonium ion.

3.2
Characterization of the Sites of Proteolytic Activation of Fusion Protein Precursors of Newcastle Disease Virus Isolates

Newcastle disease virus (NDV) is a serious avian pathogen that exists in a variety of forms with virulence ranging from highly lethal to innocuous (Waterson et al. 1967). NDV is a member of the paramyxoviridae family of viruses that infect cells by fusion of the viral lipid bilayer membrane with the plasma cell membrane. Fusion is mediated by a fusion (F) protein that consists of two disulfide-linked polypeptide (F1 and F2) chains. The F-protein is produced as an inactive single chain (F0) bioysynthetic precursor that is activated by a combination of endoproteolytic and exoproteolytic cleavages (Gorman et al; 1988, 1990a; Homma et al. 1975; Scheid et al. 1978). Variability of the virulence of NDV strains is dependent upon the susceptibilities of their various F0-proteins to proteolytic activation (Nagai et al. 1976). Susceptibility to proteolysis is dictated by the nature of the amino acids preceding the susceptible peptide bond (Glickman et al. 1988; Toyoda et al. 1987). Highly virulent strains have a cleavage signalling motif of two pairs of basic amino acids, separated by a single glutamine, preceding the susceptible peptide bond. The positions equivalent to that occupied by the first basic amino acids of these pairs are occupied by another type of amino acid, usually glycine, in avirulent and low-virulence strains. Antibodies capable of detecting these differences were produced using synthetic peptides as a means of rapid pathotyping of NDV isolates (Hodder et al. 1993). However, isolates have been encountered that do not react with the first generation of antibodies due to unusual variability in the cleavage signalling motifs such as the presence of amino acids other than glycine or basic amino acids (Hodder et al. 1994; Scanlon et al. 1999). When such isolates are encountered it is necessary to define the structural features of the cleavage sites for production of antibodies to accommodate the new sequences. The order of the polypeptide chains in the F0-protein prior to proteolysis is NH2-F2-F1-COOH. Thus the cleavage signalling motif is at the C-terminus of the F2-polypeptide of the mature F-protein. Consequently, characterization of cleavage signalling motifs using isolated mature F-proteins corresponds to characterization of the C-termini of their F2-polypeptides. A factor that must be borne in mind is that basic amino acids are exposed at the C-termini as a consequence of primary endoproteolytic cleavage of the F0-protein and these newly exposed basic amino acids are subsequently removed by a carboxypeptidaseB-like enzyme (Gorman et al. 1990a). A single basic amino acid is removed from the C-termini of the F2-polypeptides of avirulent and low-virulence strains and two basic amino acids are removed from F2-polypeptides of virulent strains.

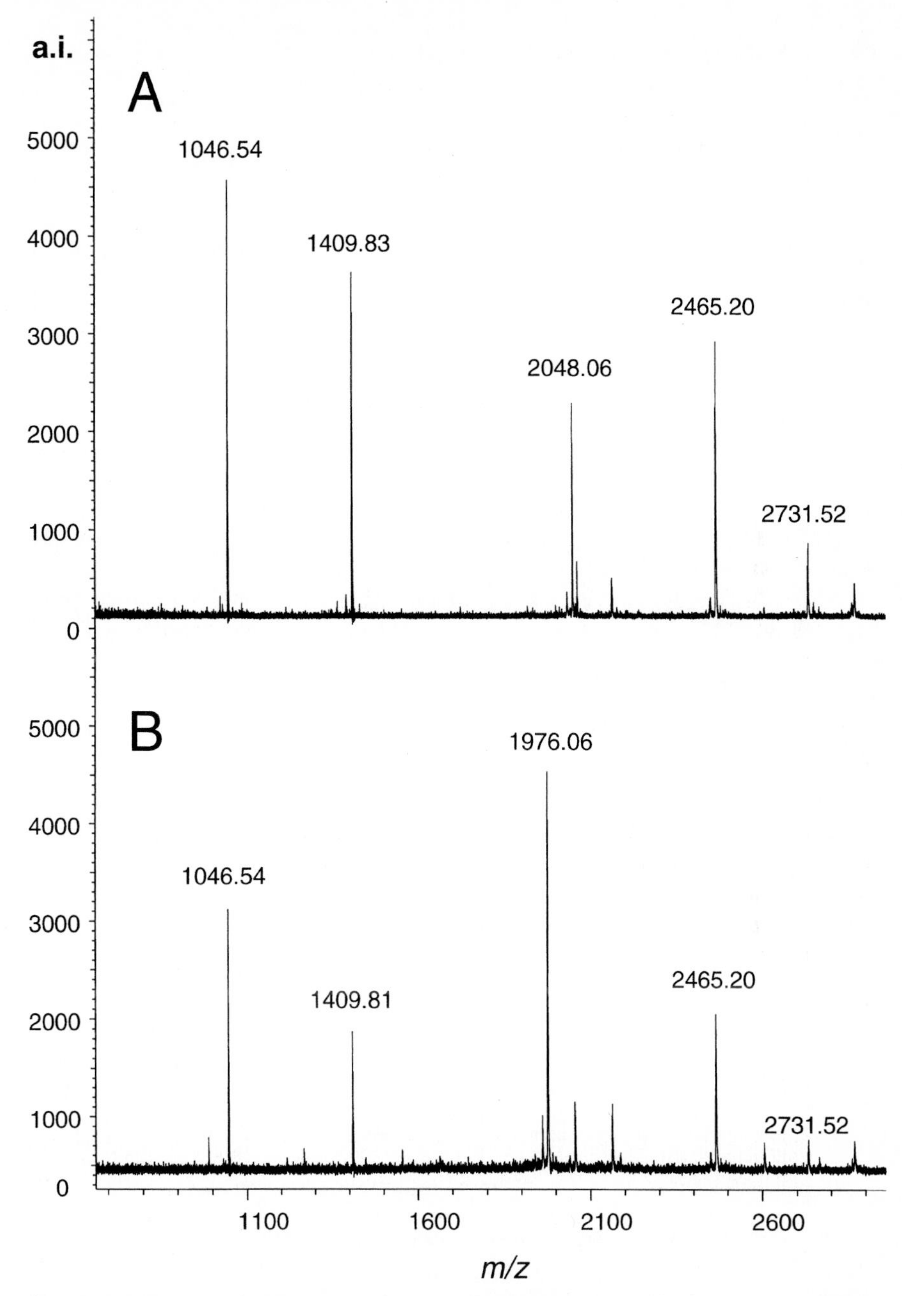

Fig. 8.4. MALDI maps of AspN protease digests of the F2-polypeptides of (A) the VRI 82-6409 and (B) V4 isolates of NDV. Ions at m/z values of 1046.54 and 2465.2 are the monoisotopic ions of the internal calibrants angiotensin II and ACTH 18-39, respectively. These spectra were all collected using delayed extraction and a digitization rate of 1GHz. These spectra were adapted with permission from (Lopaticki et al. 1998; Copyright, John Wiley + Sons)

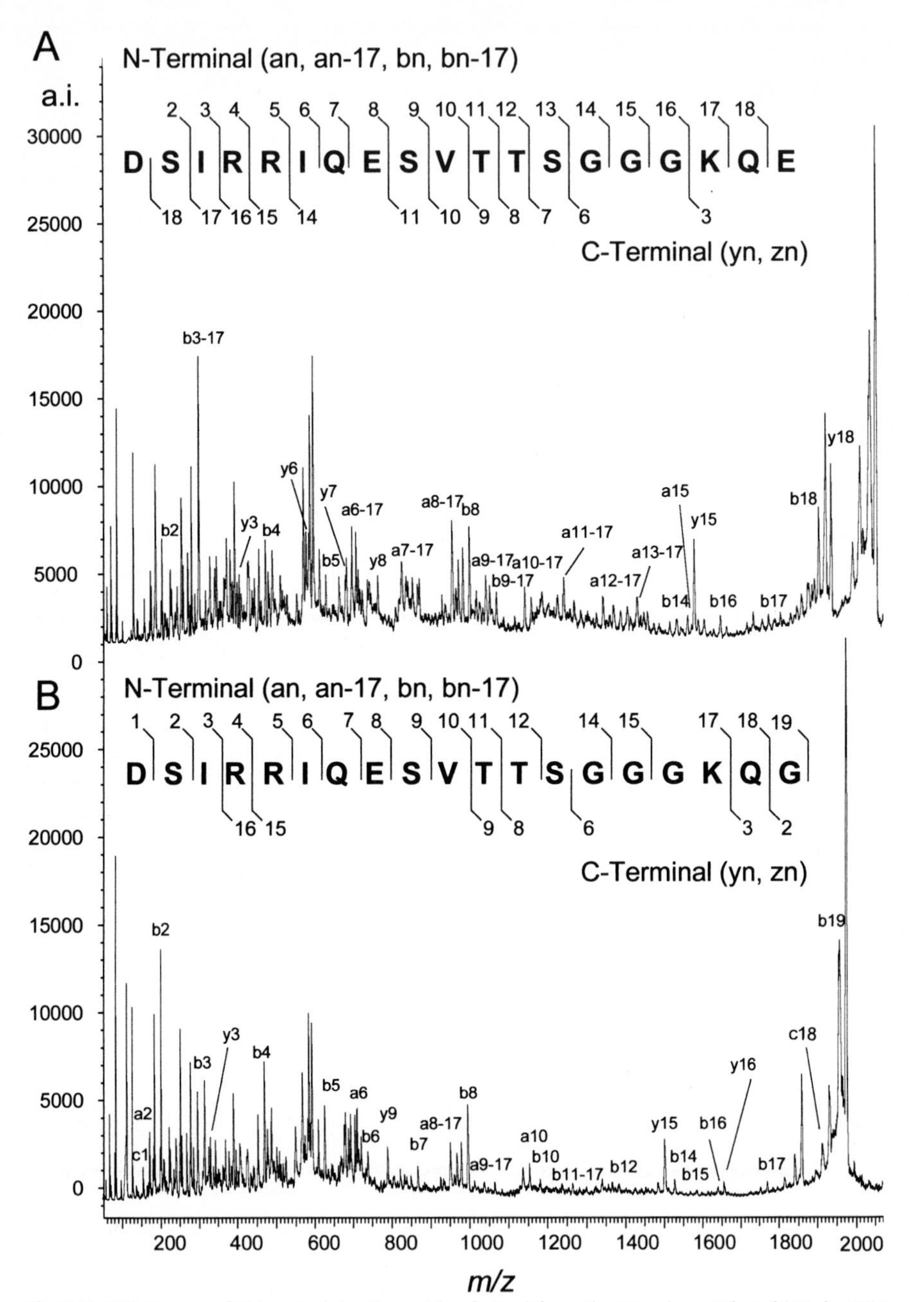

Fig. 8.5. PSD spectra of C-terminal AspN peptides derived from the F2-polypeptides of (A) the VRI 82–6409 and (B) V4 isolates of NDV. Detailed descriptions of the ions observed are presented in Table 8.2 and summarized above the respective spectra. Only representative ions are indicated on the spectra. These spectra were collected using continuous extraction and a digitization rate of 250 MHz. These spectra were adapted with permission from (Lopaticki et al. 1998; Copyright, John Wiley + Sons)

FAB-MS was used for the original characterization of the process of cleavage activation of the F0-proteins of common isolates of NDV and of some more unusual isolates (Gorman et al. 1988, 1990a; Hodder et al. 1994; Scanlon et al. 1999). However, this required considerable quantities of starting material and complementary use of Edman sequencing and amino acid analysis. In a recent study MALDI-TOF-MS was used to characterize a variant isolate of NDV that did not react with the available antipeptide antibodies (Lopaticki et al. 1998). This proved to be a much more comprehensive and sensitive approach than FAB-MS. In particular, PSD was a valuable technique for characterizing the cleavage signalling motif of this isolate. In the MALDI-TOF-MS study under current consideration (Lopaticki et al. 1998) the F2-polypeptide of an immunologically unreactive avirulent isolate (VRI 82–6409) (Hodder et al. 1994) was isolated and compared to the F2-polypeptide of a well characterized avirulent strain (Queensland or V4) (Simmons 1967) of NDV.

F2-polypeptides were isolated from both strains of NDV by SDS-PAGE followed by electroelution (Gorman et al. 1988, 1990a). These F2-polypeptides were digested with AspN protease and the unfractionated digests analysed by MALDI-TOF-MS using internal calibration (Fig. 8.4). A difference in mass between the C-terminal peptides of the two F2-polypeptides was apparent with the C-terminal peptide of the VRI 82–6409 isolate (Fig. 8.4A) being 72 Da heavier than that of the V4 isolate (Fig. 8.4B). Such mass difference is consistent with variation of one of the glycine residues at the C-terminus of the V4 F2-polypeptide to glutamic acid in the VRI isolate. These two AspN peptides were isolated by HPLC and analysed by PSD (Fig. 8.5; Table 8.2) which confirmed this difference and located the variant residues as the C-terminal glycine of the V4 isolate. Placement of the variation within the last three C-terminal residues was possible based on the fact that a series of C-terminal ions was observed from y3 to y18 for the VRI 82–6409 isolate (Fig. 8.5A; Table 8.2) with masses 72 Da greater than masses predicted and/or observed for the V4 isolate (Fig. 8.5B; Table 8.2). Definitive placement of the variation at the C-terminus was based on observation of N-terminal ions for the VRI 82–6409 isolate (Fig. 8.5A; Table 8.2), up to and including the b18 fragment, with masses the same as N-terminal fragments predicted and/or observed for the V4 isolate (Fig. 8.5B; Table 8.2).

3.3 Determination of the Disulphide Linkage Pattern in the Ectodomain of the Respiratory Syncytial Virus Attachment Protein

Respiratory syncytial virus (RSV), which is another member of the paramyxoviridae family of viruses, causes serious lower respiratory infections of humans and animals (Collins et al. 1996). Children up to 2 years of age are particularly susceptible to RSV and infection during this period is the expected norm (Collins et al. 1996). In some instances hospitalization may be required as a consequence of infection and mortality may occur in extreme cases. In order to combat RSV better it would be of benefit to understand the structural features of the RSV proteins involved in the early stages of cellular infection. Proteins of interest include the viral attachment protein and the fusion protein. The attachment protein is a

Table 8.2. Fragment Ions Formed by PSD of Residues 66-84 of VRI 82-6409 and V4 F2 Polypeptides

Fragment	m/z				
	Predicted V4 66-84	Observed VRI 82-6409 66-84	Difference	Observed V4 66-84	Difference
N-Terminal					
c1	131.1	–	–	131.2	0.1
a2	175.2	175.4	0.2	174.9	−0.3
b2	203.2	203.3	0.1	203.4	0.2
a3	288.3	288.5	0.2	288.2	−0.1
b3	316.3	316.4	0.1	316.2	−0.1
b3-17	299.3	298.5	−0.8	–	–
a4	444.5	444.6	0.1	–	–
b4	472.5	472.6	0.1	472.0	−0.5
b4-17	455.5	455.6	0.1	455.5	0
b5	628.7	629.0	0.3	628.7	0
b5-17	611.7	611.8	0.1	611.7	0
a6	713.9	713.7	−0.2	713.5	−0.4
a6-17	696.9	696.9	0	696.6	−0.3
b6	741.9	742.2	0.3	741.4	−0.5
b6-17	724.9	724.7	−0.2	–	–
a7	842.0	842.0	0	841.7	−0.3
a7-17	825.0	825.4	0.4	824.8	−0.2
b7	870.0	870.3	0.3	869.7	−0.3
b7-17	853.0	853.0	0	852.5	−0.5
a8	971.1	971.3	0.2	970.6	−0.4
a8-17	954.1	954.3	0.2	953.7	−0.4
b8	999.1	999.3	0.2	998.5	−0.6
b8-17	982.1	982.2	0.1	981.4	−0.7
a9-17	1041.2	1041.6	0.4	1040.8	−0.4
b9	1086.2	–	–	1085.8	−0.4
b9-17	1069.2	1068.6	−0.6	–	–
a10	1157.3	1157.6	0.3	1157.0	−0.3
a10-17	1140.3	1140.8	0.5	1140.5	0.2
b10	1185.3	1186.1	0.8	1184.5	−0.8
a11	1258.4	1258.1	−0.3	–	–
all-17	1241.4	1242.0	0.6	1240.4	−1
b11	1286.4	1286.0	−0.4	–	–
b11-17	1269.4	1268.4	−1	–	–
a12	1359.5	–	–	1359.6	0.1
a12-17	1342.5	1342.8	0.3	1342.9	0.4
b12	1387.5	1387.9	0.4	1386.9	−0.6
b12-17	1370.5	1370.2	−0.3	1369.3	−1.2
a13-17	1429.6	1430.3	0.7	–	–
b13-17	1457.6	1457.2	−0.4	–	–
b14	1531.7	1532.2	0.5	1531.0	−0.7
a15	1560.7	1560.4	−0.3	–	–
b15	1588.7	1588.5	−0.2	1588.4	−0.3
c15	1603.7	1603.9	0.2	–	–
b16	1645.8	1646.8	1	–	–
b17	1773.9	1773.1	−0.8	1772.8	−1.1
b18[1]	1902.1	1902.2	0.1	–	–
c18	1717.1	–	–	1916.2[1]	−0.9
b19	1959.1	–	–	1960.0	0.9
C-Terminal					
y2	204.2	–	–	204.4	0.2
y3[2]	332.4	404.7	72.3	332.4	0
y6	503.6	576.5	72.9	503.6	0

Table 8.2 (Continued)

Fragment	m/z				
	Predicted V4 66-84	Observed VRI 82-6409 66-84	Difference	Observed V4 66-84	Difference
y7	590.6	663.0	72.4	–	–
y8	691.7	764.0	72.3	691.4	−0.3
y9	792.8	865.3	72.5	792.3	−0.5
z9	775.8	–	–	775.1	−0.7
y10	892.0	965.0	73	–	–
y11	979.1	1051.4	72.3	–	–
z14	1332.4	1405.4	73	–	–
y15	1505.6	1577.6	72	1505.2	−0.4
y16	1661.8	1732.7	70.9	1660.1	−1.7
y17	1775.0	1846.4	71.4	–	–
y18	1862.1	1934.4	72.3	–	–

1. These ions are diagnostic of a C-terminal glutamic acid and C-terminal glycine for the VRI 82-6409 and V4 sequences, respectively.
2. These data were diagnostic of a difference of a glycine and glutamic acid within the last 3 residues at the C-termini of the V4 and VRI 82-6409 sequences.

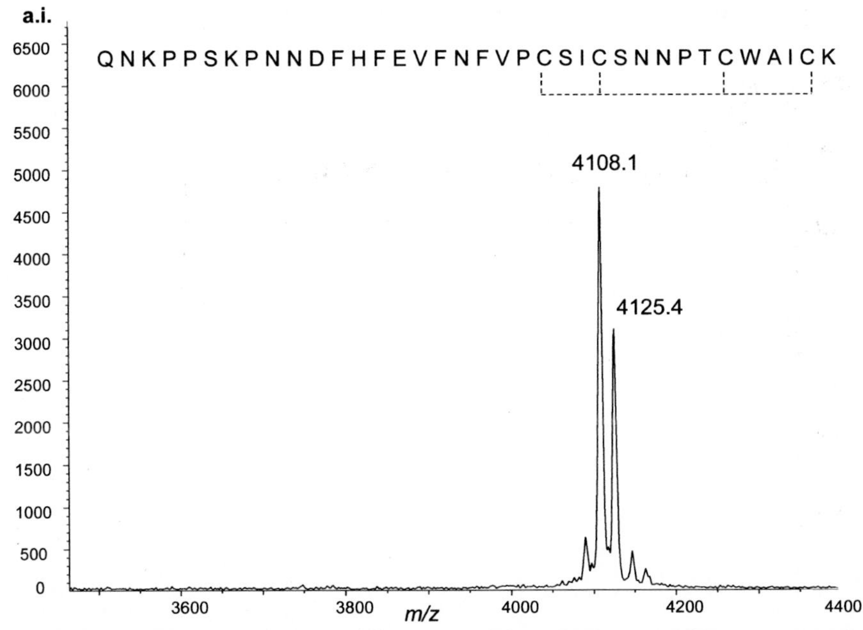

Fig. 8.6. MALDI spectrum of a major peptide obtained from the G-protein of human respiratory syncytial virus A2 strain by tryptic digestion and HPLC isolation. The theoretical average m/z for the fully reduced form of the sequence shown (residues 152–187) is 4128.7. The apparent discrepancy between the experimental (m/z = 4125.4) and theoretical masses would be accounted for if the four cysteine residues are involved in disulfide linkages, these bonds are shown as broken lines to indicate that this data does not define the linkage pattern. This was confirmed by observation of an increase in m/z of these ions by 4 upon chemical reduction. The ion at 4108.1 represents loss of NH_3 due to cyclization of the terminal Gln (Q) residue to pyroglutamic acid. This spectrum was recorded using continuous extraction and a digitization rate of 250 MHz. This figure was adapted with permission from (Gorman et al. 1997)

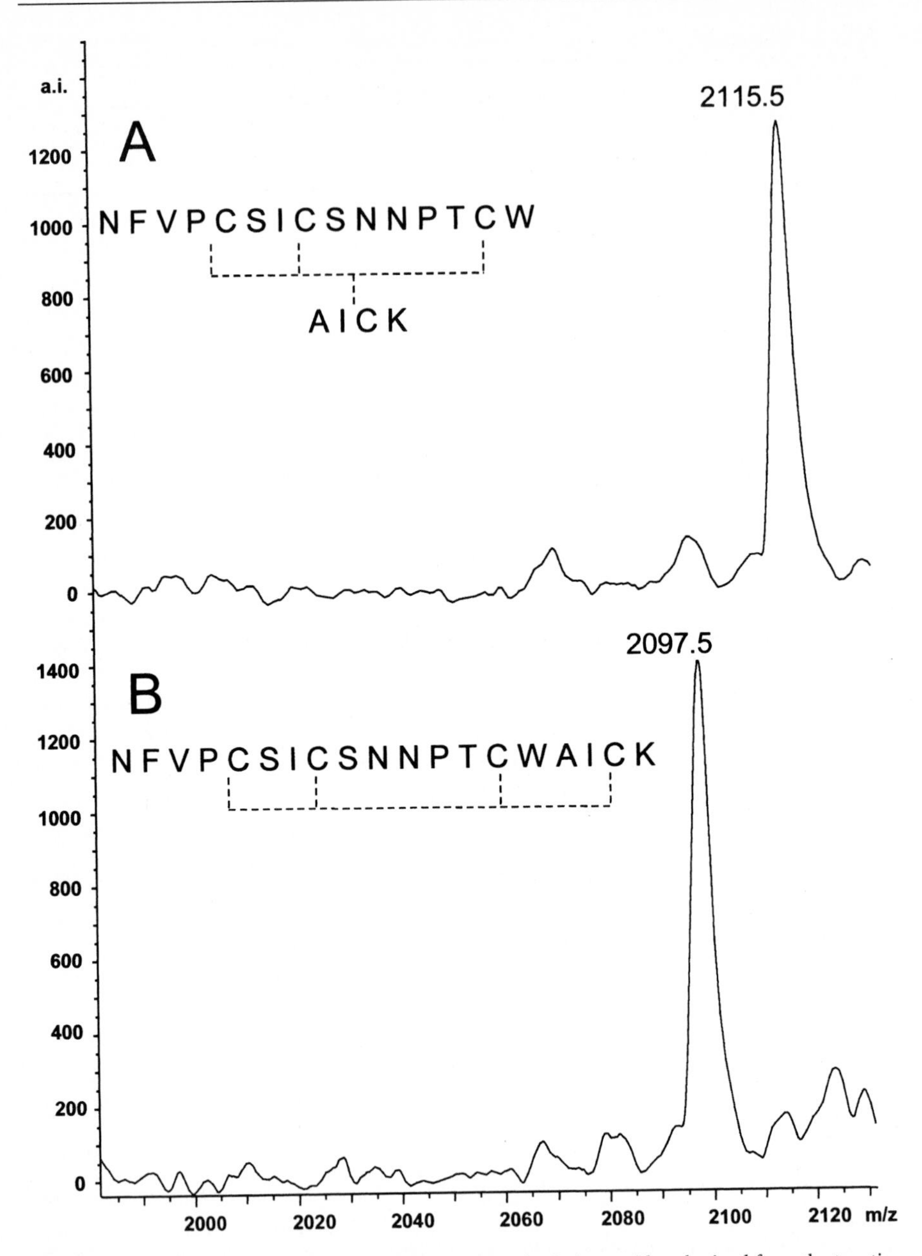

Fig. 8.7. Parent ion masses of (A) two chain and (B) single chain peptides obtained from the tryptic peptide defined in Fig. 8.5 by peptic digestion and HPLC. Broken lines are used to indicate involvement of the cysteines in undefined disulfide bonds. Chemical reduction of these peptides resulted in loss of the mass of the AICK sequence and gain of 3 for the two chain peptide and an increase of 4 in m/z for the single chain peptide. These spectra were collected using continuous extraction and a digitization rate of 250 MHz

type II integral membrane protein with a heavily glycosylated ectodomain that contains four conserved cysteine residues (Collins et al. 1996). This protein is termed the G-protein as a consequence of heavy glycosylation (Gruber and Levine 1985).

MALDI-TOF-MS was used to characterize the disulfide bond arrangement of human RSV G-protein directly isolated from virus infected cells (Gorman et al. 1997). HPLC of a tryptic digest of the G-protein revealed a predominant peak of absorbance which was shown by MALDI-TOF-MS to contain ions of m/z = 4108.1 and 4125.4 (Fig. 8.6). These ions were consistent with the sequence spanning residues 152–187 provided no glycans were attached to any amino acid side chains and the four cysteines within the sequence were in disulfide linkages. The mass difference of approximately 17 Da between these ions can be accounted for by the fact that the amino terminus of the peptide is glutamine which is susceptible to cyclization, through loss of ammonia, under the acidic conditions of separation. Further digestion of the isolated tryptic peptide with pepsin resulted in two peptides containing two disulfides each (Fig. 8.7). The difference between these two peptic peptides was due solely to an additional peptide bond cleavage in one of the peptides. This peptide consisted of two disulfide-linked peptide chains with one of the chains containing an intrachain disulfide bond in addition to the interchain disulfide. By comparison the other peptic peptide, which lacked the additional peptide bond cleavage, had two intrachain disulfide bonds.

PSD analysis of the two chain peptic peptide was used to determine that the disulfide bonds of the G-protein were in a 1 to 4 plus 2 to 3 arrangement (Gorman et al. 1997). This peptide underwent extensive post-source fragmentation that reflected cleavages along the peptide backbone (Fig. 8.8A; Table 8.3). Although some disulfide bond fission was evident (Figs. 8.8B and 8.8C), the interchain peptide bond survived during production of the majority of the PSD ions. Fragments resulting from peptide backbone cleavages were evident that effectively reflected sequential losses of amino acids along the backbone of the larger of the two chains of this peptide up to the first of the three half cystines of the larger chain. Thereafter, fragmentation reflected loss of the smaller peptide chain in addition to the amino acid sequence of the larger chain (Fig. 8.8A). Peptide backbone cleavages were apparent for the N-terminal portion of the larger peptide chain in the form of N- and C-terminal ion series but did not extend into the intrachain disulfide loop of this chain. Internal fragments were also evident (Figs. 8.8B and 8.8C; Table 8.3) that were consistent with location of the disulfides indicated by the sequential fragmentation data. The interchain disulfide was preserved within these internal fragments. Some fragmentation of the interchain disulfide also occurred via both symmetric and asymmetric fragmentation (Figs. 8.8B and 8.8C; Table 8.3).

By comparison (Fig. 8.9), the single chain peptic peptide, with two intrachain disulfide loops, only produced sequence information for the N-terminal amino acids prior to the half-cystine proposed to participate in the interchain disulfide linkage of the two chain peptic peptide (Fig. 8.9B). A single chain peptide derived by post-proline cleavage of the original tryptic peptide, which commenced with the half-cystine involved in the interchain disulfide in the two chain peptic pep-

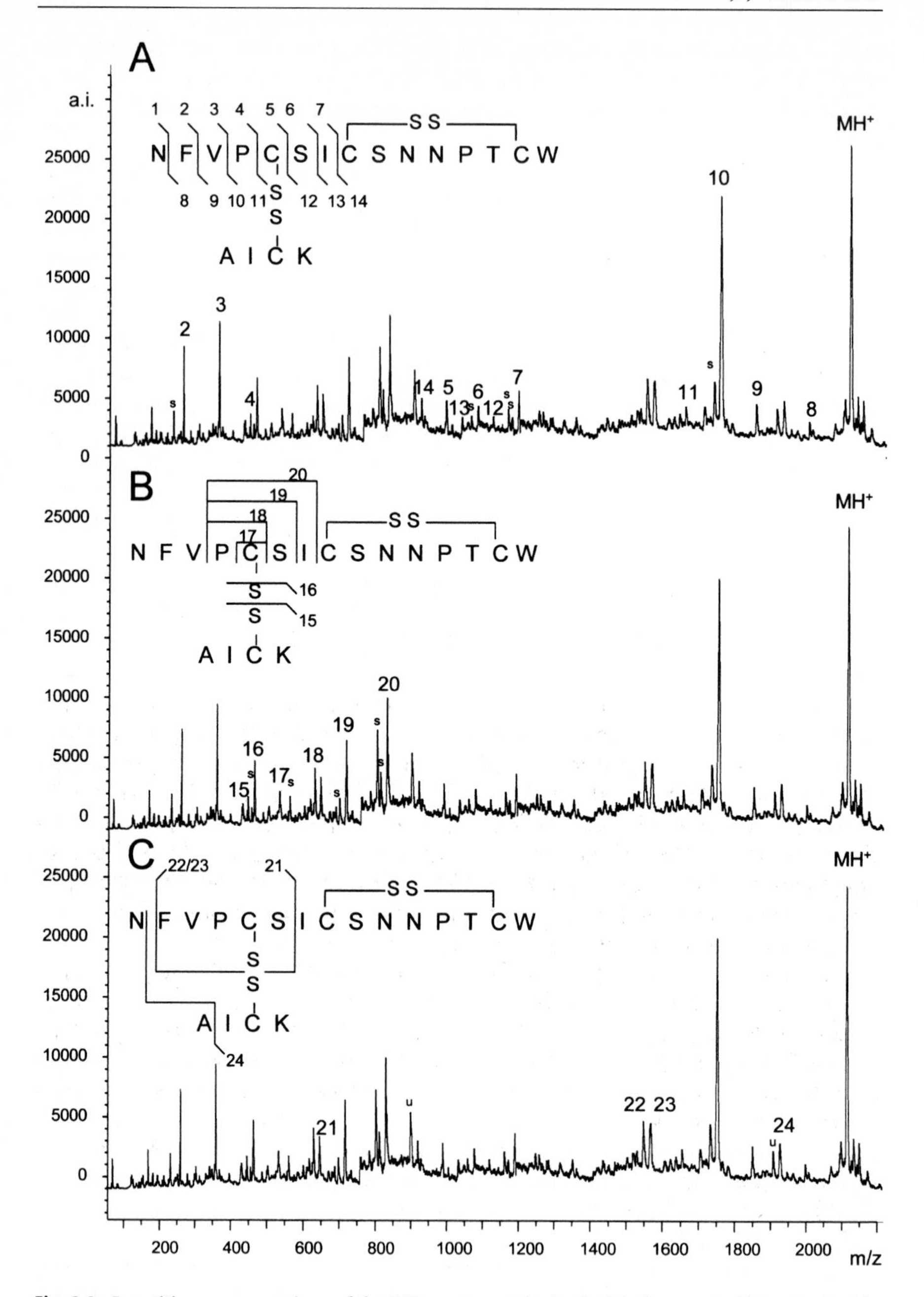

Fig. 8.8. Repetitive representations of the PSD spectrum obtained with the two chain peptic peptide described in Fig. 8.7A. Panel A depicts backbone fragmentations deduced from the fragment ion data presented in Table 8.3, panel B depicts fragmentations deduced as being internal fragments or disulfide bond fragmentation and panel C depicts combined fragmentation of peptide and disulfide bonds or combined peptide bond cleavages. S denotes satellite a type ions accompanying identified b type ions and U denotes unidentified ions. Solid lines joining cysteine residues indicate the connectivity

Table 8.3. Fragment Ions of the Two Chain Disulfide-Linked Peptic Peptide of the RSV G-Protein

Ion	m/z		Fragment type
	Observed	Predicted	
1	–	115.0	–
2	262.4+/–0.2	262.3	b2
3	361.4+/–0.3	361.4	b3
4	458.6+/–0.2	458.5	b4
5	992.9+/–0.5	993.2	b5
6	1080.6+/–0.2	1080.3	b6
7	1193.5+/–0.6	1193.5	b7
8	2001.2+/–0.8	2001.4	y14
9	1854.3+/–1.1	1854.2	y13
10	1755.2+/–1	1755.1	y12
11	1658.9+/–1	1658.0	y11
12	1123.1+/–0.3	1123.3	y10
13	1036.4+/–0.6	1036.2	y9
14	923.1+/–0.6	923.1	y8
15	434.4+/–0.4	434.6	S-S
16	466.7+/–0.1	466.6	S-CH2
17	535.5+/–0.2	535.7	Internal
18	632.7+/–0.1	632.8	Internal
19	719.9+/–0.4	719.9	Internal
20	832.9+/–0.5	833.1	Internal
21	648.9+/–0.1	648.7	b6 + S-S
22	1551.2+/–0.5	1551.8	z14 + S-S
23	1571.1+/–0.1	1570.9	y14 + S-S + 2H
24	1929.9+/–0.5	1930.3	y14 + yl(AICK)

tide, failed to produce N-terminal sequence specific ions (Fig. 8.9C). This comparative analysis served to verify the identities assigned to ions observed with the two chain peptic peptide.

3.4 Analysis of Oligosaccharide Structure

Analysis of oligosaccharides is an important component of characterization of proteins. Mass spectrometric analysis of glycans released from proteins usually involves derivatization of the reducing termini of the glycans with a reagent to facilitate ionization (Wang et al. 1984). Derivatization also generally engenders the ability to detect glycans by UV absorbance during HPLC (Wang et al. 1984). Reverse phase chromatographic performance is also generally enhanced by derivatization (Wang et al. 1984). 1-Phenyl-3-methyl-5-pyrazolone (PMP) is a very useful reagent for facile and efficient derivatization of the reducing termini of oligosaccharides (Honda et al. 1989). PMP derivatives also exhibit excellent performance during MALDI analysis (Pitt and Gorman 1997). Ionization and

◀

supported by the total data set and additional data obtained by thermolytic data of the original tryptic peptide (Gorman et al. 1997). In particular, the mass differences associated with transitions from the peptide fragmentations 4 to 5 and 11 to 12 indicate that the AICK sequence was attached to the first half-cystine of the larger chain. These spectra were collected using continuous extraction and a digitization rate of 250 MHz. Reproduced with permission from (Gorman et al. 1997)

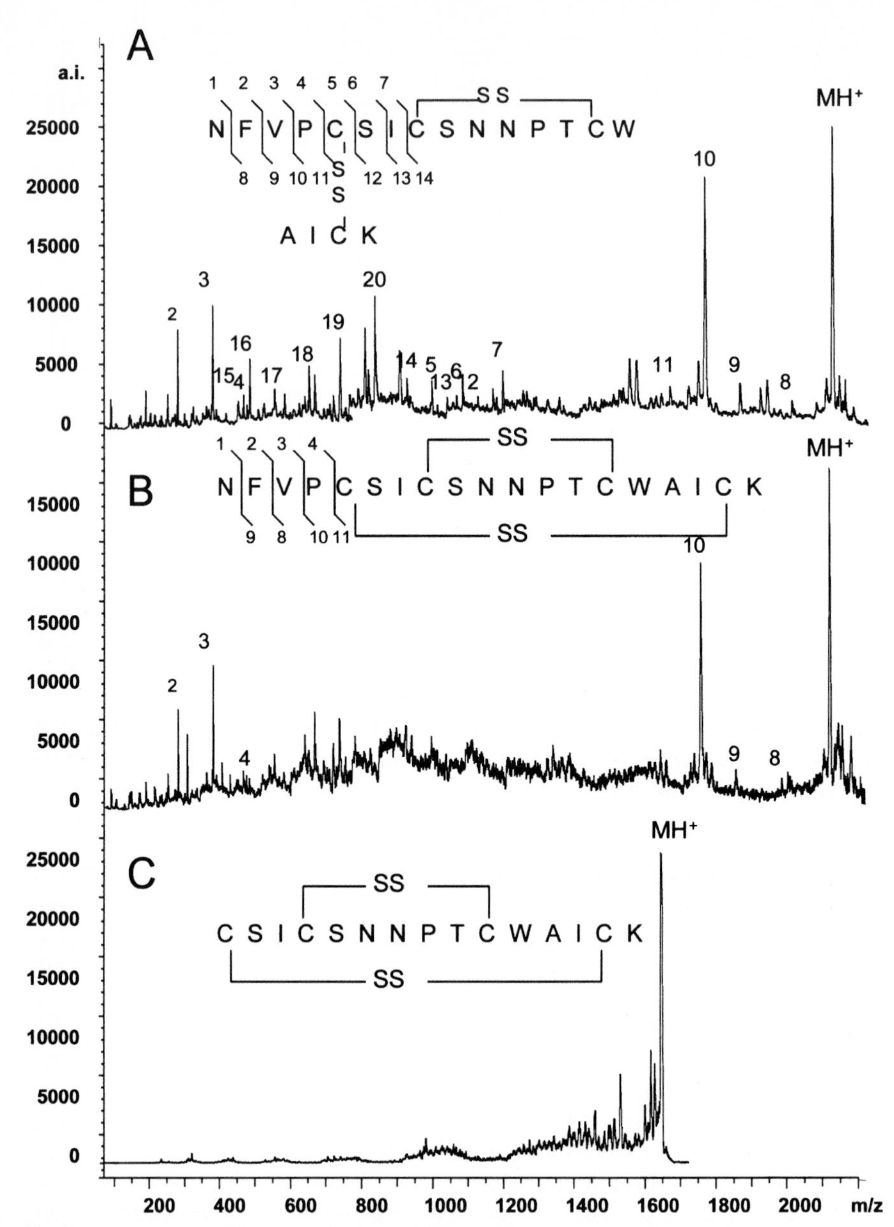

Fig. 8.9. Comparison of the PSD behaviour of the two chain (A) and single chain (B) peptic peptides described in Fig. 8.7 and a post proline cleavage enzyme fragment (C) of the original tryptic peptide described in Fig. 8.6. Observation of identical fragmentations for the two chain (A) and single chain peptic peptides (B) external to cystine loops and failure to observe intra loop fragmentation is consistent with previous observations regarding peptide bond fragmentation within disulfide loops (Bean and Carr 1992; Katta et al. 1995; Suckau et al. 1996). These comparisons add weight to the identities of the additional fragments in the two chain peptide (A and Fig. 8.8). These spectra were collected using continuous extraction and a digitization rate of 250 MHz. Panel A was adapted with permission from (Gorman et al. 1997)

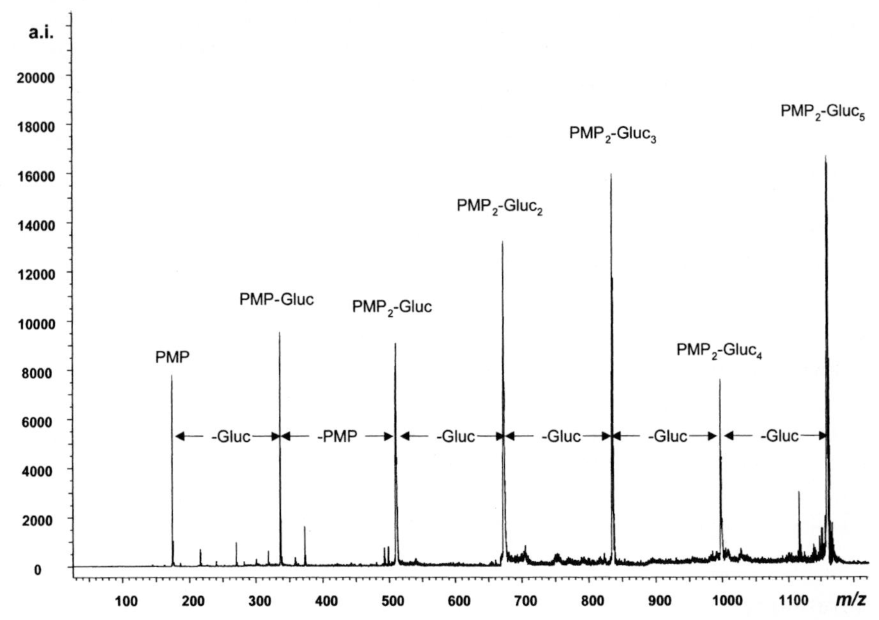

Fig. 8.10. PSD spectrum of PMP derivatized pentaglucose. Fragment ions detected are consistent with the fragments indicated on the spectrum. This spectrum was collected using delayed extraction and a digitization rate of 1 GHz

analysis of difficult glycans, for example large sialylated oligosaccharides, has been achieved by positive ion analysis of their PMP derivatives (Pitt and Gorman 1997) in conjunction with the "soft" matrix 2,6-dihydroxyacetophenone (Gorman et al. 1996; Pitt and Gorman 1996). In subsequent studies with model oligosaccharides, it has been possible to demonstrate that these derivatives are also amenable to PSD analysis (Fig. 8.10) and may provide a means of characterization of complex glycans released from proteins.

4 Conclusions

PSD has proven to be an effective technique for analysis of modified peptides. One example of this involved a superactive byproduct of peptide synthesis with Nβ-butyl substitution of an asparagine. An immonium ion characteristic of this modification (m/z = 143.21) was defined as a result of these studies.

Variation at the site proteolytic activation of the fusion protein precursor of a variant NDV isolate was defined by the use of PSD. This involved characterization of the newly generated C-terminus that was generated as a consequence of activation.

PSD has also proven to be an effective means of determination of the disulfide bond arrangement of a comparably short stretch of amino acid sequence bearing two disulfide loops in a 1 to 4 plus 2 to 3 linkage pattern. Surprisingly, an interchain disulfide bond was found to have survived MALDI ionization and reflectron based analysis under conditions whereby peptide backbone cleavages were observed. Precedence would have indicated that this interchain disulfide bond should not have been so stable (Crimmins et al. 1995; Hemling et al. 1996; Patterson and Katta 1994; Zhou et al. 1993).

Data derived with PMP derivatives of oligosaccharides suggests that MALDI-PSD of these derivatives will be of use in characterizing protein glycans.

The data reviewed herein indicate that the utility of MALDI-PSD should be assessed in a problem specific context and precedents or dogma should not rule out consideration of the use of PSD in particular circumstances.

5 Acknowledgments

Data depicted in figures 8.4 and 8.5 and table 8.2 were reproduced in adapted form with permission from (Lopaticki et al. 1998; Copyright, John Wiley + Sons) and data depicted in figures 8.6, 8.8 and 8.9 and table 8.3 were reproduced in adapted form with permission from (Gorman et al. 1997)

References

Barber M, Bordoli RS, Elliot GJ, Sedgwick RD, Tyler AN, Green BN (1982) Fast atom bombardment of bovine insulin and other large peptides. J Chem Soc Chem Commun 936–938

Barber M, Bordoli RS, Sedgwick RD, Tyler AN (1981) Fast atom bombardment of solids (FAB): a new ion source for mass spectrometry. J Chem Soc Chem Commun 325–327

Bean MF, Carr SA (1992) Characterization of disulfide positions in proteins and sequence analysis of cystine-bridged peptides by tandem mass spectrometry. Anal Biochem 201: 216–226

Biemann K (1990) The utility of mass spectrometry for the determination of the structure of peptides and proteins. In: McEwen CN, Larsen BS (eds) Mass Spectrometry of Biological Materials. Marcel Dekker Inc., New York, pp 3–24

Carr SA, Roberts GD, Hemling ME (1990) Structural analysis of posttranslationally modified proteins by mass spectrometry. In: McEwen CN, Larsen BS (eds) Mass Spectrometry of Biological Materials. Marcel Dekker Inc., New York, pp 87–136

Collins PL, McIntosh K, Chanock RM (1996) Respiratory syncytial virus. In: Fields BN, Knipe DM, Howley PM (eds) Virology. Lippincott-Raven Publishers, Philadelphia, pp 103–161

Crimmins DL, Saylor M, Rush J, Thoma RS (1995) Facile, in situ matrix-assisted laser desorption ionization-mass spectrometry analysis and assignment of disulfide bond pairings in heteropeptide molecules. Anal Biochem 226: 355–361

Desiderio DM, Katakuse I (1983) fast atom bombardment-collision activated dissociation-linked field scanning mass spectrometry of neuropeptide substance P. Anal. Biochem. 129: 425–429

Fenn JB, Mann M, Meng CK, Wong SF, Whitehouse CM (1989) Electrospray Ionization for Mass Spectrometry of Large Biomolecules. Science 246: 64–71

Glickman RL, Syddall RJ, Iorio RM, Sheehan JP, Bratt MA (1988) Quantiative basic residue requirements in the cleavage-activation site of the fusion glycoprotein as a determinant of virulence of Newcastle disease virus. J Virol 62: 354–356

Gorman JJ, Corino GL, Mitchell SJ (1987) Fluorescent labeling of cysteinyl residues: Application to extensive primary structure analysis of proteins on a microscale. Eur J Biochem 168: 169–179

Gorman JJ, Corino GL, Selleck PW (1990a) Comparison of the positions and efficiency of cleavage activation of fusion protein precursors of virulent and avirulent strains of Newcastle disease virus: insights into the specificities of activating proteases. Virol 177: 339–351

Gorman JJ, Corino GL, Shiell BJ (1990b) Role of mass spectrometry in mapping strain variation and post-translational modifications of viral proteins. Biomed Environ Mass Spectrom 19: 646–654

Gorman JJ, Ferguson BL, Nguyen TN (1996) Use of 2,6-dihydroxyacetophenone for analysis of fragile peptides, disulfide bonding and small proteins by matrix-assisted laser desorption/ionization. Rapid Commun Mass Spectrom 10: 529–536

Gorman JJ, Ferguson BL, Speelman D, Mills J (1997) Determination of the disulfide bond arrangement of human respiratory syncytial virus attachment (G) protein by matrix-assisted laser desorption/ionization time-of-flight mass spectrometry. Protein Science 6: 1308–1315
Gorman JJ, Nestorowicz A, Mitchell SJ, Corino GL, Selleck PW (1988) Characterization of the sites of proteolytic activation of Newcastle disease virus membrane glycoprotein precursors. J Biol Chem 263: 12522–12531
Gorman JJ (1987) Fluorescent labeling of cysteinyl residues to facilitate electrophoretic isolation of proteins suitable for amino-terminal sequence analysis. Anal Biochem 160: 376–387
Gruber C, Levine S (1985) Respiratory syncytial virus polypeptides. IV. The oligosaccharides of the glycoproteins. J Gen Virol 66: 417–432
Hemling ME, Mentzer MA, Capiau C, Carr SA (1996) A multifaceted strategy for the characterization of recombinant gD-2, a potent herpes vaccine. In: Burlingame AL, Carr SA (eds) Mass Spectrometry in the Biological Sciences. Humana Press, Totowa, N. J., pp 307–331
Hodder AN, Liu Z-Y, Selleck PW, Corino GL, Shiell BJ, Grix DC, Morrow CJ, Gorman JJ (1994) Characterization of field isolates of Newcastle disease virus using antipeptide antibodies. Avian Diseases 38: 103–118
Hodder AN, Selleck PW, White JR, Gorman JJ (1993) Analysis of pathotype-specific structural features and cleavage activation of Newcastle disease virus membrane glycoproteins using antipeptide antibodies. J Gen Virol 74: 1081–1091
Homma M (1975) Host-induced modification of Sendai virus. In: Mahy BWJ, Barry RDe (eds) Negative strand viruses. Academic Press, London, pp 685–697
Honda S, Akao E, Suzuki S, Okuda M, Kakehi K, Nakamura J (1989) High-performance liquid chromatography of reducing carbohydrates as strongly-absorbing and electrochemically sensitive 1-phenyl-3-methyl-5-pyrazolone derivatives. Anal Biochem 180: 351–357
Hunt DF, Shabanowitz J, Yates JR, Griffin PR (1990) Protein sequence analysis by tandem mass spectrometry. In: McEwen CN, Larsen BS (eds) Mass Spectrometry of Biological Materials. Marcel Dekker Inc., New York, pp 169–195
Karas M, Hillenkamp F (1988) Laser desortion ionization of of proteins with molecular masses exceeding 10 000 daltons. Anal Chem 60: 2299–2301
Katakuse I, Desiderio DM (1983) Positive and negative ion fast-atom bombardment–collision-activated dissociation–linked-field scanned mass spectra of leucine enkaphalin. Int J Mass Spectrom Ion Proc 54: 1–15
Kaufmann R, Kirsch D, Spengler B (1994) Sequencing of peptides in a time-of-flight mass spectrometer: evaluation of post source decay following matrix-assisted laser desoption ionisation (MALDI). International J Mass Spectrom Ion Proc 131: 355–385
Kaufmann R, Spengler B, Lutzenkirchen F (1993) Mass Spectrometric Sequencing of Linear peptides by Product-ion Analysis in a Reflectron Time-of-flight Mass Spectrometer Using Matrix-assisted Laser Desorption Ionization. Rapid Commun Mass Spectrom 7: 902–910
Kent SBH (1988) Chemical synthesis of peptides and proteins. Ann Rev Biochem 57: 957–989
Lopaticki S, Morrow CJ, Gorman JJ (1998) Characterization of pathotype-specific epitopes of Newcastle disease virus fusion glycoproteins by matrix-assisted laser desorption/ionization mass spectrometry and post-source decay sequencing. J Mass Spectrom 33: 950–960
Macfarlane RD, Thorgerson DF (1976) Californium-252 plasma desortion mass spectrometry. Science 191: 920–925
Nagai Y, Klenk H-D, Rott R (1976) Proteolytic cleavage of the viral glycoproteins and its significance for the virulence of Newcastle disease virus. Virol 72: 494–508
Patterson SD, Katta V (1994) Prompt fragmentation of disulfide-linked peptides during matrix-assisted laser desorption ionization mass spectrometry. Anal Chem 66: 3727–3732
Pitt JJ, Gorman JJ (1996) Matrix-assisted laser desorption/ionization time-of-flight mass spectrometry of sialylated glycopeptides and proteins using 2,6-dihydroxyacetophenone as a matrix. Rapid Commun Mass Spectrom 10: 1786–1788
Pitt JJ, Gorman JJ (1997) Oligosaccharide characterization and quantitation using 1-phenyl-3-methyl-5-pyrazolone derivatives and matrix-assisted laser desorption/ionization time-of-flight mass spectrometry. Anal Biochem 248: 63–75
Purcell AW, Chen W, Ede NJ, Gorman JJ, Fecondo JV, Jackson DC, Zhao Y, McCluskey J (1998) Avoidance of self-reactivity results in skewed CTL responses to rare components of synthetic immunogens. J Immunol 160: 1085–1090
Rouse JC, Yu W, Martin SA (1995) A comparison of the fragmentation obtained from a reflector matrix-assisted laser desorption-ionization time-of-flight and a tandem four sector mass spectrometer. J Am Soc Mass Spectrom 6: 822–835
Scanlon DB, Corino GL, Shiell BJ, Della-Porta AJ, Manvell R, Alexander DJ, Gorman JJ (1999) Pathotyping isolates of Newcastle disease virus using antipeptide antibodies to regions of their fusion

and hemagglutinin-neuraminidase proteins and the detection of differences in cleavage activation of fusion protein precursors of virulent isolates. Arch Virol 144: 55–72

Scheid A, Choppin PW (1974) Identification of biological activities of paramyxovirus glycoproteins. Activation of cell fusion, hemolysis, and infectivity by proteolytic cleavage of an inactive precursor protein of Sendai virus. Virology 57: 475–490

Scheid A, Graves MC, Silver SM, Choppin PW (1978) Studies on the structure and functions of paramyxovirus glycoproteins. In: Mahy BWJ, Barry RD (eds) Negative strand RNA viruses and the host cell. Academic Press, New York, pp 181–193

Simmons GC (1967) The isolation of Newcastle disease virus in Queensland. Aust Vet J 43: 29–30

Spengler B (1997) Post-source decay analysis in matrix-assisted laser desorption/ionization mass spectrometry of biomolecules. J Mass Spectrom 32: 1019–1036

Spengler B, Kirsch D, Kaufmann R (1991) Metastable decay of peptides and proteins in matrix-assisted laser-desorption mass spectrometry. Rapid Commun Mass Spectrom 5: 198–202

Spengler B, Kirsch D, Kaufmann R, Jaeger E (1992) Peptide Sequencing by Matrix-assisted Laser Desorption Mass Spectrometry. Rapid Commun Mass Spectrom 6: 105–108

Sundqvist B, Roepstorff P, Fohlmann J, Hedin A, Hakonsson P, Kamensky I, Lindberg M, Salepour M, Save G (1984) Molecular weight determination of proteins by californium plasma desorption mass spectrometry. Science 226: 696–698

Toyoda T, Sakaguchi T, Imai K, Inocencio NM, Gotoh B, Hamaguchi M, Nagai Y (1987) Structural comparison of the cleavage-activation site of the fusion glycoprotein between virulent and avirulent strains of Newcastle disease virus. Virol 158: 242–247

Wang WT, LeDonne NC, Ackerman B, Sweeley CC (1984) Structural characterization of oligosaccharides by high-performance liquid chromatography, fast-atom bombardment-mass spectrometry, and exoglycosidase digestion. Anal Biochem 141: 366–381

Waterson AP, Pennington TH, Allan WH (1967) Virulence in Newcastle disease virus: a preliminary study. Brit Med Bull 23: 138–143

Wilm M, Shevchenko A, Houthhaeve T, Breit S, Schweigerer L, Fotis T, Mann M (1996) Femtomole sequencing of proteins from polyacrylamide gels by nano electrospray mass spectrometry. Nature 379: 466–469

Zhou J, Poppe-Schriemer N, Standing KG, Westmore JB (1993) Cleavage of interchain disulfide bonds following matrix-assisted laser desorption. Int J Mass Spectrom Ion Proc 126: 115–122

CHAPTER 9

Detection of Specific Zinc Finger Peptide Complexes with Matrix-Assisted Laser Desorption/Ionization Mass Spectrometry

E. LEHMANN[1] and R. ZENOBI[1]

1 Introduction

Noncovalent complexes play an important role in biochemistry and molecular biology. Various methods have been used to investigate these complexes. Among them, size exclusion chromatography and gel electrophoresis, for example, are used for determining stoichiometry and molecular weights. These conventional methods are relatively time- and material-consuming. A recent analytical tool to overcome these problems is mass spectrometry (MS), characterized by high sensitivity and speed. Together with electrospray ionization (ESI) mass spectrometry, matrix-assisted laser desorption/ionization (MALDI) mass spectrometry is a soft ionization technique, i.e. leads to little fragmentation of the analyte. For this reason, it is a powerful method for the analysis of various biomolecules and synthetic polymers (Bahr et al. 1992; Hillenkamp et al. 1991). MALDI spectra are generated by laser irradiation of analyte molecules embedded in an excess of a crystalline matrix. This matrix is responsible for the absorption of the incident laser light and also plays an important role in analyte ionization.

A relevant question to ask is whether a correspondence exists between solution-phase behavior and the observed gas-phase ions and how soft MALDI is for the analysis of specific noncovalent complexes and complexes of biomolecules with metal ions. These species are stable under physiological conditions, but may not survive sample crystallization or laser desorption and ionization processes. Up to now, the ability of MALDI to detect specific noncovalent complexes is just starting to be explored (Cohen et al. 1997; Glocker et al. 1996; Gruic-Sovulj et al. 1997; Woods et al. 1995) and only little work has been done with MALDI (Lehmann et al. 1997; Nelson and Hutchens 1992) and also ESI (Loo 1997; Veenstra et al. 1998) to compare solution and gas-phase chemistries of noncovalent and metal ion-biomolecule complexes.

Our aim was therefore to detect the specific complexes of a zinc finger peptide with Zn^{2+}, as well as with Zn^{2+} and oligodeoxynucleotides with MALDI MS. The challenge in the second case was to transfer relatively weakly bound species intact into the gas phase. An important aim was also to distinguish between specific and nonspecific aggregation; the latter may take place in the MALDI plume

[1] Laboratorium für Organische Chemie, ETH Zentrum, Universitätstr. 16, CH-8092 Zürich, Switzerland.

(Gruic-Sovulj et al. 1997; Lehmann 1996). For this purpose, control experiments had to be designed. Furthermore, we wanted to establish a correlation between the complexation behavior in solution and in the MALDI spectra. A detailed report of the experiments described in this chapter can be found in two recent papers (Lehmann and Zenobi 1998, Lehmann et al. 1999)

2
Results and Discussion

2.1
Zinc Finger Peptide Complexation with Metal Ions

We investigated an 18-residue zinc finger peptide of CCHC type (CCHC = Cys-X_2-Cys-X_4-His-X_4-Cys, X = variable amino acid) with the sequence acVKCFN-CGKEGHIARNCRA-OH corresponding to the first zinc finger domain from the gag protein p55 of human immunodeficiency virus type 1 (HIV-1), called p55F1. The apo-peptide has no defined secondary structure. Metal ion binding induces peptide folding which is required for the interaction with nucleic acids (Lam et al. 1994). CCHC type peptides bind Zn^{2+} very tightly. In the resulting complex, Zn^{2+} is tetrahedrally coordinated by the three cysteine and one histidine residues (Fig. 9.1) (Summers et al. 1990).

The MALDI MS analysis of the metal ion-p55F1 complex was performed using the basic matrix 6-aza-2-thiothymine (ATT). This matrix was recently shown to be suited for the detection of noncovalent compounds (Glocker et al. 1996). It permits to work at a physiological pH of the sample solution, which is necessary for complex formation. Such pH conditions are in contrast to normal MALDI conditions, where acidic matrices are used.

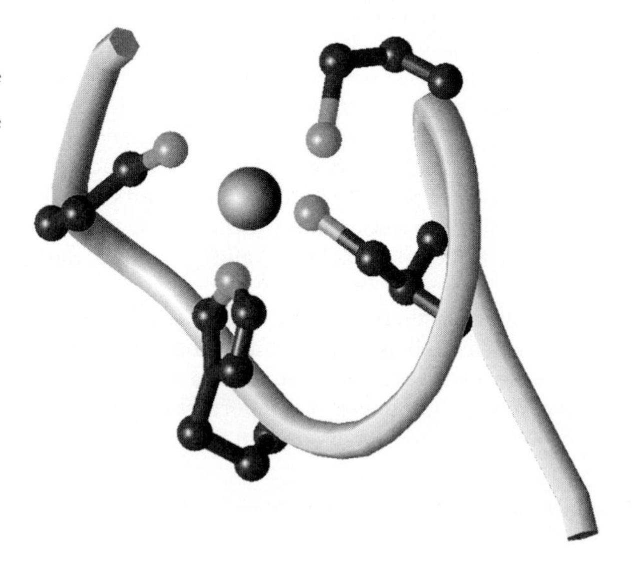

Fig. 9.1. Schematic view of the Zn^{2+}-p55F1 complex. The histidine and the three cysteine residues (black) of the CCHC motif bind to the Zn^{2+} ion. The binding heteroatoms are depicted in grey

The complexation behavior in solution was probed by circular dichroism (CD) spectroscopy because the peptides secondary structure elements show characteristic CD spectra (Fasman 1996).

Fig. 9.2 depicts the MALDI mass spectra of p55F1 with ATT as a matrix. A signal of the protonated molecular ion is observed (Fig. 9.2A). Upon addition of 10 molar equivalents of Zn^{2+}, a signal corresponding to the peptide adduct with Zn^{2+} appears (Fig. 9.2B). The reasons why a large excess of Zn^{2+} is needed to see a significant zinc complex signal in MALDI, whereas CD spectra already show a complete formation of the Zn^{2+}-peptide complex in solution in a 1:1 molar ratio, are explained in detail in the paper of Lehmann et al. 1998.

In the following, the ratio of the peak integrals of the Zn^{2+}-adduct and the protonated peptide will be called relative Zn^{2+}-adduct intensity. The Zn-p55F1 signal may either represent a specific 1:1 complex or a nonspecific adduct. To distinguish between both possibilities, the peptide gramicidin S, which does not form specific adducts with Zn^{2+}, was added to the sample (Fig. 9.2C). The relative Zn^{2+}-adduct intensity of p55F1 is the same as without addition of gramicidin S. The Zn^{2+}-adduct of gramicidin S is negligible. Therefore, this experiment indicates a specific complex of Zn^{2+} with p55F1.

The association reaction of the Zn^{2+}-p55F1 complex in solution can be written as follows:

$$\text{p55F1} + Zn^{2+} \rightleftharpoons [(\text{p55F1}) + Zn]^{2+} \tag{1}$$

The formation constant is 10^{10} M^{-1}. The affinities of p55F1 for other metal ions are assumed to be lower, but are not known. Therefore, they were qualitatively

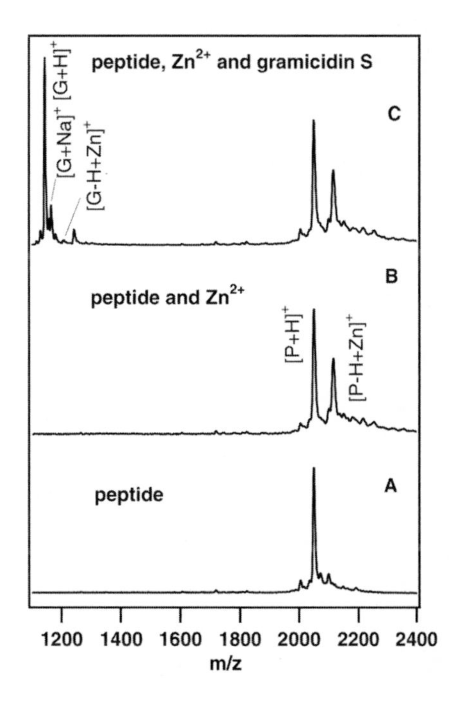

Fig. 9.2. MALDI mass spectra with ATT matrix of (A) peptide p55F1 alone, (B) peptide and Zn^{2+} in molar ratio 1:10, and (C) peptide and Zn^{2+} in molar ratio 1:10 with addition of gramicidin S. Solvent: water, pH 5. The spectra are normalized to the $[P+H]^+$ signal (adapted with permission from Lehmann et al. 1999)

determined using CD spectroscopy. The degree of peptide folding was found to decrease from Zn^{2+} over Cd^{2+} to Ni^{2+}. Since complexation induces folding, the CD spectra give a qualitative estimate of the affinity of p55F1 for the different metal ions.

If MALDI spectra reflect this behavior in solution, then the use of the different metal ions should have an influence on the relative intensities of the metal ion-peptide complex. It can clearly be seen in Fig. 9.3 that the relative signal intensities decrease from Zn^{2+} over Cd^{2+} to Ni^{2+}. These results indicate a correlation between the behavior of the metal ion-peptide complexes in solution and in the gas phase.

As CD spectroscopy and MALDI mass spectrometry rely on completely different principles, the results cannot be compared quantitatively, but only qualitatively. In this sense, talking about a correlation between the behavior of the metal ion-peptide complex in solution and in the gas phase means that changes in solution conditions lead to corresponding changes in the MALDI spectra. Absolute complex abundances in solution and complex peak intensities in the MALDI spectra can, of course, not be directly compared.

The complexation of p55F1 and Zn^{2+} in solution is pH dependent:

$$[(\text{p55F1}) + 2\text{H}]^{2+} + \text{Zn}^{2+} \rightleftharpoons [(\text{p55F1}) + \text{Zn}]^{2+} + 2\text{H}^{+} \qquad (2)$$

At neutral pH, the peptide is doubly protonated. Upon formation of the tetrahedral metal ion-zinc finger peptide complex, at least two of the three cysteine -SH groups are deprotonated. Addition of base leads to an increase of the concentration of Zn-p55F1, as protons are eliminated from the equilibrium. CD experi-

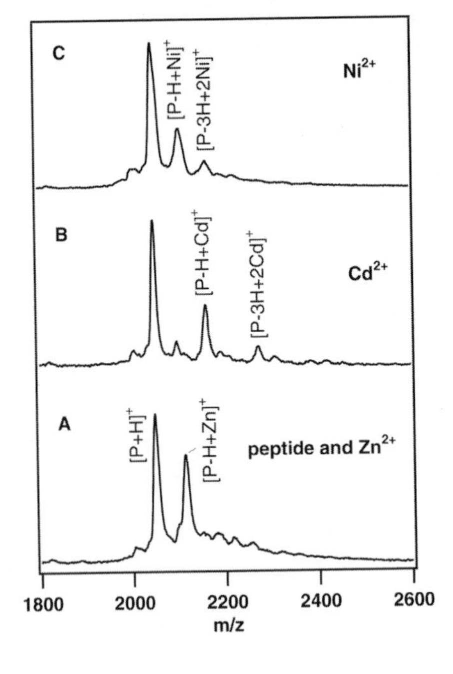

Fig. 9.3. MALDI mass spectra of p55F1 alone and with different metal ions in molar ratio 1:10 at pH 5. Matrix: ATT. Solvent: water. Adducts of two metal ions to p55F1 most probably correspond to nonspecific gas-phase products. The spectra are normalized to the $[P+H]^{+}$ signal (adapted with permission from Lehmann et al. 1999)

ments show that the peptide complex is completely formed in solution above pH 6 and that it is not stable at a pH < 6. When studying the influence of the pH on the MALDI spectra of Zn-p55F1, we found that the relative intensity of Zn-p55F1 increases with pH. This behavior correlates with the one in solution.

Note that the MALDI spectra at a pH below 6 exhibit Zn-p55F1 complex signals, whereas CD spectra of the solutions at this pH do not. This is most probably due to the fact that MALDI spectra are taken from crystalline samples. We have shown in previous work that the amount of complexes present in the solid phase is higher than in solution if volatile byproducts, such as HCl, form upon complexation (Lehmann et al. 1997). This is the case here, because $ZnCl_2$ was used as metal salt. The reason for this is that HCl is eliminated during crystallization, resulting in a shift of equilibrium (2) to the right.

2.2 Zinc Finger Peptide Complexation with Oligodeoxynucleotides

The complexation of Zn-p55F1 to single-stranded nucleic acids is essential for the replication of HIV-1 and has been extensively studied by various methods in solution (Lam et al. 1994; South and Summers 1993). In the resulting noncovalent triple complex the oligodeoxynucleotide binds within a hydrophobic cleft onto the peptide surface. CCHC zinc fingers bind to ribonucleic acids, but NMR studies in solution, to which our MALDI experiments can be compared, have been performed using single-stranded oligodeoxynucleotides as structural probes (South and Summers 1993). Therefore, the MALDI results presented here were carried out using oligodeoxynucleotides, mainly d(TTGTT).

2-Amino-4-methyl-5-nitropyridin (AMNP) was chosen as a MALDI matrix for the analysis of the triple complexe (Fitzgerald et al. 1993). This basic matrix allowed the simultaneous detection of both p55F1 and oligodeoxynucleotide in positive ion mode. It also permits work at physiological conditions of the sample solution, using ammonium bicarbonate to adjust the pH.

When analysing the oligodeoxynucleotide-containing samples with MALDI, undesired alkali ion adducts were first observed. They were eliminated using the drop dialysis method (Lehmann and Zenobi 1998), where the alkali ions are exchanged against ammonium ions. Upon sample crystallization, ammonia evaporates and the overall result is that alkali ions are exchanged against protons.

Fig. 9.4A shows the MALDI mass spectrum of a mixture of d(TTGTT) and p55F1 at neutral pH. The protonated signals of the oligodeoxynucleotide and p55F1 as well as a small nonspecific adduct of both are detected. Upon Zn^{2+} addition, the specific triple complex of p55F1, Zn^{2+} and d(TTGTT), which we expect from solution chemisty, is observed. Less intense p55F1-d(TTGTT) adducts without and with two Zn^{2+} ions are also detected. We also succeeded in detecting the triple complex between p55F1, Zn^{2+}, and an oligodeoxynucleotide 11-mer, d(TTTTTGTTTTT) using a similar sample preparation for MALDI MS (Fig. 9.5).

The signals of both the triple complex and the Zn-p55F1 complex are less intense than those of the individual components. The reasons in the case of Zn-p55F1 have already been discussed (Lehmann et al. 1999). Similar arguments can also be made for the triple complex.

The investigation of Zn-p55F1 with CD spectroscopy in solution showed that this complex is only stable at pH > 6. The same is expected to be valid for the triple complex with d(TTGTT) since the oligodeoxynucleotide only binds to the metal-complexed peptide. If MALDI spectra reflect solution-phase behavior, then decreasing the pH to a value below 6 should lead to a significant decrease of the MALDI signal of the specific triple complex. This is exactly what was observed experimentally (Fig. 9.4C). Besides a strong decrease of the Zn-p55F1 signal compared to that of p55F1, the triple complex signal is absent. Instead, a distribution

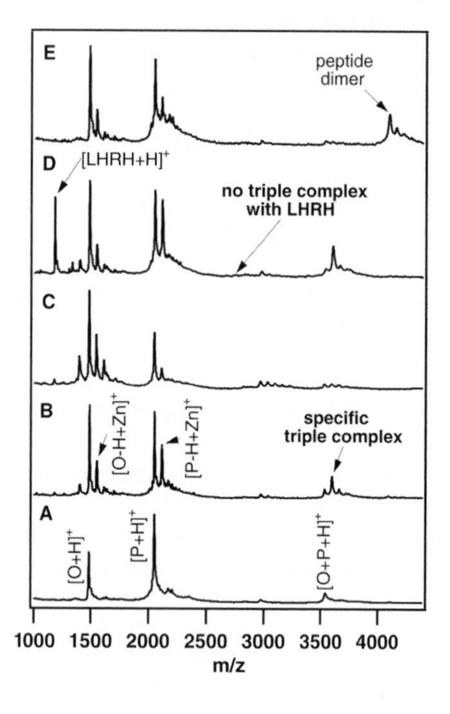

Fig. 9.4. MALDI mass spectra with AMNP matrix of p55F1 and d(TTGTT) in a 33:1 molar ratio (A) without addition of Zn^{2+} at pH 6.5–7, (B) with Zn^{2+} added (Zn^{2+}:p55F1 molar ratio = 5:1) at pH 6.5–7, (C) same as (B) at pH 5–5.5, (D) same as (B) with addition of LHRH at pH 7, (E) same as (B) with addition of Cu^{2+} in the same molar amount as Zn^{2+} (adapted with permission from Lehmann and Zenobi 1998)

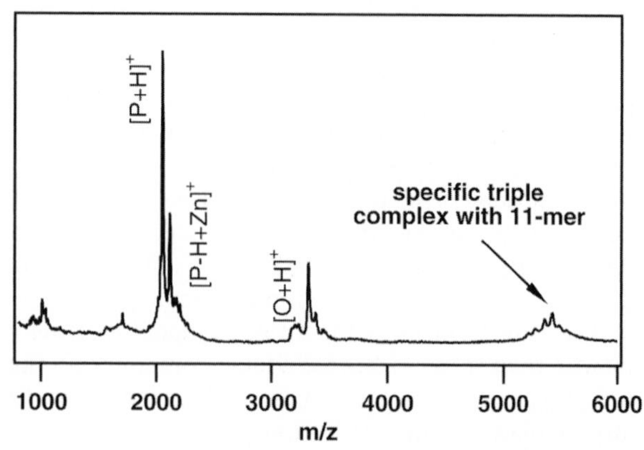

Fig. 9.5. MALDI mass spectrum with AMNP matrix of the triple complex Zn-p55F1-d(TTTTTGTTTTT). Molar ratio Zn^{2+} : p55F1 : d(TTTTTGTTTTT) = 165:33:1

of multiple Zn^{2+}-adducts to p55F1-d(TTGTT) is observed. We assume that these are nonspecific adducts of p55F1 with the oligodeoxynucleotide, in which the oligodeoxynucleotide binds multiple Zn^{2+} ions through its phosphodiester backbone. This hypothesis is confirmed by the observation of multiple Zn^{2+}-adducts to the oligodeoxynucleotide alone and to its dimer.

To further confirm the specificity of the triple complex, another peptide, luteinizing hormone releasing hormone, LHRH, was added to the sample (Fig. 9.4D). LHRH contains a histidine residue in its sequence and thus potentially binds to Zn^{2+} ions. It also contains arginine with a positively charged side chain, which may bind through electrostatic forces to the oligodeoxynucleotide backbone. Both interactions would be nonspecific. The spectrum shows that neither a Zn^{2+}-adduct to this peptide nor a triple complex with Zn^{2+} and d(TTGTT) is detected. The corresponding complex signals of p55F1 are as intense as in the absence of LHRH. This experiment supports a specific complex formation of Zn^{2+} with p55F1 and d(TTGTT) which can be followed with MALDI.

An experiment demonstrating the effect of a chemical modification of the zinc finger peptide on its complexation to the oligodeoxynucleotide was also performed. Since the binding of p55F1 to oligodeoxynucleotides is important for HIV replication, such a modification mimics the action of an antiviral agent. For this purpose, Cu^{2+} was added to a sample containing peptide, Zn^{2+} and the oligodeoxynucleotide. It is known from solution experiments that this metal ion oxidizes the peptide's thiol functions on the zinc complexing cysteine residues to an intramolecular disulfide bond by simultaneous reduction of Cu^{2+} to Cu^{+}. The modified p55F1 forms complexes neither with Zn^{2+} nor with oligodeoxynucleotides. This is also reflected in the MALDI spectrum in Fig. 9.4E: if Cu^{2+} is added to the sample, the cysteine residues are oxidized and the signal of the triple complex disappears. A strong p55F1 dimer is detected instead, because peptide oxidation can also lead to intermolecular disulfide bonds involving the third cysteine thiol. This experiment therefore indicates that MALDI MS may be a potential method for rapidly screening antiviral HIV agents.

Besides the complexation of Zn-p55F1 with d(TTGTT), other sequences, like d(TTATT) and d(ACGCC) were also investigated. Our MALDI experiments with d(TTATT) instead of d(TTGTT) support the findings of Lam et al. 1994 and Gorelick et al. 1993 who report that Zn-p55F1 is not sequence-specific in solution. We found that the complex of d(TTATT) with Zn-p55F1 gave signals of about the same intensity as with d(TTGTT). MALDI experiments with oligodeoxynucleotide sequences lacking thymidine are, unfortunately, not very informative. If d(ACGCC) was used as the binding partner hardly any MALDI signal could be seen at all, although the corresponding triple complex should be stable in solution (South and Summers 1993). It is common knowledge that oligodeoxynucleotides that do not contain thymidines are much less stable as ions and thus much more difficult to detect by MALDI MS. Also, they may crystallize less favorably with the matrix, using the same sample preparation, than those containing mostly thymidines (Schneider and Chait 1993).

The experiments probing the sequence specificity of p55F1 reveal a current limitation of MALDI: a "MALDI window" in which the target complex is observed has to be found. The MALDI window is given by the experimental sample preparation conditions, such as the nature of the matrix or the matrix analyte ratio. Once this window is found, experiments involving, for example, pH or oligodeoxynucleotide sequence variations can be performed. However, if a change of the experimental conditions prohibits observation, no conclusions can be drawn. An example is the above described experiment involving an oligodeoxynucleotide sequence change: the experimental conditions adequate for the sequences d(TTGTT) and d(TTATT) were not adequate for d(ACGCC). No conclusion on the sequence-specificity of Zn-p55F1 can therefore be drawn. Future work has to concentrate on finding more generally usable experimental conditions.

3 Conclusions

In the present study on zinc finger peptide complexation, we demonstrated that MALDI MS is suitable for the detection of specific biomolecule-metal ion and noncovalent complexes. The specificity can be proven by carefully designed chemical controls, such as the variation of pH or competition experiments. We could also show that MALDI mass spectra reflect solution-phase chemistry. In this sense, we showed that it is possible to qualitatively study metal-binding properties of peptides and the pH dependence of the corresponding complexes. Furthermore, the experiment on the effect of chemical modifications of p55F1 on the formation of the Zn-p55F1-oligodeoxynucleotide complex revealed that MALDI MS is a potential method for rapidly screening antiviral HIV agents. In order to establish MALDI as a reliable method for detecting noncovalent complexes consisting of molecules of different chemical classes, future work has to focus on finding more generally usable experimental conditions, such as MALDI matrices allowing simultaneous detection of the different classes of compounds.

4 Acknowledgements

EL gratefully acknowledges a Kekulé-stipend from the Fonds der Chemischen Industrie (Germany). The authors thank Stefan Vetter for the synthesis of p55F1 and many helpful discussions as well as the Kommission für Technologie und Innovation (project 3165.1), Switzerland, for financial support.

References

Bahr U, Deppe A, Karas M, Hillenkamp F, Giessmann U (1992) Mass spectrometry of synthetic polymers by UV MALDI. Anal. Chem. 64: 2866–2869

Cohen LRH, Strupat K, Hillenkamp F (1997) Analysis of quarternary protein ensembles by MALDI MS. J Am. Soc. Mass Spectrom. 8: 1046–1052

Fasman GD (1996) Circular Dichroism and the Conformational Analysis of Biomolecules. Plenum Press, New York

Fitzgerald MC, Parr GR, Smith LM (1993) Basic matrices for MALDI MS of proteins and oligonucleotides. Anal Chem 65: 3204–3211

Glocker MO, Bauer SHJ, Kast J, Volz J, Przybylski M (1996) Characterization of specific noncovalent protein complexes by UV MALDI MS. J Mass Spectrom 31: 1221–1227

Gorelick RJ, Chabot DJ, Rein A, Henderson LE, Arthur LO (1993) The two zinc fingers in the HIV-1 nucleocapsid protein are not functionally equivalent. J Virol 67: 4207–4036
Gruic-Sovulj I, □Lüdemann H-C, Hillenkamp F, Weygand-Durasevic I, Kucan Z, Peter-Katalinic J (1997) Dection of noncovalent tRNA aminoacyl-tRNA synthease complexes by MALDI MS. J Biol Chem 272: 32084–32091
Hillenkamp F, Karas M, Beavis RC, Chait BT (1991) MALDI MS of biopolymers. Anal Chem 63: 1193A-1203A
Lam W-C, Maki AH, Casas-Finet JR, Erickson JW, Kane BP, Sowder II RC, Henderson LE (1994) Phosphorescence and optically detected magnetic resonance investigation of the binding of the nucleocapsid protein of the HIV virus type 1 and related peptides to RNA. Biochemistry 33: 10693–10700
Lehmann E, Zenobi R (1998) Detection of specific noncovalent zinc finger peptide-oligodeoxynucleotide complexes by MALDI MS. Angew Chem Int Ed 37: 3430–3432
Lehmann E, Knochenmuss R, Zenobi R (1997) Ionization Mechanisms in MALDI MS: Contribution of pre-formed ions. Rapid Commun Mass Spectrom 11: 1483–1492
Lehmann E, Zenobi R, Vetter S (1999) MALDI mass spectra reflect solution-phase zinc finger peptide complexation. J Am Soc Mass Spectrom 10: 27–34
Lehmann WD (1996) Massenspektrometrie in der Biochemie. Spektrum Akad Verl, Heidelberg
Loo J (1997) Studying noncovalent protein complexes by ESI MS. Mass Spectrom Rev 16: 1–23
Nelson RW, Hutchens TW (1992) Mass spectrometric analysis of a transition metal binding peptide using MALDI TOF. A demonstration of probe tip chemistry. Rapid Commun Mass Spectrom 6: 4–8
Schneider K, Chait BT (1993) MALD MS of homopolymer oligodeoxyribonucleotides. Influence of base composition on the mass spectrometric response. Org Mass Spectrom 28: 1353–1361
South TL, Summers MF (1993) Zinc- and sequence-depending binding to nucleic acids by the N-terminal zinc finger of the HIV-1 nucleocapsid protein: NMR structure of the complex with the psi-site analog dACGCC. Protein Sci 2: 3–19
Summers MF, South TL, Kim B, Hare DR (1990) High resolution structure of an HIV zinc fingerlike domain via a new NMR-based distance geometry. Biochemistry 29: 329–340
Veenstra TD, Johnson KL, Tomlinson AJ, Kumar R, Naylor S (1998) Correlation of fluorescence and circular dichroism spectroscopy with ESI MS in the determination of tertiary conformational changes in calcium-binding proteins. Rapid Commun Mass Spectrom 12: 613–619
Woods AS, Buchsbaum JC, Worrall TA, Berg JM, Cotter RJ (1995) MALDI of noncovalently bound compounds. Anal Chem 67: 4462–4465

CHAPTER 10

Epitope Mapping for the Monoclonal Antibody that Inhibits Intramolecular Flavin to Heme Electron Transfer in Flavocytochrome b_2 from Baker's Yeast (L-Lactate Dehydrogenase)

K. H. Diêp Lê[1], M. Mayer[1] and F. Lederer[1]

1 Introduction

Flavocytochrome b_2 is a homotetramer, each subunit of which carries one FMN and one protoheme IX. The FMN oxidizes lactate to pyruvate in a reaction involving the transfer of two reducing equivalents. These are then transferred in two successive steps to heme b_2 in the same subunit. Heme b_2 in turn is reoxidized by cytochrome *c*, the physiological acceptor, in the intermembrane space of yeast mitochondria. The enzyme enables yeast to grow on lactate as sole carbon source (for review, see Lederer 1991).

The enzyme can now be expressed under a recombinant form in *E. coli* (Black et al. 1989). Its crystal structure has been refined to 2.4 Å resolution (Xia and Mathews 1990). The asymmetric unit is composed of two subunits. Subunit S1 shows two domains (Fig. 10.1); residues 1 to 99 encompass the heme-binding domain. Its fold is similar to that of cytochrome b_5, as predicted from sequence data (Guiard et al. 1974). Residues 100 to 511 encompass the flavodehydrogenase domain (FDH). It presents the typical $\beta_8\alpha_8$ barrel fold, also present in glycolate oxidase, another member of the family of the FMN-dependent α-hydroxy acid oxidizing enzymes (Lindqvist et al. 1991). The planes of the two prosthetic groups are nearly parallel to each other, and the distance between flavin N5 and the closest porphyrin ring atom is 10 Å. There is no intervening protein matter between flavin N5 and the opening of the heme crevice. In contrast, subunit S2 only shows the FDH domain, with pyruvate, the reaction product, bound at the active site.

Thus, the crystal structure shows evidence for mobility of the heme domain in the crystal. Subsequently, NMR data suggested that it is also mobile in solution (Labeyrie et al. 1988). The interdomain contacts comprise a number of hydrophobic interactions, a single electrostatic interaction between K296 and a heme propionate carboxylate, as well as a single direct hydrogen bond between Y143 and the other heme propionate carboxylate, and a few water-mediated hydrogen bonds (Xia and Mathews 1990). Altogether, these interactions do not provide strong cohesive forces, since, once the covalent bond between the domains is broken, the domains do not recognize each other and are incapable of exchanging

[1] Laboratoire d'Enzymologie et Biochimie Structurales, Centre National de la Recherche Scientifique, 91198 Gif-sur-Yvette Cedex, France.

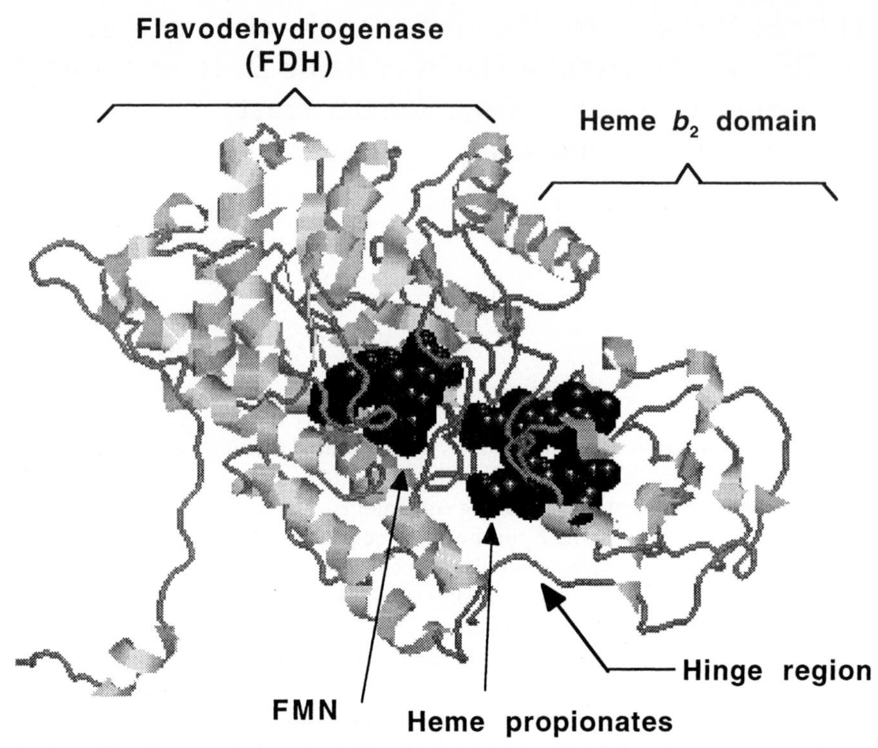

Fig. 10.1. Three-dimensional structure of flavocytochrome b_2 subunit S1. The space-filling representation was used for the heme and flavin prosthetic groups

electrons (Gervais and Tegoni 1980; Balme et al. 1995). The covalent link between the domains thus must limit the space the heme domain has to search for finding a docking position on the FDH which would be competent for electron transfer. One may therefore wonder if the flavin to heme electron transfer rates determined in kinetic experiments are actual transfer rates, or underestimated rates due to limitations arising from heme domain movements and the fraction of time it spends in a productive contact with the FDH.

Site-directed mutagenesis has been used in order to shed light on this question. Mutations of interface residues (Miles et al. 1992; Rouvière et al. 1997) and manipulations of the sequence linking the two domains (Sharp et al. 1994, 1996; White et al. 1993) showed the importance of the hydrogen bond between Y143 and the heme propionate, as well as of the structural integrity of the linker region. We have been using another approach, based on the use of a monoclonal antibody directed against the heme-binding domain. We have recently described some properties of this antibody (Miles et al. 1998). The enzyme complexed with this antibody shows a normal flavin reduction by lactate, but the reduced FMN is incapable of transferring electrons to heme b_2. It can only be reoxidized by ferricyanide, a small non physiological acceptor (Fig. 10.2). As a consequence of the inhibition of heme reduction, cytochrome *c* reduction is also inhibited. In view of

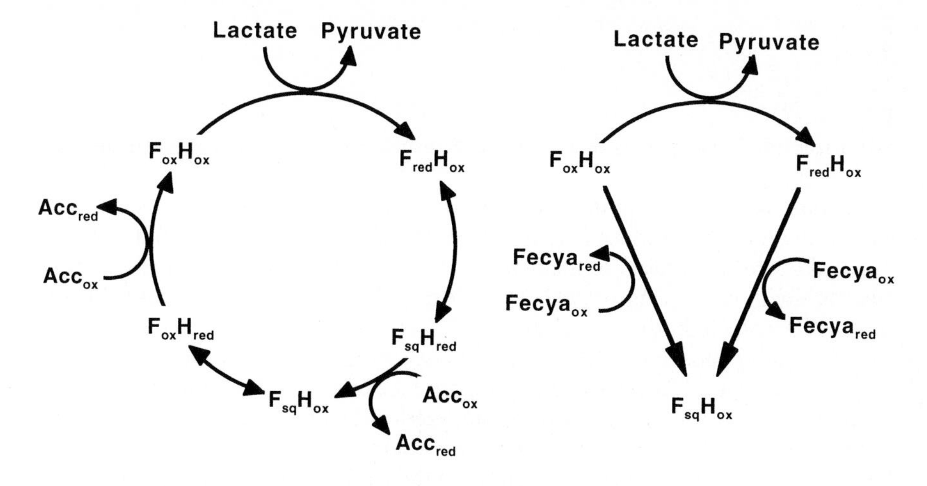

Fig. 10.2. The flavocytochrome b_2 catalytic cycle for the free (left) and the complexed enzyme (right)

the three-dimensional structure of the enzyme (Fig. 10.1; Xia and Mathews 1990), it appeared difficult to rationalize these results without invoking the heme-domain mobility. Antibody binding would freeze the heme domain in a wrong orientation with respect to the FDH by preventing movements necessary for the catalytic cycle, or by acting as a wedge between the domains. The first experiments designed to locate the antibody epitope on the heme-binding domain indicated that it is a conformational epitope, and that the linker region between the domains (residues 86 to 116; White et al. 1993) is not part of the epitope. We describe here further efforts, using chemical and site-directed mutagenesis, designed for locating the binding area of the antibody on the heme-binding domain.

2
Materials and Methods

Wild-type enzyme, IgG and Fab fragment preparations have been described before, as well as the competitive ELISA procedure for determining dissociation constants and the protocols for kinetic assays of enzyme activity in the presence and absence of antibody (Miles et al. 1998). Site-directed mutagenesis was carried out using PCR with appropriate oligonucleotides. The fragment to be mutagenized was excised from the plasmid coding for the WT enzyme with *Bam* HI (after TCC corresponding to Ser 87) and *Eco* RI (plasmid polylinker site), and was religated after mutagenesis, amplification and sequencing, using the same restriction sites. Surface plasmon resonance (SPR) analyses were carried out using a Biacore instrument model 1000. The Fab fragment was covalently attached to the chip through its amino groups. All experiments were carried out in 0.1 M phosphate buffer pH 7 at 20°C. The tetrameric enzyme or the isolated heme-binding domain were run in the mobile phase at 30 μl/min.

3
Results

3.1
Site-Directed Mutagenesis of the Heme-Binding Domain in Flavocytochrome b_2

For a start we chose to mutate charged residues or polar ones. Four mutations were individually introduced: R38E, E63K, N69K and D72A. The first position mutated lies at the "waist" of the heme-binding domain, the three others are very close to the interface (Fig. 10.3A). The effect of the mutations was assessed using three different methods. In steady-state assays of cytochrome *c* reduction, the lactate dehydrogenase activity at saturating substrate and acceptor concentrations was compared in the presence and absence of a fixed IgG concentration. In parallel, dissociation constants were analyzed in competitive ELISA assays and comparatively with the Biacore. The results are reproduced in Table 10.1. Clearly, R38E and D72A mutant enzymes behave as the wild type enzyme: they are inhibited to the same extent at the same antibody concentration and they exhibit, within experimental error, the same affinity for the antibody. In contrast, E63K and N69K mutant enzymes were not inhibited in the presence of the IgG, and it was very difficult to accurately determine affinities. In the ELISA tests, no competition with the WT enzyme adsorbed on the plates could be determined up to 1.5 μM mutant enzymes and the amplitudes on the Biacore sensorgrams up to 1.7 μM enzyme in the mobile phase were very small. It can be estimated that the

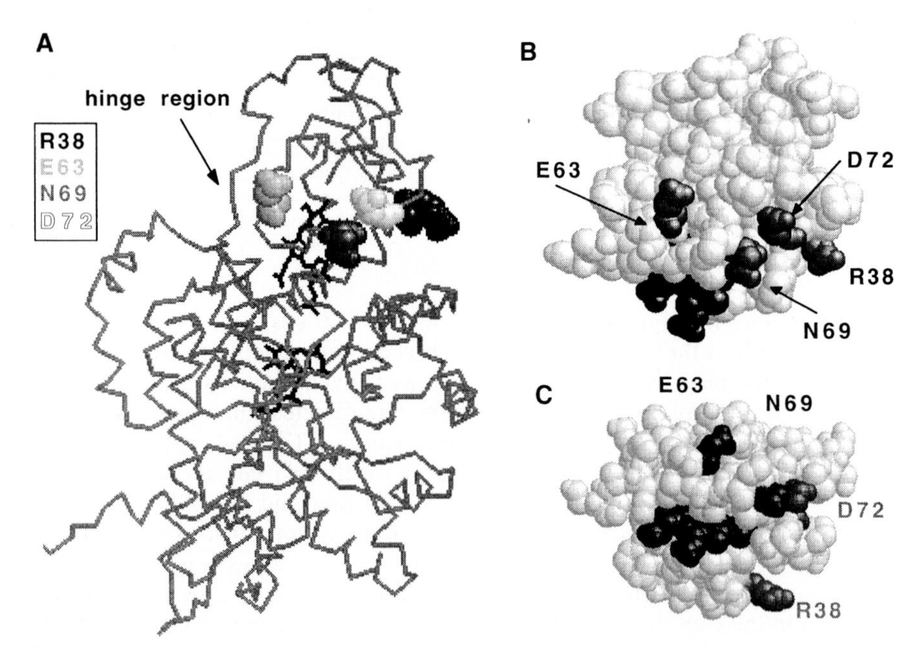

Fig. 10.3. Location of the positions mutated. A: in the subunit; B: in the isolated heme domain (same orientation as in A); C: in the isolated heme domain after a rotation

Table 10.1. Effect of point mutations on inhibition of cytochrome *c* reduction by the antibody and on its affinity for the enzyme. Activity assays were carried out in the presence of 20 mM L-lactate and 500 μM cytochrome *c* in 0.1 M phosphate buffer pH 7, at 30 °C. The enzyme (0.26 μM) was preincubated for 5 minutes with the IgG (0.85 μM) at 30 °C, then kept on ice before a 60-fold dilution in the assay cuvettes. Competition ELISA tests were carried out as described in Miles et al. (1998), using the Fab fragment. BIACORE affinities were determined by flowing the enzyme over covalently-coupled Fab followed by NaOH regeneration. Buffer conditions were 0.1 M phosphate pH 7, 0.01 % Tween at a flow rate of 30 μl/min at 20 °C

	WT	R38E	D72A	E63K	N69K
Cytochromse *c* reduction:					
$K_{cat}(s^{-1})$					
-IgG	362±32	380±44	377±40	278±27	241±22
+IgG	23±13	18±7	21±6	271±30	238±19
ELISA					
K_d (nM)	24±15	49±19	21±17	>2000	>2000
BIACORE					
K_d (nM)	209±97	282±22	161±18	not measurable	not measurable

affinity of the mutant enzymes decreased by 2 to 3 orders of magnitude. These results indicate that E63 and N69 are important determinants in the epitope. It can be noted that the values obtained by competition ELISA assays and with the Biacore present a five to ten fold difference. These may possibly be ascribed to several of the factors which have been discussed in the literature (Schuck 1997): in particular the antigen in the mobile phase is a tetramer and the antibody is not bound to the chip in a single orientation. Nevertheless, the two methods give results that are qualitatively coherent.

3.2 A Chemical Mutation

As can be seen in Fig. 10.3A, positions 63 and 69 are adjacent to the interface. We decided to probe residues in the interface, but the first attempt was actually directed at the heme. If one considers the structure of the isolated heme-binding domain (Fig. 10.3B), the heme propionates clearly stick out from the heme crevice and are fully accessible; they become buried in the interface and make the interactions mentioned above. We used the acid acetone procedure to reversibly remove the heme (Teale 1959); the apo heme-binding domain was reconstituted in parallel with normal protoheme IX and its dimethyl ester, following the reconstitution procedure described for cytochrome b_5 (Reid et al. 1984). Heme removal was not attempted with the holoenzyme, because the acid conditions would also remove part or all of the more labile flavin, in a possibly irreversible manner. The effect of the propionate groups esterification was analysed using the competive ELISA assays and the Biacore. The results are presented in Table 10.2; with the ELISA tests, the domain reconstituted with normal heme exhibited the same affinity as the native one, whereas the one reconstituted with the dimethyl ester showed an affinity drop by a factor of about 50; with the Biacore, despite some

Table 10.2. Effect of heme replacement by its dimethylester derivative (DME) on the affinity of the Fab for the heme-binding domain. For ELISA tests, the protocol described in Miles et al. (1998) was followed with two modifications: the 96-well plates were coated with 0.2 µg flavocytochrome b_2 instead of 1 µg, and all incubations were carried out at 20 °C. The Biacore experiments were carried out as described under Table 10.1, with the heme-binding domain in the mobile phase

	Native heme domain	Apodomain + normal heme	Apodomain + DME heme
ELISA K_d (nM)	8.5±3.1	7.5±0.6	408±103
BIACORE K_d (nM)	67±4.2	18.1±0.1	1820±50

small differences in absolute values, the decrease was of the same order of magnitude. Thus, both methods indicate that one or both of the heme propionates are also determinants in the epitope.

4 Discussion

The results presented here yield a preliminary definition of the heme-binding domain epitope of the monoclonal antibody which inhibits flavin to heme electon transfer in flavocytochrome b_2. It minimally encompasses a surface defined between residues 63, 69 and the heme propionates (Fig. 10.3C). It is uncertain whether only one or both heme propionates are involved. Nevertheless, it is clear that the epitope overlaps part of the domains interface. This is the first direct proof of the heme domain mobility in solution, since it is clear that the fixed structure shown in Fig. 10.1 cannot give access to the heme propionates. Another conclusion of the present results is that the domain movements must have a rather large amplitude, since the size of an Fab is somewhat larger than that of the flavodehydrogenase domain. It is now clear that electron transfer between flavin and heme is prevented by the antibody acting indeed as a wedge between the domains and preventing close approach and correct docking. Further experiments are designed to elucidate whether or not an overlap between the antibody epitope and the cytochrome *c* binding area on the heme-binding domain , as well as to obtain information about the frequency of the domain movements.

References

Balme A, Brunt C E, Pallister R, Chapman S K. & Reid G A (1995) Isolation and characterization of the flavin-binding domain of flavocytochrome b_2, expressed independently in *Escherichia coli*. Biochem J 309: 601–605

Black M T, White S A, Reid G A & Chapman S K (1989) High-level expression of fully active yeast flavocytochrome b_2 in *Escherichia coli*. Biochem J 258: 255–259

Gervais M & Tegoni M (1980) Spontaneous dissociation of a cytochrome core and a biglobular flavoprotein after mild trypsinolysis of the bifunctional *Saccharomyces cerevisiae* flavocytochrome b_2. Eur J Biochem 111: 357–367

Guiard B, Groudinsky O & Lederer F (1974) Homology between baker's yeast cytochrome b_2 *and liver microsomal cytochrome* b_5. Proc Natl Acad Sci USA. 71: 2539–2543

Labeyrie F, Beloeil J C & Thomas M A (1988) Evidence by NMR for mobility of the cytochrome domain within flavocytochrome b_2. Biochim Biophys Acta 953: 134–141

Lederer F (1991) Flavocytochrome b_2. In: Chemistry and Biochemistry of Flavoenzymes (F. Müller, ed.) CRC Press, Boca Raton, FL, Vol. 2: pp 153–242.
Lederer F (1994) The cytochrome b_5-fold: an adaptable module. Biochimie 76: 674–692
Lindqvist Y, Brändén C I, Mathews F S & Lederer F (1991) Spinach glycolate oxidase and yeast flavocytochrome b_2 are structurally homologous and evolutionarily related enzymes with distinctly different function and flavin mononucleotide binding. J Biol Chem 266: 3198–3207
Miles C S, Rouvière-Fourmy N, Lederer F, Mathews F S, Reid G A., Black M T & Chapman S K (1992) Tyr-143 facilitates interdomain electron transfer in flavocytochrome b_2. Biochem J 285: 187–192
Miles C S, Lederer F & Lê K H D (1998) Probing intramolecular electron transfer within flavocytochrome b_2 with a monoclonal antibody. Biochemistry 37: 3440–3448
Reid L S, Mauk M R & Mauk A G (1984) Role of heme propionate groups in cytochrome b_5 electron transfer. J Am Chem Soc 106: 2182–2185
Rouvière N, Mayer M, Tegoni M, Capeillère-Blandin C & Lederer F (1997) Molecular interpretation of inhibition by excess substrate in flavocytochrome b_2: a study with wild-type and Y143F mutant enzymes. Biochemistry 36: 7126–7135
Sharp R E, White P, Chapman S K & Reid G A (1994) The role of the interdomain hinge of flavocytochrome b_2 in intra- and inter-protein electron transfer. Biochemistry 33: 5115–5120
Sharp R E, Chapman S K & Reid G A (1996) Deletions in the interdomain hinge region of flavocytochrome b_2: Effects on intraprotein electron transfer. Biochemistry 35: 891–899
Schuck P (1997) Use of surface plasmon resonance to probe the equilibrium and dynamic aspects of interactions between biological macromolecules. Annu Rev Biophys Biomol Struct 26: 541–566
Teale F W J (1959) Cleavage of the haem-protein link by acid methylethylketone. Biochim Biophys Acta 35: 543
White P, Manson F D C, Brunt C E, Chapman S K & Reid G A (1993) The importance of the interdomain hinge in intramolecular electron transfer in flavocytochrome b_2. Biochem J 291: 89–94
Xia Z X & Mathews F S (1990) Molecular structure of flavocytochrome b_2 at 2.4 Å resolution. J Mol Biol 212: 837–863

CHAPTER 11

A MALDI-TOF Mass Spectrometry Approach to Investigate the Defense Reactions in *Drosophila melanogaster*, an Insect Model for the Study of Innate Immunity

Ph. Bulet[1] and S. Uttenweiler-Joseph[2]

1 Summary

A protocol consisting of matrix-assisted laser desorption/ionization mass spectrometry (MALDI-MS), high performance liquid chromatography (HPLC), Edman degradation and cDNA cloning has been optimized for the rapid identification and structural characterization of peptides induced during the humoral immune response of *Drosophila*. MALDI-MS analysis led to the detection of at least 24 immune-induced molecules (DIMs) in the hemolymph (body fluid, 0.1 μl) of a single fruitfly in response to an experimental challenge. The use of micropurification by HPLC, peptide sequencing and molecular biology techniques allowed the characterization of three out of the 24 DIMs from a hemolymph sample collected from less than 140 flies. This study clearly shows that MALDI-MS differential mapping constitutes a powerful tool to detect compounds involved in inducible physiological processes.

2 Introduction

The present interest for insect immunity was stimulated by the recent discovery of striking parallels that exist between insect immunity and innate immunity in vertebrates (Ezekowitz and Hoffmann 1998). Similarly, studies on vertebrate innate immunity have been encouraged by the perception that it plays a role in the induction of adaptive immunity and it directs particular effector mechanisms in this response (for review, see Fearon and Locksley 1996). The discovery of similarities between vertebrate and insect innate immunity suggests that *Drosophila melanogaster*, through its powerful genetics, is a suitable model for evolutionary studies on innate immunity (for review, see Meister et al. 1997).

The immunity of *Drosophila*, like that of other insects, relies on both cellular and humoral reactions. Cellular defense includes the phagocytosis and encapsulation of intruders by blood cells. This aspect has mostly been described at the

[1] Institut de Biologie Moléculaire et Cellulaire, Unité Propre de Recherche 9022 du Centre National de la Recherche Scientifique, 15, rue René Descartes, 67084 Strasbourg Cedex, France.
[2] Present address: European Molecular Biology Laboratory (EMBL), Protein & Peptide Group, Meyerhofstrasse, 1, 69117 Heidelberg, Germany.

morphological level (Rizki and Rizki 1984) while the molecular aspects have only been recently investigated (Franc et al. 1996, Braun et al. 1997). The humoral defense involves, as first line, the activation of proteolytic cascades leading to coagulation and melanization at the site of injury. Little information is available on insect coagulation. Melanization, which has been largely investigated in Lepidopera (for review, see Ashida and Brey 1997), is poorly understood in *Drosophila* as only enzymes acting at the end of this cascade were characterized (Fujimoto et al. 1993, Chosa et al. 1997). The second line of humoral defense in insects is the rapid synthesis by the fat body (an equivalent of the mammalian liver) and by certain blood cells of antimicrobial peptides and their release into the hemolymph (insect body fluid). This latter aspect of insect immunity is well documented in *Drosophila* where seven distinct antimicrobial peptides/polypeptides (plus isoforms) have been characterized (for review, see Hoffmann and Reichhart 1997). In general, insect antimicrobial peptides have a broad range of activity and are not cytotoxic. They share common features such as a low molecular mass (below 5 kDa), a positive net charge at physiological pH and for most of them, amphiphilic α-helices or hairpin-like β-sheets or mixed structures. The existence of several antimicrobial peptides with different functional properties in association with the genetic potential of *Drosophila* allowed to demonstrate the existence of two signaling pathways, referred to as the imd (for immune deficiency) and Toll pathways, leading to the expression of the antibacterial or antifungal genes, respectively (Lemaitre et al. 1995, Lemaitre et al. 1996). However, many of the components of these two signaling cascades have yet to be identified.

Although significant progress has been made in recent years in the field of *Drosophila* immunity, essential questions remain to be answered. First, no information is available to date on the mechanisms of recognition of the pathogens that induce the immune response. At a second level, the interconnections between the different reactions of the insect immune response described above are unknown. Furthermore, many effectors involved in these different immune mechanisms remain to be characterized. Finally, we cannot eliminate the possible existence of yet unknown aspects of the immune response.

The components of *Drosophila* immunity which have been characterized to date were isolated either by a biochemical approach on the basis of their activities (e.g. antimicrobial peptides) or by molecular biology (cloning of genes by homology) or by genetics (e.g. genes involved in the imd and Toll pathways) (for review, see Hoffmann and Reichhart 1997). Our aim was to develop a methodology to detect novel molecules involved in *Drosophila* humoral immunity at the protein level, regardless of a precise biological activity and omitting any purification step that could introduce a certain selectivity. As the immune reactions can be activated by a septic injury, a differential analysis between the hemolymph of unchallenged and experimentally immune-challenged *Drosophila* can be performed to detect molecules induced during the humoral immune response. This strategy could be operated by two-dimensional gel electrophoresis but as collecting *Drosophila* hemolymph is labor-intensive, a more sensitive technology is required for analyzing the components present in the body fluid (0.1 μl) of individual flies.

During the last years, matrix-assisted laser desorption/ionization time-of-flight mass spectrometry (MALDI-TOF MS) has appeared as a powerful technique to analyze complex mixtures of peptides within different biological samples. The performance of MALDI-TOF MS is explained by its sensitivity (femtomolar to attomolar concentrations), its tolerance towards buffers and salts, and by the ionization process that chiefly produces single-charged ions leading to clear spectra (for review, see Burlingame et al. 1998). Consequently, MALDI-TOF MS is the most suitable MS technique for the direct analysis of complex biological samples such as cell lines (van Adrichem et al. 1998), thin-layer preparations of tissues (Caprioli et al. 1997), single giant neurons (Jimenez et al. 1994, Li et al. 1994), procaryotic organisms (Claydon et al. 1996, Easterling et al. 1998), or eucaryotic cells (Stahl et al. 1997, Redeker et al. 1998). In these studies, MALDI-TOF MS allowed the establishment of peptide profiles which served to obtain characteristic fingerprints or to follow the synthesis or the maturation processing of bioactive peptides.

In the present contribution, MALDI-TOF MS differential display of the hemolymph of individual immune-challenged versus unchallenged *Drosophila* led to the detection and molecular mass characterization of new molecules involved in insect immunity. This strategy, as developed for *Drosophila*, can be applied to other biological systems for the direct identification from complex biological samples of compounds induced or repressed during a given physiological process, regardless of their activities.

3 Measuring Molecular Masses in Complex Biological Samples: Establishment of Experimental Conditions

3.1 Critical Points for MALDI-TOF MS Analysis of Complex Mixtures

Briefly, for MALDI analysis, the analyte is mixed with an excess (10^3 to 10^5-fold) of "matrix" to form co-crystals. The matrix serves to isolate the analyte molecules from each other and to absorb the intense laser radiation used to vaporize and propel the analyte molecules into the gas phase where they are ionized. Matrix is clearly the most important factor to obtain a mass signal, but many other parameters influence the quality of the mass spectrum, like the sample preparation (how the co-crystals are formed), the matrix to analyte ratio and the laser power (Cohen and Chait 1996, Kussmann et al. 1997, Roepstorff et al. 1998).

In a previous study, we observed that all these parameters are critical when working on complex mixtures of peptides and proteins (Uttenweiler-Joseph et al. 1997). With regard to this observation, our first step in establishing the protocol for MALDI-TOF MS analysis of *Drosophila* hemolymph was to determine the optimal conditions to obtain high quality spectra.

3.2
Sample Preparation and Experimental Conditions of MALDI-TOF MS Analysis of *Drosophila* hemolymph

Drosophila hemolymph was collected from a single fly using a glass capillary (Fig. 11.1A) and directly deposited onto the MALDI target pretreated as described below (Fig. 11.1B). There are many different sample preparations to incorporate a protein mixture into matrix crystals such as the dried droplet, the thin polycrystalline film, the seeded microcrystalline film or the thick layer methods (for reviews, see Beavis and Chait 1996 and Kussmann et al. 1997). A sample preparation including nitrocellulose (for further experimental details, see Uttenweiler-Joseph et al. 1998) was preferred because nitrocellulose has the property to enhance the ionization yield of molecules as already described (Preston et al. 1993, Liu et al. 1995, Uttenweiler-Joseph et al. 1997). In addition, this preparation allows an intensive washing of the final preparation before analysis resulting in a better resolution and a higher mass accuracy due to the elimination of salts adducts. Two different matrices were tested with the nitrocellulose preparation: α-cyano-4-hydroxycinnamic acid (4-HCCA) usually used for the analysis of peptides, and sinapinic acid which is more selective for proteins. During this first approach, all mass spectra were acquired in the positive linear mode with an instrument equipped with a continuous extraction (no delayed-extraction). The same pattern was obtained in the negative mode, but with a lower ionization efficiency (data not shown). Comparison of the mass spectra obtained with 4-HCCA or with sinapinic acid as matrix reveals that sinapinic acid enhanced the signal

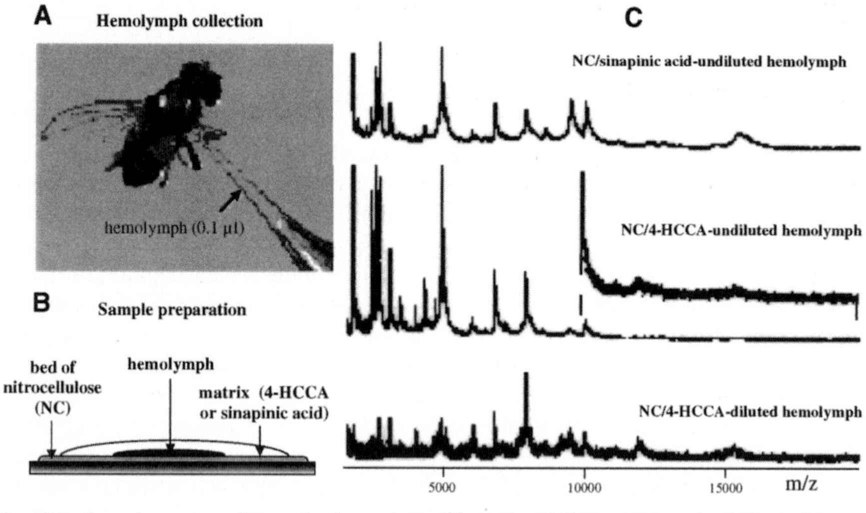

Fig. 11.1. Experimental conditions for the analysis of hemolymph from an immune-challenged *Drosophila*. The hemolymph (0.1 µl), collected from a single fly with a glass capillary (A), is directly analyzed by MALDI-TOF MS with a sample preparation using nitrocellulose (B) (for further details, see Uttenweiler-Joseph et al. 1998). In order to obtain the most qualitative mass spectrum, different parameters were tested like the nature of the matrix [α-cyano-4-hydroxycinnamic acid (4-HCCA) or sinapinic acid] and the matrix/analyte ratio modified by hemolymph dilution (C)

intensity of the compounds with a molecular mass higher than 9 kDa (Fig. 11.1C). Nevertheless, with sinapinic acid, some compounds with a low molecular mass were undetectable, and resolution as well as mass accuracy were decreasing compared to the results observed with 4-HCCA. To be as qualitative as possible, the matrix 4-HCCA was preferred to sinapinic acid.

For MALDI MS, the matrix/analyte ratio is important as it influences the formation of the co-crystals on which depends the quality of the mass spectra. To analyze this parameter, different dilutions were performed on *Drosophila* hemolymph (10 and 100-fold in acidic water). Compared to crude hemolymph, a 10-fold dilution had the effect of dramatically decreasing the signal to noise ratio resulting in a loss of resolution and mass accuracy (Fig. 11.1C). Furthermore, no additional peak was visible. Consequently, our subsequent investigations were performed directly on undiluted *Drosophila* hemolymph with a sample preparation including nitrocellulose and 4-HCCA as a matrix. Finally, only the mass range from 1.5 to 11 kDa was investigated because the mass signal intensity was very low above this mass range resulting in a lack of reproducibility.

We would also like to stress the importance of the laser power intensity. In fact, we clearly observed from these preliminary experiments, that quality of mass spectra is dependent on laser power intensity. A higher laser power intensity increased the number of peaks detectable. However, when laser power intensity is higher than the ionization threshold, resolution and mass accuracy decrease. For the purpose of this study, laser power intensity was first adjusted to a level allowing the detection of the highest number of compounds as illustrated in figure 11.1C and secondly to a level corresponding to the ionization threshold of any individual compound detected for the determination of their most accurate mass.

The best conditions of MALDI-TOF MS analysis must be found for each complex biological mixture with regard to the aims (qualitative analysis versus detection of a specific compound) and the requirements for reproducibility, mass resolution and mass accuracy.

4 Differential Analysis of Hemolymph from Immune-Challenged and Unchallenged *Drosophila* by MALDI-TOF MS

The mass spectrum obtained from the hemolymph of a 24h-immune-challenged fly was compared to that of a control (unchallenged) fly (Fig. 11.2). In the conditions mentioned above and whatever the sex of the fly, the mass spectra were highly reproducible for *m/z* values between 1,500 and 11,000 (more than 50 control and immune-challenged flies were individually analyzed).

Mass spectrum of the hemolymph of immune-challenged *Drosophila* is more complex than that of controls. More precisely, in addition to compounds already found in control flies, 24 additional molecules (numbered from 1 to 24 in Fig. 11.2) were clearly present in immune-challenged hemolymph.

The 24 molecules detected after injury could be systemic molecules or cytoplasmic components released by the *Drosophila* blood cells which can be lysed by the acidity of the MALDI matrix solution used. To distinguish between these two

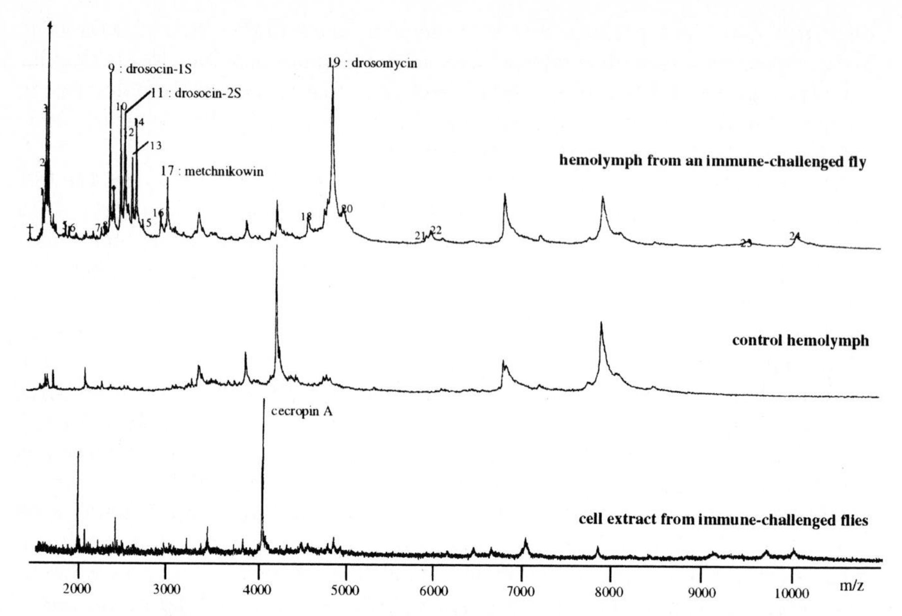

Fig. 11.2. Differential analysis by MALDI-TOF MS of hemolymph from a control and a 24h bacteria-challenged adult *Drosophila* and from a blood cell extract. The crude hemolymph collected from a single fly was analyzed by MALDI-TOF MS in the positive linear mode using the sample preparation presented in Fig. 11.1, including 4-HCCA as matrix. The singly charged ions of the molecules induced in the crude hemolymph after bacterial challenge are numbered from 1 to 24. Four of them correspond to already characterized antimicrobial peptides which are designated by their names. 1S and 2S refer to the monosaccharide and the disaccharide glycoforms of drosocin, respectively. + and ◆ are the doubly charged ions of metchnikowin and drosomycin, respectively. The mass spectrum of blood cells extract from immune-challenged flies does not contain the induced molecules found in crude hemolymph but mostly presents a compound with a molecular mass corresponding to the antibacterial cecropin A

possibilities, *Drosophila* hemolymph was collected in a physiological buffer (*Drosophila* Ringer) and the blood cells were separated by gentle centrifugation. The cell-free hemolymph and the acidic extract of blood cells were then analyzed by MALDI-TOF MS. Mass spectrum from cell-free hemolymph was identical to that obtained for native hemolymph suggesting that all 24 molecules were systemic (data not shown). In sharp contrast, the mass spectrum of the acidic extract of blood cells was much more simple with one major single-charged ion detected at *m/z* 4156.9 (Fig. 11.2). The measured molecular mass (4155.9 Da) corresponds to the mass of a compound already characterized from an acidic extract of thousands of *Drosophila* flies, namely the antibacterial peptide cecropin A (calculated molecular mass of 4155.8 Da) whose expression in blood cells had been established at the mRNA level (Samakovlis et al. 1992). This result confirms the presence of mature cecropin within *Drosophila* blood cells.

As the immune response is induced by pricking adult *Drosophila* with a needle previously dipped into a bacterial pellet, we checked whether the 24 molecules might be of bacterial origin. The same qualitative mass spectra were obtained after an injury with a septic or an aseptic needle (data not shown) suggesting that the 24 induced molecules do not result from bacteria. This result was confirmed by the analysis of *Drosophila* mutants (see below).

In summary, none of the 24 induced molecules detected by MALDI-TOF MS in *Drosophila* hemolymph are of bacterial origin or coming from the lysis of the *Drosophila* blood cells. Therefore they represent molecules involved in the systemic immune response of *Drosophila* and are hereafter called DIMs for "*Drosophila* Immune-induced Molecules". DIMs are revealed by direct MALDI-TOF MS analysis of the hemolymph of single flies, without any purification step. Their precise function in this process remains to be established. For this, we first determined the most accurate molecular mass of the DIMs to see if some of these compounds correspond to already described molecules.

MALDI-TOF MS allows differential mass analysis of complex biological mixtures in order to compare the qualitative pattern of compounds present before and after a physiological process.

5
Molecular Mass Characterization of the DIMs

At their ionization threshold, where the best resolution and mass accuracy are obtained (Jensen et al. 1997), and with an external calibration, the masses of peaks 9, 11, 17 and 19 correlate with the masses of already identified antimicrobial peptides which are secreted into the hemolymph after immune challenge. (i) Peaks 9 and 11 (*m/z* measured at 2402.9 and 2566.1) correspond to the monocharged ions of the antibacterial *O*-glycosylated drosocin (Bulet et al. 1993) carrying an *N*-acetylgalactosamine (drosocin-1S, calculated molecular mass of 2401.9 Da) and an *N*-acetylgalactosamine-galactose (drosocin-2S, calculated molecular mass of 2564.4 Da), respectively. (ii) Peak 17 at *m/z* 3046.6 corresponds to one form of the antimicrobial metchnikowin with a calculated mass of 3045.4 Da (Levashina et al. 1995). (iii) Peak 19 at *m/z* 4890.6 corresponds to drosomycin (calculated mass of 4889.5 Da), which exhibits potent antifungal activity (Fehlbaum et al. 1994).

However, not all the systemic antimicrobial peptides already characterized in immune challenged *Drosophila* were detectable in the mass spectra recorded from the hemolymph of immune-challenged flies (Fig. 11.2). Several hypotheses can explain this discrepancy. For example, we did not observe the mass signal of defensin, an antibacterial peptide of 4354 Da (Dimarcq et al. 1994), certainly due to its very low concentration (<2 μM) in the hemolymph of immune-challenged flies compared to the concentration of 100 μM for drosomycin, 40 μM for the two glycoforms of drosocin and 10 μM for metchnikowin. This hypothesis was verified since a clear mass signal was observed when purified *Drosophila* defensin was added to the hemolymph of immune-challenged flies at a final concentration of 10 μM (data not shown). This result was surprising since MALDI-TOF MS is considered to be highly sensitive (femtomole range). However, such a high sensitivity appears to be accessible only on purified or partially purified compounds. In fact, we clearly observed that efficiency of MALDI-TOF MS, from the point of view of sensitivity, decreases when a compound is analyzed in more and more complex mixtures. This alteration in sensitivity can be the result of suppression effects that often occur when mixtures are analyzed (Beavis and Chait 1990, Kratzer et al. 1998). In addition, suppression effects can also affect the detection of compounds of neighboring molecular masses. The antibacterial peptide cecropin A (calculated molecular mass of 4156 Da) was not detected in the mass spectrum presented in Fig. 11.2 although its concentration is around 20 μM in the hemolymph of immune-challenged flies. We believe that this absence of signal results from the suppression effects due to the vicinity of the drosomycin peak (DIM 19). To confirm this hypothesis, we performed the following series of experiments: we added the equivalent of 20 μM of cecropin A to the hemolymph of an unchallenged fly (where no drosomycin is present) and observed a clear mass signal at the expected *m/z* of 4157. When we further added drosomycin at increasing concentrations, the signal of cecropin A decreased to become undetectable when the final concentration of drosomycin reached the 100 μM concentration observed *in vivo* in *Drosophila* hemolymph (data not shown).

The absence of detection of some of the already characterized *Drosophila* antimicrobial peptides (namely defensin and cecropin A) suggests that suppression effects can also minimize the exact number of DIMs detected in our experimental conditions. This has been verified by the experiments performed on different *Drosophila* mutants (see below).

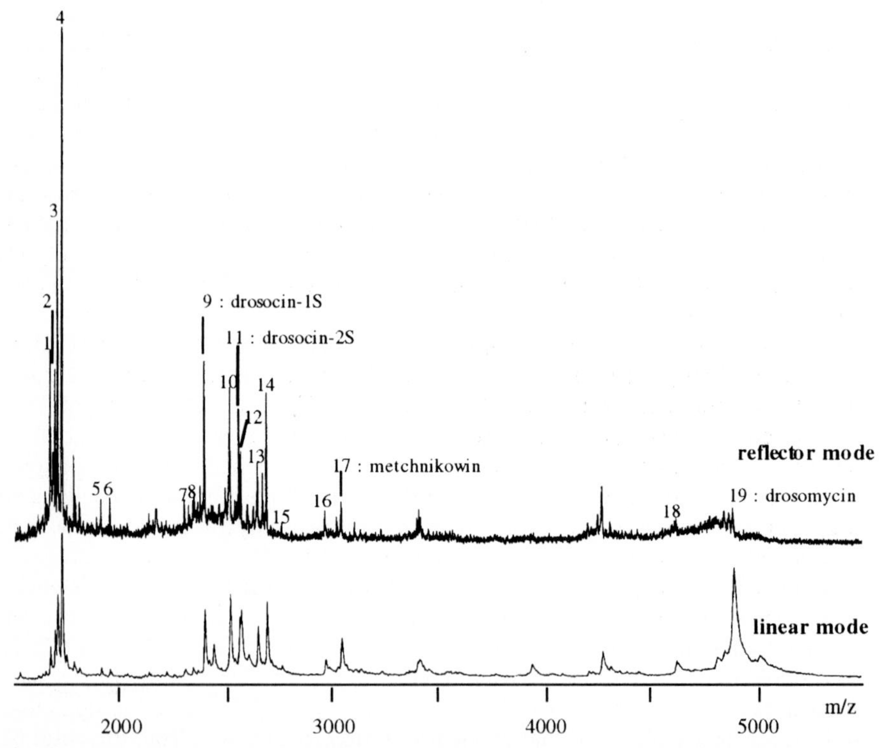

Fig. 11.3. MALDI-TOF mass spectra of hemolymph from an immune-challenged *Drosophila* in linear and reflector mode. The conditions used for the sample preparation are identical to these presented in Fig. 11.2 (for details see Uttenweiler-Joseph et al. 1998)

As we were able to detect already characterized antimicrobial peptides in the hemolymph of immune-challenged flies (namely the two glycoforms of drosocin, metchnikowin and drosomycin) we used them as internal calibrants. With this internal calibration, we assigned more accurate *m/z* to the unknown DIMs by measuring them in reflector mode (Fig. 11.3). Although we could not reach monoisotopic resolution, a better resolution was obtained with reflectron compared to the linear mode, but only for DIMs with a molecular mass below 5 kDa, as fragmentations occurred for compounds of higher mass, as illustrated for drosomycin (see Fig. 11.3). The average masses were obtained with the following internal calibrants: DIM 9 (drosocin-1S), DIM 11 (drosocin-2S), DIM 17 (metchnikowin) and DIM 19 (drosomycin), in reflector mode for DIMs 1–19 and in linear mode for DIMs 20–24 (Table 11.I).

We hypothesized that all DIMs are peptides/polypeptides as, referring to other studies performed on complex biological mixtures, MALDI MS preferentially ionized this type of substance compared to nucleic acids, oligosaccharides and phospholipids which are less detectable using the same conditions of analysis. Following this assumption, we scanned protein data banks (Swiss Prot and Protein Information Resource) to possibly correlate peptide *m/z* information data obtained for the DIMs with known sequences of *Drosophila* hemolymph peptides/proteins. No match was found, except for the antimicrobial peptides referred to above. We also controlled that none of the DIMs are degradation products of the known antimicrobial peptides.

Table 11.1 Molecular mass characterization of the DIMs in linear and reflector modes

DIMs	Mode	m/z	Peptide identity
1		1667.6	
2	R	1690.0	
3		1701.7	
4	E	1723.0	
5		1915.0	
6	F	1956.2	
7		2308.1	
8	L	2349.2	
9		2402.9	drosocin-1S
10	E	2521.6	
11		2565.0	drosocin-2S
12	C	2574.0	
13		2652.0	
14	T	2695.0	
15		2768.6	
16	O	2972.1	
17		3046.6	metchnikowin
18	R	4626.0	
19		4890.8	drosomycin
	L		
20	I	50.23	
21	N	5939	
22	E	5984	
23	A	9521	
24	R	10064	

To summarize, 4 out of the 24 DIMs could be identified by their molecular mass as already known antimicrobial peptides (the two glycoforms of drosocin, metchnikowin and drosomycin). The 20 other DIMs represent novel molecules involved in the immune response of *Drosophila* which have been characterized by their molecular mass.

MALDI-TOF MS analysis of complex biological mixtures often permits the detection of peptide compounds. The mass accuracy of MALDI-TOF MS is sufficient to allow the identification of molecules by their molecular mass if they are already known. Identified peptides can then serve as internal calibrants to measure a more accurate molecular mass for unknown peptides. Consequently, these accurate molecular masses characterize the unknown compounds.

6
Time Course of Induction and Degradation Process of the DIMs

In order to know if the DIMs are elements of immediate defense reactions such as coagulation and melanization (see Introduction) or involved in slower reactions, we investigated by MALDI MS their appearance in the hemolymph of immune-induced *Drosophila* at various time intervals post infection (from 1h up to 3 weeks). The most significant results were reported on Fig. 11.4 (for further details see Uttenweiler-Joseph et al. 1998). All the DIMs begin to be detectable in the hemolymph 6h after immunization and reach a maximum of intensity after 24h in a very reproducible fashion. In contrast, individual variations appear between flies after one day of immunization. However, as a general rule, the level of DIMs slowly decreases after 24h and they become undetectable 2 or 3 weeks after immune-challenge (Fig. 11.4).

Regarding the antimicrobial peptides (drosocin, metchnikowin and drosomycin), their stability in the hemolymph of infected flies could be determined precisely as their transcription profiles have already been reported (Charlet et al. 1996, Levashina et al. 1995, Fehlbaum et al. 1994). Briefly, Northern blot analysis has revealed a decrease in transcriptional activity of genes encoding the antimicrobial peptides 24–36h after challenge. However, MALDI-TOF MS analysis shows that drosocin-2S (DIM 11) disappears 2 weeks after the immune-challenge, whereas drosocin-1S (DIM 9) is undetectable only after 3 weeks. The stability course observed for metchnikowin (DIM 17) is identical to the one observed for DIM 9 while drosomycin (DIM 19) remains detectable for up to 3 weeks in most flies (Fig. 11.4). The *in vivo* stability of the antimicrobial peptides can be explained by two main structural features: (1) an overrepresentation in proline residues for drosocin and metchnikowin and (2) a compact three-dimensional structure for drosomycin (Landon et al. 1997). The changes in the relative ratio between the two glycoforms of drosocin suggest that a circulating exoglycosidase cleaves the distal sugar moiety (namely galactose).

MALDI-TOF MS analysis of complex biological mixtures at different time intervals after stimulation of a physiological process allows to directly study the life time (induction, persistence and disappearance) of a molecule that has been characterized by its molecular mass. If the transcription profile of the gene encoding a certain peptide is known, the in vivo stability of this peptide can be investigated by direct MALDI analysis of the biological sample.

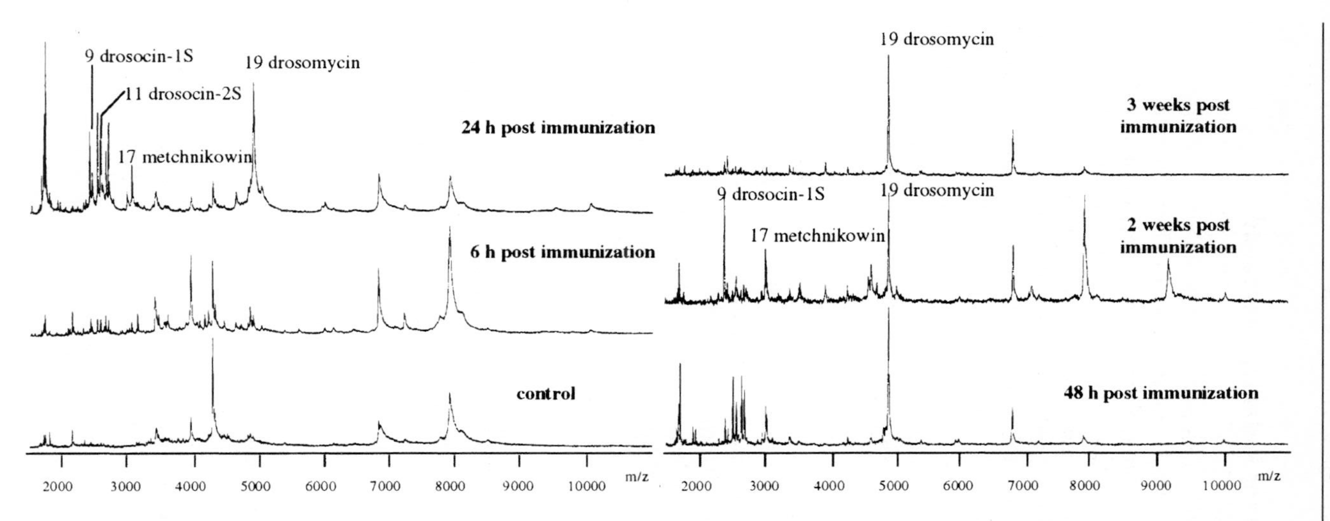

Fig. 11.4. Kinetics of induction and disappearance of the DIMs in the hemolymph of bacteria-challenged and control *Drosophila*. For the conditions of sample preparation, referred to legend of Fig. 11.2. The mass spectra were recorded at different time intervals: 6h, 24h, 2 and 3 weeks post infection. Complementary information is available in Uttenweiler-Joseph et al. 1998

7 Induction of the DIMs in Different Mutant Backgrounds

The presence of the DIMs can also be followed in the hemolymph of *Drosophila* carrying different mutations. As already discussed in the Introduction, recent studies based on genetic analysis of the humoral immune response in different mutant strains of *Drosophila* has led to the definition of two distinct pathways regulating the antimicrobial peptide gene expression (Lemaitre et al. 1995, Lemaitre et al. 1996). In adult *Drosophila*, the synthesis of the antifungal peptide drosomycin is predominantly under the control of the Toll pathway, which includes various genes also involved in the dorso-ventral patterning of *Drosophila* embryos (for review, see Morisato and Anderson 1995). The expression of the genes encoding the antibacterial peptides (drosocin, cecropin, attacin, defensin and diptericin) is either controlled by the imd pathway or by both pathways (for review, see Hoffmann and Reichhart 1997). The regulation of metchnikowin, which exhibits both antibacterial and antifungal activities, is unique compared to that of the other antimicrobial peptides, in that it appears to be regulated independently by the Toll and/or the imd pathways (Levashina et al. 1998).

MALDI-TOF MS analysis of the hemolymph of *Drosophila* carrying a mutation in genes involved in these immune pathways confirms at the peptide level the results obtained by Northern blot analysis and demonstrates that expression of the unidentified DIMs is also controlled by these two pathways. The MALDI-TOF MS spectra obtained from the hemolymph of unchallenged and immune-

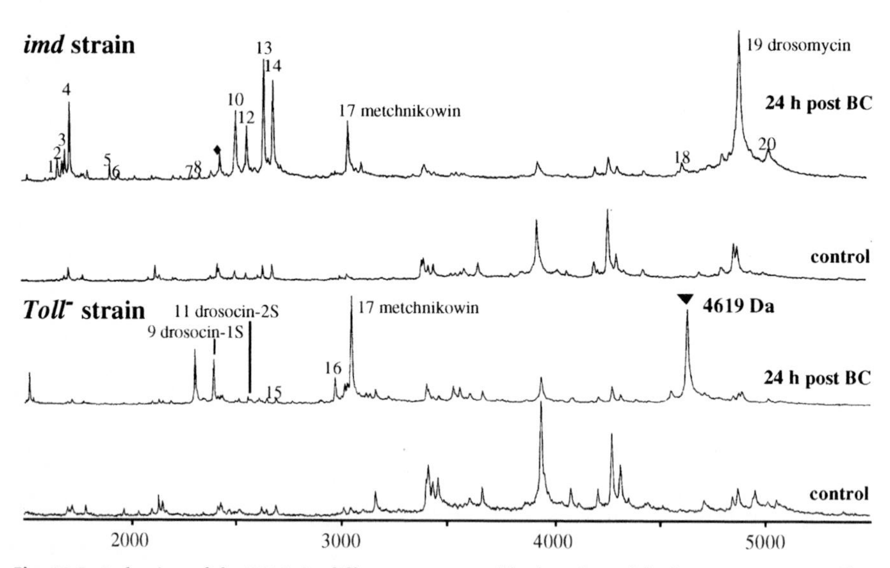

Fig. 11.5. Induction of the DIMs in different mutants with alteration of the immune response. The induction of the DIMs was monitored in *Drosophila* mutants carrying a loss-of-function mutation for the receptor *Toll* (*Toll⁻*) and in *Drosophila* with an imd mutation background (*imd*) before (control) and after a 24h bacterial-challenge (24h post BC). An additional DIM (molecular mass at 4619 Da) was detected in the *Toll⁻* mutant

challenged *Drosophila* carrying a loss-of-function mutation of either the *Toll* (*Toll*$^-$) or the *imd* gene are illustrated in Fig. 11.5. For the imd strain, all the DIMs are induced after bacterial challenge (BC), with the exception of the two glycoforms of drosocin (DIMs 9 and 11), as expected, and the DIMs 15 and 16 (Fig. 11.5, top). Therefore, it seems that, like drosocin, the genes encoding DIMs 15 and 16 are under the control of the imd pathway. In the case of *Drosophila* carrying a *Toll*-mutation, only six molecules are induced after the infection: the two glycoforms of drosocin (DIMs 9 and 11), metchnikowin (DIM 17), DIMs 15 and 16 and a molecule with a molecular mass of 4619 Da close to the molecular mass of DIM 18 (4625 Da) (Fig. 11.5, bottom). Interestingly, this molecule at 4619 Da, detected in the *Toll*-strain where no drosomycin is induced, has not been clearly detected in the wild-type *Drosophila* strain because of the proximity of the drosomycin mass signal at *m/z* 4890.8. This confirms that suppression effects affect the quality of the fingerprint of the *Drosophila* hemolymph. Furthermore, in double mutants (*imd/Toll*$^-$) no DIMs are detectable, suggesting that only the two distinct pathways already reported (imd and Toll) control the gene expression of the DIMs. We also analyzed the DIM expression in *Drosophila* mutants carrying a dominant mutation of the *Toll* gene (*Toll*^{10b}; where the *Toll* receptor is constitutively activated; Lemaitre et al. 1996) and compared the mass spectra obtained before and after immune challenge with the one recorded from wild-type flies (data not shown). In unchallenged *Toll* gain-of-function mutants (*Toll*^{10b}), most of the DIMs are constitutively expressed confirming our first assumption that DIMs are not products of bacterial origin (see above and for the mass spectra Uttenweiler-Joseph et al. 1998).

In conclusion, all the unknown DIMs are controlled by the Toll pathway, with the exception of DIMs 15–16 and DIM at 4619 Da which are controlled by the imd pathway. In addition, all these experiments on *Drosophila* mutants gave valuable information on the place of the DIMs in the general scheme of *Drosophila* immunity. However, the precise role of the DIMs remains to be clarified.

MALDI-TOF MS allows to follow, in a genetic model such as the fruitfly Drosophila melanogaster, *the presence of molecules in different contexts of mutations affecting a physiological process.*

8
Identification of the DIMs by Sequencing

In an attempt to understand the precise function of the DIMs in the *Drosophila* defense reactions, the induced molecules have to be structurally characterized. On such minute amounts of molecules available from hemolymph samples, methods for protein sequence analysis directly from complex mixtures can be done by tandem MS (MS/MS) but only with MALDI ionization. One MS/MS strategy involving ionization by MALDI is MALDI-Post Source Decay (PSD). However limitations exist such as difficulties in interpreting a mass spectrum resulting from this kind of fragmentation and also at the level of the mass precision for the selection of the ion for isolation, a prerequisite parameter in our complex sample (for review, see Yates 1998). For this reason, we developed a methodology based on the prepurification of the peptides by HPLC. However this methodology has the disadvantage of imposing a relative selectivity resulting in a possible loss of certain DIMs compared to the MALDI-MS mapping achieved on the crude hemolymph.

In a pilot experiment, the hemolymph of 20 unchallenged and 20 bacteria-challenged flies, collected in acidified water, were subjected separately to microbore (1 mm of internal diameter) reversed-phase HPLC separation with a linear gradient of acetonitrile (for details, see Uttenweiler-Joseph et al. 1998). All the HPLC fractions were screened by UV spectroscopy and MALDI-TOF MS in order to detect the DIMs initially characterized by their molecular mass in the fingerprint performed on the crude hemolymph. This strategy allowed to detect DIMs 1 to 19 in the fractions collected from the HPLC analysis of hemolymph from infected *Drosophila* (data not shown). The peptide nature of these DIMs was authenticated by subjecting the hemolymph sample to pronase, a treatment that totally abolished all UV and mass signals corresponding to DIMs 1–19 (data not shown). Following this first step of purification we were unable to detect DIMs 20 to 24.

Different methodologies can be applied to identify purified peptides/proteins depending on the sequence strategy which can be used: *in silico* or *de novo* sequencing. If the genome of the biological model is already known, *in silico* characterization (using mostly nucleotide sequences as source of information) can be performed even if only partial information on the amino acid sequence (obtained by ESI-MS/MS or Edman degradation) or tryptic fragment masses (MALDI peptide mapping) are available (for review, see Roepstorff 1997). If genome information is not available, *de novo* sequencing must be performed in order to get accurate peptide sequence information necessary to unambiguously identify the protein by complete sequencing or/and after cDNA cloning. *De novo* sequencing by MS is not yet straightforward but the new technology of quadrupole-TOF is promising (Shevchenko et al. 1997).

As the *Drosophila* genome is only expected to be fully sequenced by the turn of the century, we choose to sequence the DIMs by a more classical technique: Edman chemistry. However such an approach has different limitations to unambiguously define a primary amino acid sequence: (i) the peptide must have a non blocked N-terminus and (ii) the quantity of purified peptide must be sufficient (approximately 1–10 picomoles). After purification to homogeneity from an hemolymph sample collected from less than 140 immune-challenged flies, we fully sequenced DIM 4 and partially DIMs 1 and 2 that appeared to be isoforms. The full primary structure of DIM 2 was obtained by molecular cloning of cDNA prepared from RNA of immune-challenged adult *Drosophila*. The information obtained after cDNA sequencing established that DIM 2 does not result from the degradation of a larger protein (for more details on the molecular biology analysis see Uttenweiler-Joseph et al. 1998). Unfortunately, database searching using the amino acid sequences obtained on DIMs 1, 2 and 4 did not reveal any significant similarity with already known peptides that could be used to understand the precise role of these compounds in *Drosophila* immunity. Three additional DIMs (DIMs 10, 13 and 16) were purified to homogeneity but found to be resistant to Edman degradation.

The structural characterization of peptides detected by MALDI-TOF MS directly from a biological complex mixture is challenging without any purification step.

Although a purification step introduces selectivity, the isolation of a specific product can be followed with the help of its molecular mass. The structural characterization can be performed using different techniques depending on the amount of peptide available, the necessity to have a partial or a full sequence and the possibility to conduct molecular cloning experiments.

9 MALDI-TOF MS as a Tool to Investigate Inducible Tissue-Specific Expression of the Drosomycin Gene

We have also investigated the possible expression of *Drosophila* antimicrobial peptides in barrier epithelia using the antifungal peptide drosomycin as a model. Using a transgenic reporter system based on the green fluorescent protein (GFP from the jellyfish *Aequorea victori*, Chalfie et al. 1994), it has been shown that the reporter gene drosomycin-GFP was expressed in different epithelial tissues, for example in tracheal trunks of some experimentally uninfected *Drosophila* larvae (for more details, see Ferrandon et al. 1998). However, to be certain that the GFP expression visualized at the level of this epithelial tissue corresponded to the synthesis of endogenous drosomycin, fluorescent and non-fluorescent pieces of tracheal trunks were dissected under a fluorescence microscope and separately analyzed by MALDI-TOF MS.

To analyze these tissue portions, we tested several sample preparations, and the best results were obtained with the sample preparation established for the analysis of *Drosophila* hemolymph (for further details, see Ferrandon et al. 1998).

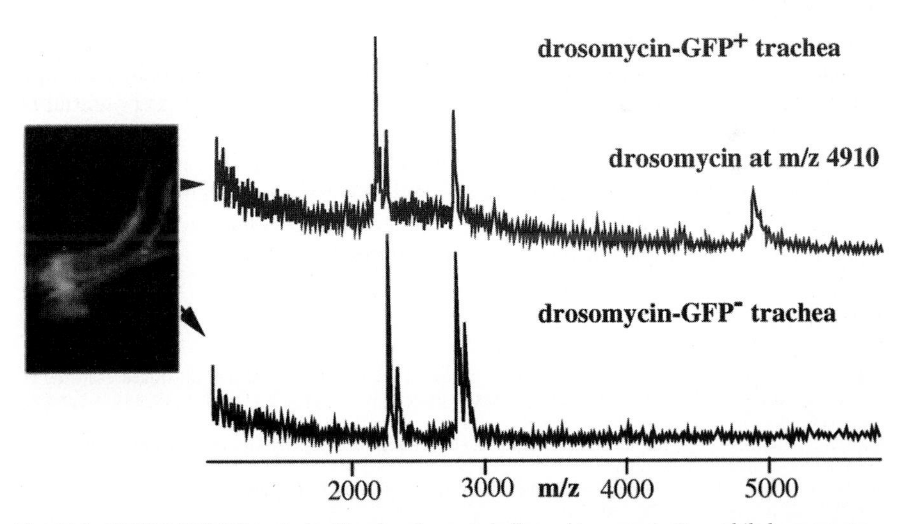

Fig. 11.6. MALDI-TOF MS analysis of trachea from unchallenged transgenic *Drosophila* larvae carrying a green fluorescent protein (GFP) reporter gene under the control of the drosomycin promoter. Fluorescent (drosomycin-GFP^{+}) and non-fluorescent (drosomycin-GFP^{-}) pieces of tracheal trunks were dissected under a fluorescence microscope and separately analyzed by MALDI-TOF MS. The sample preparation used is identical to the method established for the peptide mapping of the crude hemolymph sample (for experimental details see Ferrandon et al. 1998)

The MALDI mass spectra obtained on GFP^+ and GFP^- pieces of tracheal tissue were relatively simple and quite similar, with the exception of a peak at *m/z* 4910 only detectable in the mass spectrum from fluorescent trachea (Fig. 11.6). It should be noticed that these mass spectra were acquired at high laser power, that can result in (i) a loss of resolution and (ii) a shift of the measured mass to a higher mass than expected. The peak detected at *m/z* 4910 in the mass spectrum of GFP^+ trachea corresponds to endogenous drosomycin as controlled with recombinant drosomycin in the same conditions of analysis. This observation strongly suggests that the respiratory tract of *Drosophila* is not only acting as a physical barrier but as the potency to develop a local immune response in order to fight off a natural microorganism penetration.

The detection of an already characterized protein in a biological tissue can be rapidly and easily performed by MALDI-TOF MS without the need of any purification step. This technology can be applied to all biological models as already shown in other publications (see for example references in the Introduction).

10 Acknowledgments

The authors are indebted to Dr. Jules A. Hoffmann for his continuous interest in this study and to Dr. Alain Van Dorsselaer for providing the facilities in mass spectrometry.

11 Conclusion

The data obtained by differential MALDI-TOF MS analysis of crude hemolymph from bacteria-challenged and control *Drosophila* illustrate the great potential of this method for the detection of molecules induced in a complex biological process. However, this MALDI-TOF MS approach appeared to be more appropriate for the analysis of low molecular mass peptides. For the complementary analysis of higher molecular mass proteins, 2-D gel electrophoresis associated to mass spectrometry techniques is certainly more adapted.

References

Ashida M, Brey P (1997) Recent advances in research on the insect prophenoloxidase cascade. In Molecular mechanisms of immune responses in insects, P. T. Brey and D. Hultmark, eds.: Chapman & Hall: pp 135–172

Beavis R, Chait B (1990) Rapid, sensitive analysis of protein mixtures by mass spectrometry. Proc Natl Acad Sci USA 87: 6873–6877

Beavis R, Chait B (1996) Matrix-assisted laser desorption ionization mass spectrometry of proteins. Methods in Enzymol 270: 519–551

Braun A, Lemaitre B, Lanot R, Zachary D, Meister M (1997) *Drosophila* immunity: analysis of larval hemocytes by P-element-mediated enhancer trap. Genetics 147: 623–634

Bulet P, Dimarcq J, Hetru C, Lagueux M, Charlet M, Hegy G, Van Dorsselaer A, Hoffmann J (1993) A novel inducible antibacterial peptide of *Drosophila* carries an O-glycosylated substitution. J Biol Chem 268: 14893–14897

Burlingame P, Boyd R K, Gaskell S J (1998) Mass Spectrometry. Anal Chem 70: 647R-716R

Caprioli R, Farmer T, Gile J (1997) Molecular imaging of biological samples: localization of peptides and proteins using MALDI-TOF MS. Anal Chem 69: 4751–4760
Chalfie M, Tu Y, Euskirchen G, Ward W W, Prasher D C (1994) Green Fluorescent Protein as a marker for gene expression. Science 263: 802–805
Charlet M, Lagheux M, Reichhart JM, Hoffmann JA (1996) Cloning of the gene encoding the antibacterial peptide drosocin involved in *Drosophila* immunity. Expression studies during the immune response. Eur J Biochem 241: 699–706
Chosa N, Fukumitsu T, Fujimoto K, Ohnishi E (1997) Activation of prophenoloxidase A1 by an activating enzyme in *Drosophila melanogaster*. Insect Biochem Mol Biol 27: 61–68
Claydon M, Davey S, Edwards-Jones V, Gordon D (1996) The rapid identification of intact microorganisms using mass spectrometry. Nat Biotech 14: 1584–1586
Cohen S, Chait B (1996) Influence of matrix solution conditions on the MALDI-MS analysis of peptides and proteins. Anal Chem 68: 31–37
Dimarcq J L, Hoffmann D, Meister M, Bulet P, Lanot R, Reichhart J M, Hoffmann J A (1994) Characterization and transcriptional profiles of a *Drosophila* gene encoding an insect defensin. A study in insect immunity. Eur J Biochem 221: 201–209
Easterling M, Colangelo C, Scott R, Amster I (1998) Monitoring protein expression in whole bacterial cells with MALDI time-of-flight mass spectrometry. Anal Chem 70: 2704–2709
Ezekowitz R, Hoffmann J (1998) Innate immunity. The blossoming of innate immunity. Curr Opin Immunol 10: 9–11
Fearon D, Locksley R (1996) The instructive role of innate immunity in the acquired immune response. Science 272: 50–53
Fehlbaum P, Bulet P, Michaut L, Lagueux M, Broekaert W, Hetru C, Hoffmann J (1994) Insect immunity. Septic injury of *Drosophila* induces the synthesis of a potent antifungal peptide with sequence homology to plant antifungal peptides. J Biol Chem 269: 33159–33163
Ferrandon D, Jung A C, Criqui M, Lemaitre B, Uttenweiler-Joseph S, Michaut L, Reichhart J, Hoffmann J A (1998) A drosomycin-GFP reporter transgene reveals a local immune response in *Drosophila* that is not dependent on the Toll pathway. EMBO J 17: 1217–1227
Franc N C, Dimarcq J L, Lagueux M, Hoffmann J A, Ezekowitz A B (1996) Croquemort, a novel *Drosophila* hemocyte/macrophage receptor that recognizes apoptotic cells. Immunity 4: 434–444
Fujimoto K, Masuda K, Asada N, Ohnishi E (1993) Purification and characterization of prophenoloxidases from pupae of *Drosophila melanogaster*. J Biochem 113: 285–291
Hoffmann J, Reichhart J (1997) *Drosophila* immunity. Trends in Cell Biol 7: 309–316
Jensen O N, Mortensen P, Vorm O, Mann N (1997) Automation of matrix assisted laser desorption/ionization-mass spectrometry using fuzzy logic feedback control. Anal Chem 69: 1706–1714
Jimenez C, van Veelen P, Li K, Wildering W, Geraerts W, Tjaden U, van der Greef J (1994) Neuropeptide expression and processing as revealed by direct matrix-assisted laser desorption/ionization mass spectrometry of single neurons. J Neurochem 62: 404–407
Kratzer R, Eckerskorn C, Karas M, Lottspeich F (1998) Suppression effects in enzymatic peptide ladder sequencing using ultraviolet-matrix assisted laser desorption/ionization mass spectrometry. Electrophoresis 19: 1910–1919
Kussmann M, Nordhoff E, Rahbek-Nielsen H, Haebel S, Rossel-Larsen M, Jakobsen L, Gobom J, Mirgorodskaya E, Kroll-Kristensen A, Palm L, Roepstorff P (1997) Matrix-assisted laser desorption/ionization mass spectrometry sample preparation techniques designed for various peptides and protein analytes. J Mass Spectrom 32: 593–601
Landon C, Sodano P, Hetru C, Hoffmann J, Ptak M (1997) Solution structure of drosomycin, the first inducible antifungal protein from insects. Protein Sci 6: 1878–1884
Lemaitre B, Kromer-Metzger E, Michaut L, Nicolas E, Meister M, Georgel P, Reichhart J, Hoffmann J (1995) A recessive mutation, *immune deficiency (imd)*, defines two distinct control pathways in the *Drosophila* host defence. Proc Natl Acad Sci USA 92: 9465–9469
Lemaitre B, Nicolas E, Michaut L, Reichhart J, Hoffmann J (1996) The dorsoventral regulatory gene cassette *spätzle/Toll/cactus* controls the potent antifungal response in *Drosophila* adults. Cell 86: 973–983
Levashina E, Ohresser S, Bulet P, Reichhart J, Hetru C, Hoffmann J (1995) Metchnikowin, a novel immune-inducible proline-rich peptide from *Drosophila* with antibacterial and antifungal properties. Eur J Biochem 233: 694–700
Levashina E A, Ohresser S, Lemaitre B, Imler J L (1998) Two distinct pathways can control expression of the gene encoding the *Drosophila* antimicrobial peptide metchnikowin. J Mol Biol 278: 515–527
Li K W, Hoek R M, Smith F, Jimenez C R, van der Schors R C, van Veelen P A, Chen S, van der Greef J, Parish D C, Benjamin P R, Geraerts W P M (1994) Direct peptide profiling by mass spectrometry of single identified neurons reveals complex neuropeptide-processing pattern. J Biol Chem 269: 30288–30292

Liu Y, Bai J, Liang X, Lubman D, Venta P (1995) Use of a nitrocellulose film substrate in matrix-assisted laser desorption/ionization time-of-flight mass spectrometry for DNA mapping and screening. Anal Chem 67: 3482–3490

Meister M, Lemaitre B, Hoffmann J A (1997) Antimicrobial peptide defense in *Drosophila*. BioEssays 19: 1019–1026

Morisato D, Anderson K (1995) Signaling pathways that establish the dorsal-ventral pattern of the *Drosophila* embryo. Annu Rev Genetics 29: 371–379

Preston L, Murray K, Russell D (1993) Reproducibility and quantitation of matrix-assisted laser desorption ionization mass spectrometry: effects of nitrocellulose on peptide ion yields. Biol Mass Spectrom 22: 544–550

Redeker V, Toullec J, Vinh J, Rossier J, Soyez D (1998). Combination of peptide profiling by matrix-assisted laser desorption/ionization time-of-flight mass spectrometry and immunodetection on single glands or cells. Anal Chem 70: 1805–1811

Rizki T, Rizki R, (1984) The cellular defense system of *Drosophila melanogaster*. In Insect ultrastructure, R. King and H. Akai, eds.: Plenum Publishing Corporation: pp 579–604

Roepstorff P (1997) Mass spectrometry in protein studies from genome to function. Curr Opin Biotechnol 8: 6–13

Roepstorff P, Larsen M, Rahbek-Nielsen H, Nordhoff E (1998) Sample preparation methods for matrix-assisted laser desorption/ionization mass spectrometry of peptides, proteins, and nucleic acids. In cell Biology: a laboratory Handbook, J E Celis eds.: San Diego, Academic Press: pp 556–565

Samakovlis C, Asling B, Boman H, Gateff E, Hultmark D (1992) *In vitro* induction of cecropin genes: an immune response in a *Drosophila* blood cell line. Biochem Biophys Res Commun 188: 1169–1175

Shevchenko A, Chernushevich I, Ens W, Standing K G, Thomson B, Wilm M, Mann M (1997) Rapid *de novo* peptide sequencing by a combination of nanoelectrospray, isotopic labeling and a quadrupole/time-of-flight mass spectrometer. Rapid Commun Mass Spectrom 11: 1015–1024

Stahl B, Linos A, Karas M, Hillenkamp F, Steup M (1997) Analysis of fructans from higher plants by matrix-assisted laser desorption/ionization mass spectrometry. Anal Biochem 246: 195–204

Uttenweiler-Joseph S, Moniatte M, Lagueux M, Van Dorsselaer A, Hoffmann J, Bulet P (1998) Differential display of peptides induced during the immune response of *Drosophila*: a matrix-assisted laser desorption ionization time-of-flight mass spectrometry study. Proc Natl Acad Sci USA 95: 11342–11347

Uttenweiler-Joseph S, Moniatte M, Lambert J, Van Dorsselaer A, Bulet P (1997) Matrix-assisted laser desorption ionization time-of-flight mass spectrometry approach to identify the origin of the glycan heterogeneity of diptericin, an *O*-glycosylated antibacterial peptide from insects. Anal Biochem 247: 366–375

van Adrichem J, Börnsen K, Conzelmann H, Gass M, Eppenberger H, Kresbach G, Ehrat M, Leist C (1998) Investigation of protein patterns in mammalian cells and culture supernatants by matrix-assisted laser desorption/ionization mass spectrometry. Anal Chem 70: 923–930

Yates J R (1998) Mass spectrometry and the age of the proteome. J Mass Spectrom 33: 1–19

CHAPTER 12

A New Method for the Isolation of the C-Terminal Peptide of Proteins by the Combination of Selective Blocking of Proteolytic Peptides and Cation Exchange Chromatography

L. J. González[1], E. Torres[1], Y. García[1], L. H. Betancourt[1], G. Moya[2], V. Huerta, V. Besada[1] and G. Padrón[1]

1 Introduction

The sequence information on the *C*-terminal end of proteins is highly appreciated. This information is used for the quality control of recombinant proteins to ensure the correct transcription of the gen of interest and also provide an exact idea about the C-terminal processing of the protein. In the case of unknown proteins, this information is very useful because it allows designing oligonucleotides for the selective isolation of encoding gen and their further DNA sequencing.

The *C*-terminal sequencing has followed two separate trends, one of them has been the development of methods that allow the direct sequencing of the *C*-terminal end of proteins, such as the development of the *C*-terminal sequencers, the use of carboxypeptidases (CPases) and a chemical hydrolysis that mimics the CPase degradation.

The protein chemists have devoted great efforts to the development of a *C*-terminal sequencer (Boyd et al. 1991, Bailey and Shively 1990, Shenoy et al. 1993, Miller and Baley 1995) and considerable progress has been made in this direction, in fact, there is a C-terminal sequencer commercially available today. However, the performances of the *C*-terminal sequencers in terms of sensitivity, and initial and repetitive yields are still inferior in comparison with the *N*-terminal sequencers, which use the well-established Edman chemistry.

The use of CPases to release the amino acids from the C-terminus of the intact proteins (Nguyen et al. 1995, Thiede et al. 1995) has an important limitation: Cpases are substrate-dependent and some proteins are resistant to CPase digestion.

The use of perfluorated organic acids and its corresponding anhydrides to yield CPase mimetic degradation (Tsuguita et al. 1992, Takamoto et al. 1995) works mainly in short peptides because, in the case of intact proteins, internal degradation in several parts of the proteins can also take place.

These are the reasons why another trend has been followed in developing strategies to isolate the C-terminal peptide of the protein from all the proteolytic peptides, and subsequently perform its sequencing by well-established methods such as Edman degradation or Mass Spectrometry (MS).

[1] Division of Physical-Chemistry and
[2] Quality Control Division of the Center for Genetic Engineering and Biotechnology. P.O.Box 6162. Havana Cuba.

The first method reported for this purpose was the diagonal electrophoresis (Naughton and Hagopian 1962). This involves the separation of the tryptic peptides on a paper electrophoresis, their treatment with carboxypeptidase B (CPB) and their further separation in a second dimension. The *C*-terminal peptide is not affected by the CPB treatment unless the protein has a basic amino acid at the *C*-terminal (Lys/Arg). Therefore, the *C*-terminal peptide should be the only species that does not change its mobility in either dimension of the electrophoresis, and will be detected just in the diagonal. It should be noticed that the success of this methodology depends on the high purity of the CPB preparation, because contamination with other CPases could also degrade the C-terminal peptide and affect its mobility in the second dimension.

Another approach is based on the comparative rp-HPLC peptide mapping between the Cpase-treated and the native protein (Isobe K et al.1986). The *C*-terminal peptide should be the only fraction to shift its retention time when both chromatograms are compared. However, this method has limitations similar to the strategies that use Cpases and on the other hand, it is difficult to estimate the extent of the C-terminal degradation when the protein has a C-terminal ragged end.

Other authors (Rose et al. 1988) have used the stable ^{18}O-labeling of proteolytic peptides to identify the *C*-terminal peptide of the protein. All internal peptides incorporate ^{18}O at their *C*-termini during the proteolytic digestion, while the C-terminal peptide doesn't. This difference can be observed in the shift of the signals when the mass spectra of the fractions obtained in presence and in absence of $H_2{}^{18}O$ are compared. One limitation of this methodology is that the C-terminal peptide of the protein can also be labeled if the protease have affinity for binding to the C-terminal amino acid of the protein, and this is much more probable with longer digestion times.

This strategy has additional advantages for the sequencing of all internal peptides (^{18}O-labeled species), since the *N*- and the *C*-terminal ions can be easily differentiated in the daughter ion spectra (Takao et al. 1991, Takao et al. 1993). However the above mentioned advantages cannot be utilized to obtain an easier and more reliable sequencing by mass spectrometry of the *C*-terminal peptide because it is a non-labeled species.

Kumazaki *et.al* in 1986 proposed the affinity chromatography for the isolation of the *C*-terminal peptide. The immobilized anhydrotrypsin binds all the tryptic peptides (end in Lys or Arg) except the C-terminal peptide. In this strategy, the internal peptides originated by non-specific cleavages of trypsin will also be collected in the same fraction where the *C*-terminal peptide appears and if the protein ends in lysine or arginine, the C-terminal peptide would also be retained on the column.

Another strategy has been also developed for the isolation of the *C*-terminal peptide after the cyanogen bromide treatment of the protein (Murphy and Fenselau, 1995). It comprises the determination of the increment in mass of all cyanogen bromide peptides after methylation (CH_3OH/HCl). The *C*-terminal peptide should be the only peptide that increases its molecular weight by a multiple of 14 Da because it has a free carboxyl terminal group, while other peptides are transformed into homoserine lactone. This strategy has the disadvantage that it is cyanogen bromide dependent, and it requires special handling. On the other hand, peptides originated by the non-specific cleavages of the cyanogen bromide, or others that could be originated by the drastic acidic conditions used in this chemical cleavage, will exhibit the same behavior as the C-terminal peptide during the methylation analysis.

The cation exchange chromatography could be used to isolate the *C*-terminal peptide in the complex mixture generated by a proteolytic digestion. However, it is not easy to find unique and reproducible conditions that would allow the isolation of the C-terminal peptide in all proteins. In 1993, Gorman and Shiell published the use of cation exchange chromatography to successfully isolate, simultaneously, the *C*- and the *N*-terminal peptides of *N*-terminal blocked proteins.

In this paper, we report on a new strategy for the selective isolation of the *C*-terminal peptide of proteins by using the combination of the highly specific chemical derivatization of the *N*-terminal end of the tryptic peptides (Wetzel et al. 1990, Stults et al. 1993) and cation exchange chromatography. By this procedure, the tryptic peptides are transformed at acid pH, into a mixture of single-charge and neutral peptides, which can be easily separated by cation exchange chromatography.

2 Procedures

2.1 Selective Blocking of NH_2-Terminal Groups

The peptide mixture was dissolved in buffer MES (300 mM, pH = 6.0) containing Glycine 60 mM to reach a final concentration of 130–150 μM.

Fifty-fold molar excess of the organic acid anhydride (dissolved in dry THF) with respect to the amount of amino terminal groups generated during the proteolytic digestion, was added. The reaction time for acetic anhydride was only 3 minutes and for the other anhydrides, 10 to 15 minutes were required. The reaction was maintained at 0 °C using an ice bath with an efficient stirring inside the reaction vial.

2.2 Desalting of N-Terminal Blocked Peptides

The peptide mixture was desalted using a mini-column (2 × 0.2 cm) packed with C4 widepore material (VYDAC). A linear gradient from 0 to 80 % of CH_3CN (0.05 % of TFA) in 10 minutes was used for desalting the derivatized peptides. The pool of blocked peptides was detected by monitoring the absorbance at 226 nm.

2.3
Cation Exchange Chromatography

The cation exchange chromatography was performed on a centricon tube with 0.1 μm pore size (MILLIPORE), containing approximately 50–100 μL of the anionic resin EMD-650 (S) SO^{3-} (MERCK) or DEAE Sepharose Fast-Flow (Pharmacia). The N-terminal blocked peptides were dissolved in few μL of the same equilibrium buffer (H_2O/TFA 0.05 % containing 0.5 % of octyl glucoside) and applied onto the resin. A short spin (10 secs.) was done in a small centrifuge (~ 6000 rpm) and the non-retained fraction was collected and analyzed by rp-HPLC. The other peptides retained into the resin were eluted after passing a solution of H_2O/TFA 0.05 % containing NaCl (1 mol/L).

3
Results and Discussion

The strategy proposed for the isolation of the *C*-terminal peptide of proteins is shown in Fig. 12.1. In the first step the protein is digested with trypsin and all proteolytic peptides end with basic amino acids (lysine or arginine) except the *C*-terminal peptide.

After this, by using any of the three anhydrides shown in Fig. 12.1 (acetic, succinic or maleic) all the tryptic peptides are selectively blocked at their amino terminal group, leaving the ε amine group of lysine residues unmodified. The *N*-terminal blocked peptides are desalted by using a small rp-C4 column and, at acid pH, they are transformed into a mixture of single-charge and neutral peptides. When they are analyzed by cation exchange chromatography, the charged peptides are retained on the column. However, once transformed as neutral species the *C*-terminal peptide will elute in the non-retained fraction.

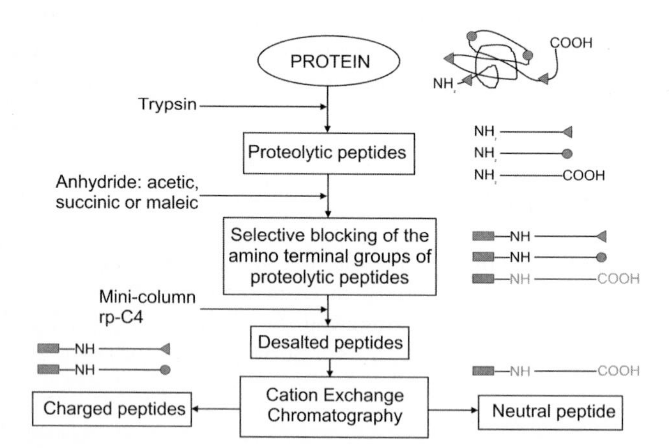

Fig. 12.1. General strategy for the isolation of the C-terminal peptide. The triangles and circles represent the basic amino acids (lysine and arginine) in the hypothetical protein. The rectangles represent the blocking groups (acetyl, maleyl or succinyl) introduced during the selective blocking of the N-terminal end of the proteolytic peptides

In 1990, Wetzel et al. reported the best conditions to achieve the highly selective blocking of the amino terminal groups of peptides. The selectivity of this reaction is based on the difference between the pK of the ε amino groups of the side chain of lysine and the amino terminal groups of the peptides. At pH 6.0, due to its higher basicity, the ε amino groups of lysine residues are predominantly protonated, while the amino terminal groups are not. This is the reason why the reaction occurs preferably through the amino terminal group of peptides.

We selected the TAB-9 protein as a model to evaluate the efficacy of the methodology. It is a 16 kDa chimeric protein containing six lysine and 15 arginine residues. In the tryptic map of this protein, the fraction labeled with an arrow contains the *C*-terminal peptide (Fig. 12.2A). The proteolytic peptides were selectively acetylated at their amino terminal group in three minutes and were desalted prior the cation exchange chromatography with a mini-column packed with reverse phase (C4). The *N*-terminal blocked peptides were eluted with a fast gradient of acetonitrile and the pool of blocked peptides was collected and concentrated (data not shown).

In the cation exchange chromatography a non-retained fraction was obtained. After increasing the ionic strength of the buffer, the remaining peptides retained

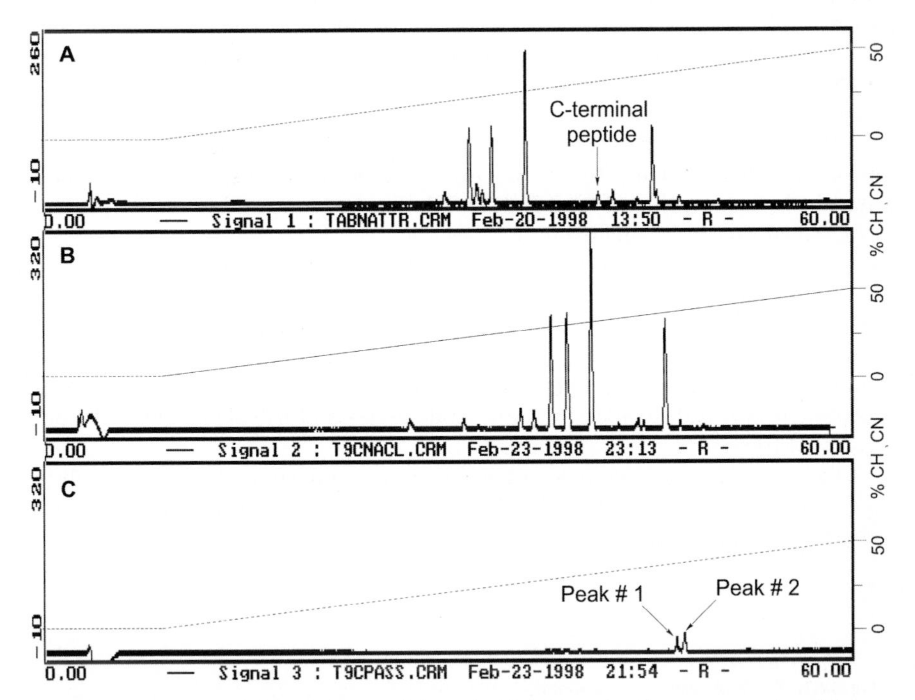

Fig. 12.2. (A) RP-HPLC of the tryptic map of the native TAB-9 protein. Chromatograms B and C correspond to the rp-HPLC profiles of the retained and non-retained fractions obtained by cation exchange chromatography, respectively. In each chromatogram the gradient is indicated by broken lines

on the column were eluted. Both retained and the non-retained fractions were analyzed by rp-HPLC (Fig. 12.2B and 12.2C).

By comparing the chromatograms of the tryptic map of the native protein (Fig. 12.2A) with the one obtained for the retained fraction (Fig. 12.2B), it is evident that the reaction has been completed, since all the *N*-terminal blocked peptides (Fig. 12.2B) have increased their retention time. Also, in the chromatogram of the non-retained fraction (Fig. 12.2C), two peaks appeared with higher retention time than that corresponding to the *C*-terminal peptide.

The results of this methodology were not affected by adding different quantities of acetic acid anhydride, at least within a reasonable range of concentrations (600–1500 μM), because the rp-HPLC profiles of the non-retained fractions were very similar (data not shown). Amino acid analysis of peak 1 and 2 (Fig. 12.2C) showed ten fold higher amounts of peak # 2 as well as the absence of lysine residues (data not shown). Therefore, non-specific acetylation at the ε amino group of the lysine residues had not occurred.

In the FAB mass spectra of peak # 1 (Fig. 12.3A) a signal appeared, matching very well with peptides corresponding to five different regions in the whole protein. The sequencing by CID-linked scan (Fig. 12.3B) revealed that this is an internal peptide (Ac-Gln^{69}-Tyr^{80}), originated by a non-specific cleavage of the trypsin. It should be noticed that the non-specific cleavages of the trypsin should be minimized, since those peptides will be also transformed as neutral species after acetylation, and will co-elute with the C-terminal peptide in the non-retained fraction. Therefore, trypsin sequencing grade and the gentlest digestions should be used to avoid ambiguous results when an unknown protein is analyzed.

In the FAB mass spectrum of peak # 2 (Fig. 12.3C), a signal at m/z 592.2 was observed which matches very well with the theoretical value expected for the C-terminal pentapeptide of TAB-9 (AFVTI). The sequencing by CID-linked scan confirmed this assignment (Fig. 12.3D).

However, it is clear that the presence of a basic amino acid at the *C*-terminus of the protein could be a limitation for the methodology, since the *C*-terminal peptide could also be retained on the column in the cation exchange chromatography. This disadvantage can be overcome if the basic residues (lysine or arginine), are blocked before the tryptic digestion. By this procedure, the *C*-terminal peptide is also transformed into a neutral species and will also elute in the non-retained fraction.

The modified methodology was evaluated with the streptokinase (SKr). It is a 47 kDa recombinant protein with 32 lysine residues, one of them located just at their *C*-terminal end.

Fig. 12.3. The FAB mass spectra A and C were obtained from the peaks 1 and 2 shown in the Fig. 12.2C. The CID-linked scan spectra shown in B and D correspond to the main molecular ions (1289.5 and 592.2 Da) observed in the FAB mass spectra shown in A and C, respectively. The peaks labeled with asterisks in the spectra A and C are signals corresponding to the matrix (glycerol). Ac- represents a blocking acetyl group at the N-terminal end of the peptides

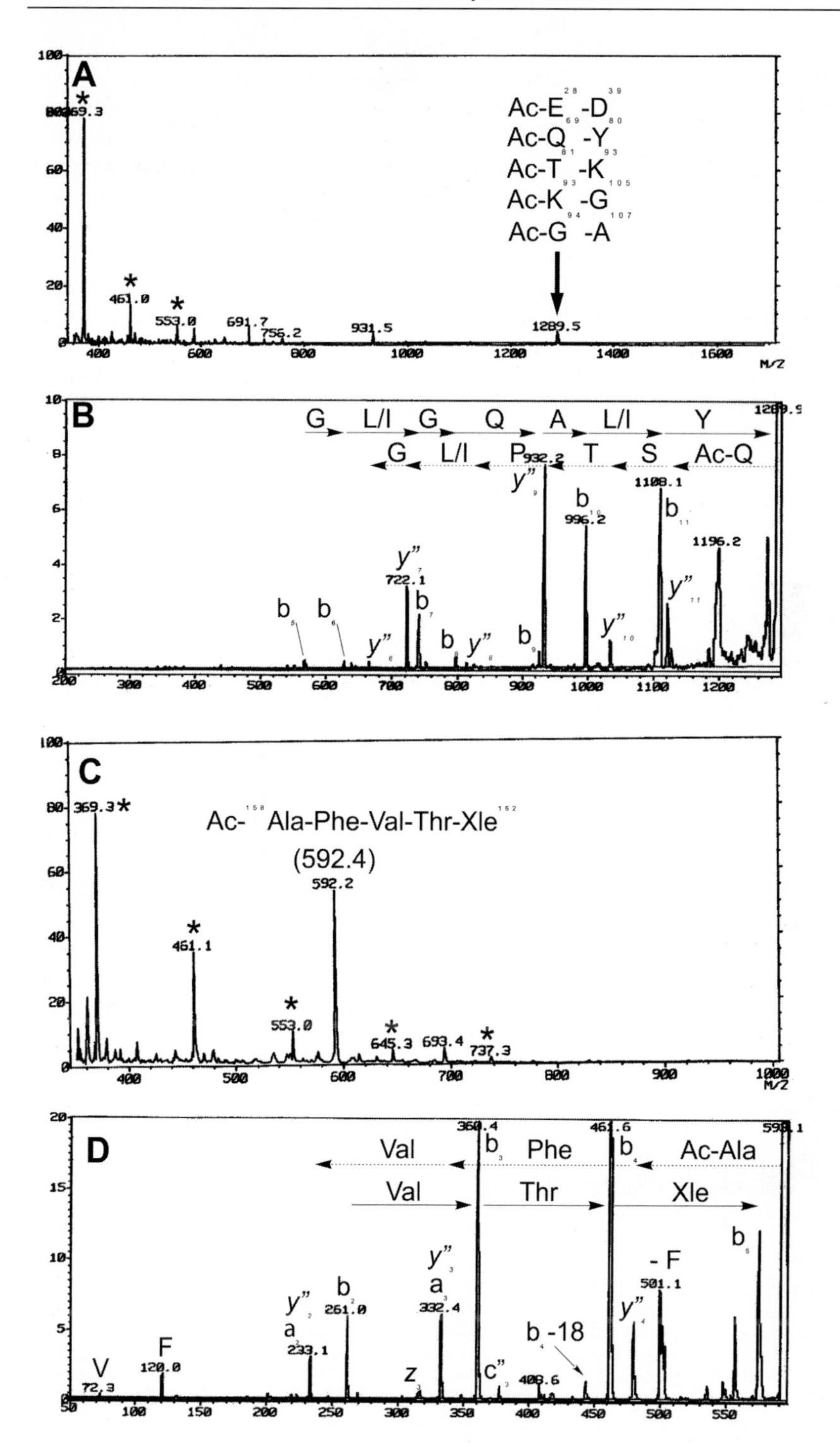
A
B
C
D
Ac-Ala-Phe-Val-Thr-Xle
(592.4)
Val
Phe
Ac-Ala
Thr
Xle
M/Z

The intact protein was succinylated and its molecular weight increased in comparison with the native protein by approximately 3 kDa (SDS PAGE, Fig. 12.4). The protein was digested with trypsin and the proteolytic peptides were selectively succinylated. The rp-HPLC of the retained fraction obtained in the cation exchange chromatography showed several peaks (Fig. 12.4A). However, in the chromatogram of the non-retained fraction, a major peak was obtained (Fig. 12.4B). It was analyzed by FAB mass spectrometry (Fig. 12.4C) and the molecular mass obtained experimentally was 1 Da higher than expected for the *C*-terminal peptide of streptokinase, taking into account the presence of two succinyl lysine residues, one of them located at the C-terminal end. This result suggests the occurrence of deamidation in one of the two Asn residues contained within the peptide sequence. Furthermore, the amino acid analysis of this peptide (data not shown) confirms that the *C*-terminal peptide of the SKr was successfully isolated.

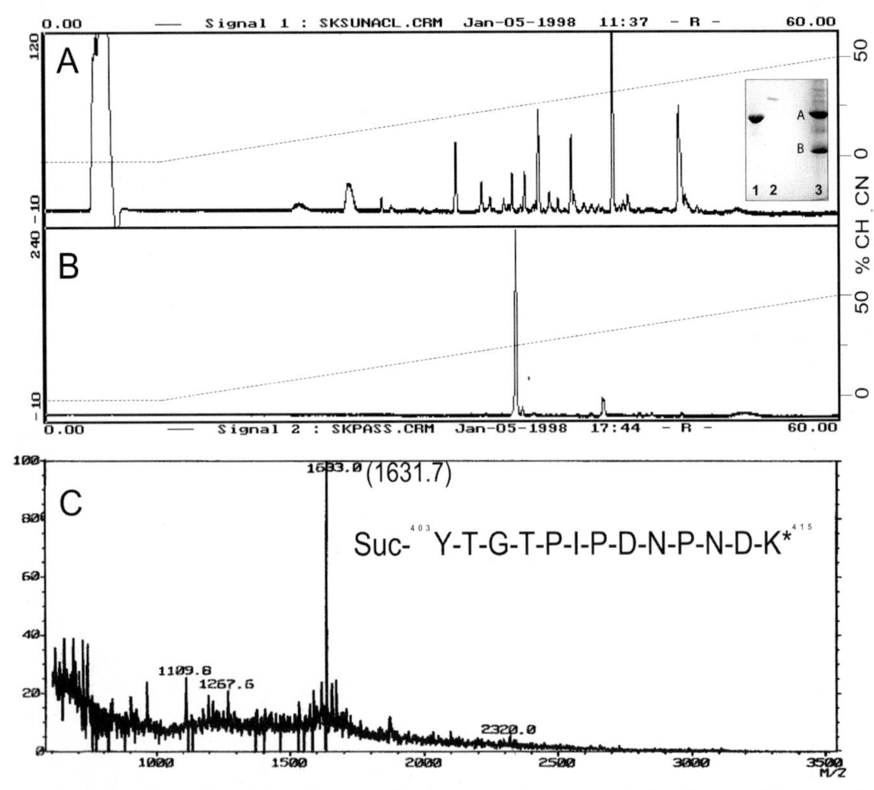

Fig. 12.4. The SDS-PAGE shows the recombinant streptokinase, succinylated streptokinase and molecular weight markers [(B) 32 and (A) 43 kDa] in lanes 1, 2 and 3, respectively. The chromatograms A and B show a comparison between the rp-HPLC profiles for the non-retained and the retained fractions obtained in the cation exchange chromatography during the isolation of the C-terminal peptide of the SKr. The FAB mass spectrum shown in C corresponds to the major fraction obtained in chromatogram B. The theoretical mass value of the C-terminal peptide is indicated in parenthesis and Suc- indicates a succinyl residue at the N-terminal end. K* indicates a succinyl lysine residue

Upon reaching this point, however it is clear that this methodology is not compatible with the Edman sequencing because the *C*-terminal peptide is obtained as an N-terminal blocked species. In 1969, Butler et al. reported that the ε amino group of lysine residues in polypeptide chains could be reversibly blocked when maleic anhydride is used. Since the chemical nature of the ε amino group of lysine and the amino terminal group of peptides are very close, in similar conditions, the maleyl group at the amino terminal group of the C-terminal peptide should be also released. By this procedure, this strategy could be compatible with the automatic sequencing.

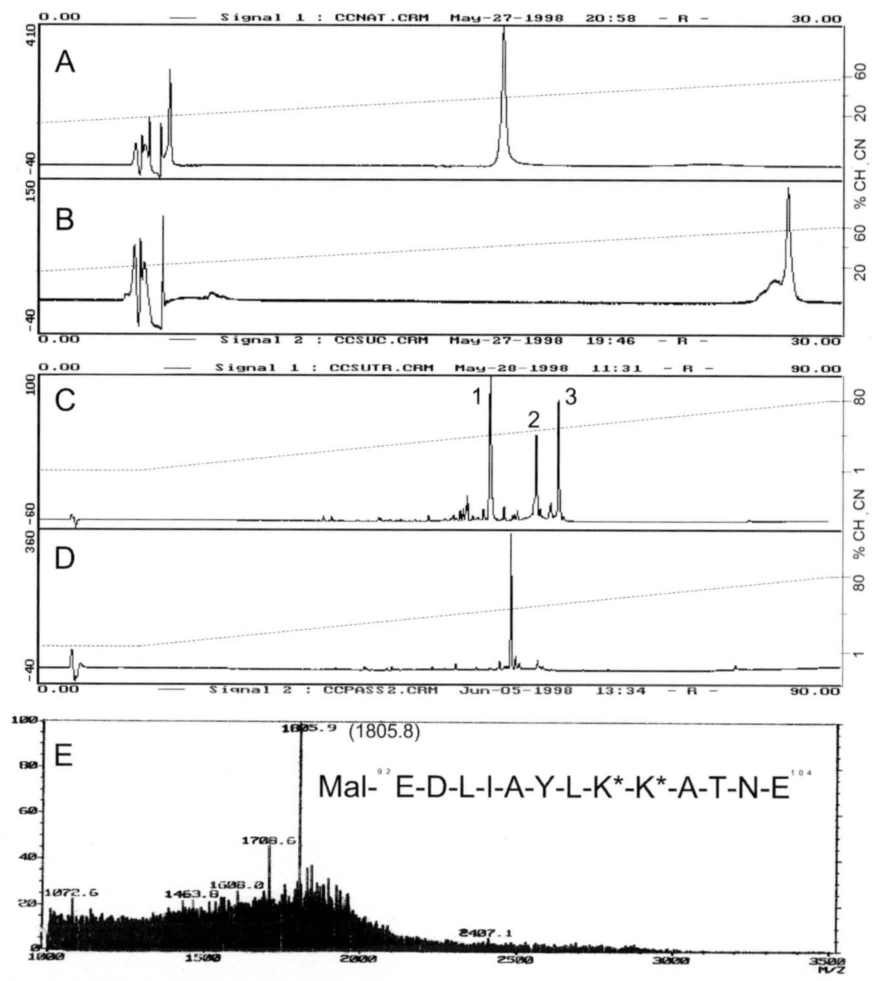

Fig. 12.5. Chromatograms A and B show a comparison between the rp-HPLC profile of the native and the succinylated cytochrome-C. The chromatograms C and D show the tryptic map of the succinylated Cytochrome C and the chromatogram obtained for the non retained fraction, respectively. E is the FAB mass spectra of the major fraction obtained in D. Mal- and K* indicate the maleyl group at the N-terminal end of the peptide and two succinyl residues within the sequence, respectively

This compatibility was evaluated with the cytochrome C. It is a small protein (11.7 kDa) with a great number of lysine residues (nineteen) to be modified. A comparison between the rp-HPLC profiles of the native (Fig. 12.5A) and succinylated (Fig. 12.5B) proteins showed that the reaction has occurred due to the increase of retention time of the succinylated protein. However the lysine residues seem to be partially blocked due to the presence of a shoulder in the main peak observed in Fig. 5B. This result was also confirmed by SDS-PAGE when a band of the modified protein migrated very close to the native one (data not shown).

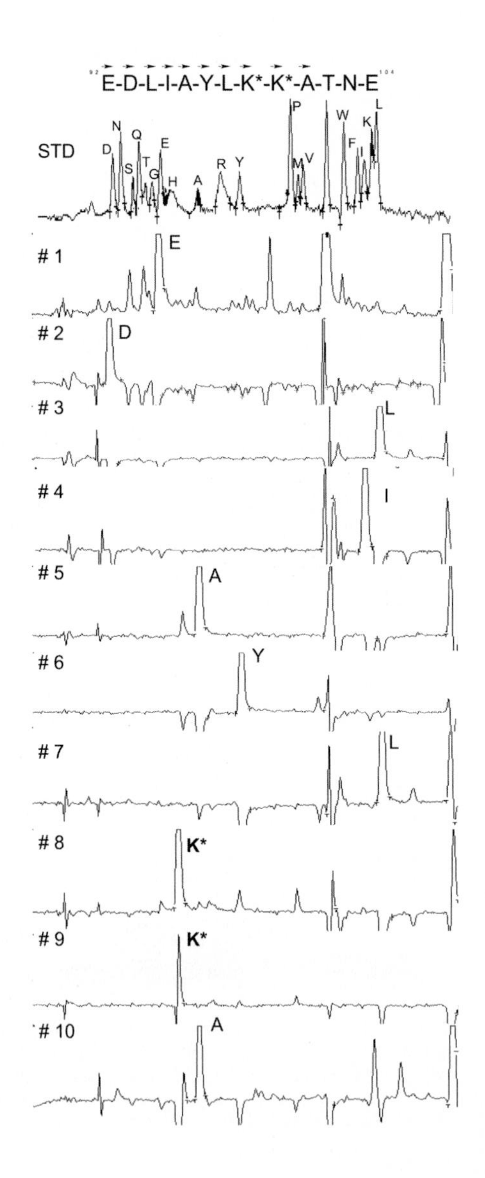

Fig. 12.6. The first ten cycles obtained in the automatic N-terminal sequencing of the modified C-terminal peptide of cytochrome-C. The identified amino acid is indicated in each cycle. In the cycles 8 and 9, the K* indicates a succinyl lysine residue

If all lysine residues are completely blocked, only three tryptic peptides should be expected. In fact, three major peaks were obtained (Fig. 12.5C). However, it is also evident that other minor peaks have also been generated by the cleavage of trypsin in lysine residues that were not completely blocked.

In the selective blocking of proteolytic peptides, the maleic anhydride was used. The rp-HPLC of the non-retained fraction showed a major peak (Fig. 12.5D). When it was analyzed by FAB mass spectrometry (Fig. 12.5E), a signal appeared at 1805.9 Da. It matches very well with the theoretical mass value expected (1805.8) for the C-terminal peptide of Cytochrome C, taking into account the presence of two succinyl lysine residues and a maleyl residue at the N-terminus end. This result indicates that incomplete blocking of the basic residues in the whole protein is not a limitation to successful isolation of the C-terminal peptide of the analyzed proteins.

The previous maleyl-blocked peptide was treated with a mixture of acetic and formic acids and analyzed by automatic Edman degradation (Fig. 12.6). The first ten amino acids from the *N*-terminus were clearly determined, including the succinyl-lysine residues at the cycles 8 and 9 that eluted between His and Ala in the PTH amino acid standard. This result confirms that a free amino terminal group was generated after the acid treatment.

The FAB mass spectrum of the acid-treated peptide showed a signal at 1707.9 Da that matched very well with the theoretical mass value expected (1707.8 Da), taking into account the presence of two succinyl-lysine residues and a free amino terminal group (data not shown). Since no signal of the blocked peptide was observed in the mass spectra, it indicates that a quantitative de-blocking of the amino terminal group was obtained. Therefore, our methodology is also compatible with the Edman sequencing particularly when maleyl anhydride is used in the selective and reversible blocking of the proteolytic peptides.

This methodology was successfully applied to three other proteins myoglobin, mutated IL-2 and bovine mucorpepsin (Fig. 12.7). In the chromatogram of the retained fraction several peaks were obtained however in the chromatogram of the non-retained fraction a major peak was obtained and it always contained the C-terminal peptide. Table 12.1 summarized the FAB-MS analysis of the major fraction obtained in the rp-HPLC analysis of the non-retained fraction. Amino acid analysis of the C-terminal peptides was performed and good agreement was found between the theoretical and experimental amino acid composition (data not shown).

4 Conclusions

This strategy is very easy to implement in any protein chemistry lab. The derivatization reaction, the desalting of the blocked peptides and cation exchange chromatography for the isolation of the C-terminal peptide can be easily performed. All the reagents and materials required for the isolation of the C-terminal peptide are not expensive and are commercially available.

This strategy is compatible with the automatic Edman sequencing and all the proteolytic peptides can also be recovered for their further analysis or sequencing.

Once the tryptic peptides are selectively derivatized, it is only a matter of separating charged from the neutral species. Using two different chromatographic

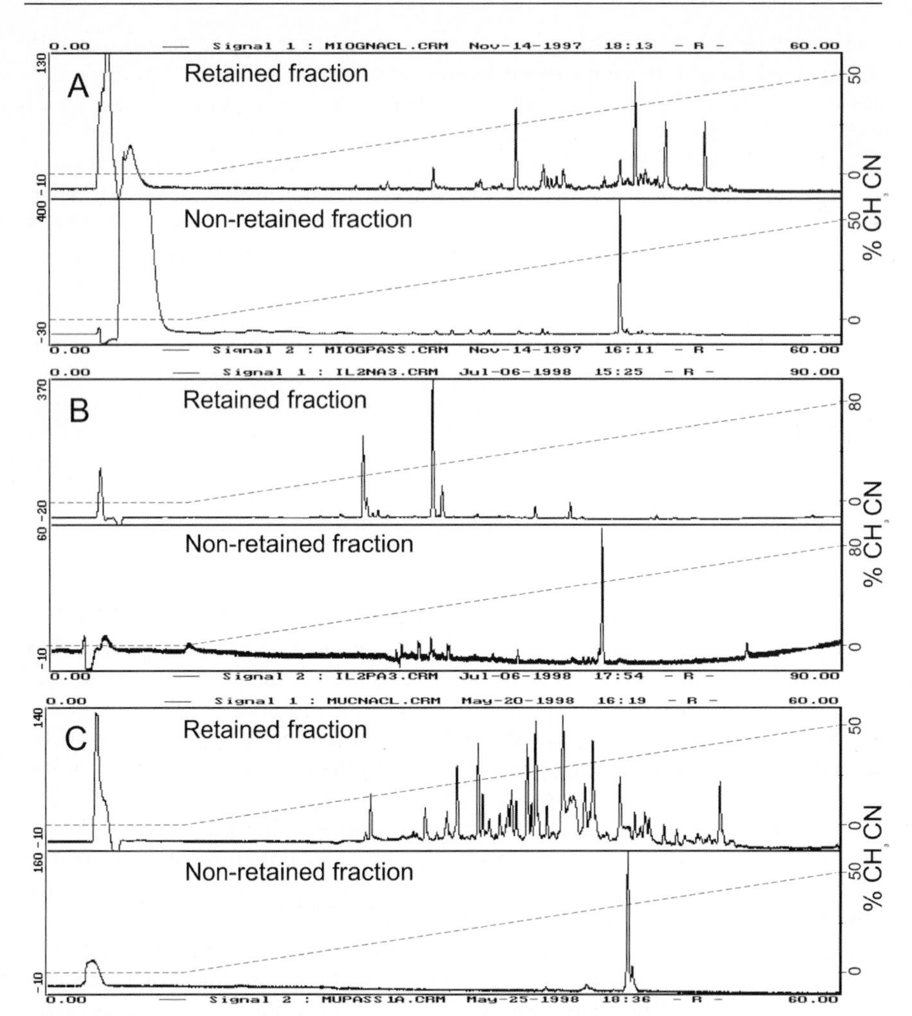

Fig. 12.7. The pairs of chromatograms shown in A, B and C are the rp-HPLC analysis of the retained and the non-retained fractions obtained in the isolation of the C-terminal peptides of the proteins Myoglobin, mutated IL-2 and bovine mucorpepsin, respectively

matrices and the same chromatographic conditions the results in the isolation of the C-terminal peptide have been the same. Therefore our strategy can be considered as a method of choice for the isolation of C-terminal peptide of proteins.

It should be noted that the presence of several histidine residues within the C-terminal peptide might be a limitation of the methodology. It could yield abundant positive charge in the C-terminal peptide and, of course, it would be retained in the column. This means, for example, that proteins with a polihistidine tail at the C-terminal cannot be analyzed with this methodology.

We have also applied our strategy for the isolation of the C-terminal peptide of proteins on the SDS-PAGE with successful results and this will be published elsewhere.

Table 12.1. Summary of the experimental masses observed in the isolated C-terminal peptide from the different proteins characterized in this study

Protein	Molecular Weight (kDA)	$(M+H)^+$ theor.	$(M+H)^+$	Sequence
Cytochrome C	11.7	1805.8	1805.9	Mal-^{92}EDLIAYLK*K*ATNE^{104}
IL-2m	15.6	1522.9	1522.8	Ac-^{122}WITFAQSIISTLT^{134}
TAB-9	16.6	592.3	592.2	Ac-^{158}AFVTI^{162}
Myoglobin	17.2	1983.8	1984.1	Suc-^{140}K*DIAK*YK*ELGYQG^{153}
Streptokinase	47.3	1631.7	1633.0	Suc-^{403}YTGTPIPDNPNDK*^{415}
Mucorpepsin	57.2	1394.6	1395.7	Ac-^{122}IGFAPLASGYENN^{134}

References

Bailey JM, Shively JE (1990) Carboxy-terminal sequencing: Formation and hydrolysis of C-terminal Peptidylthiohydantoins. Biochemistry 29: 3145–3156.

Boyd VL, Bozzini M, Zon G, Noble RL, Mattaliano R (1992) Sequencing of peptides and proteins from the carboxy terminus. Anal. Biochem. 206: 344–352

Butler PJG, Harris JI, Hartley, Bsleberman R (1969) The use of maleic anhydride for the reversible blocking of amino groups in polypeptide chains. Biochem J 112: 679–689

Gorman JJ, Shiell BJ (1993) Isolation of carboxyl-termini and blocked amino-termini of viral proteins by high-performance cation-exchange chromatography J. Chrom. 646: 193–205.

Isobe T, Ichimura T, Okuyama T (1986) Identification of the C-Terminal Portion of a Protein by Comparative Peptide Mapping. Anal Biochem. 155: 135–140

Kumazaki T, Nakako T, Arisaka F, Ishii S (1986) A novel method for selective isolation of C-terminal peptides from tryptic digests of proteins by immobilized anhydrotrypsin: application to structural analyses of the tail sheath and tube proteins from bacteriophage T4. Proteins 1: 100 -107.

Miller CG, Bailey JM (1995) Biotechnology applications of C-terminal sequence analysis. Genetic Engineering News 15: 12–16

Murphy CM, Fenselau C (1995) Recognition of Carboxy-Terminal Peptide in Cyanogen Bromide Digest of Proteins. Anal Chem. 67: 1644–1647

Naughton, J and Hagopian, K. (1962) Some Applications of Two-Dimensional Ionophoresis. Anal Biochem. 3: 276–284

Nguyen DN, Becker GW, Riggin RM (1995) Protein mass spectrometry: applications to analytical biotechnology. J. Chromatogr. 705: 21–45

Rose K, Savoy LA, Offord RE, Wingfield P (1988) C-terminal peptide identification by fast atom bombardment mass spectrometry. Biochem J. 250: 253–259

Shenoy NR, Shively JE, Bailey JM (1993) Studies in C-terminal Sequencing: New reagents for the synthesis of peptidylthiohydantoins. J. Prot. Chem. 12: 195–205

Stults JT, Lai J, McCune S, Wetzel R (1993) Simplification of high-energy collision spectra of peptides by amino terminal derivatization. Anal Chem 65: 1703–1708

Takamoto K, Kamo M, Kubota K, Satake K, Tsugita A (1995) Carboxy-terminal degradation of peptides using perfluoroacyl anhydrides. Eur. J. Biochem 228: 362–372.

Takao T, Gonzalez LJ, Yoshidome K, Sato K, Asada T, Kammei Y, Shimonishi Y. (1993) Automatic precursor-ion switching in a four sector tandem mass spectrometer and its application to acquisition of the MS/MS product ions derived from a partially ^{18}O-labeled peptide for their facile assignments. Anal Chem. 65: 2394–2399

Takao T, Hori H, Okamoto K, Harada A, Kamachi M, Shimonishi Y. (1991) Facile assignment of sequence ions of a peptide labelled with ^{18}O at the carboxyl terminus. Rapid Commun. Mass Spectrom. 5: 312–315.

Tsugita A, Takamoto K, Kamo M, Iwadate H (1992) C-terminal sequencing of protein. A novel partial acid hydrolysis and analysis by mass spectrometry. Eur. J. Biochem. 206 : 691–696

Wetzel R, Halualani R, Stults JT, Quan C (1990) A general method for highly selective cross-linking of unprotected polypeptides via pH controlled modification of N-terminal α-amino groups. Bioconjugate Chem 1: 114–122

CHAPTER 13

The Folding Pathway of Disulfide Containing Proteins

J. Y. Chang[1]

1 Summary

The pathway of oxidative folding of four single domain, 3-disulfide containing proteins has been studied in our laboratory. These four proteins are hirudin core domain (Hir, 49 amino acids), potato carboxypeptidase inhibitor (PCI, 39 amino acids), human epidermal growth factor (EGF, 58 amino acids) and tick anticoagulant peptide (TAP, 60 amino acids). Their folding pathways were analyzed by characterization of the acid and iodoacetate trapped folding intermediates. The results demonstrate a high degree of heterogeneity of the 1- and 2-disulfide intermediates and the presence of 3-disulfide scrambled species along the folding pathway of all these four proteins. Their folding mechanism differs significantly from the well documented case of bovine pancreatic trypsin inhibitor (BPTI).

2 Introduction

The elucidation of "Protein folding pathway" is a major issue and a very much debated subject in the field of protein chemistry (Baldwin, 1989; Kim & Baldwin, 1990; Creighton, 1990; Richards, 1991; Matthews, 1993). There are limited methodologies available to conduct such analysis. A newly developed technique of pulsed-label NMR (Roder et al., 1988; Udgaonkar & Baldwin, 1988; Jennings & Wright, 1993) permits trapping and identification of amide groups that are engaged in the structured elements during the process of folding. However, one of the best established techniques is so called "disulfide folding pathway" pioneered by Creighton (1978 &1990). The disulfide folding pathway is defined by the kinetics of formation of native disulfide bonds during the process of oxidative folding (Creighton, 1986). Extensive application of this technique in the past 20 years has produced one major model of protein folding, the disulfide folding pathway of bovine pancreatic trypsin inhibitor (BPTI) (Creighton, 1990; Weissmann & Kim, 1991), a kunitz-type protease inhibitor which comprises 58 amino acids and 3 disulfides. Despite the lingering debate about its detailed mechanism

[1] Research Center for Protein Chemistry, Institute of Molecular Medicine The University of Texas, Houston, Texas 77030.

(Creighton, 1992; Weissmann & Kim, 1992a), the folding pathway of BPTI has been established as one of the central dogma of protein folding.

For the past 5 years, our lab analyzed the disulfide folding pathways of four single domain proteins that have the size and disulfide numbers similar to that of BPTI. These four proteins are hirudin core domain (Hir, 49 amino acids) (Chatrenet & Chang, 1993), potato carboxypeptidase inhibitor (PCI, 39 amino acids) (Chang et al., 1994), human epidermal growth factor (EGF, 58 amino acids) (Chang et al., 1995) and tick anticoagulant peptide (TAP, 60 amino acids) (Chang, 1996). We found that these four proteins refold *through* a similar mechanism. But the mechanism is very different from what has been described in the case of BPTI. There are two major differences; (1) One is that the folding intermediates are much more heterogeneous than in the case of BPTI (Creighton, 1990; Weissmann & Kim, 1991). (2) Another important difference is that scrambled 3-disulfide species were found as folding intermediates in all these four proteins. Some of our data will be presented here to illustrate these major differences. It is important to mention that TAP is structurally homologous to BPTI in terms of 3-D conformation and disulfide pattern (Autuch et al., 1994; Lim-Wilby et al., 1995). Yet, the folding mechanism of TAP is simply incompatible with those described in the case of BPTI.

3
Experimental Procedures

Materials

All four proteins, hirudin core domain (Hir. residues 1–49), tick anticoagulant peptides (TAP), potato carboxypeptidase inhibitor (PCI), and human epidermal growth factor (EGF), are recombinant proteins. Their purity was greater than 95% as judged by HPLC, mass analysis and N-terminal sequence analysis. Reduced glutathione (GSH), oxidized glutathione (GSSG), cysteine (Cys), cystine (Cys-Cys), β-mercaptoethanol were obtained from Sigma.

Folding experiments performed in the absence of redox agent

The native proteins (1.5 mg) were first reduced and denatured in 0.5 ml of Tris-HCl buffer (0.5 M, pH 8.5) containing 5 M of GdmCl and 30 mM of dithiothreitol. Reduction and denaturation was carried out at 22oC for 90 mins. To initiate the folding, the sample was passed through a PD-10 column (*Pharmacia*) equilibrated in 0.1 M Tris-HCl buffer (pH 8.5). Desalted and unfolded protein was recovered in a volume of 1.1 ml, which was immediately diluted with the same Tris-HCl buffer to a final protein concentration of 1 mg/ml, both in the absence (control -) and presence (control +) of 0.25 mM 2-mercaptoethanol. Folding intermediates were trapped in a time-course manner by mixing aliquots of the sample with equal volume of 4% trifluoroacetic acid in water or with 0.4 M iodoacetic acid in the Tris-HCl buffer (0.5 M, pH 8.5). In the case of iodoacetate trapping, carboxymethylation was performed at 22oC for 30 mins, followed by desalting using the PD-10 column. Trapped folding intermediates were analyzed by HPLC.

Folding in the presence of redox agents
The procedures of unfolding and refolding are as those described in the control folding experiments. Selected concentrations of redox agents were introduced immediately after reduced and denatured proteins were collected from the PD-10 column. Folding intermediates were similarly trapped by acidification or carboxymethylation as those described above.

Carboxymethylation of acid trapped intermediates isolated by HPLC
Acid trapped intermediates were separated and isolated by HPLC. The samples were dried in a speedvac and immediately treated with 1M of iodoacetic acid in 0.1 ml of Tris-HCl buffer (0.5 M, pH 6.5) containing 40 % (by volume) of dimethylformamide (Chang, 1993). The reaction was allowed for 20 mins and the carboxymethylated intermediates were removed from the excess reagent and salt by a NAP-5 column (*Pharmacia*).

Amino acid analysis, amino acid sequencing and MALDI mass spectrometry
Amino analysis was performed with the dabsyl chloride precolumn derivatization method which permits direct evaluation of the disulfide (cystine) content (Chang & Knecht, 1991). Amino acid sequencing was done with a Hewlett-Packard G-1000A sequencer. The MALDI mass spectrometer was a home-built time of flight (TOF) instrument with a nitrogen laser of 337 nm wavelength and 3ns pulse width (Boernsen et al., 1990). The apparatus has been described in detail elsewhere. The calibration was performed either externally or internally, by using standard proteins (Hypertensin, M.W. 1031.19; Synacthen, 2934.50 and Calcitonin, 3418.91).

4 Results and Discussion

4.1 The Folding Intermediates of Hirudin and TAP are Highly Heterogeneous

Reduced and denatured hirudin and TAP were first allowed to refold in the Tris-HCl buffer in the absence and presence of thiol catalyst (0.25 mM β-mercaptoethanol). These two folding experiments were designated as "control minus" (without β-mercaptoethanol) and "control plus" (with β-mercaptoethanol). Acid trapped intermediates were analyzed by HPLC and the results are shown in Fig. 13.1 (Hirudin) and Fig. 13.2 (TAP). The HPLC profiles demonstrate the heterogeneity of the intermediates, their progression along the folding pathway and the formation of native structure. In the two cases of "control plus" experiments, the recovery of the native structure is nearly quantitative. However, in the two cases of "control minus" experiments, about 50 % of the protein becomes trapped as non-native species, unable to convert to the native species even during prolonged folding (e.g up to 72 hours).

Each of these time-course trapped intermediates was extensively characterized for the disulfide content (by dabsyl chloride method) and the composition of 1-, 2- and 3-disulfide species (by MALDI mass spectrometry). The data reveals that there are three groups of intermediates, namely 1-disulfide, 2-disulfide and 3-

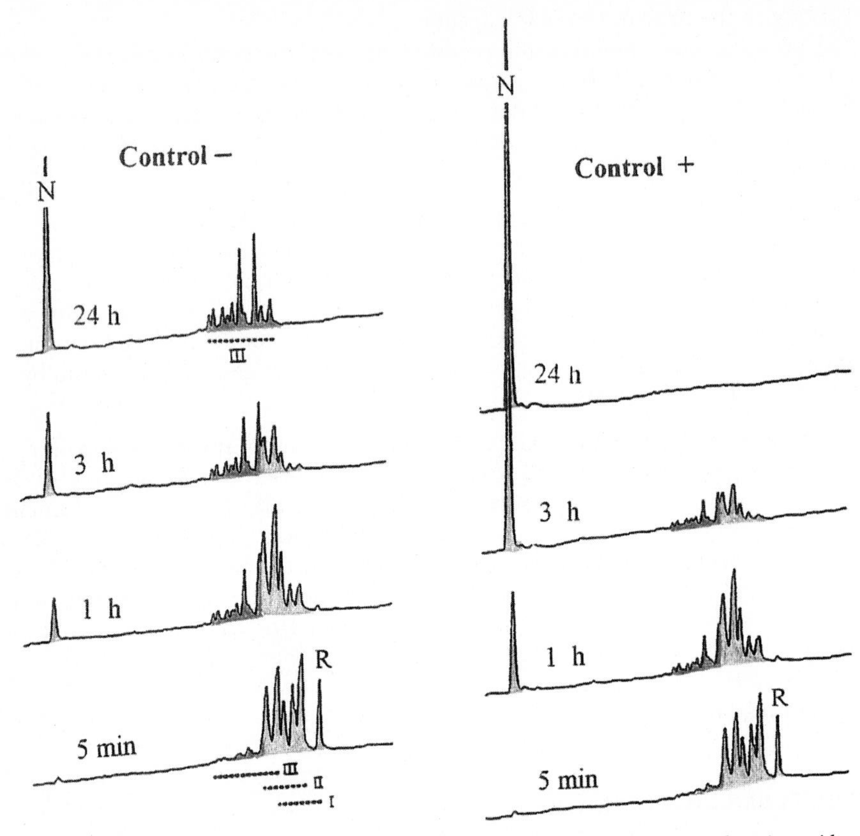

Fig. 13.1. HPLC profiles of acid-trapped folding intermediates of Hirudin core domain residues 1–49. Folding was performed in the Tris-HCl buffer (0.1 M, pH 8.4) in the absence (control minus) and presence (control plus) of b-mercaptoethanol (0.25 mM). The samples were analyzed by HPLC using the following conditions. Solvent A was water containing 0.1 % trifluoroacetic acid. Solvent B was acetonitrile/water (9:1, by volume) containing 0.1 % trifluoroacetic acid. The gradient was 14 % to 32 % solvent B linear in 50 mins. Column was Vydac C-18 for peptides and proteins, 4.6 mm, 10 mm. Column temperature was 23oC. **R** (orange color) and **N** (blue color) indicate the elution positions of the fully reduced and the native species. Fractions containing 1- and 2-disulfide species are colored with green. Those containing 3-disulfide species are colored with red

disulfide (scrambled) species, which overlap and advance sequentially along the folding pathway. Those which become trapped in the case "control minus" folding are 3-disulfide, scrambled species, unable to convert to the native species because of the depletion and absence of free thiol as catalyst. β-mercaptoethanol is not the only reagent that is useful for "control plus" folding experiment, other thiol reagents, such as cysteine and reduced glutathione are equally effective. If one compares the patterns of folding intermediates performed in the absence and presence of β-mecaptoethanol, it becomes very clear that the only difference is the relative level of accumulation of scrambled 3-disulfide species and the recovery of the native species. We have found this phenomenon in all four proteins that have been analyzed in our lab.

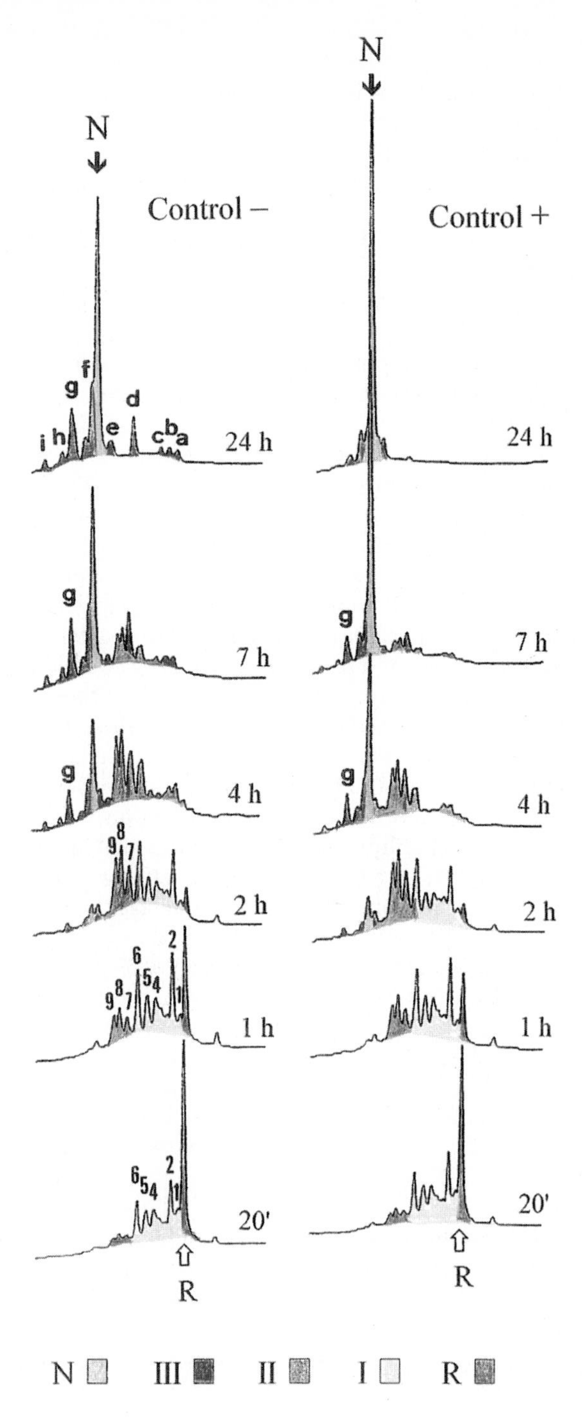

Fig. 13.2. HPLC profiles of acid-trapped folding intermediates of TAP. Folding was performed in the Tris-HCl buffer (0.1 M, pH 8.4) in the absence (control minus) and presence (control plus) of b-mercaptoethanol (0.25 mM). HPLC conditions are as those described in the legend of Fig. 13.1, except for using a different gradient which was 28 % to 45 % solvent B linear in 40 mins. **R** (orange color) and N (blue color) indicate the elution positions of the fully reduced and the native species. Major fractions of 1-disulfide (yellow color) and 2-disulfide (green color) intermediates are numbered from 1 to 6 and 7 to 9, respectively. Scrambled 3-disulfide species (red color) are marked alphabetically (24 h sample, left column)

The folding intermediates of hirudin and TAP are actually far more heterogeneous than what has appeared on HPLC as shown in Fig. 13.1 and Fig. 13.2. Take the 5 min trapped intermediates from the hirudin “control minus” folding as an example (Fig. 13.1). Each HPLC fraction, when isolated and further analyzed by capillary electrophoresis, was found to contain a mixture of complex species. If all these species are added up, it approaches the possible numbers of the disulfide isomers (there are 60 possible 1- and 2-disulfide isomers for a protein containing 3 disulfides). This high heterogeneity was similarly observed in the cases of TAP (Chang, 1996), PCI (Chang et al., 1994), and EGF (Chang et al., 1995).

4.2 Redox Agents Promote the Efficiency of Folding, but do not Alter the Composition of Folding Intermediates

Redox reagents are common ingredients for the oxidative folding of disulfide containing proteins (Sexena & Wetlaufer, 1970; Creighton, 1986; Lyles & Gilbert, 1991). A systematic study has been performed to examine the effect of GSH/GSSG and Cys/Cys-Cys on the folding mechanism of hirudin (Chang., 1994), TAP, PCI and EGF. These studies include evaluation of single redox agent as well as combined redox components at varying concentrations. Our results allow us to reach two important conclusions (Chang., 1994). First, the redox agents accelerate the kinetics and promote the efficiency of folding, but do not alter the composition of folding intermediates. For each of these four proteins, the composition of folding intermediates, as judged by their HPLC patterns, remains indistinguishable under various combinations of redox agents. Second, redox agents promote the folding of hirudin, TAP, PCI and EGF in a two-stage manner. GSSG and Cys-Cys accelerate the disulfide formation, whereas GSH and Cys catalyze the disulfide reshuffling. As a consequence, the kinetic of the flow of folding intermediates and the level of their accumulation along the folding pathway are dependent upon the concentration of redox agents. When the folding of PCI was carried out in the buffer containing GSSG or Cys-Cys, rapid formation of the disulfide bonds leads to the accumulation of 3-disulfide scrambled PCI as the folding intermediates (Fig. 13.3) (Chang et al., 1994). For instance, under 2 mM of Cys-Cys, the only detectable folding intermediates after 1 min of folding are scrambled PCI. The inclusion of both Cys-Cys and Cys (or both GSSG and GSH) resulted in a simultaneous disulfide reshuffling of the scrambled species and the rapid recovery of the native structure. This two-stage mechanism has been observed with all four proteins analyzed in our laboratory.

4.3 Nearly all Possible Isomers of Scrambled Species of Hirudin and TAP were Shown to Exist Along the Folding Pathway

The complexity of the folding intermediates of hirudin and TAP suggests that nearly all possible disulfide isomers may exist along the folding pathway. Among the three groups of folding intermediates, however, the 3-disulfide scrambled species represents the only class of intermediates that can be purified to homoge-

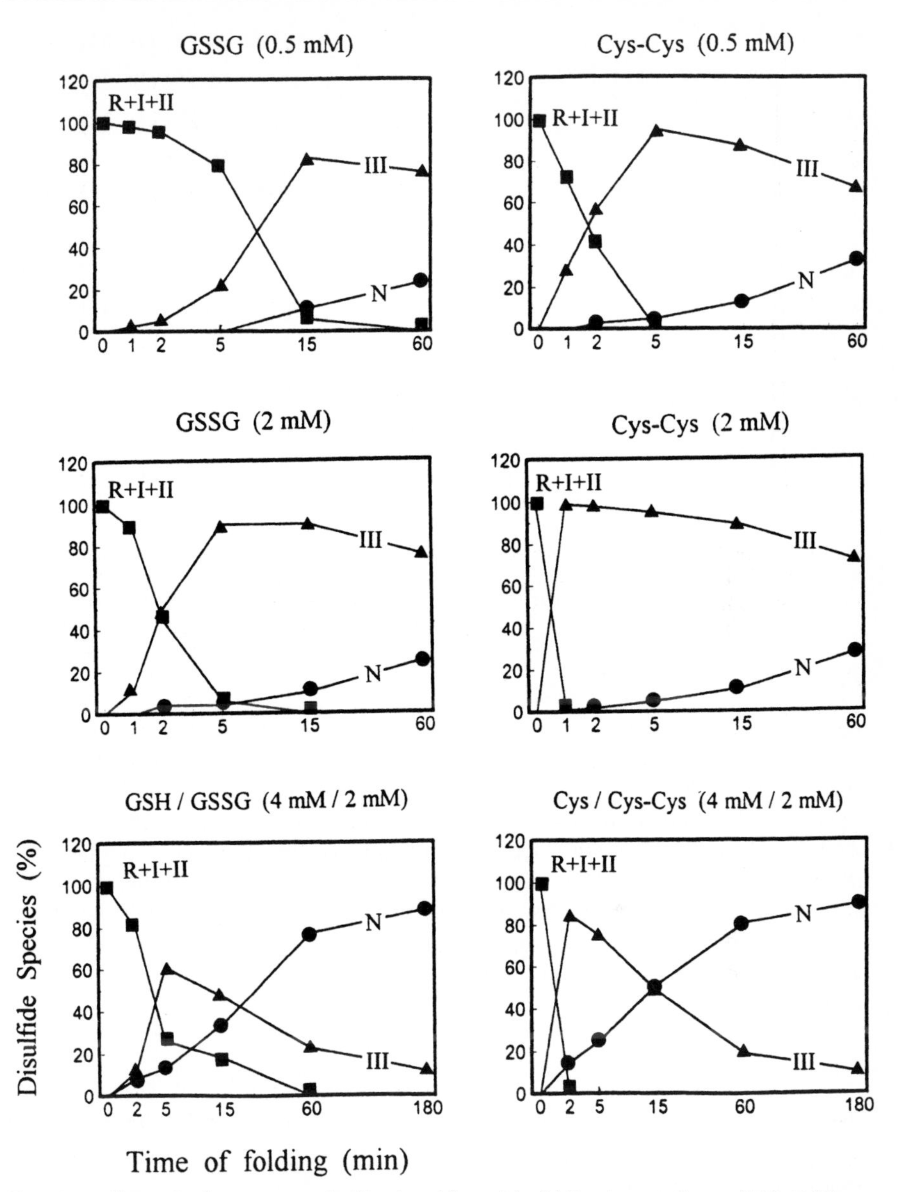

Fig. 13.3. Effects of redox agents on the kinetics of flow of the folding intermediates of PCI.. Foldings were performed in the Tris-HCl buffer containing the indicated redox agents. **I, II** and **III** stand for 1-disulfide, 2-disulfide and 3-disulfide scrambled species of PCI, respectively. [**R+I+II**], [**III**] and [**N**] were followed quantitatively in each time-course trapped sample analyzed by HPLC. [**R**], [I] and [**II**] overlap extensively, therefore, their recoveries were quantitated collectively

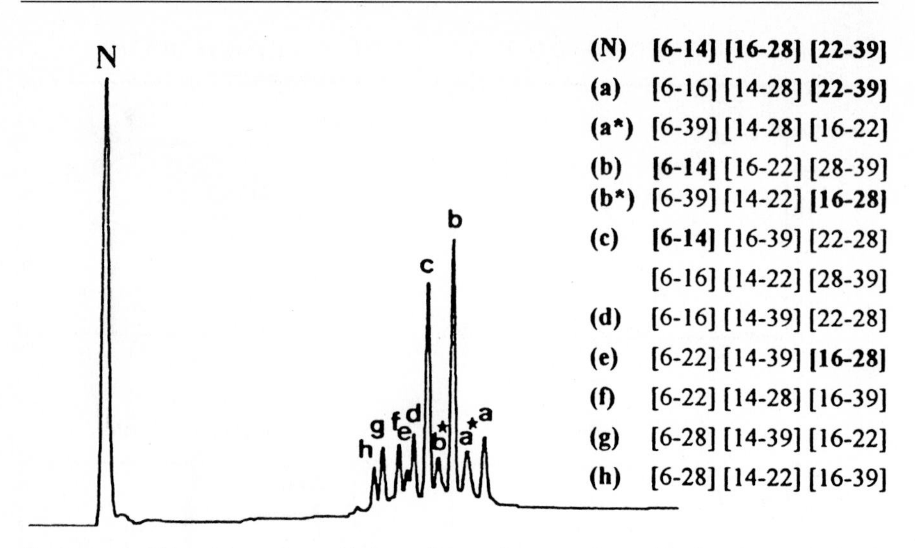

Fig. 13.4. Disulfide pairings of scrambled Hirudin. Ten fractions of scrambled hirudin were isolated from HPLC. Each was shown to contain one single species of scrambled hirudin, except for fraction "c" that comprises two species. All together, 11 species of scrambled hirudins were identified (Chang, 1995)

neity for further structural characterization. Their disulfide structures were deduced from the analysis of thermolytic peptide by both Edman sequencing and MALDI mass spectrometry and the results are shown in Fig. 13.4 and Fig. 13.5. In the case of hirudin, 11 out of the 14 possible scrambled isomers were characterized (Fig. 13.4) (Chang 1995). The three hirudin scrambled isomers that have not been found are those containing Cys14-Cys16, presumably due to the steric constraint. In the case of TAP, 7 scrambled isomers were isolated and structurally characterized (Fig. 13.5) (Chang, 1996). An additional 4 fractions of scrambled TAP were found to contain multiple species. All together, at least 11 species of scrambled TAP were observed.

Fig. 13.5. Disulfide pairings of scrambled TAP. Ten fractions of scrambled TAP were separated by HPLC (see Fig. 13.2). Seven of them were isolated and structurally characterized

(N)	**[5-59] [15-39] [33-55]**
(a)	[5-15] [33-39] [55-59]
(d)	[5-15] [33-59] [39-55]
(e)	[5-33] **[15-39]** [55-59]
(f)	[5-55] [15-33] [39-59]
(g)	[5-39] [15-33] [55-59]
(h)	[5-39] [15-55] [33-59]
(i)	[5-55] **[15-39]** [33-59]

The structures of scrambled hirudin and TAP imply that their 1- and 2-disulfide intermediates may be equally heterogeneous. The presence of significant amount of scrambled species along the folding pathway of hirudin, TAP, PCI and EGF also indicate that they could not be simply dismissed as by-product or abortive structures of mis-folding. Indeed, scrambled species have also been observed during the productive folding of ribonuclease A (Creighton, 1979) and pro-BPTI (Weissmann & Kim, 1992b) that involved no denaturants.

5 Acknowledgements

I would like to acknowledge the collaboration with Dr. F. X. Aviles and Dr. P. H. Lai in the analysis of folding mechanism of potato carboxypeptidase inhibitor and human epidermal growth factor.

References

Autuch, W., Güntert, P., Billeter, M., Hawthorne, T., Grossenbacher, H., and Wüthrich, K. (1994) *FEBS Lett.* 352, 251–257.
Baldwin, R. L. (1989) *Trends Biochem.* Sci. 14, 292–294.
Boernsen, K. O., Schaer, M., and Widmer, M. (1990) *Chimia* 44, 412–416.
Chang, J.-Y. (1993) *J. Biol. Chem.* 268, 4043–4049.
Chang, J.-Y. (1994) *Biochem. J.* 300, 643–650.
Chang, J.-Y. (1995) *J. Biol. Chem.* 270, 25661–25666.
Chang, J.-Y. (1996) *Biochemistry* 35, 11702–11709.
Chang, J.-Y. and Knecht, R. (1991) *Anal. Biochem.* 197, 52–58. J. Biol
Chang, J.-Y., Canals, F., Schindler, P., Querol, E., and Aviles, F. X. (1994). Chem. 269, 22087–22094.
Chang, J.-Y., Schindler, P., Ramseier, U., and Lai, P.-H. (1995) *J. Biol. Chem.* 270, 9207–9216.
Chatrenet, B., and Chang, J.-Y. (1993) *J. Biol. Chem.* 268, 20988–20996.
Creighton, T. E. (1978) *Prog. Biophys. Mol.* Biol. 33, 231–297.
Creighton, T. E. (1979) *J. Mol. Biol.* 129, 411–431.
Creighton, T. E. (1986) *Methods Enzymol.* 131, 83–106.
Creighton, T. E. (1990) *Biochem.* J. 270, 1–16.
Creighton, T. E. (1992) *Science* 256, 111–112.
Jennings, P. A., and Wright, P. E. (1993) *Science* 262, 892–896.
Kim, P. S., and Baldwin, R. L. (1990) *Annu. Rev. Biochem.* 59, 631–660.
Lim-Wilby, M. S. L., Hallenga, K., De Maeyer, M., Lasters, I., Vlasuk, G. P., and Brunck, T. K. (1995) *Protein Sci.* 4, 178–186.
Lyles, M. M., and Gilbert, H. F. (1991) *Biochemistry* 30, 613–619.
Matthews, C. R. (1993) *Annu. Rev. Biochem.* 62, 653–683.
Richards, F. M. (1991) *Sci. Am.* 34–41.
Roder, H., Elöve, G., Englander, S. W. (1988) *Nature* 335, 700–704.
Sexena V. P., and Wetlaufer, D. B. (1970) *Biochemistry* 9, 5015–5023.
Udgaonkar, J. B., and Baldwin, R. L. (1988) *Nature* 335, 694–699.
Weissman, J. S., and Kim, P. S. (1991) *Science* 253, 1386–1393.
Weissman, J. S., and Kim, P. S. (1992a) *Science* 256, 112–114.
Weissman, J. S., and Kim, P. S. (1992b) *Cell* 71, 841–851.

CHAPTER 14

Cavity Supported HPLC of *Cis/Trans* Isomers of Proline Containing Peptides Using Cyclodextrins and Calixarenes

S. MENGE[1], S. GEBAUER[1] and GERD-J. KRAUSS[1]

1 Introduction

In biological systems the cellular metabolism is decisively controlled by the dynamics of folding and refolding of peptidic structures. Essential processes of protein biogenesis such as protein synthesis, and translocation of proteins into intracellular compartments require the protein to exist temporarily in unfolded or partially folded conformation (Beissinger and Buchner 1998). Therefore, information on in vivo and in vitro folding and refolding of proteins is of actual interest in biochemistry and biotechnology and the knowledge of these processes is of remarkable importance for the production of biological active recombinant proteins.

The amino acid proline plays an outstanding role in such processes. Proline represents a molecular switch in the polypeptide chain, which controls the conformational changes of the prolyl bond and, therefore, the alignment of the polypeptide backbone.

In peptides containing proline in a C-terminal position the free rotation around the peptide bond is hindered due to the formation of a pyrrolidin ring system with the prolyl nitrogen (Brandts et al. 1975). The C-N bond is marked by a partial double bond character leading to energy barriers for the rotation around the peptidyl-prolyl bond, resulting in two lowest energy arrangements for the prolyl peptide bond: *cis* and *trans*. *Cis* and *trans* conformers are defined by the dihedral angle ω with $\omega \sim 0\,°C$, *cis* and $\omega \sim 180\,°C$, *trans*. They exist in the aqueous peptide solution in comparable amounts due to comparable thermodynamic stabilities, but the *trans* conformer is a little energetically favored. In principle, polypeptides with n Xaa-proline bonds can form 2^n bond isomers. Frequently, the dynamic *cis/trans* isomerization of prolyl bond turned out to be the rate limiting step in folding/refolding processes of proteins (Brandts et al. 1997, Lin and Brandts 1987). The interconversion of both energetic arrangements can be catalyzed by peptidyl-prolyl *cis/trans* isomerases (PPI'ases), which accelerate the *cis* to trans isomerization by lowering the rotation barriers (Lang et al. 1987, Fischer 1994).

[1] Martin-Luther-University Halle-Wittenberg, Department of Biochemistry/Biotechnology, Institute of Biochemistry, Kurt-Mothes-Str. 3, D-06120 Halle/S., Germany.

The partial double bond character and the energy barriers result in a delay of the *cis/trans* isomerization with relaxation times of 10–1000 sec^{-1} so that analytical methods with corresponding time scales of measurement are able to detect both isomers in their aqueous solution. Quantitative data, describing this type of conformational changes, can be determined by spectroscopic and kinetic methods (Fischer et al. 1984, Hübner et al. 1990).

Several papers report on the use of proline containing oligopeptides as model substrates for the determination of the catalytic activity of PPI'ases from different natural sources. Peptide isomers can be used as models to study different biological processes including protein folding/refolding (Brandts et al. 1975), immune response (Fischer et al. 1989, Schreiber 1991) or the assembly of hormon receptor complexes (Schmidt et al. 1995, Owens-Grillo et al. 1995).

New studies are published on the introduction of a protease-free assay for the determination of the time course of *cis/trans* isomerization on the basis of different UV/VIS spectra of the conformational states (Küllertz et al. 1998, Janowski et al. 1997). Special methods of NMR spectroscopy were used to study the *cis/trans* ratio of different peptide structures and rates of isomerization depending on different conditions such as pH, as well as the ionization grade of substrates, the role of phosphorylation, the content of organic solvent in the aqueous solution or the interaction to micelles (Hübner et al. 1990, Kramer and Fischer 1997, Schutkowski et al. 1998).

In order to study whether rate and ratios of *cis/trans* isomerization may effect biological processes high resolution chromatographic and electrophoretic methods have been applied, too. HPLC and CZE are suitable separation techniques allowing the isolation and identification of peptide conformers.

2 HPLC of *cis/trans* Isomers of Proline Containing Peptides

2.1 Commonly Applied Separation Principles

In the early eighties, Melander and Horvath (1982) first reported on the appearance of a secondary equilibrium during the RP-HPLC of proline containing peptides. They observed that peak splitting and deformity of unprotected peptides including proline in a C-terminal position are attributed to the dynamic interconversion of peptide bond isomers. During the chromatographic run the *cis/trans* isomerization of prolyl peptides interferes with the second dynamic equilibrium: the distribution of both conformers between the mobile and the stationary phases (Kalman et al. 1996). This interference of the two dynamic processes leads not only to an insufficient resolution of the isomer peaks, but also to an influence on the rate of interconversion and the *cis* to *trans* ratio depending on the operation conditions.

In several papers the fundamentals for the RP-HPLC of interconverting species have been studied. They succeeded in the isomer separation of oligopeptides only on the basis of slight differences in the overall hydrophobicity of both conformational species. Optimum separation of *cis/trans* isomers for given prolyl peptides can be achieved considering the fact that the time scale of the chromatographic run is in a faster/similar scale compared to the relaxation time of the *cis/trans* isomerization. The relaxation rate is affected by temperature (low temperatures delay the conformer interconversion), mineral acids and organic solvents which increase the rate of isomerization. Based on these fundamental studies general principles can be deduced for the chromatographic resolution of peptide isomers:

- short columns with high resolution packings;
- high flow rates of mobile phase or steeply rising gradients, respectively;
- low working temperatures;
- pH values stabilizing the zwitterionic state of the peptide.

However, low temperatures and high flow rates of mobile phase result in a diminished permeability of the column and in poorer diffusion coefficients leading to insufficient mass transfers between mobile and stationary phases. These characteristic features can be overcome for instance by the application of unporous micropellicular supports and solvent additives which diminish the viscosity of the mobile phase.

A simple re-chromatography of the fractionated isomer peaks should be performed to check that the split peaks are really conformers and not peptide impurities. After a certain time of re-equilibration the two isomer peaks can be monitored again.

On-line and off-line NMR measurements of the fractionated isomers from several prolyl peptides confirmed the higher hydrophobicity of *cis* isomers leading to an elution order *trans* before *cis* (Kalman et al. 1996, Gebauer et al. 1998). Table 14.1 summarizes some examples for the RP-HPLC of biologically active peptides bearing one or more Xaa-Pro bonds in their sequence, including the opioid peptides ß-casomorphin and morphiceptin, the blood pressure controlling hormones angiotensin and bradykinin, dermorphin and tryptophilin. Of special interest are present studies on the immunosupressiva cyclosporin A, C, and D and FK 506 (Joshua 1991).

2.2 Cyclodextrin Bonded Stationary Phases

In recent years macrocyclic compounds including cyclodextrins (CDs), crown ethers and calixarenes gained increasing interest in host-guest chemistry due to their extraordinary potential for the molecular recognition of neutral and ionic molecules. These compounds are characterized by well defined cavities, referring to size and shape as well as ionic and hydrophobic properties. Therefore, they are preferentially used as selectors in chromatography and electrophoresis for separations exploiting stereochemical interactions. Cyclodextrin bonded silicas as stationary phases in HPLC were developed for separating enantiomers by forming diastereomers with the chiral centers of the CDs, but they were used too for the steric discrimination of geometric and positional isomers.

Cyclodextrins of different ring sizes (α-, β-, γ-CD) offer the opportunity to exploit a variety of selective interactions with peptides, such as inclusion com-

Table 14.1. Reversed phase HPLC of *cis/trans* conformational isomers of biologically active peptides with Xaa-Pro Bonds

Peptides	Column	Mobile phase	Temp./Ref.
Xaa-Pro (Xaa = Ala, Leu, Ile, Val, Phe)	**LiChrosorb RP 18,** (10µm)m 250 × 4 mm	0.05 M phosphate, pH 6	25 °C Melander et al. (1982)
diprotin B (Val-Pro-Leu) **bradykinin:** (Arg-Pro-Pro-Ala-Phe-Ser-Pro-Phe-Arg) **morphiceptin:** (Tyr-Pro-Phe-Pro-NH_2) **β-casomorphin:** (Tyr-Pro-Phe-Pro-Gly-Pro-Ile) **Proctolin:** (Arg-Tyr-Leu-Pro-Thr)	**Pecospher C18,** 3 × 3 mm, 33 × 4,6 mm	methanol/ 0.05 phosphate, (pH 7), (15:85), 2 mL/min	0–11 °C Henderson et al. (1990)
tryptophyllin(TPH 13)	**µBondapak C18,** 300 × 3.9 mm	acetonitrile/0.1 % HCOOH (40:60), 1 mL/min	0 °C Rusconi et al. (1985)
bradykinin, synthetic oligopeptides	**Nucleosil C18,** (3 µm), 100 × 4.6 mm	acetonitrile/phosphate (pH 3.5)-gradient, 0.7 mL/min	–10–20 °C Gesquiere et al. (1989)
dermorphin-analogs (Tyr-DL-Ala-Phe-Gly-Tyr-Pro-Ser-NH_2)	**1. LiChrospher 100 RP 18** (5 µm), 125 × 4 mm, **2. Ultraspher ODS,** 5 µm, **3. microparticular C18-silica gel**	acetonitrile/TFA gradient, MeOH/0.05 M phosphate, pH 2.5	–5 °C Schmidt et al. (1995)
ramipril	Nucleosil C18, (3µm), 100 × 4.6 mm	a Acetonitrile/THF/ phosphate 25:5:70, pH 2, pH 6, 1 mL/min	5–40°C Gustafsson et al. (1990)
cyclosporin A, C, D	**Inertsil ODS-2,** 250 × 4 mm	acetonitrile/H_2O (75:25), 1 mL/min	0–60 °C Nishikawa et al. (1994)
colecystokinin-derivate (Gly-Trp-MeNle-Asp-Phe-NH_2)	**Vydac C18-Peptides and Proteins,** 250 × 4 mm	methanol/acetonitrile/ 0.1 % TFA (50:13:37), 1 mL/min	23–17 °C Lebl et al. (1991)

plexation, steric discrimination, hydrophobic interactions in the interior of the CD toroid, dipol-dipol interactions and hydrogen bonding with the primary and secondary hydroxy groups at the lower and upper rim. Considering the stereochemical differences of the conformational states in proline containing peptides we applied α-, β-, and γ-CD bonded stationary phases in the reversed-phase mode. Applying a phosphate buffer of low ionic strength, acetonitrile as mobile phase modifier and a pH forcing the unprotected peptides into their zwitterionic state, the chromatographic separation of *cis/trans* isomers of several di- to oligopeptides succeeded on β-CD stationary phases. Low working temperatures cause

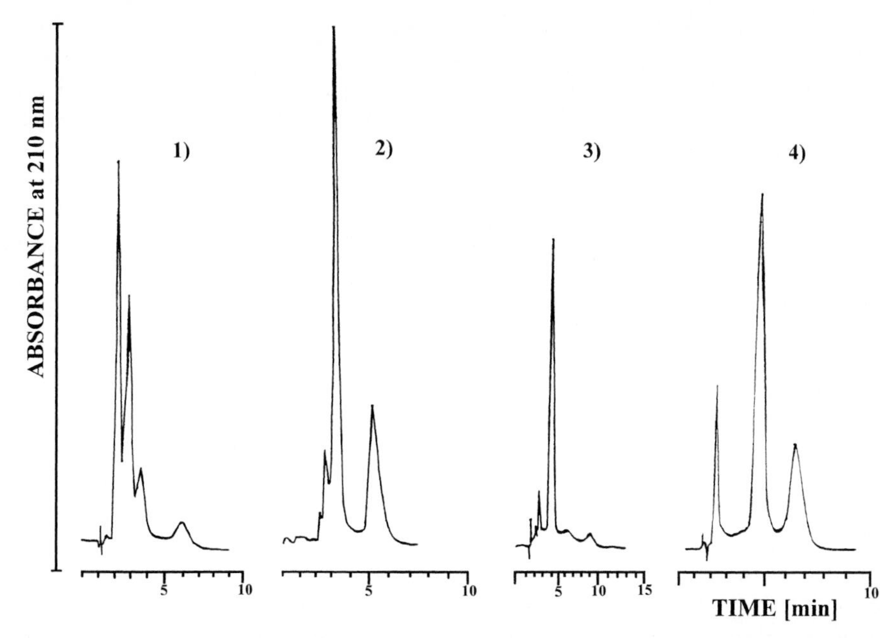

Fig. 14.1. HPLC profiles of oligopeptides bearing two Xaa-Pro bonds: Tyr-Pro-Phe-Pro-Gly (Casomorphin-5) (1), Phe-Pro-Phe-D-Pro-Gly (2), Tyr-D-Pro-Phe-Pro-NH_2 (3) and Tyr- Pro-Phe-Pro-Gly (4) on two connected 125 × 4.6 mm chiral ß-cyclodextrin Si-100 (10 μm) columns at 2 °C. Mobile phase: 0.02 M ammonium dihydrogenphosphate (pH 6.2)-acetonitrile (70:30); flow rate: 3 mL/min (Friebe et al. 1994)

a decrease of the interconversion rate and contribute to an improvement of the chromatographic resolution, probably also by an entropy effect of the stereochemical interaction (Fig. 14.1.).

One type of the investigated proline containing peptides were casomorphins representing proline containing peptides with high opioid activity, central nervous effects and immune regulating properties (Neubert et al. 1990). The investigated peptides may be released from the milk protein ß-casein by proteolytic fragmentation. The introduction of D-amino acids gives derivatives that are characterized by an increased proteolytic stability and a prolonged opioid activity. Obviously, a stereochemical discrimination is responsible for the separation of the *cis/trans* conformers as shown for the tetrapeptides Pro-D-Phe-Pro-Gly (DL) and Pro-Phe-Pro-Gly (LL). Comparing the corresponding isomer separations on a β-CD column a reversed elution order was observed for the *cis* and *trans* conformers (Friebe et al. 1994).

Comparative studies on α-, β-, and γ-CD bonded silicas have been performed in order to investigate how the molecular dimensions of the cyclodextrins can be adapted to the size and shape of the analyte. Only a peak splitting was achieved with an α-CD bonded stationary phase for the resolution of conformers of Xaa-Pro dipeptides (Xaa = Ala, Leu, Ile), whereas on γ-CD phases the chromatographic resolution is sufficient for cyclic or larger peptides, like the pentapeptide ß-casomorphin-5. Generally, β-CD bonded phases show the highest efficiency in separating peptide bond isomers, especially if an aromatic amino acid is N-terminal bonded to the proline residue. These results are in agreement with other findings which describe the high selectivity of β-CD phases for analytes with an aromatic ring system in the vicinity of the chiral, geometrical or structural center (Ward and Armstrong 1988).

The high resolution power of β-CD bonded silicas for the conformer separation of proline containing peptides was demonstrated for oligopeptides which include two Xaa-Pro bonds. That means, theoretically four *cis/trans* isomers could be detected. As shown in Fig. 14.1. we really succeeded in separating these peptide isomers.

2.3 Calix[n]arenes

Calix[n]arenes are cyclic oligomeric condensates of phenols and formaldehyde. As calixarenes have a cavity-shaped architecture, they are frequently used as building blocks for host-guest receptors in supramolecular chemistry (Gutsche 1989, Vicens and Böhmer 1991, Böhmer 1995). Based on molecular recognition the use of calixarenes as selectors in chromatography for the study of *cis/trans* isomer separations of proline containing dipeptides seems to be possible (Friebe et al. 1995). More recently, we introduced a series of new calix[n]arene bonded stationary phases (n = 4, 5, 6, 8) for HPLC based on silica gel (Gebauer et al. 1998a, 1998b) and called these supports Si[n]Arenes. Applying compounds of

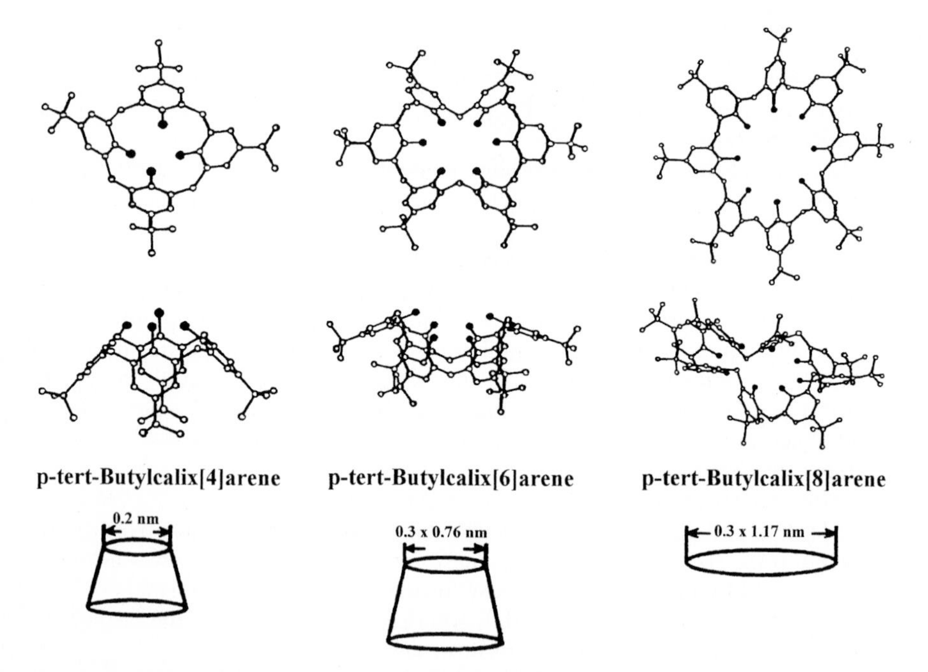

Fig. 14.2. Molecular dimensions of p-tert-butylcalix[n]arenes (n = 4, 6, 8) (Vicens and Böhmer 1991)

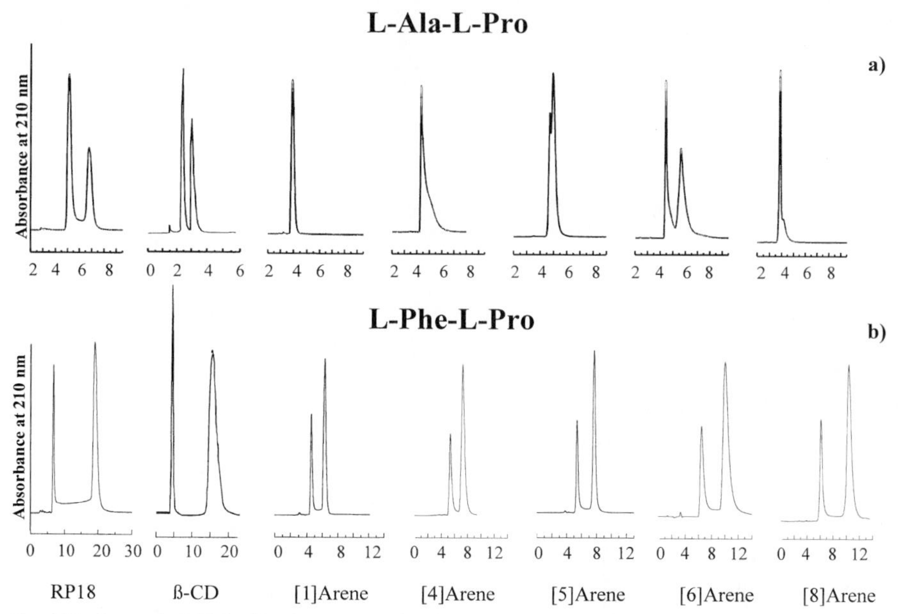

Fig. 14.3. Isocratic HPLC of L-Ala-L-Pro and L-Phe-Pro at 5 °C on different stationary phases using (a) 0.02 M $NH_4H_2PO_4$ (pH 6.2) and (b) 0.02 M $NH_4H_2PO_4$ (pH 6.2)–acetonitrile (93:7) as mobile phase; flow rate 1 mL/min (Gebauer et al. 1998)

different structural features we demonstrated the resolution power and chromatographic selectivity of the new calixarene phases. It has been shown that the chromatographic resolution of regio- and stereoisomers depend on the shape and size of calixarene cavities. The geometrical extents of p-*tert*-butylcalix[n] arenes (n = 4, 6, 8) are given in Fig. 14.2.

The results of low-temperature HPLC studies of L-Ala-L-Pro and L-Phe-L-Pro on calixarene silica gels are summarized in Fig. 14.3. and compared with those on RP 18 and β-CD phases. As demonstrated the isomers of L-Ala-L-Pro are only separated on RP 18, β-CD and calix[6]arene. A peak splitting is observed for the calix[5]arene phase. The broad, deformed peaks for calix[4]arene and calix[8]arene point to the existence of an isomeric equilibrium. On [1]arene, representing the corresponding monomeric phase, the isomers were co-eluted as a single symmetric peak.

In general, the retention factors for all separations are very low and show that the interaction of the analytes with the stationary phases is very small. Hence, sophisticated resolution factors of 1.1 and 1.2 were obtained for β-CD and calix[6]arene. For the RP 18 phase a relative high resolution factor was monitored. This high resolution value reduced the baseline separation considerably. The plateaus between the isomeric peaks represent the transition state of *cis/ trans* isomerism because of mass transfer problems and on-column isomerization of the conformers. The similar elution pattern of peptides on β-CD and calix[6]arene silica gels as well as their comparable cavity sizes support the assumption of host-guest complex formation.

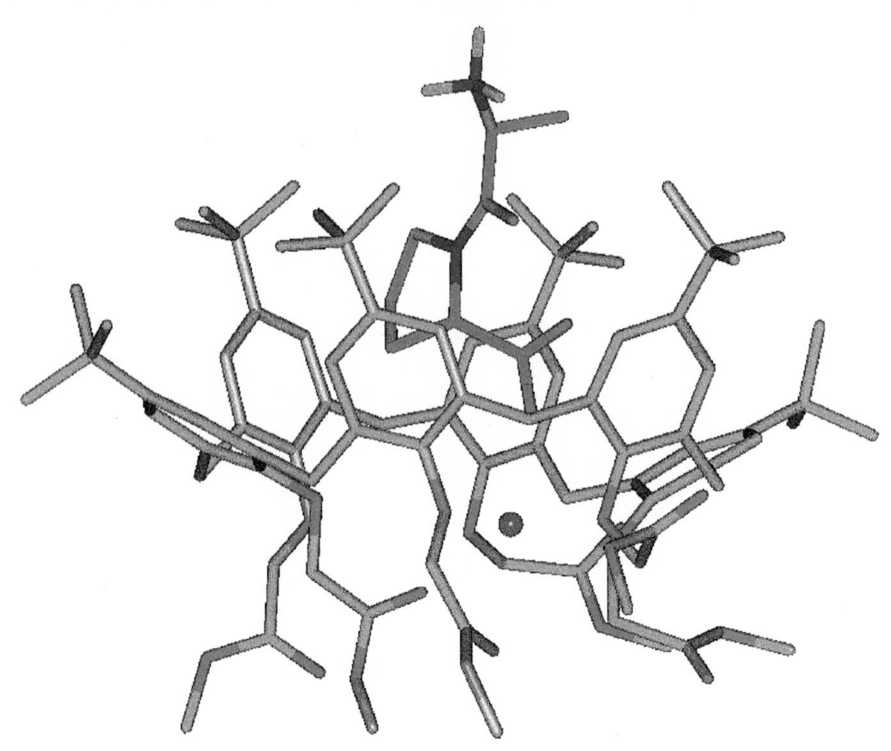

Fig. 14.4. Side view of the inclusion complex of the *p-tert*-butylcalix[6]arene ester pre-organized by a sodium ion and the *trans* isomer of L-Ala-L-Pro. For clarity hydrogen atoms of the calixarene are omitted (Gebauer 1998, unpublished)

Molecular modeling studies confirm our assumption. Using the force field AMBER 4.1 molecular dynamics (MD) simulations were carried out with *p-tert*-butylcalix[n]arene derivatives (n = 4, 5, 6) and *cis* and *trans* isomers of the dipeptides in a theoretical waterbox. Our findings show that only *p-tert*-butylcalix[6]arene is able to form stable inclusion complexes (Fig. 14.4.). The driving forces base on electrostatic energies due to the interaction of the isomers and a complexed cation (Na^+) at the lower rim of calix[6]arene. Furthermore, the cavity size of calix[6]arene is a well preorganized binding site to include these molecules. On the other hand the cavities of calix[4]arene and calix[5]arene are too small for selective host-guest interactions. Only *p-tert*-butylcalix[5]arene is able to include the *trans* isomer of L-Ala-L-Pro because of its relative small stretch shape. These findings are in good agreement with our HPLC results. These modelling experiments explain that the selective separations on [6]Arene and the peak splitting on [5]Arene are results of inclusion complex formations. The elution profiles of L-Phe-L-Pro presented in Fig. 14.3. show baseline separations on all stationary phases due to the hydrophobic surface of this dipeptide. The retention factors are higher than those of L-Ala-L-Pro and increase with the calixarene ring size. The typical plateau (rising baseline between the peaks) is monitored only when the RP 18 phase was used. Similar plateaus are also observed on the calix[n]arene phases. That means, that the chromatographic process is only slowly going on, the interconversion of the conformers takes place during the chromatographic run and no complete discrimination is achieved compared to the separation on β-CD phase. However, not to be excluded is that hydrophobic moieties of the dipeptides interact with the cavity of calixarenes showed by MD simulations.

3
Capillary Zone Electrophoresis (CZE)

The capillary electrophoresis as a time resolving method represents a new method which allows the detection of conformational changes in peptides. CZE has an advantage over HPLC methods in so far as the analytes remain in solution and no interaction exists to a solid phase. The mobilities of peptides, however, depend on parameters which do not vary with the conformation, such as molecular weight and the number of amino acid residues (Hilser and Freire 1995). However, Kilar and Hjerten (1993) observed changed electrophoretic mobilities for unfolded compared to native proteins which can be attributed to changes in surface polarity and surface charges, but also to an alteration of the geometry of the whole molecule. Referring to the fundamentals of CZE the effective mobility of a compound depends strongly on the experimental conditions with a definitive charge and ionic radius, but also on the frictional resistance during the migration in an electric field. Therefore, it is possible that sterical differences such as the hydrodynamic radius or the symmetry of the molecule can contribute to the electrophoretic discrimination of *cis/trans* conformers of prolyl peptides. Compared to the presented HPLC conditions which allow the resolution of dynamically interconverting species, the commonly prevailing conditions for the CZE of *cis/trans* isomers are to be described as follows (Ma et al. 1995, 1998):
- short capillaries,
- low working temperatures,
- high voltage,
- low viscosity of the electrolyte system.

The main advantage of CZE compared to HPLC represents the possibility of obtaining direct interrelationships to kinetic and thermodynamic parameters of the *cis/trans* interconversion of the peptide bond, such as *cis/trans* isomer ratio or the rate constants of the *cis* to *trans* interconversion and the corresponding back reaction.

In recent years several studies on capillary electrophoretic separations of *cis/trans* peptide bond isomers have been published (Thunecke et al. 1998, Brandsch et al. 1998, Molle et al 1994). The rotational isomers of enalapril maleate, an alanyl-proline derivative which is used as an angiotensin inhibitor, were resolved by micellar CZE with SDS (Qin et al. 1992). In addition to the differences in hydrodynamic radii in this case hydrophobic interactions with the SDS micelles supported the separation of the peptide conformers. Meyer et al. (1994) succeeded in the electrophoretic conformer separation of proline containing thioxopeptides of the type Ala-Xaa-ψ [CS-N]-Pro-Phe-4-nitroanilides. These compounds show up to 100 fold delayed isomerization rates compared to the corresponding oxopeptides, while the isomer ratio should not be affected by the substitution of oxygen by sulfur. The particularly extreme requirements of CZE of peptide bond isomers resulted in the development of special methods. Moore and Jorgenson (1995) describe the *cis/trans* isomer separation of proline containing di- and oligopeptides (ß-casomorphins) on extremely short capillaries with a small inner diameter (6μm) after derivatization with fluorescinisothiocyanate and ultra short injection times (5–50 ms) and called this approach Fast-CZE. The application of subzero temperatures was published by Ma et al.(1995). The separation of proline containing di- to hexapeptides was performed down to −17 °C with glycerol diminishing the viscosity of the borate buffer at such very low temperatures. Peptides with two Xaa-Pro bonds could be resolved into a conformer pattern of four peaks. The migration order *cis* before *trans* was confirmed by accompanying RP-HPLC and molecular modeling studies.

The application of carboxymethylated ß-Cyclodextrin (CMBCD) as additive to the bulk electrolyte represents an approach to separate the *cis/trans* conformers of prolyl peptides by inclusion complexation in CZE (Rathore and Horvath 1998). In this way the complete resolution of angiotensin I conformers succeeded within few minutes. A decrease of isomer mobilities with increasing CMBCD concentrations and a slight shift in the *cis/trans* ratio hint at the formation of peptide-cyclodextrin complexes.

In conclusion, low temperature HPLC and CE supported by cavity interaction has been found to be applicable in the detection of conformational changes in proline containing peptides, for the determination of thermodynamic and kinetic parameters of *cis/trans* interconversion and for the fractionation of *cis* and *trans* conformers as valuable substrates for advanced studies. Shifts in equilibrium distribution of conformers by inclusion complexation or interactions to a hydrophobic, solid support have to be taken into consideration.

References

Beissinger M, Buchner J (1998) How chaperons fold proteins. Biol. Chem. 379: 245–259

Böhmer V (1995) Calixarene-Makrocyclen mit (fast) unbegrenzten Möglichkeiten. Angew. Chem. 107: 785–818

Brandsch M, Thunecke F, Küllertz G, Schutkowski M, Fischer G, Neubert K (1998) Evidence for the absolute conformational specificity of the intestinal H^+/peptide symporter, PEPT1. J. Biol. Chem. 273: 3861–3864

Brandts JF, Halvorson HR, Brennan M (1975) Consideration of the possibility that the slow step in protein denaturation is due to cis-trans isomerism of proline residues. Biochemistry 14: 4953–4963

Fischer G, Bang H, Mech C (1984) Nachweis einer Enzymkatalyse für die cis-trans Isomerisierung der Peptidbindung in prolinhaltigen Peptiden. Biomed. Biochem. Acta 43: 1104–1111

Fischer G (1994) Peptidyl-prolyl *cis/trans* isomerases and their effectors. Angew. Chem. Int. Ed. Engl. 33: 1415–1436

Friebe S, Krauss GJ, Nitsche H (1992) High-performance liquid chromatographic separation of *cis-trans* isomers of proline containing peptides. I. Separation on cyclodextrin silica. J. Chromatogr. 598: 139–145

Friebe S, Hartrodt B, Neubert K, Krauss GJ (1994), High-performance liquid chromatographic separation of *cis-trans* isomers of proline containing peptides. II. Fractionation in different cyclodextrin systems. J. Chromatogr. A 661: 7–12

Friebe S, Gebauer S, Krauss GJ, Goermar G, Krueger J (1995) HPLC on calixarene bonded silica gels. I. Characterization and applications of the p-tert-butyl-calix[4]arene. J. Chromatogr. Sci. 33: 281–284

Gebauer S, Friebe S, Gübitz G, Krauss GJ (1998a) HPLC on calixarene bonded silica gels. II. Separation of regio- and stereoisomers on *p-tert*-butylcalix[n]arene phases. J. Chromatogr. Sci. 36:383–387

Gebauer S, Friebe S, Scherer G, Gübitz G, Krauss GJ (1998) HPLC on calixarene bonded silica gels. III. Separations of *cis/trans* isomers of proline containing peptides. J. Chromatogr. Sci. 36: 388–394

Gesquiere JC, Diesis E, Cung MT, Tartar A (1989) Slow isomerization of some proline-containing peptides inducing peak splitting during RP-HPLC. J. Chromatogr. 478: 121–129

Gustafsson S, Eriksson B, Nilsson I (1990) Mutiple peak formation in reversed-phase liquid chromatography of ramipril and ramiprilate. J. Chromatogr. 506: 75–83

Gutsche CD (1989) Calixarenes. Monographs in Supramolecular Chemistry, ed. J. F. Stoddard, The Royal Society of Chemistry, Cambridge

Henderson DE, Mello JA (1990) Physico-chemical studies of biologically active peptides by low temperature RP-HPLC. J. Chromatogr. 499: 79–88

Hilser VJ, Freire E (1995) Quantitative analysis of conformational equilibrium using capillary electrophoresis: applications to protein folding. Anal. Biochem. 224: 465–485

Hübner D, Fischer G, Ströhl D, Kleinpeter E, Hartrodt B, Brandt W (1990) Conformational analysis of proline containing peptides by multinuclear NMR-spectroscopy. J. Anal. Chem. 337: 131–132

Janowski B, Wollner S, Schutkowski M, Fischer G (1997) A protease-free assay for peptidyl-prolyl *cis/trans* isomerases using standard peptide substrates. Anal. Biochem. 252: 299–307

Joshua H (1991) Temperature-dependent chromatographic fractionation of FK-506 conformational isomers. Rainin Dynamax Review

Kalman A, Thunecke F, Schmidt R, Schiller P, Horvath C (1996) Isolation and identification of peptide conformers by RP-HPLC and NMR at low temperature. J. Chromatogr. A 729: 155–177

Kilar F, Hjerten S (1993) Unfolding of serum transferrin in urea studied by high performance capillary electrophoresis. J. Chromatogr. 638: 269–276

Kramer ML, Fischer G (1997) FKBP-like catalysis of peptidyl-prolyl bond isomerization by micelles and membranes. Biopolymers 42: 49–60

Küllertz G, Luthe S, Fischer G (1998) Semiautomated microtiter plate assay for monitoring peptidyl-prolyl *cis/trans* isomerase activity in normal and pathological human sera. Clinic. Chem. 44: 502–508

Lang K, Schmid F, Fischer G (1987) Catalysis of protein folding by prolyl isomerase. Nature 329: 268–270

Lebl M, Fang S, Hruby VJ (1991) HPLC of peptides at reduced temperatures: Separation of isomers. J. Chromatogr. 586: 145–148

Lin LN, Brandts JF (1978) Further evidence that the slow phase in protein unfolding and refolding is due to proline isomerisation. Biochemistry 17: 4102–4110

Ma S, Kalman S, Kalman A, Thunecke F, Horvath C (1995) Capillary zone electrophoresis at subzero temperatures I. Separation of the *cis* and *trans* conformers of small peptides. J. Chromatogr. A 716: 167–182

Ma S, Horvath C (1998) Capillary zone electrophoresis at subzero temperatures. III. Operating Conditions and separation efficiency. J. Chromatogr. A 825: 81–88

Melander WR, Jacobson J, Horvat C (1982) Effect of molecular structure and conformational change of proline containing dipeptides in reversed phase chromatography. J. Chromatogr. 234: 269–276

Meyer S, Jabs A, Schutkowski, M Fischer G (1994) Separation of *cis/trans* isomers of a prolyl peptide by capillary zone electrophoresis. Electrophoresis 15: 1151- 1157

Molle D, Leonil J, Bouhallab S (1995) Separation of two proline-containing peptides by capillary elelctrophoresis. Anal. Biochem. 225:161–162

Moore AW, Jorgenson JW (1995) Resolution of *cis* and *trans* isomers of peptides containing proline using capillary electrophoresis. Anal. Chem. 67: 3464–3475

Nishikawa T, Hasumi H, Suzuki S, Kubo H, Ohtani H (1994) Interconversion of cyclosporin molecular form inducing peak broadening , tailing and splitting during reversed-phase liquid chromatography. Chromatographia 38: 359–364

Neubert K, Hartrodt B, Born I, Barth A, Rüthrich HL, Grecksch G, Schrader U, Liebmann C (1990) Structural Modifications of β-Casomorphin-5 and Related Peptides. Proceedings of the 1st International Symposium on β-Casomorphins and Related Peptides, eds F Nyberg, V Brantl, Fyris-Tryck, Uppsala.

Owens-Grillo JK, Hoffmann K, Hutchinson KA, Yem AW, Deibel MR, Handschumacher RE, Pratt WB (1995) The cyclosporin A-binding immunophilin Cyp-40 and the FK506-binding immunophilin hsp56 bind to a common site on hsp90 and exist in independent cytosolic heterocomplexes with the untransformed glucocorticoid receptor. J. Biol. Chem. 270: 20479–20484

Pearlman DA, Case DA, Caldwell JW, Ross WS, Cheatham III TE, Ferguson DM, Seibel GL, Singh UC, Weiner PK and Kollman PA (1995) AMBER 4.1. University of California, San Francisco

Rathore AS, Horvath C (1997) Capillary zone electrophoresis of interconverting *cis-trans* conformers of peptidyl-proline dipeptides: Estimation of the kinetic parameters. Electrophoresis 18: 2935–2943
Rathore AS, Horvath C (1998) Cyclodextrin aided separation of peptides and proteins by capillary zone electrophoresis. J. Chromatogr. A 796: 367–373
Rusconi L, Perseo L, Frantoi L, Montecucci P (1985) RP-HPLC characterization of a novel proline rich tryptophyllin. J. Chromatogr. 349: 117–130
Schmidt R, Kalman A, Chung N, Lemieux C, Horvath C, Schiller P (1995) Structure-activity relationships of dermorphin analogues containing N-substituted amino acids in the 2-position of the peptide sequence. J. Peptide Protein Res. 46: 47–55
Schreiber SL (1991) Chemistry and biology of the immunophilins and their immunosuppressive ligands. Science 251: 283–287
Schutkowski M, Bernhardt A, Zhou XZ, Shen M, Reimer U, Rahfeld JU, Lu KP
Fischer G (1998) Role of phosphorylation in determining the backbone dynamics of the serine/threonine-proline motif and Pin1 substrate recognition. Biochemistry 37: 5566–5575
Thunecke F, Fischer G (1998) Separation of *cis/trans* conformers of human and salmon calcitonin by low temperature capillary electrophoresis. Electrophoresis 19: 288–294
Vicens J, Böhmer V (1991) Topics in inclusion science. Calixarenes–a versatile class of macrocyclic compounds. Ed. J. Vicens , Böhmer V., Kluwer Academic Publishers, Dordrecht
Ward TM, Armstrong DW (1988) Cyclodextrin-stationary phases. In: Zief M, Cranein LY (eds.) Chromatographic chiral separation (Chromatographic science Series, Vol 40), Marcel Dekker, New York, pp 131–163
Qin XZ, Dominic P, Tsai, PI, Tsai EW (1992) Determination and rotamer separation of enalapril and enalapril maleate by capillary electrophoresis. J.Chromatogr. 626: 251–258

CHAPTER 15

The Perception of Hydrophobic Clusters in the Native and Partially Unfolded States of a Protein

G. VANDERHEEREN[1] and I. HANSSENS[1]

1 Introduction

1.1 Hydrophobic Interactions as the Driving Force in Protein Folding

The folding mechanism of globular proteins is dominated by hydrophobic interactions. They are believed to play a main role in the formation of α-helices and β-structures and also in the further folding into a partially folded protein with fluctuating tertiary structure (Miranker and Dobson 1996). This fluctuating protein is called a 'molten globule'. The final transition to the native structure with tight close-packed contacts between the amino acid side-chains is governed by short-range electrostatic interactions (Levitt et al. 1997). The residues in the hydrophobic core of the partially (un)folded protein are accessible to external agents and can effect binding interactions with hydrophobic probes, such as 1,1'-bis(4-anilino-5-naphthalenesulfonate) (bis-ANS). As the fluorescence quantum yield of those molecules greatly increases upon binding to hydrophobic sites, this interaction with molten globule-like intermediates of the protein leads to a large fluorescence increase. This property has been used to estimate the population of the molten globule state in protein folding studies (Ptitsyn et al. 1990; Teschke et al. 1993; Das and Surewicz 1995). The use of these hydrophobic probes for the characterization of the molten globule state has been questioned by other research groups since they are suspected of inducing changes in the native protein conformation. In this way Coco and Lecomte (1994) revealed perturbations in the structure of native-like apomyoglobin upon ANS-binding. Shi et al. (1994) observed that bis-ANS, which preferentially binds to the molten globule state of DnaK, shifts the equilibrium from the native to the molten globule state. Engelhard and Evans (1995) observed that ANS, used in a kinetic study of the α-lactalbumin refolding process, stabilizes the dye-bound intermediates.

With this contribution we demonstrate that bis-ANS is helpful in obtaining information on the degeneration of the hydrophobic domains at different stages of the thermal unfolding of α-lactalbumin. Furthermore we discuss the interaction of the hydrophobic probe with the native protein state of α-lactalbumin and

[1] Interdisciplinary Research Center, Katholieke Universiteit Leuven, Campus Kortrijk, B-8500 Kortrijk, Belgium.

explore the structural perturbations of the native protein as a consequence of the above mentioned interaction.

1.2 α-Lactalbumin

α-Lactalbumin, the protein of this study, has a single polypeptidechain of 123 amino acids. The amino acid sequence and the conformation of α-lactalbumin resembles that of c-type lysozyme. However, both homologous proteins have totally different biological functions and stabilities. The shape of both native proteins resembles a slightly elongated sphere nearly cut in two by a large crevice. One half of the molecules mainly consists of four α-helices, the other half contains a β-structured sheet. In the domain connecting the halves, the α-lactalbumins from different animals contain a site able to bind one Ca^{2+} in a characteristic way. Most c-type lysozymes do not bind Ca^{2+}. The high affinity Ca^{2+}-binding site has been identified as a short loop structure, consisting of ten residues (79–88), five of which contribute liganding atoms (Stuart et al. 1986). At room temperature in the absence of Ca^{2+}, α-lactalbumin is in the above mentioned molten globule state. To regain the native protein state, the protein solution needs to be cooled to almost 0 °C or an excess of Ca^{2+} should be added. Ca^{2+}-bound α-lactalbumin only starts to unfold near 60 °C. In protein folding studies α-lactalbumin has received considerable attention because the transitions among the compact native state, the molten globule state and more unfolded states can easily be realized.

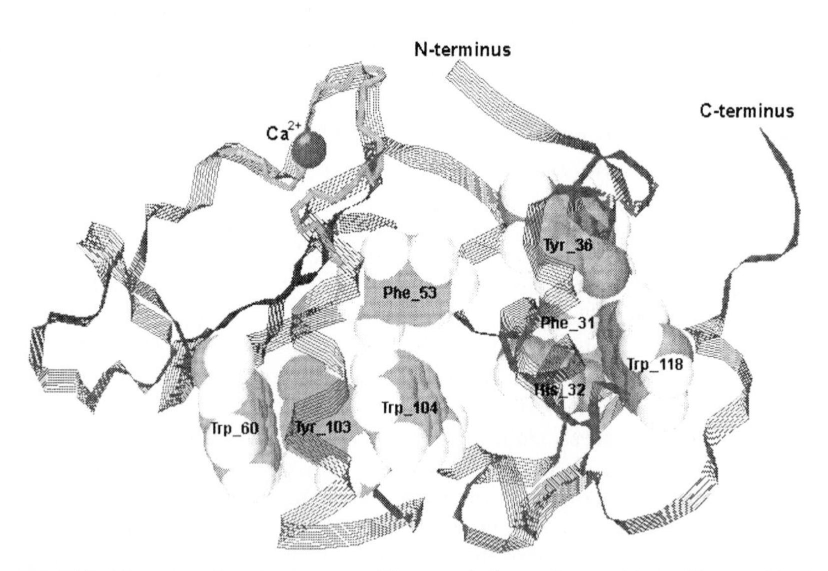

Fig. 15.1. The α-lactalbumin structure. The strand shows the peptide backbone with the Ca^{2+}-binding site that stabilizes the native structure and the two domains with the different helices and β-sheet. The clusters of aromatic residues are space-filled. The picture was created using the structure of baboon α-lactalbumin (PDB code 1ALC)

For this hydrophobic interaction study it is worth mentioning that an α-lactalbumin molecule possesses two clusters of aromatic residues. The planes of these aromatic rings are approximately perpendicular (Fig. 15.1). One of the clusters (Phe-53, Trp-60, Tyr-103, Trp-104) is situated in the crevice and contains amino acids belonging to the two structural halves of the protein, the other one (Phe-31, His-32, Tyr-36, Trp-118) is completely situated in the half with the α-helices (Acharya et al. 1989).

2 Materials and Methods

2.1 Materials

Bovine α-lactalbumin (BLA) is from Sigma (St Louis, Missouri). The protein is decalcified on a Sephadex G-25 column in 10 mM HCl, lyophilized, and stored at −20 °C until use. Goat α-lactalbumin (GLA) is prepared from fresh milk whey (Vanderheeren et al. 1998) and stored as described for BLA. In solutions, the concentration of both proteins is determined from the absorption at 280 nm (ε_{280} = 28 500 M^{-1} cm^{-1}).

The hydrophobic probe 1,1'-bis(4-anilino-5-naphthalenesulfonate) (bis-ANS) is from Molecular Probes Inc. (Eugene, Oregon). Its concentration is determined from its absorption at 385 nm (ε_{385} = 16 790 M^{-1} cm^{-1}).

All experiments are performed in 10 mM Tris-HCl buffer (pH 7.5) containing 2 mM Ca^{2+} or 2 mM EGTA, for Ca^{2+}-bound protein and apo-protein (Ca^{2+}-free), respectively. Generally, mixtures of α-lactalbumin and bis-ANS are incubated overnight at the desired temperature.

2.2 Circular Dichroism

The CD measurements are carried out on a Jasco J-600 spectropolarimeter (Tokyo, Japan). Cuvettes of 5 (or 10) mm and 1 mm are used for the near-UV and far-UV regions, respectively. The α-lactalbumin concentration is about 25 μM. Circular dichroism signals are monitored as ellipticity changes (expressed as deg cm^2 $dmol^{-1}$). Near 220 nm (far-UV) the ellipticity changes are generally dominated by peptide groups in helical structures, while the CD measurements near 270 nm (near-UV) monitor aromatic groups fixed in a specific orientation due to tertiary structure.

The protein-bis-ANS mixtures for thermal transition studies are preincubated overnight at the lowest temperature. At each temperature of the transition curve, the measurements are started 5 minutes after temperature equilibration of the sample. Evidence that equilibrium states are obtained is presented by the fact that the ellipticity values are identical in heating and cooling runs, provided they have not been exposed to more than 70 °C for several minutes.

2.3 Fluorescence Measurements

The fluorescence measurements are performed with an Aminco SPF-500 spectrofluorimeter (Rochester, New York). The equilibrium binding data for the interaction of bis-ANS with α-lactalbumin is obtained from two fluorescence titration curves, both at equal pH (pH 7.5), temperature and Ca^{2+}-content. First, the limit fluorescence increase of 1 μM bis-ANS (ΔI_{max}) is determined in the presence of increasing amounts of protein. Next, this value is used to calculate the concentration of bound bis-ANS from the fluorescence increase on titrating 1 μM protein with an excess of bis-ANS. It is assumed that the ΔI_{max}-values obtained from the former titrations represent the extra fluorescence increase (ΔI) of 1 μM bound bis-ANS in the latter one (Vanderheeren and Hanssens 1994).

For both fluorescence titrations the excitation is at 465 nm, which is situated on the edge of the bis-ANS absorption peak. At this wavelength, even at 400 μM bis-ANS, the absorbance does not exceed the value of 0.2. The fluorescence emission intensity between 515 and 550 nm is integrated. The background fluorescence of free bis-ANS is subtracted and the resulting fluorescence increase is corrected for inner filter effects.

If tryptophan and bis-ANS are close to each other, they constitute a good fluorescence energy-transfer donor-acceptor pair. For these energy-transfer experiments the excitation is at 290 nm, the absorption maximum for tryptophan. The fluorescence emission intensity of tryptophan residues between 300 and 420 nm is measured and corrected for inner filter effects at the emission and excitation wavelengths.

2.4 Photolabelling of bis-ANS to α-Lactalbumin

It has been shown that bis-ANS can be covalently photobound to various proteins (Gorovits et al. 1995; Seale et al. 1995). For some measurements we needed a large association of bis-ANS to native Ca^{2+}-bound goat α-lactalbumin with small free dye concentration. To obtain this, a mixture of 35 μM GLA and 150 μM bis-ANS in Ca^{2+} buffer, thermostated at 37 °C, is irradiated at a 366 nm wavelength for 2 hours using a UV 131 000 lamp from Desaga (Heidelberg, Germany). Next the irradiated sample is applied to and eluted from a PD-10 column containing Sephadex G-25 (Pharmacia, Sweden). The protein with the covalently bound bis-ANS passes through the column together with the void volume while the free and non-covalently bound bis-ANS is retarded. In the eluate obtained in this way, on average 3.3 bis-ANS molecules are covalently bound to 1 Ca^{2+}-GLA molecule.

3
The Degeneration of the Hydrophobic Clusters During the Thermal Unfolding of Bovine α-Lactalbumin

3.1
Aim

It is generally accepted that the progressive accumulation of hydrophobic interactions plays the main role in protein folding. Nevertheless no systematic studies of the hydrophobic behaviour of globular proteins during (un)folding are available. Therefore, we have related the thermal unfolding of apo- and Ca^{2+}-bound bovine α-lactalbumin (BLA), determined by CD measurements at 220 and 270 nm, with the ability of the hydrophobic groups to interact with bis-ANS. The measurements allow to follow the progressive degeneration of the hydrophobic regions during the thermal unfolding of Ca^{2+}-free and Ca^{2+}-bound BLA.

3.2
Thermal Unfolding of BLA Monitored by Circular Dichroism

In Fig.s 15.2 A and B the temperature dependence of the mean residue ellipticity at 220 and 270 nm of BLA in 2 mM EGTA and in 2 mM Ca^{2+} (pH 7.5) is presented. Upon heating BLA to 25 °C in 2 mM EGTA (Fig. 15.2, filled squares) the ellipticity in the near-UV region changes strongly while the far-UV ellipticity is poorly

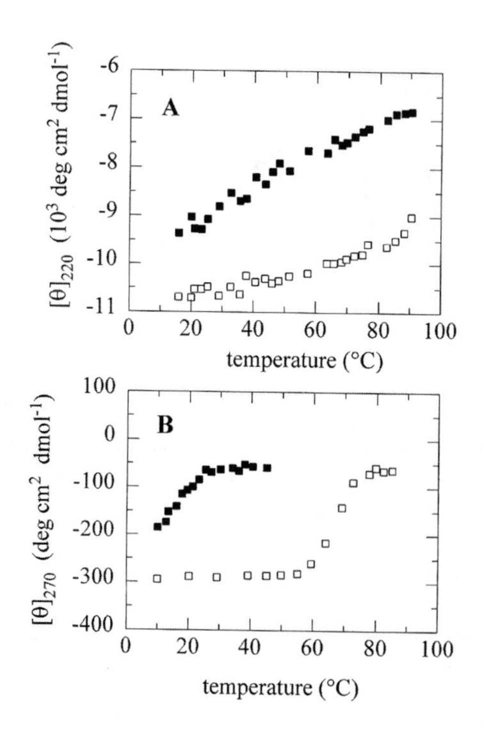

Fig. 15.2. Thermal transition curves of BLA measured by the ellipticity change at 220 (A) and 270 nm (B) and in 2 mM EGTA (filled squares) or 2 mM Ca^{2+} (open squares) (Vanderheeren and Hanssens 1994)

effectuated. It indicates that the tertiary protein structure is loosened while the secondary structure elements are largely conserved. By this partial unfolding, the so called, molten globule state is obtained. Above 25 °C, the numeric value of the ellipticity at 220 nm decreases gradually. The temperature dependence of the ellipticity is not known. But the fact that the ellipticity decrease is more pronounced in 2 mM EGTA than in 2 mM Ca^{2+} proves that the conformation in 2 mM EGTA changes more as a function of the temperature than in 2 mM Ca^{2+}.

In 2 mM Ca^{2+} (Fig. 15.2, open squares) the ellipticity changes indicate that up to 60 °C the compact tertiary structure of BLA is conserved. The aromatic groups determining the ellipticity at 270 nm are kept in fixed orientations and the ellipticity changes at 220 nm indicate that, if any change of secondary structure occurs at all, it will be small. Between 60 and 75 °C the tertiary structure loosens its compact packing while no pronounced unfolding of the secondary structure is observed. At 80 °C the tertiary structure of Ca^{2+}-bound BLA is unfolded. However the ellipticity at 220 nm indicates that the secondary structure of BLA is more preserved in 2 mM Ca^{2+} than in 2 mM EGTA. This observation indicates that Ca^{2+} remains associated to BLA after thermal destabilization of its tertiary structure. Furthermore, the associated ion stabilizes elements of secondary structure in the partially unfolded protein. By X-ray diffraction it has been demonstrated that the Ca^{2+}-binding loop of (baboon) α-lactalbumin is at the same time a part of the amino-terminal side of the 3_{10}-helix (residues 76–82) in the β-domain and a part of the carboxyl-terminal side of the α-helix C (residues 86–99) in the α-domain (Acharya et al. 1989). It is clear that at least these elements of the secondary structure will be stabilized by Ca^{2+}-binding.

3.3
Determination of the Binding Properties of bis-ANS to BLA

The binding properties for the interaction of bis-ANS with apo- and Ca^{2+}-BLA are obtained from two series of fluorescence titration curves, as described in Materials and Methods. The titrations in the absence of Ca^{2+} at 25, 50, 70 and 80 °C are presented in Fig.s 15.3 A and B. The corresponding Scatchard plots for the bis-ANS binding to 1 μM apo-BLA are presented in Fig. 15.3 C. At 25 °C, the strong fluorescence increase at the start of the titrations of 1 μM bis-ANS with apo-GLA and of those of 1 μM apo-GLA with bis-ANS (Fig.s 15.3 A and B, squares) is characteristic for the strong binding of fluorophore to the protein. The limited fluorescence increase obtained in the second titration is nearly twice that obtained in the first titration indicating that two probe molecules bind to apo-BLA. This is also deduced from the X-intercept in the corresponding Scatchard plot (Fig. 15.3 C, squares). In addition, the linear shape of that plot indicates that the two probe molecules independently bind to apo-BLA with nearly equal strength. As an α-lactalbumin molecule possesses two distinct aromatic clusters, it seems reasonable to suppose that one bis-ANS molecule can penetrate into each cluster. At 55 °C (Fig. 15.3 C, down triangles), the Scatchard plot is hyperbolic. The straight line slopes drawn to the extremities of the curve near the X- and Y-axes, intersect the X-axis at about 3 and 1 μM bis-ANS, respectively. This indicates that one molecule BLA can bind 3 bis-ANS molecules of which one

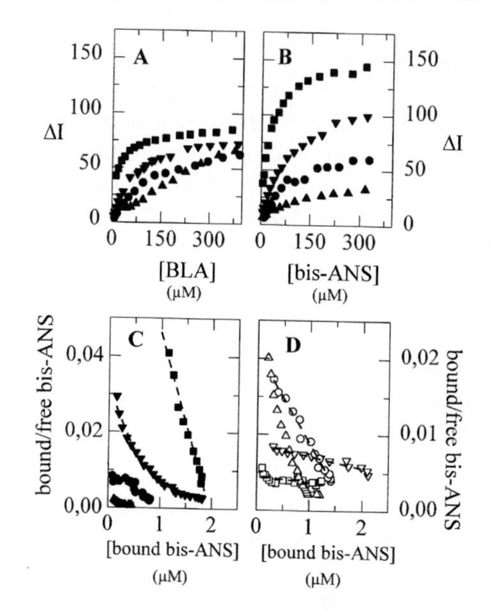

Fig. 15.3. Fluorescence titrations of 1 μM bis-ANS with BLA (A) and 1 μM BLA with bis-ANS (B) at four different temperatures. Scatchard plots for the titrations of 1 μM BLA with bis-ANS in 2 mM EGTA (C) or 2 mM Ca^{2+} (D). The temperatures are 25 °C (squares), 55 °C (down triangles), 70 °C (circles) and 80 °C (up triangles)

molecule binds with clearly stronger affinity than the two others. At 70 and 80 °C (Fig. 15.3 C, circles and up triangles) linear Scatchard plots with X-intercepts at about 1 and 0.5 μM bis-ANS are obtained. Their slopes indicate that the binding becomes weaker at higher temperature.

As hydrophobic interactions are endothermic, the bis-ANS-binding constant for the interaction with an invariable hydrophobic site should increase with increasing temperature. Therefore, the weaker binding of bis-ANS can only be explained if heating causes the apolar clusters in the BLA molecule to reduce their hydrophobic character. The temperature-dependent loss of protein conformation, observed by the loss of ellipticity at 220 nm (Fig. 15.2 A, filled squares), implies dissipation of the hydrophobic clusters resulting in the temperature-dependent reduction of the bis-ANS-binding constants. In the first stage, the loss of protein conformation may cause a splitting of one of the clusters into two smaller regions with weak binding constants for the hydrophobic probe. Such phenomenon may explain the transformation of one of the strong binding sites observed at 25 °C to two weaker binding sites at 55 °C. The weakening of the bis-ANS binding at higher temperatures is indicative of the further degeneration of the hydrophobic clusters.

In 2 mM Ca^{2+} the tertiary structure of Ca^{2+}-BLA unfolds between 60 °C and 80 °C (Fig. 15. 2 B, open squares), while the secondary structure (Fig. 15. 2 A, open squares) is well conserved. The Scatchard plot at 70 °C and 2 mM Ca^{2+}(Fig. 15.3 D, circles) is representative for the independent interaction of two bis-ANS molecules with one BLA molecule in a similar way as was observed for apo-BLA near 25 °C. Consequently, a kind of Ca^{2+}-loaded molten globule state of BLA is obtained. By further unfolding of Ca^{2+}-BLA to 80 °C (Fig. 15. 3 D, up triangles) one of the strong binding sites transforms to one or two weaker binding sites, as was observed for apo-BLA at 55 °C, and presumably results from the splitting of one of the two hydrophobic clusters.

3.4 Conclusion

This study demonstrates that a more advanced research of the interaction of hydrophobic probes offers information on the degeneration of the hydrophobic domains during the protein denaturation.

In the molten globule state of α-lactalbumin, the two clusters with strong hydrophobic character seem to be accessible. By heating the hydrophobic domains degenerate, but the temperature of destabilization is different for the two domains.

A similar evolution of the hydrophobic behavior is observed for apo- and Ca^{2+}-BLA, but the destabilization steps of Ca^{2+}-BLA occur at higher temperatures than those of apo-BLA. The results imply that Ca^{2+} remains associated with BLA after the thermal induced destabilization of its native tertiary structure.

4 The Perturbations Induced by the Hydrophobic Probe in the Native State of Goat α-Lactalbumin

4.1 Aim

It has been shown that bis-ANS is helpful in following the behavior of the hydrophobic domains along the thermal unfolding pathway of an α-lactalbumin.

In this chapter we deal with the ability of bis-ANS to penetrate into the hydrophobic core of native α-lactalbumin inducing perturbations in the native protein structure. Again we distinguish the apo- and the Ca^{2+}-bound protein. For this study, goat α-lactalbumin is preferred to bovine α-lactalbumin because at low temperatures a clear native state of the goat apo-protein is obtained with spectroscopic properties similar to those of the Ca^{2+}-bound protein.

4.2 Thermal Unfolding of GLA

The squares in Fig. 15.4 represent the ellipticity at 270 nm of GLA at different temperatures in the presence and absence of Ca^{2+}. From the Fig. it can be deduced that in a solution containing 10 mM Tris-HCl and 2 mM EGTA (filled squares) the protein is in the fully native conformation only at temperatures below 10 °C. By replacing EGTA for Ca^{2+} (open squares), the native protein conformation remains stable up to 55 °C. The increased thermal stability resulting from Ca^{2+} binding is a general property of all α-lactalbumins (Acharya et al. 1989).

Next, we concentrate on the comparison of the effects of bis-ANS interaction with native GLA at 4 °C in 2 mM EGTA with those at 37 °C in 2mM Ca^{2+}. In both conditions the protein is in the native state close to the thermal transition and shows the same spectroscopic properties. Therefore, although the temperatures are different, the protein can be expected to be equally susceptible to conformational changes.

4.3 Equilibrium Binding of bis-ANS to Native GLA

We determined the binding properties of bis-ANS to native GLA in the previously selected conditions (4 °C and 2 mM EGTA or 37 °C and 2 mM Ca^{2+}) by performing two series of fluorescence titrations, as described in Materials and Methods.

The strong fluorescence increase at the start of the titrations of 1 μM bis-ANS with apo-GLA and 1 μM apo-GLA with bis-ANS (Fig.s 15.5 A and B, filled squares) are characteristic for the strong binding of (one or more) fluorophore molecules to the Ca^{2+}-free protein. As can be expected for such a strong binding, the fluorescence increase for the addition of bis-ANS to apo-GLA (Fig. 15.5 B, filled squares) readily tends to a limit. However, this limit is not obtained by the addition of apo-GLA to bis-ANS (Fig. 15.5 A, filled squares). It is judged that the most plausible explanation for this continued fluorescence increase at higher

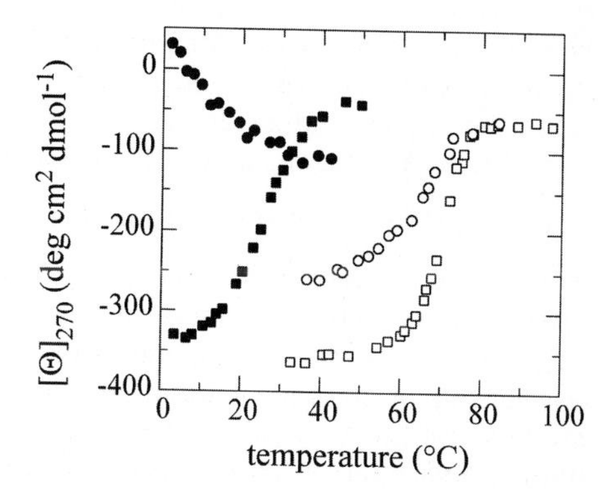

Fig. 15.4. Thermal transition curves of GLA measured by the ellipticity change at 270 nm. Conditions: 25 μM GLA in Tris-HCl buffer containing 2 mM EGTA (filled squares) or 2 mM Ca^{2+} (open squares), after equilibration with 60 μM bis-ANS in 2 mM EGTA (filled circles), containing 3.3 photo-labeled bis-ANS molecules per GLA in 2 mM Ca^{2+} (open circles) (Vanderheeren et al. 1998)

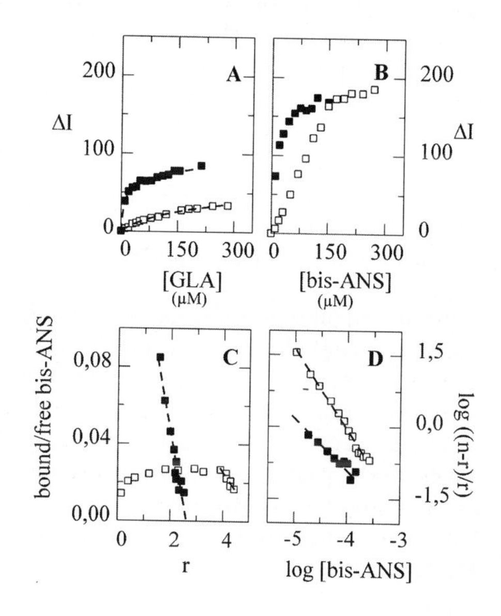

Fig. 15.5. Fluorescence titrations of 1 μM bis-ANS with GLA (A) and 1 μM GLA with bis-ANS (B). Scatchard plots (C) and Hill plots (D) for the titration of 1 μM GLA with bis-ANS. Conditions: 4 °C and 2 mM EGTA (filled squares) or 37 °C and 2 mM Ca^{2+} (open squares) (Vanderheeren et al. 1998)

apo-GLA concentrations is self-association of the protein. By such self-association bis-ANS is expected to be further protected from the water phase. This provokes a further increase of the fluorescence at protein concentrations for which the probe is already completely bound. The maximum fluorescence increase of 1 μM bis-ANS at a low protein concentration (as used in the experiments presented in Fig. 15.5 B, filled squares) is estimated 66 ± 13 and 38 ± 7 (arbitrary units) for apo-GLA and Ca^{2+}-GLA, respectively. These maximum values are used to calculate the concentration of bound bis-ANS from the fluorescence increase observed on titrating 1 μM GLA with bis-ANS (Fig. 15.5 B). The titration curve for the titration of 1 μM native Ca^{2+}-GLA with bis-ANS at 37 °C (Fig. 15.5 B, open squares) is slightly S-shaped. This is indicative of positive cooperativity for the dye binding which is confirmed by the clear concave-downward shape of the according Scatchard plot (Fig. 15.5 C, open squares). On the other hand, neither the hyperbolic titration curve, nor the linear Scatchard plot gives any indication of cooperative behavior for the bis-ANS interaction with native apo-GLA at 4°C (Figs. 15.5 B and C, filled squares). From the abscis intercept of the Scatchard plot the total number of bis-ANS binding sites is found to be about 5.3 for native Ca^{2+}-GLA and 2.6 for native apo-GLA. In order to determine the corresponding degree of cooperativity, the binding data are plotted according to the Hill equation (Fig. 15.5 D).

$$\log((n-r)/r) = -m \log K_b - m \log [\text{bis-ANS}]$$

where K_b is the average binding constant of bis-ANS to GLA, n is the number of binding sites, r is the number of sites occupied and m is the Hill coefficient which expresses the degree of cooperativity. The data of both titrations fit the linear relationship of the Hill equation. The Hill coefficients determined from the slopes are 1.08 and 1.67 for apo- and Ca^{2+}-GLA, respectively. The values confirm that for the bis-ANS interaction with native apo-GLA no cooperativity can be detected ($m \approx 1$), whereas for the dye interaction with native Ca^{2+}-GLA a positive cooperativity has been found ($m > 1$). The positive cooperativity as well as the fact that more dye molecules, as expected from the number of available hydrophobic sites, bind to Ca^{2+}-GLA, may imply that several adsorbed bis-ANS molecules bind at the same site. A similar behavior has been observed for the interaction of bis-ANS with tubulin (Ward and Timasheff 1994).

4.4 Conformational Changes Induced by bis-ANS in Native GLA

Information on the conformational changes that occur in the GLA-bis-ANS complexes is gathered from protein ellipticity changes in the far- and near-UV wavelength regions (Figs. 15.6 A and B,respectively).

In the absence of bis-ANS the far- and near-UV CD-spectra of native apo-GLA at 4°C and Ca^{2+}-GLA at 37 °C are quite similar, in agreement with their similar states. In the presence of 60 μM bis-ANS, the absolute values for the far-UV ellipticity changes of apo-GLA at 4 °C (Fig. 15.6 A) are decreased at wavelengths near 225 nm and increased near 208 nm as compared with the native state. Comparable behavior is observed in the molten globule state of GLA, as shown by the far-

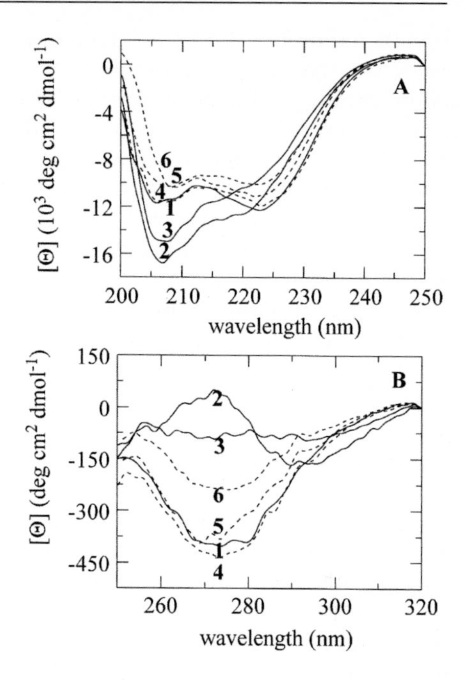

Fig. 15.6. Circular dichroism spectra of GLA under different conditions in the far (A) and near (B) ultraviolet regions. In 2 mM EGTA at 4 °C (1), in 2 mM EGTA and 60 μM bis-ANS at 4 °C (2), in 2 mM EGTA at 42 °C (3), in 2 mM Ca^{2+} at 37 °C (4), in 2 mM Ca^{2+} and 60 μM bis-ANS at 37 °C (5), containing 3.3 photolabeled bis-ANS molecules per GLA in 2 mM Ca^{2+} at 37 °C (6) (Vanderheeren et al. 1998)

UV spectrum of apo-GLA at 42 °C. The similar shape of both far-UV CD-spectra indicates that the penetration of the dye in native apo-GLA causes changes in the secondary structure comparable to the formation of a molten globule. It is likely that the changes are local unfoldings which must allow some expansion of the protein. Also the near-UV properties of apo-GLA change substantially by the addition of 60 μM bis-ANS. However, the resulting near-UV spectrum is clearly more structured than the spectrum of the molten globule state obtained at 42 °C (Fig. 15.6 B). This indicates that the aromatic side-chains of apo-GLA at 4 °C, preferentially conserve more anisotropic interactions in 60 μM bis-ANS than in the thermally induced molten globule state and thus remain relatively immobile. The immobility of the aromatic residues within the complex is also proven by the temperature dependence of their ellipticity changes at 270 nm (Fig. 15.4, filled circles). The ellipticity value progressively decreases as a function of the temperature and approaches the value of the thermal induced molten globule. Visibly, this curve represents a part of the thermal unfolding of the apo-GLA-bis-ANS complex, for which the transition is shifted to a lower temperature compared with the original apo-GLA.

At 37 °C and 2 mM Ca^{2+}, the far- and near-UV CD-spectra hardly change in the presence of 60 μM bis-ANS, indicating that the conformation of the compact native Ca^{2+}-GLA remains intact. From Fig. 15.6 B it is obvious that this limited influence may be related to the weak dye interaction under those circumstances. In order to obtain saturation, high dye concentrations (> 100 μM) are required. Because of the strong light absorption of bis-ANS at 270 nm, CD measurements are compromised in such conditions. To circumvent this problem we have photolabeled the protein with the interacting dye, which allows observations at relatively low free dye concentrations. As explained in Materials and Methods, by photolabelling and chromatography, we obtained a sample with 3.3 bis-ANS molecules irreversibly bound per Ca^{2+}-GLA molecule. Although the number of bound probe molecules is larger than for apo-GLA at saturation, the far-UV CD spectrum of the labeled Ca^{2+}-GLA.bis-ANS complex (Fig. 15.6 A) shows the typical shape observed for native GLA with conserved secondary structure elements. Similarly, the residual ellipticity in the near-UV (Fig. 15.6 B) indicates that the bound dye molecules only partially disrupt the tertiary structure. The stability of the tertiary structure of this bis-ANS-labeled Ca^{2+}-GLA is lowered as shown by the decrease in the transition temperature from 70 °C to 62 °C (Fig. 15.4, open circles).

4.5 Fluorescence Energy Transfer

Because of the endothermic character of the hydrophobic interaction, the smaller effect that bis-ANS has on the conformation in the presence of Ca^{2+} than in its absence, directly results from the higher temperature at which the data are obtained. The smaller perturbations of Ca^{2+}-GLA rather result from a reduction of the probe access to the hydrophobic regions of the protein. This is confirmed by energy-transfer experiments in the absence and presence of Ca^{2+}. Fig. 15.7 shows the influence of bis-ANS on the tryptophan fluorescence spectra of origi-

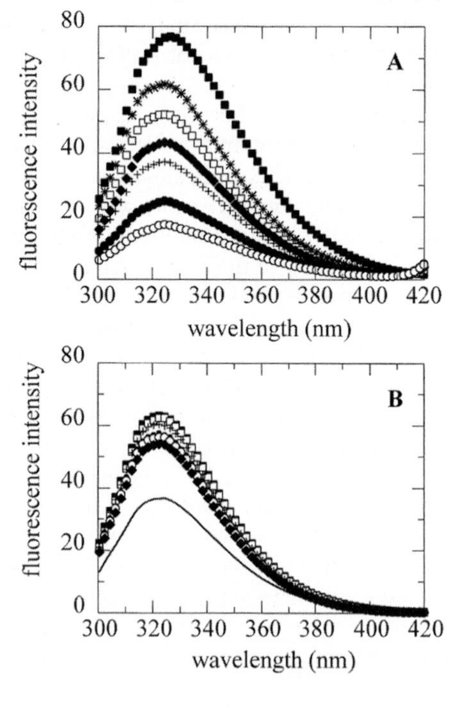

Fig. 15.7. Reduction of the tryptophan emission fluorescence as a consequence of energy transfer to bis-ANS. (A) 10 μM GLA in 2 mM EGTA at 4 °C and increasing bis-ANS concentrations: 0 μM (■), 5 μM (★), 10 μM (□), 15 μM (◆), 20 μM (+), 30 μM (●) and 40 μM (○). (B) 10 μM GLA in 2 mM Ca^{2+} at 37 °C and increasing bis-ANS concentrations: 0 μM (■), 10 μM (□), 20 μM (+), 30 μM (●), 40 μM (○) and 50 μM (◆) or containing 3.3 photolabeled bis-ANS molecules per GLA (solid line) (Vanderheeren et al. 1998)

nally compact native GLA. The pronounced decrease of the tryptophan fluorescence intensity obtained for apo-GLA in the presence of bis-ANS at 4 °C (Fig. 15.7 A) is accompanied by a fluorescence increase of bis-ANS near 500 nm (not shown). It indicates that an important part of the tryptophan fluorescence energy is transferred to the dye. The interaction of bis-ANS with Ca^{2+}-GLA at 37 °C does not result in a comparable energy-transfer (Fig. 15.7 B). Even the tryptophan fluorescence of the bis-ANS-photolabeled-Ca^{2+}-GLA complex is not reduced to the same degree as for apo-GLA.

4.6
Conclusion

In 10 mM Tris-HCl, pH 7.5 and at 4 °C in 2 mM EGTA as well as at 37 °C in 2 mM Ca^{2+}, native GLA is close to its thermal transition. Nevertheless, there are a number of interesting differences in the interaction of the dye with the apo- and Ca^{2+}-form of the protein.

Native apo-GLA binds 2 bis-ANS molecules. The interactions with both probe molecules can be described by independent equilibria. Therefore, it is likely that both interacting bis-ANS molecules bind to a different hydrophobic cluster. The efficient quenching of the tryptophan fluorescence also indicates that the binding of bis-ANS to the apo-protein occurs at locations in the vicinity of these aromatic groups. In addition, the far-UV CD spectrum of the complex becomes similar to that of the protein in its molten globule state. In contrast, 5 bis-ANS molecules are bound in a cooperative way to native Ca^{2+}-GLA. The far-UV CD spectrum of native Ca^{2+}-GLA is conserved for the complex. The limited energy-transfer confirms that in the presence of Ca^{2+}, the interacting bis-ANS molecules only approach a part of the tryptophan residues or even remain relatively distant from them.

As native apo-GLA and Ca^{2+}-GLA are close to their thermal destabilization at the considered temperatures, the reduction of the perturbation in the presence of Ca^{2+} is not due to the stabilization of GLA as a whole. It is rather due to stiffening of local structure by Ca^{2+}. Close inspection of the protein structure proves the ability for such local effects. As described earlier, one of the hydrophobic clusters of α-lactalbumin, Phe-53, Trp-60, Tyr-103 and Trp-104 (Fig. 15.1), is situated in the crevice that divides the protein in two structural halves and contains aromatic groups that belong to both halves. The Ca^{2+}-binding loop of α-lactalbumins (with ligand residues Lys-79, Asp-82, Asp-84, Asp-87 and Asp-88) contributes to the amino-terminal side of a 3_{10}-helix (residues 76–82) and to the carboxyl-terminus of the α-helix C (residues 86–99) (Acharya et al. 1989). Ca^{2+} binding closes the crevice and may prevent access of bis-ANS to these hydrophobic residues.

This local stabilization may have as a consequence that the bis-ANS molecules interacting with native Ca^{2+}-GLA are only situated at the cluster distant from the specific Ca^{2+}-binding loop. In our further research we concentrate on the determination of the exact binding sites of the dye in the GLA-bis-ANS complexes in the presence and absence of Ca^{2+}.

References

Acharya KR, Stuart DI, Walker NPC, Lewis M, Philips DC (1989) Refined structure of baboon α-lactalbumin at 1.7 A resolution. J Mol Biol 208: 99–127.

Coco MJ, Lecomte JT (1994) The native state of apomyoglobin described by proton NMR spectroscopy: Interaction with the paramagnetic probe HyTEMPO and the fluorescent probe ANS. Protein Sci 3: 267–281.

Engelhard M, Evans PA (1995) Kinetics of interaction of partially folded proteins with a hydrophobic dye: evidence that the molten globule character is maximal in early folding intermediates. Protein Sci 4: 1553–1562.

Das KP, Surewicz WK (1995) Temperature induced exposure of hydrophobic surfaces and its effect on the chaperone activity of α-Crystallin. FEBS Lett 369: 321–325.

Gorovits BM, Seale JW, Horowitz PM (1995) Residual structure in urea-denaturated chaperonin GroEL. Biochemistry 34: 13928–13933.

Levitt M, Goestein M, Huang E, Subbiah S, Tsai J (1997) Protein folding: the endgame. Ann Rev Biochem 66: 549–579.

Miranker AD, Dobson CD (1996) Collapse and cooperativity in protein folding. Curr Opin Struct Biol 6: 31–42.

Ptitsyn OB, Pain RH, Semisotnov GV, Zerovnik E, Razgulyaev OI (1990) Evidence for a molten globule state as a general intermediate state in protein folding. FEBS Lett 262: 20–24.

Seale WS, Martinez JL, Horowitz PM (1995) Photoincorporation of 4,4'-bis(1-anilino-8-naphthalenesulfonic acid) into the apical domain of GroEL. Biochemistry 35: 7443–7449.

Shi L, Palleros DR, Fink AL (1994) Protein conformational change induced by 1,1'-bis(4-anilino-5-naphthalenesulfonic acid): preferential binding to the molten globule of DnaK. Biochemistry 33: 7536–7546.

Stuart DI, Acharya KR, Walker NPC, Smith SG, Lewis M, Phillips DC (1986) α-lactalbumin possesses a novel calcium binding loop. Nature 324: 84–87.

Teschke CM, King J, Prevelige PE (1993) Folding of the phage P22 coat protein in vitro. Biochemistry 32: 10839–10847.

Vanderheeren G, Hanssens I (1994) Thermal unfolding of bovine α-lactalbumin. Comparison of circular dichroism with hydrophobicity measurements. J Biol Chem 269: 7090–7094.

Vanderheeren G, Hanssens I, Noyelle K, Van Dael H, Joniau M (1998) The perturbations of the native state of goat α-lactalbumin induced by 1,1'-bis(4-anilino-5-naphthalenesulfonate) are Ca^{2+}-dependent. Biophys J 75: 2195–2204.

Ward LD, Timasheff SN (1994) Cooperative multiple binding of bis-ANS and daunomycin to tubulin. Biochemistry 33: 11891–11899.

CHAPTER 16

Structure, Dynamics and Function of the Proton Pump Bacteriorhodopsin

G. Büldt[1], J. Heberle[1], R. Schlesinger[1] and H.-J. Sass[1]

1 Introduction

The integral membrane protein bacteriorhodopsin (bR) is an excellent example to show how information is obtained from different biophysical methods to understand the function of this protein on the basis of its structure and dynamics. There are only a few such proteins, which attracted the interest of a large number of research groups.

1.1 The Photocycle and Proton Translocation

Bacteriorhodopsin is densly packed in hexagonal two dimensional lattices the so-called purple membranes (PM) in the plasma membrane of the *Halobacterium salinarum*. It was discovered by Oesterhelt and Stockenius (1973) that bR contains the chromophore, retinal, covalently bound via a Schiff base to Lys 216, which enables this protein to pump protons out of the cell (Fig. 16.1). The photocycle of bR (Fig. 16.2) is linked to the active transport of protons over the membrane by delivering part of the energy of the absorbed photon to this process. In the ground state (bR_{568})of the photocycle the retinal is in an *all-trans* conformation, the Schiff's base is protonated (pK13) and positively charged. Upon absorption of a photon the retinal goes via an excited state to a 13-*cis* retinal conformation, which results in a dramatic pK down-shift to 3. Hence the Schiff's base proton is taken up by Asp 85 (Fig. 16.1) and another proton appears at the extracellular surface of bR. These steps pass intermediate states of the protein which can be monitored by colour changes of the chromophore (Fig. 16.2), which originates from the altered electric field at the retinal position and its change in the isomerisation state. At this point of the photocycle bR has reached the M_{412} intermediate which is easily detected by the blue shift of the absorption maximum to 412 nm. In order to complete the cycle and to return to the ground state, the Schiff's base has to be reprotonated from the protonated Asp96 (Fig. 16.1). This amino acid will then be reprotonated and the retinal is driven back to its *all-trans* conformation. (For recent reviews see Lanyi, 1997 and Oesterhelt, 1998.)

[1] Forschungszentrum Jülich, IBI-2: Biologische Strukturforschung, D-52425 Jülich.

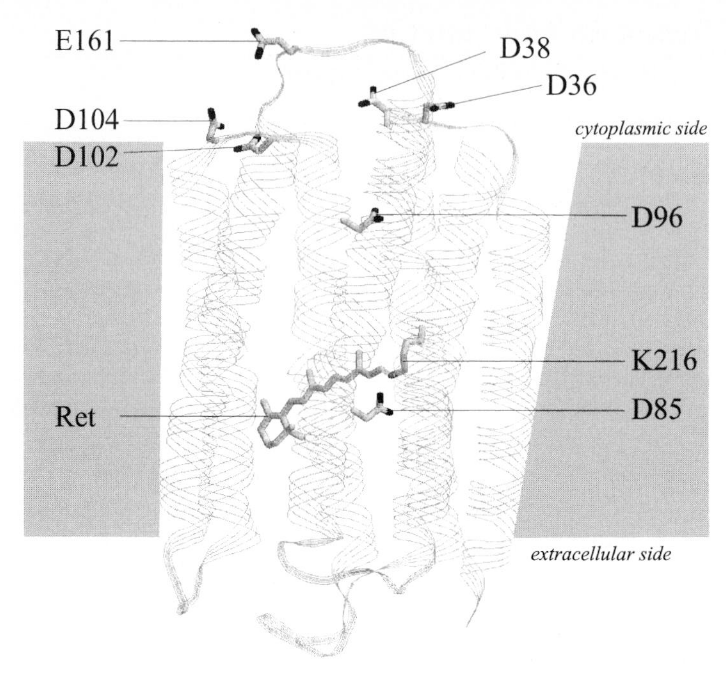

Fig. 16.1. Schematic picture of the seven transmembrane α-helices of bacteriorhodopsin displaying the retinal in its all-*trans* conformation connected to K216 and several negatively charged amino acids

It should be noted that this active proton translocation can take place against a proton gradient of a certain magnitude. It seems clear that the vectoriality of the pump is achieved by controlled changes of the pK's from different groups, the accessibility of certain positions in the structure for protons and the position of water molecules to bridge gaps in the proton translocation pathway. What drives these changes in pK's, accessibility and water positions? So far, experiments reveal three observations: The retinal isomerisation is followed by a charge redistribution, which then drives conformational changes within the protein.

In the following we would like to present some structural details which give an understanding of these processes. Diffraction methods using X-rays, neutrons and electrons have contributed a lot to the current understanding of the proton pumping mechanism. Further important details were gained, especially from FTIR-spectroscopy and site-directed mutagenesis.

1.2 Characterisation of Intermediate States by Infrared Spectroscopy

At the very beginning of bR research, photocycle intermediates were detected by UV/Vis-spectroscopy. For more than 15 years vibrational spectroscopy has (in particular resonance Raman scattering (Althaus et al., 1995) and infrared absorption (Siebert, 1990)) added molecular information to the understanding of the

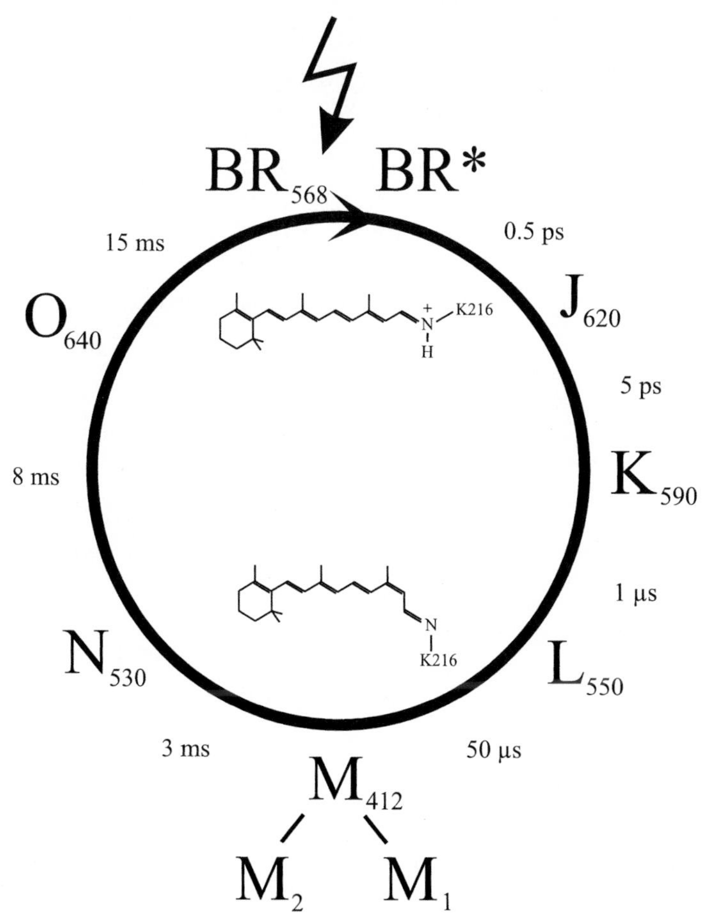

Fig. 16.2. The photocycle of bR showing the light-adapted ground state bR_{568} and several intermediates with their absorption maxima and life times. The conformation of all-*trans* and 13-cis retinal is also depicted

proton pump. However, strong overlap of the 3N-6 vibrations (with N = number of atoms) makes it impossible to record highly resolved infrared spectra of large proteins. Absorption of water in the mid-infrared is an additional complication. A powerful technique to overcome this obstacle is difference spectroscopy where only those vibrations are selected that change during the action of the protein. With the advent of Fourier-transform infrared (FTIR) spectroscopy this procedure makes it possible to record single vibrations in front of the whole protein. Even dynamics of single vibrations can be followed with high temporal resolution. Nanosecond difference spectra over a broad spectral range have been obtained by state-of-the-art FT-techniques (Weidlich et al., 1993; Rödis et al., 1999; Rammelsberg et al., 1997).

Infrared spectroscopy was particularly helpful in elucidating the role of certain amino acids in proton translocation across bR. A major advantage of IR-

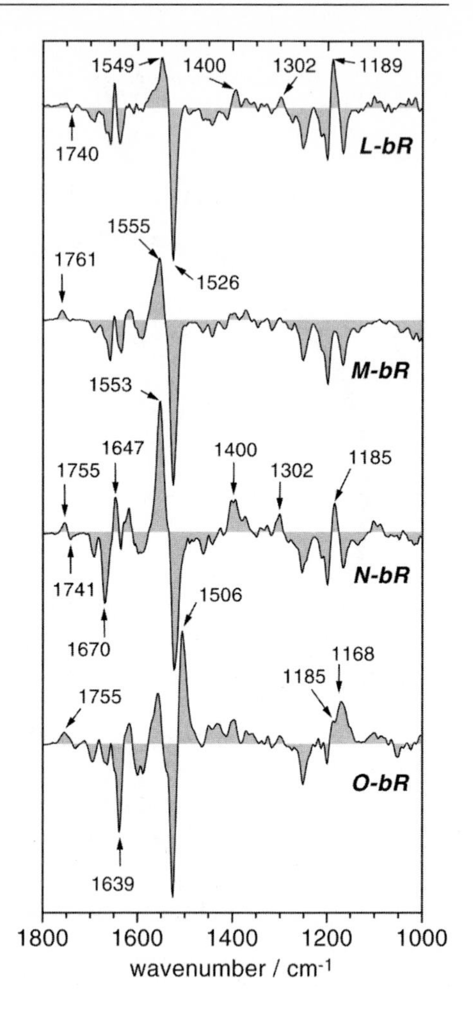

Fig. 16.3. Difference spectra of the photocycle intermediates of bR. Spectra have been extracted from time-resolved ATR/FT-IR experiments under various conditions: L-BR (10 μs, pH 6.6, 20 °C), M-BR (300–400 μs, pH 8.4, 20 °C), N-BR (80–100 ms, pH 8.4, 20 °C), and O-BR (5–10 ms, pH 4.0, 40 °C). Spectra were scaled to yield identical difference absorbance at 1252 cm^{-1}

spectroscopy is that the protonation state of acidic amino acids can be observed by the detection of the C=O double bond vibration. As an example, the positive amplitude at 1761 cm^{-1} (Fig. 16.3) is due to the C=O vibration that appears when the aspartate at position 85 accepts the proton from the retinal Schiff base during the L to M transition to form the corresponding aspartic acid (R-COOH).

Fig. 16.3 displays difference spectra between ground-state bR (negative bands) and the intermediates L, M, N, and O (positive bands). The range selected from 1800–1000 cm^{-1} is particularly useful because of the richness in band features. Difference spectra have been obtained with time-resolved attenuated total reflection (ATR) FTIR spectroscopy (Heberle and Zscherp, 1996; Zscherb and Heberle, 1997; Zscherp et al. 1999). This evanescent wave technique allows the precise control of parameters like pH, temperature, ionic strength. Conditions have been chosen that maximize the transient concentration of the respective intermediate state.

The strongest band in Fig. 16.3 is located at 1526 cm^{-1} and has been assigned to the C=C double bond of retinal in the unphotolyzed state. The corresponding bands of the photoproducts are shifted to 1549 cm^{-1} in L, 1555 cm^{-1} in M, 1553 cm^{-1} in N, and 1506 cm^{-1} in O but are overlapped by amide II difference bands. The C-C single bonds of the chromophore are absorbing in the region between 1150 and 1280 cm^{-1}. The negative band at 1639 cm^{-1} has been attributed to the protonated Schiff base in the unphotolyzed state of bR. It is of equal intensity in L and M, much smaller in N and most pronounced in O. Other bands involving the Schiff base can be found at 1302 cm^{-1} (C-H in-plane vibration) and 1400 cm^{-1} (N-H in-plane). Typical for the 13-*cis* chromophore, they are strong in L and N.

Conformational changes of the protein backbone can be observed between 1500 and 1600 cm^{-1} (C=N-H, amide II) and between 1600 and 1700 cm^{-1} (C=O, amide I). They already show up in L and M, reach maximum intensity in N and are mostly reversed in O.

Bands above 1700 cm^{-1} are due to the C=O stretching vibration of protonated carboxylic acids coupled with the in-plane bending vibration of the O-H. The band at 1761 cm^{-1} in M has been assigned to the primary acceptor of the Schiff base proton, D85 (Braiman et al., 1988; Fahmy et al., 1992). This band shifts to 1755 cm^{-1} in N, interpreted as a change of environment of D85. It shows the same amplitude in the O-bR difference spectrum. Therefore, it can be concluded that D85 is still protonated in O. D96, which is the proton donor of the Schiff base, is protonated in the ground state and transiently deprotonated in N. The corresponding negative band in the N-bR difference spectrum is located at 1741 cm^{-1}.

In conclusion, infrared spectroscopy is capable of monitoring the dynamics of retinal isomerization, proton transfer as well as structural changes of the protein backbone in one experiment. Determination of the protonation state of single residues by crystallographic methods is very difficult. In this respect, IR spectroscopy is superior as it also delivers the dynamics of proton transfer.

2 bR Structure Obtained from 2D-Lattices

2.1 The Electron Microscopic Structure of bR and the Refinement of this Structure by Neutron and X-Ray Diffraction

The characterization of the aggregation state of bR in the purple membrane was given by X-ray diffraction on these natural two-dimensional lattices (Blaurock, 1975; Henderson 1975). The basic picture of the tertiary structure, describing the three-dimensional arrangement of α-helices spanning the membrane, was given in several papers on low dose electron microscopy (e.g., Unwin and Henderson, 1975; Henderson and Unwin, 1975). The knowledge of the amino acid sequence (Ovchinnikov et al., 1979) allowed the prediction of a folding model by determining seven α-helical stretches in the polypeptide chain, which were assigned to the seven α-helices seen in the electron microscopical structure (Engelman et al., 1980). In the following years neutron diffraction was the method of choice to prove this model using bR molecules containing certain perdeuterated amino acids (Engelman and Zaccai, 1980; Trewhella et al., 1983) and partially deuterated α-helices (Trewhella et al., 1986; Popot et al., 1989). This technique was also successful in the localization of perdeuterated or partially deuterated retinal within the protein (Jubb et al., 1984; Heyn et al., 1988). Heavy-atom labelled retinal analogues incorporated into the protein confirmed these results by X-ray diffraction (Büldt et al., 1991). In early neutron diffraction experiments the hydration of purple membranes was studied (Zaccai and Gilmore, 1979) and more recently the location of the proton channel in bR was determined by neutron diffraction in combination with H_2O/D_2O exchange experiments (Papadopoulos et al., 1990).

In 1990, cryo-electron microscopy provided an additional breakthrough providing a structure of bR at 3.5 Å resolution in the X-Y plane and 7 Å in the Z-direction. This picture allowed for the first time, to construct a molecular model of bR including all the information from other methods mentioned above (Henderson et al., 1990). This model was further improved by cryo-electron microscopy data of Grigorieff et al. (1996) and Kimura et al (1997).

2.2 Structural Investigations on Photocycle Intermediates

For a complete understanding of the function of a protein like bR it is desirable to detect structural changes in space and time with high resolution parallel to the working cycle. Since the intensity scattered from a single molecule is too low, taking into account the limits set by radiation damage, an ensemble of molecules has always to be considered. In this situation information about structural changes during the working cycle can be obtained in two ways, either by trapping intermediate states or by time-resolved detection of the scattered intensity after excitation. Both alternatives were successfully carried out in the case of bR.

2.2.1 Trapping of the M-State in Wild-Type bR

After the observation that a proton is releaced by bR at the extracellular side during the transition from L to M, the M-intermediate was considered to be a strategic state for the pumping process. It seemed that the knowledge of the M-state structure would give some insight into the mechanism of proton translocation. One of the first trapping experiments of the M-intermediate at low temperatures was performed by Glaser et al. (1986) using electron diffraction. These experiments showed no intensity changes in the resolution region from 60–5 Å and small changes between 5 and 3.5 Å. Therefore neutron diffraction experiments were undertaken by Dencher et al. (1989) with the aim of observing changes in the distribution of water molecules in comparison to the ground state. As it was known that GuaHCl at high pH slows down the decay of the M state, a stack of several PM-films was soaked in a buffer containing GuaHCl at pH 9.4 and illuminated at +8 °C. The films became yellow indicating the complete transformation to the M state, which was then preserved at liquid-nitrogen temperatures in a cryostat. Neutron diffraction patterns of the ground state and the M-intermediate showed clear differences in the reflection intensities of up to 9 % in $\Sigma|\Delta I|/\Sigma I$ in the resolution region of 60 to 7 Å. By comparing diffraction patterns between films in H_2O and D_2O it was derived that at least 80 % of the differences resulted from changes in the protein structure and only a minor contribution could originate from a redistribution of water molecules. These observations indicate that changes in the tertiary structure of bR take place during the photocycle. These results are in contradiction to the electron diffraction experiments of Glaser et al. (1986).

2.2.2
Trapping the M-State in the Mutant Asp96Asn

It was observed by optical spectroscopy that the bR-mutant Asp96Asn is characterized by a large decrease in the decay rate of the M-state with increasing pH. Koch et al. (1991) performed X-ray diffraction experiments on films of this mutant under continuous illumination at room temperature and pH 9.6. They found similar intensity changes between the bR_{568} and M_{412} state structures (Fig.16.4B, 100 % r.h.) as observed in the neutron diffraction measurements.

2.2.3
Time-Resolved X-Ray Diffraction Experiment on the bR Mutant Asp96Asn

In order to examine how these structural changes correlate with relaxation processes in the photo- and pumping-cycle of bR, the structural transition from the M-state to the ground state was followed by time-resolved X-ray diffraction using intense synchrotron radiation (Koch et al., 1991). The time course of flash-induced changes for three reflections at neutral pH is illustrated in Fig. 16.5. The changes in individual reflections before and immediately after the light flash are consistent both in amplitude and in direction with the steady-state experiments.

A comparison of these structural relaxation times with optical decay rates of intermediate states in the photocycle indicated that the observed structural changes decay with the transition from the N state to the ground state. In functional terms this means that the structural changes relax after the reprotonation of the Schiff's base.

2.2.4
M Splits into Two States M_1 and M_2

So far the relaxation of tertiary structural changes was followed by time-resolved X-ray experiments. The onset of these changes could not precisely be attributed to photocycle intermediates. A first hint for an answer to this problem was obtained from X-ray diffraction experiments on bR mutant Asp96Asn (pH 9.6) at different hydration levels (Sass et al., 1997). PM-films equilibrated at different relative humidities (15, 57, 75 and 100 % r.h.) were transformed to the M-state by continuous illumination. Films, equilibrated at relative humidities above 60 % showed the known changes in the tertiary structure, whereas films below 60 % r.h. displayed only very small changes in their diffraction pictures (Fig. 16.4B). The corresponding difference density maps show at high humidity positive difference density at helices B, F and G (Fig. 16.4C). No significant difference peaks

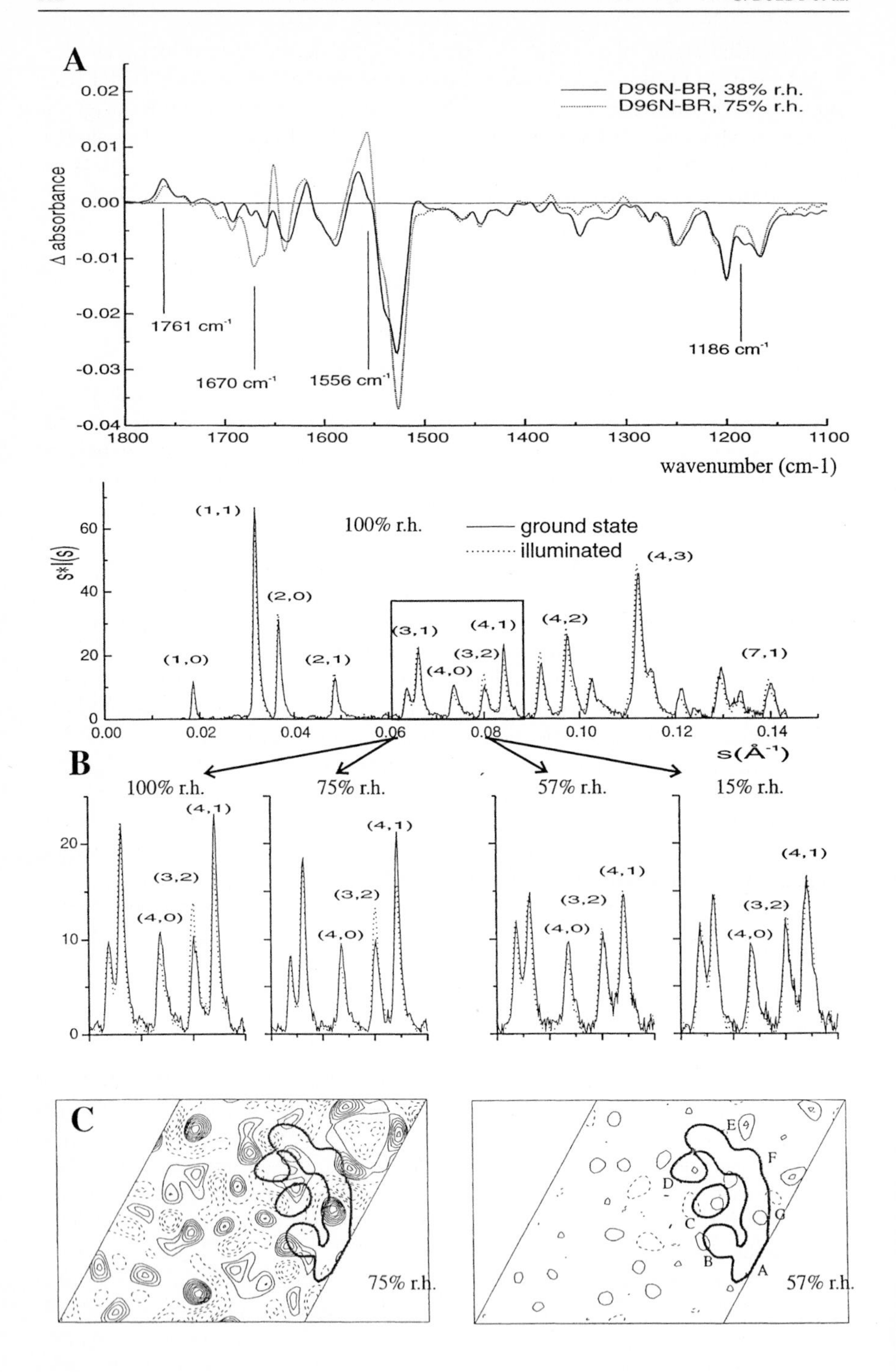
A
D96N-BR, 38% r.h.
D96N-BR, 75% r.h.
Δ absorbance
1761 cm⁻¹
1670 cm⁻¹
1556 cm⁻¹
1186 cm⁻¹
wavenumber (cm-1)
100% r.h.
ground state
illuminated
S*I(S)
(1,1)
(2,0)
(1,0)
(2,1)
(3,1)
(4,0)
(3,2)
(4,1)
(4,2)
(4,3)
(7,1)
s(Å⁻¹)
B
100% r.h.
75% r.h.
57% r.h.
15% r.h.
C
75% r.h.
57% r.h.

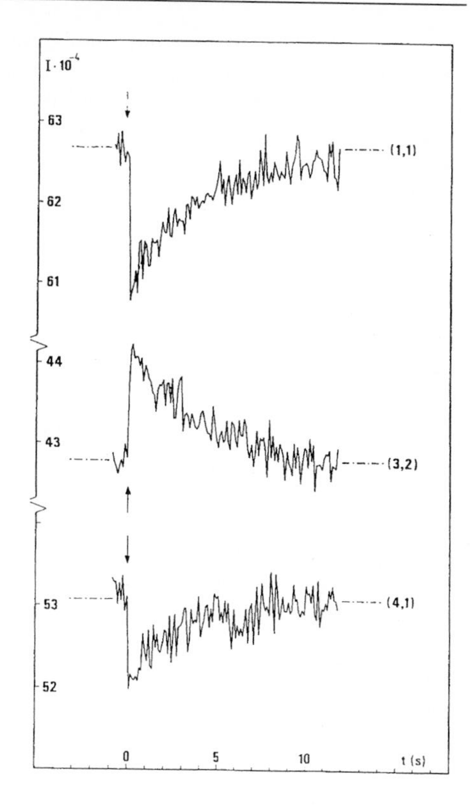

Fig. 16.5. Time course after a short light flash (~ 0.75 ms FWHM) at t = 0 s of the integrated intensity including background under the (1,1), (3,2) and (4,1) reflections for purple membranes with mutated bR D96N at pH 9.6 (100 mM carbonate buffer, 500 mM KCl). The data were collected with a time resolution of 100 ms

can be seen below 60 % r.h.. These experiments demonstrated that two types of M-states with and without structural changes compared to the ground state exist. If one assumes that both M-states would in the photocycle also at high hydration in a sequential order, the changes in the tertiary structure would develop within the M intermediate between M1 and M2 and would relax after the N-state.

FTIR measurements under identical conditions, although with much thinner PM films verify that for all hydration levels the films were trapped in the M intermediate. The distinction between the M and L intermediate as well as the N intermediate was made upon the appearance of the finger print region of the retinal in the FTIR difference spectra, the most obvious criteria for such a differentiation

Fig. 16.4. (A) FTIR difference spectra of bR mutant D96N at 38 % and 75 % relative humidity (r.h.). (B) X-ray diffraction patterns of bR-D96N light-adapted purple membranes (pH 9.6, room temperature) at different hydration levels in the absence of light (solid line) and under steady-state illumination (dotted line). $s = 2\sin\theta/\lambda$, with Bragg angle θ and wavelength $\lambda = 1.5$ Å. The lower panels represent the reflection range (2,2) to (4,1) at different r.h. on an expanded scale.
(C) Difference electron density maps (M state minus light-adapted ground state) of the bR-D96N purple membranes at different hydration levels. Left: 57 % r.h., M1 state. Right: 75 % r.h., M2 state. The bold contour outlines the bR monomer, individual helices are marked by upper case letters from A to G. Continuous lines correspond to positive, dashed lines to negative electron density levels. Contour levels are scaled to each other in both maps

(Fig. 16.4A). The largest differences between the differently hydrated samples were found in the amide regions. It appears that samples at hydration levels greater than 60 % r.h. display the structural changes in the diffraction experiment and at the same time show in the amide I region a larger difference band at 1670 cm^{-1} than at 1660 cm^{-1}. On the other hand the samples which do not show the changes in the tertiary structure (r.h. less than 60 %) display a larger difference band at 1660 cm^{-1} than at 1670 cm^{-1}. These differences in the amide I region are also found in the amide II region, where the difference band at 1556 cm^{-1} is much larger in the more hydrated samples.

The concept of two M states was brought up from the evaluation of time-resolved spectroscopy data in the visible wavelength region (Varo and Lanyi, 1990). An irreversible step was assumed at this position in the photocycle between M_1 and M_2 acting as a switch by changing the proton accessibility of the Schiff base from the extracellular to the cytoplasmic side and thus creating the vectoriality of the proton pump.

With respect to the function of bR, Thiedemann et al. (1992) were able to show that proton pumping was only found in samples with hydration levels above 60 % r.h. These results indicate that the observed structural changes are necessary for proton translocation and that at least part of these changes may form the switch which changes the accessibility to the Schiff's base.

2.2.5
Charge-Controlled Conformational Changes

Further support for this interpretation of the proton translocation mechanism was given through the investigation on the bR mutant Asp38Arg. X-ray diffracton experiments on samples at pH 6.7 do not show changes in the tertiary structure of bR whereas measurements at pH 9.6 display a diffraction pattern, showing the characteristic large structural changes (Sass et al., 1998). The interpretation of these experiments is, that at pH 6.7 the M_1-state is trapped under illumination whereas at pH 9.6 the M_2-state is accumulated. Assuming a sequential appearence of M_1 and M_2 independent of pH, but with individual pH-dependent relaxation times, it seems that also for this mutant the observed large structural changes are necessary for vectorial proton pumping. In addition, these results give a clear indication that the changes in the tertiary structure are driven by alterations in the charge distribution of the protein, which follow photoisomerization.

The substitution of an aspartic acid by an arginine makes the charge pattern at the cytoplasmic side more positive either directly by the positive charge of the arginine or indirectly but more effectively since another positive charge is no longer compensated by the interaction with the aspartate. This new charge pattern, more positive at the cytoplasmic side than in wt bR, could interfere with the charge variation resulting from the deprotonation of the Schiff base and therefore slows down the large structural rearrangements. This would result in the accumulation of the M_1 state. Since no large structural changes are detectable under this condition and if one assumes a sequential order of M_1 and M_2 the general conclusion for wt bR can be drawn that a charge redistribution around the Schiff base results in an altered force field within bR which drives the large structural changes.

At a pH above 9 another group on the cytoplasmic side might be deprotonated and compensate at least partly the added positive charge of the arginine. This would allow the structural changes associated with the transition to the M_2 state to take place.

Reconsidering the earlier experiments of Glaeser et al. (1986) it seems now very probable that the M_1 intermediate was accumulated and therefore no changes in the tertiary structure were observed.

3
The Crystal Structure

3.1
Crystallization of bR

bR was one of the first integral membrane proteins that was crystallized in three dimensions (Michel and Oesterhelt, 1980). Considerable progress was made by Schertler et al. (1993) using the surfaces of freshly formed benzamidine crystals as nucleation sites. These plate-shaped crystals gave diffraction to 3.6 Å in a and b direction and 4.2 Å in the c direction.

A novel concept for crystallization of membrane proteins in a lipidic cubic phase was reported by Landau et al. (1996). They crystallized bR in monoolein-water cubic phase and obtained microcrystals of space group $P6_3$ with a unit cell of a = b = 61.76 Å; c = 104.16 Å, $\alpha = \beta = 90°$ and $\gamma = 120°$. Although the exact mechanism of crystallization is still unknown, it seems probable that bR, once inserted into the continuous 3D curved lipid bilayer of the cubic phase, diffuses within this bilayer to form a nucleation particle from which well-ordered crystals can grow. They started with very small microcrystals of up to 30 μm in diameter and 5 μm in thickness. The crystallization conditions were continuously improved so that crystals of 150 μm in diameter are available now. The cell constant in a,b-direction of 61.76 Å together with the hexagonal space group gave an indicates an arrangement of bR molecules similar to the purple membrane. In accordance with the screw axis along the c direction these purple membrane like 2D lattices are stacked on each other, where two adjacent lattices are rotated by 60 degrees (Pebay-Peyroula et al., 1997). Using a highly focused beam with a diameter of about the same size as the crystal, the signal to noise ratio was strongly increased, so that diffraction patterns of up to 2 Å resolution could be obtained.

3.2
Vibrational Spectroscopy to Assess the Functionality of bR in Microcrystals

After solving the problem of crystallizing a membrane protein the proper functionality of the protein within the crystal lattice has to be proven. Common tests on functionality of membrane proteins usually call for compartmentation. This strategy is apparently impossible with crystals. Thus, we employed vibrational spectroscopy as a non-invasive tool to assess the functionality of bR microcrystals (Helerle et al., 1998).

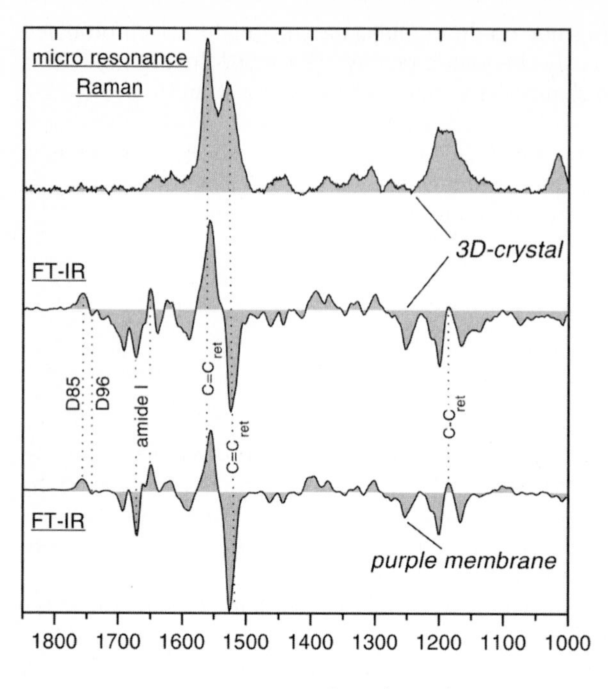

Fig. 16.6. Upper spectrum: Resonance Raman microspectrum of a single BR crystal at room temperature. The high power density of the exciting laser leads to accumulation of intermediate states during the measurement. The frequency of the strong bands at 1528 and 1563 cm^{-1} are assigned to the C=C stretching vibration of retinal in the ground and M state, respectively.
Middle spectrum: Light-induced FTIR difference spectrum of an ensemble of BR microcrystals. Spectra were obtained under continuous illumination with yellow light at 5 °C with the light-adapted ground state as reference. Marker bands are indicated by dotted lines and are discussed in the text.
Bottom spectrum: FTIR difference spectrum of BR within the native purple membrane. The spectrum was scaled by a factor of 0.07 to facilitate comparison with the spectrum of the microcrystals obtained under identical conditions (middle spectrum)

Microcrystals within the crystallization matrix of the lipidic cubic phase were used directly from the crystallization batch. Since the diameter of the probing laser spot is about 1 μm a single crystal can be investigated. With the wavelength of the Ar-laser used (488 nm) the Raman spectrum of bR is resonance enhanced (Fig. 16.6, top spectrum). Bands of the chromophore retinal are selectively observable. However, the expected Raman spectrum of retinal in the ground state of bR is overlaid by additional bands. The frequencies of those bands argue for the presence of the M intermediate. This mixture of states is due to the high power density of the laser. Directing the exciting laser into the lipidic phase reveals only small (non-resonant) bands from the lysolipid monoolein (data not shown) which was used as crystallization matrix. Yet, no vibrations from retinal are discernable. This provides evidence that the lipidic phase is completely devoid of bR.

As demonstrated in a previous section, FTIR difference spectroscopy allows the light-induced changes to be resolved at particular vibrations of bR. However, the information content goes beyond the chromophore as compared to the resonance Raman spectrum, and includes bands from the apoprotein. With regard to the microcrystals being embedded in the lipidic cubic phase the difference approach is even more advantageous since the strong IR-absorber monoolein does not respond to light. Its vibrational contribution is thus cancelled. A spectrum of an ensemble of bR crystals within the lipidic cubic phase was recorded under continuous illumination with yellow light at 5 °C (Fig. 16.6, middle spectrum) and ratioed against a spectrum in the dark. The band pattern of the resulting difference spectrum is almost identical to the one obtained with purple membrane (Fig. 16.10, bottom spectrum). The frequencies of the bands indicate that a mixture of the M with the N intermediate is accumulated under the applied conditions. Comparison with the intermediate spectra (Fig. 16.3) allows detailed understanding of the processes during the photocycle; thus, the chromophore retinal undergoes the typical isomerization reactions, as detected by the characteristic stretching modes of the C-C single bonds (at 1254, 1200 and 1167 cm^{-1}) and double bonds (at 1525 cm^{-1}) of the depleted ground state bR. Retinal bands due to the photointermediates are observed at around 1560 and at 1186 cm^{-1}, respectively. Active, vectorial proton transfer is demonstrated (see the positive band at 1755 cm^{-1}) by the protonation of residue D85. The subsequent deprotonation of D96 is in turn shown from the negative band at 1743 cm^{-1}. Changes in secondary structure of the protein, presumed to allow efficient proton transfer, are evident from the shift of the amide I vibration to lower frequencies resulting in a negative difference band at 1670 cm^{-1} and a positive band at 1650 cm^{-1}. The high light stability of the bR crystals allowed time-resolved FTIR spectroscopy to be performed. The millisecond kinetics of bR in the 3D crystal tally with those obtained in the natural 2D arrays of the purple membrane (data not shown). Taken together, these results demonstrate that the function of bR is not impaired in the 3D crystals.

In conclusion, it is shown that vibrational spectroscopy is a powerful technique to assess the functionality of a membrane protein in the crystallized state. Isomerization of the co-factor retinal and structural changes of the protein backbone are monitored. Finally, active proton transfer which is the physiological task of bR, was detected on the single residue level. This is clear evidence that the high resolution structure of bR presented in the next section is biologically relevant.

3.3 The Ground State Crystal Structure of bR

The crystal structure of bR was solved by molecular replacement on the basis of the electron microscopic structure (Grigorieff et al. 1996) to 2.5 Å resolution (Pebay-Peyroula et al., 1997). The interactions between the purple membrane like protein layers are limited. Protein-protein contacts exist only between loops AB (loop between helix A and helix B) and loop BC. The membrane spanning helices are numbered from A to G going from the N- to the C-terminal end. Intertrimer protein contacts were not found. These specific interactions between adjacent protein layers result in markedly different conformations in loops AB and BC in comparison to the EM structure. After improving the crystallization conditions, larger crystals became available showing a high degree of merohedral twinning which had to be corrected for (Luecke et al., 1998). Meanwhile two additional crystal structures of bR were published resulting from different crystallization strategies (Essen et al., 1998, Takeda et al., 1998). The structure of Essen et al. nicely shows structural features of how lipids in the protein boundary interact with amino acid side chains. The most important new features of all these structures are that the conformations of functionally important amino acid side chains become more and more reliable and that several water molecules were localized in the proton translocation channel. However, also these structures do not give a satisfactory answer how a proton is translocated. At least more water molecules have to be identified in the channel before a more probable model can be presented. This situation already demonstrates that only the high resolution structures of a few important intermediates like the L, M1, M2 and N state will give the desired information for modelling a more probable mechanism for proton translocation through bR.

4 Summary

bR is an excellent example of the mosaic like picture in the contributions of different methods for the understanding of proton pumping. The following methods were especially fruitful in the case of bR:

- EM and X-ray crystal structure analysis to obtain a structural basis for the mechanism.
- Time-resolved UV-Vis spectroscopy to establish the photocycle of bR with respect to the identification of intermediates and their rise and decay times.
- Time-resolved and steady state FTIR to localize proton transfer steps to certain amino acids and to characterize intermediate states of the protein and the chromophore.
- Site-directed mutagenesis to select functionally important amino acids to assign FTIR difference bands to certain amino acids and to manipulate the kinetics of the photocycle for a better analysis of various intermediate states.

References

Althaus T, Eisfeld W, Lohrmann R, Stockburger M (1995) Application of Raman Spectroscopy to Retinal Proteins. Israel J. Chem. 35:227–252.

Blaurock A E (1975) Bacteriorhodopsin: A Trans-membrane Pump Containing α-helix. J. Mol. Biol. 93, 139–158

Braiman M S, Mogi T, Marti T, Stern L J, Khorana H G, Rothschild K J (1988) Vibrational spectroscopy of bacteriorhodopsin mutants: light-driven proton transport involves protonation changes of aspartic acid residues 85, 96, and 212. *Biochemistry* 27:8516–8520.

Büldt G, Konno K, Nakanishi K, Plöhn H J, Rao B N, Dencher N A (1991) Heavy-atom labelled retinal analogues located in bacteriorhodopsin by X-ray diffraction. Photochem. Photobiol. 54, 873–879

Dencher N A , Dresselhaus D, Zaccai G, Büldt,G (1989) Structural changes in bacteriorhodopsin during proton translocation revealed by neutron diffraction. Proc. Natl. Acad. Sci. USA 86, 7876–7879

Engelman D M , Henderson R, McLachlan A D, Wallace B A (1980) Path of the polypeptide in bacteriorhodopsin. Proc. Natl. Acad. Sci. USA 77, 2023–2027

Engelman D M, Zaccai G (1980) Bacteriorhodopsin is an inside-out protein. Proc. Natl. Acad. Sci.USA 77, 5894–5898

Essen L, Siegert R, Lehmann W D, Oesterhelt D (1998) Lipid patches in membrane protein oligomers: crystal structure of the bacteriorhodopsin-lipid complex. Proc. Natl. Acac. Sci. USA 95, 11673–11678

Fahmy K, Weidlich O, Engelhard M, Tittor J, Oesterhelt D, Siebert F (1992) Identification of the Proton Acceptor of Schiff Base Deprotonation in Bacteriorhodopsin: A Fourier-Transform-Infrared Study of the Mutant Asp85Glu in its Natural Lipid Environment. *Photochem.Photobiol.* 56:1073–1083.

Glaeser R M, Baldwin J, Ceska T A, Henderson R (1986) Electron Diffraction Analysis of the M412 Intermediate of Bacteriorhodopsin. Biophys. J. 50, 913–920

Grigorieff N, Ceska T A, Downing K H, Baldwin J M, Henderson R. (1996) Electron-crystallographic refinement of the structure of bacteriorhodopsin. J. Mol. Biol. 259, 393–421

Heberle, J. and C. Zscherp. 1996. ATR/FT-IR difference spectroscopy of biological matter with microsecond time resolution. *Appl.Spectrosc.* 50:588–596.

Heberle J, Büldt G, Koglin E, Rosenbusch J P, Landau E M (1998) Assessing the functionality of a membrane protein in a three-dimensional crystal. *J Mol Biol* 281:587–592.

Henderson R (1975) The Structure of the Purple Membran from *Halobacterium halobium*: Analysis of the X-ray Diffraction Pattern. J. Mol. Biol. 93, 123–138

Henderson R, Baldwin J M , Ceska T A, Zemlin F, Beckmann E, Downing K H (1990) Model for the structure of bacteriorhodopsin based on high-resolution electron cryo-microscopy. J. Mol. Biol. 213, 899–929

Henderson R, Unwin P N T (1975) Three-dimensional model of purple membrane obtained by electron microscopy. Nature 257, 28–32

Heyn M P, Westerhausen, J, Wallat I and Seiff F (1988) High-sensitivity neutron diffraction of membranes: Location of the Schiff base end of the chromophore of bacteriorhodopsin. Proc. Natl. Acad. Sci.USA 85, 2146–2150

Jubb J S, Worcester D L , Crespi H L, Zaccai G (1984) Retinal location in purple membrane of *Halobacterium halobium*: a neutron diffraction study of membranes labelled in vivo with deuterated retinal. EMBO J. 3, 1455–1461

Koch M H J , Dencher N A , Osterhelt D, Plöhn H J, Rapp G, Büldt G. (1991) Time-resolved X-ray diffraction study of structural changes associated with the photocycle of bacteriorhodopsin. EMBO J.,10, 521–526

Kimura Y, Vassylyev D G , Miyazawa, A, Kidera A, Matsushima M, Mitsuoka K, Murata K, Hirai T, Fujiyoshi, Y. (1997) Surface of bacteriorhodopsin revealed by high-resolution electron crystallography. Nature 389, 206–211

Landau E M, Rosenbusch J P (1996) Lipidic cubic phases: A novel concept for the crystallization of membrane proteins. Proc. Natl. Acad Sci. USA 93, 14532–14535

Lanyi J K (1997) Mechanism of ion transport across membranes. Bacteriorhodopsin as a prototype for proton pumps. *J Biol Chem* 272:31209–31212.

Luecke H, Richter H Th, Lanyi J K (1998) Proton Transfer Pathways in Bacteriorhodopsin at 2.3 Angstrom Resolution. Science 280, 1934–1937

Michel H, Oesterhelt D. (1980) Three-dimensional crystals of membrane proteins: bacteriorhodopsin. Proc. Natl. Acad. Sci. USA 77, 1283–1285

Oesterhelt D, Stoeckenius W (1973) Functions of a new photoreceptor membrane. Proc. Natl. Acad Sci. USA 70, 2853–2857

Oesterhelt, D. 1998. The structure and mechanism of the family of retinal proteins from halophilic archaea. *Curr Opin Struct Biol* 8:489–500.

Ovchinnikov Yu, Abdulaev N, Fergira M, Kiselev A, Lobanov N (1979). The structural basis of the functioning of bacteriorhodopsin: an overview. FEBS Lett. 100, 219–224

Sass H J, Schachowa I W, Rapp G, Koch M H, Oesterhelt D, Dencher N A, Büldt G (1997) The tertiary structural changes in bacteriorhodopsin occur between M states: X-ray diffraction and Fourier transform infrared spectroscopy. *EMBO J* 16:1484–1491.

Sass H J, Gessenich R, Koch M H, Oesterhelt D, Dencher N A, Büldt G, Rapp G (1998) Evidence for charge-controlled conformational changes in the photocycle of bacteriorhodopsin. *Biophys J* 75:399–405.

Papadopoulos G, Dencher N A, Zaccai G, Büldt G (1990) Water molecules and exchangeable hydrogen ions at the active centre of bacteriorhodopsin localized by neutron diffraction. J. Mol. Biol. 214, 15–19

Pebay-Peyroula E, Rummel G, Rosenbusch J P, Landau E M (1997) X-ray Structure of Bacteriorhodopsin at 2.5 Angstroms from Microcrystals Grown in Lipidic Cubic Phases. Science 277, 1676–1681

Popot J L, Engelman D M, Gurel O, Zaccai G. (1989) Tertiary Structure of Bacteriorhodopsin. Positions and Orientations of Helices A and B in the Structure Map Determined by Neutron Diffraction. J. Mol. Biol. 210, 829–847

Rammelsberg R, Heßling B, Chorongiewski H, Gerwert K (1997) Molecular Reaction Mechanism of Proteins Monitored by Nanosecond Step-Scan FT-IR Difference Spectroscopy. *Appl.Spectrosc.* 51:558–562.

Rödig C, Chizhov I, Weidlich O, Siebert F (1999) Time-Resolved Step-Scan Fourier Transform Infrared Spectroscopy Reveals Differences between Early and Late M Intermediates of Bacteriorhodopsin. *Biophys J* 76:2687–2701.

Schertler G F X, Bartunik H D, Michel H, Oesterhelt D (1993) Orthorhombic crystal form of bacteriorhodopsin nucleated on benzamidine diffracting to 3.6 Å. J. Mol. Biol. 234, 156–164

Siebert F (1990) Resonance Raman and infrared difference spectroscopy of retinal proteins. *Methods Enzymol* 189:123–136.

Takeda K, Sato H, Hino T, Kono M, Fukuda K, Sakurai I, Okada T, Kouyama T (1998) A Novel Three-dimensional Crystal of Bacteriorhodopsin Obtained by Successive Fusion of the Vesicular Assemblies. J. Mol. Biol. 283, 463–474

Thiedemann G, Heberle J, Dencher N A (1992) Structures and Functions of Retinal Proteins. Colloque ENSERM/John Libbey Eurotext Ltd, Vol. 221, 217

Trewhella J, Popot J L, Zaccai G, Engelman D M (1986) Localization of two chymotryptic fragments in the structure of renatured Bacteriorhodopsin by neutron diffraction. EMBO J. 5, 3045–3049

Trewhella J, Anderson S, Fox R, Gogol E, Khan S, Engelman D, Zaccai G. (1983) Assignment of Segments of the Bacteriorhodopsin to Positions in the Structural Map. Biophys. J. 42, 233–241

Unwin P N T, Henderson R J (1975) Molecular Structure Determination by Electron Microscopy of Unstained Crystalline Specimens. J. Mol. Biol. 94, 425–440

Varo G, Lanyi J (1990) Pathways of the rise and decay of the M photointermediate(s) of bacteriorhodopsin. Biochemistry 29, 2241–2250

Weidlich O, Siebert F (1993) Time Resolved Step-Scan FTIR Investigations of the Transition from KL to L in the Bacteriorhodpsin Photocycle: Identification of Chromophore Twists by Assigning Hydrogen-Out-Of-Plane (HOOP) Bending Vibrations. *Appl.Spectrosc.* 47:1394–1400.

Zaccai G, Gilmore D J (1979) Areas of hydration in the purple membrane of *Halobacterium halobium*: a neutron diffraction study. J. Mol. Biol. 132, 181–191

Zscherp C, Heberle J (1997) Infrared difference spectra of the intermediates L, M, N, and O of the bacteriorhodopsin photoreaction obtained by time-resolved attenuated total reflection spectroscopy. *J.Phys.Chem.B* 101:10542–10547.

Zscherp C, Schlesinger R, Tittor J, Oesterhelt D, Heberle J (1999) In situ determination of transient pKa changes of internal amino acids of bacteriorhodopsin by using time-resolved attenuated total reflection Fourier-transform infrared spectroscopy. *Proc Natl Acad Sci U S A* 96:5498–5503.

CHAPTER 17

Structural Characterisation of Porcine Seminal Plasma Psp-I/Psp-II, a Paradigm Spermadhesin Molecule Built by Heterodimerization of Glycosylated Subunits

J. J. CALVETE[1] and L. SANZ[1]

1 Introduction: Lectin-Carbohydrate Binding Mechanism in Mammalian Fertilisation

The mechanism of initial binding between mammalian gametes involves recognition of glycan moieties attached to glycoprotein components of the egg acellular coat, the zona pellucida, by carbohydrate-binding proteins (lectins) on the surface of acrosome-intact, capacitated spermatozoa (reviewed by Benoff 1997, Sinowatz et al. 1997; Tulsiani et al. 1997; Töpfer-Petersen et al. 1998). A considerable number of molecules have been proposed to participate in this complementary, carbohydrate-based binding process in the six mammalian species (mouse, rat, boar, guinea pig, rabbit, and human) most thoroughly investigated by different laboratories using diverse techniques, like

(i) incubation of Western blots of sperm proteins with labelled zona pellucida glycoproteins or specific poly- and monoclonal antibodies which inhibit sperm-zona pellucida binding;
(ii) affinity chromatography of detergent-extracted sperm proteins on immobilised zona pellucida glycoproteins;
(iii) cross-linking of sperm surface proteins to zona pellucida ligands; and
(iv) correlation with infertility.

On the other hand, mammalian *zonae pellucidae* have conserved structure and simple protein composition. They typically consist of only 3 glycoproteins, encoded by three different genes termed ZPA, ZPB, and ZPC, arranged into a loose 2–25 μm thick meshwork which encases the oocyte. However, zona pellucida glycoproteins display species-specific glycosylation patterns which contribute to the non-conserved sperm receptor activity among species (Green 1997). The large diversity of both the structure of glycans of zona pellucida glycoproteins and the receptor molecules on spermatozoa, strongly suggests that several glycan-lectin recognition events may be involved before productive gamete interaction can occur. Noteworthy, little is known about the hierarchy of interactions and the cross-talking between different receptor/ligand systems, which may collectively dictate the species-specificity of gamete interaction. Thus, although the

[1] Instituto de Biomedicina de Valencia, C.S.I.C., Jaime Roig 11, E-46010 Valencia, Spain; Phone, +34 96 339 1775; Fax, +34 96 369 0800; e-mail: jcalvete@ibv.csic.es.

concept that sperm-egg binding involves complementary molecules was postulated more than 80 years ago (Lillie 1913), our understanding of the molecular basis of fertilisation is still in its infancy.

In the pig, sperm-binding activity resides in the 55 kDa ZPα (ZPB) glycoprotein (Yurewicz et al. 1991). However, conflicting results have been reported regarding the identity of the glycans possessing the sperm-zona pellucida binding capability, e.g. O-linked (Yurewicz et al. 1991) *versus* N-linked tri- and/or tetraantennary (Kudo et al. 1998) carbohydrate structures. On the sperm side, both integral plasma membrane proteins (APz, zonadhesin) and peripherally associated proteins (spermadhesins, P47) have been described as potential components of the primary gamete recognition and/or signalling sperm machinery (Ensslin et al. 1998). Among the variety of sperm proteins known to possess zona pellucida glycoprotein- and carbohydrate-binding capabilities, spermadhesins emerge as a novel protein family of animal lectins.

2 Spermadhesins: An Overview

Spermadhesins are a group of homologous polypeptides (Fig. 17.1), which represent over 90 % of the total boar seminal plasma proteins and coat the sperm surface at ejaculation (reviewed by Calvete et al. 1995a, 1996a; Töpfer-Petersen and Calvete, 1996; Töpfer-Petersen et al. 1998). Although the major biological source of porcine spermadhesins is the secretion of the seminal vesicle epithelium, where the concentration of various spermadhesins (AQN-1, AQN-3, AWN, PSP-I, PSP-II) ranges from 0.6–7 mg/ml, spermadhesin AWN is also synthesised by the *tubuli recti* and *rete testis*. It is the only member of its protein family found on the surface of porcine epididymal spermatozoa. About 6 million AWN molecules have been quantitated on the surface of a single boar epididymal spermatozoon. After ejaculation, 12–60 million of each spermadhesin AQN-1, PSP-I, AQN-3, and AWN become coated on the apical third of the sperm acrosomal cap. Most of this material, however, is released during in vitro capacitating conditions, and the level of individual spermadhesins drops to 5–6 million molecules per spermatozoon. This suggests that the loosely attached spermadhesin population may serve as decapacitation or acrosome-stabilizing factors, which protect the acrosomal membrane from premature acrosome reactions.

Increasing evidence indicates that the population of spermadhesin molecules tightly bound to the surface of capacitated sperm may serve as primary zona pellucida-binding molecules. Thus, spermadhesins AQN-1, AQN-3, and AWN display cation-independent binding affinity (Kd in the μM range) for both zona pellucida glycoproteins and β-galactosides in Galβ(1–3)-GalNAc and Galβ(1–4)-GlcNAc sequences of mono- (O-linked) and trianntenary (N-linked) oligosaccharide structures of standard glycoproteins (Dostálova et al. 1995; Calvete et al. 1996b). Interestingly, the smallest O-linked-type, intact zona pellucida-derived, synthetic oligosaccharide possessing sperm-zona pellucida binding inhibitory activity has the structure Galβ(1–4)[HSO_3-6]GlcNAcβ(1–3)-Galβ(1–4)-GlcNAcβ (1–3)-Galβ(1–3)-GalNAc-ol (Spijker et al. 1996). On the other hand, Kudo et al.

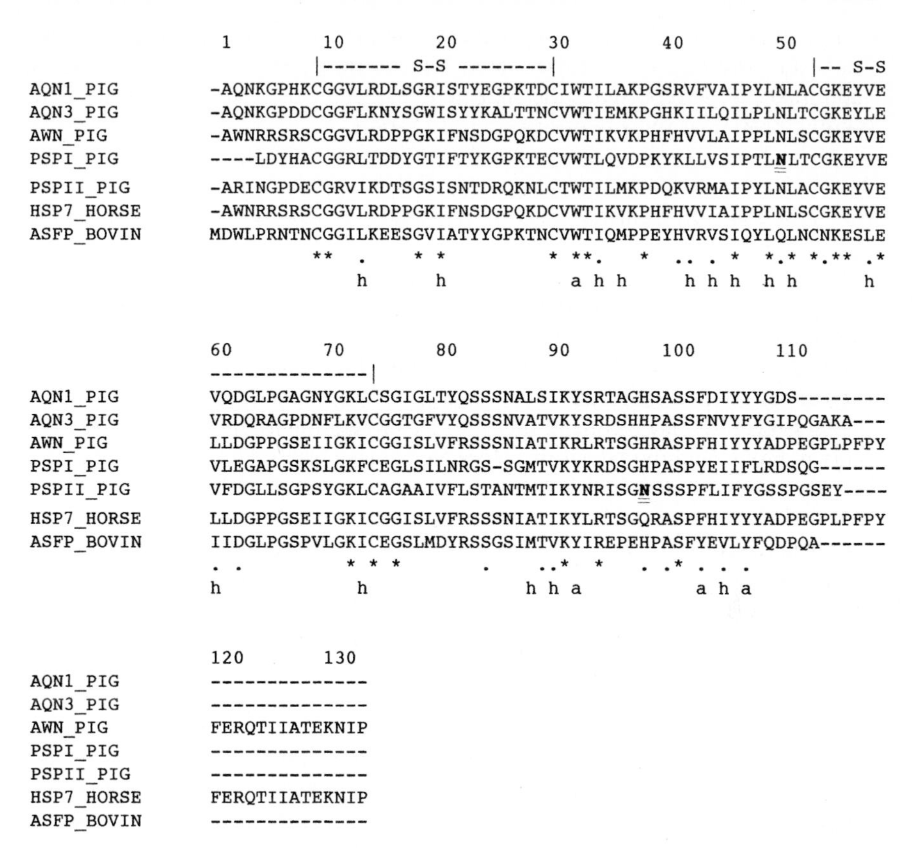

1 10 20 30 40 50
|------- S-S --------| |-- S-S
AQN1_PIG -AQNKGPHKCGGVLRDLSGRISTYEGPKTDCIWTILAKPGSRVFVAIPYLNLACGKEYVE
AQN3_PIG -AQNKGPDDCGGFLKNYSGWISYYKALTTNCVWTIEMKPGHKIILQILPLNLTCGKEYLE
AWN_PIG -AWNRRSRSCGGVLRDPPGKIFNSDGPQKDCVWTIKVKPHFHVVLAIPPLNLSCGKEYVE
PSPI_PIG ----LDYHACGGRLTDDYGTIFTYKGPKTECVWTLQVDPKYKLLVSIPTLNLTCGKEYVE
PSPII_PIG -ARINGPDECGRVIKDTSGSISNTDRQKNLCTWTILMKPDQKVRMAIPYLNLACGKEYVE
HSP7_HORSE -AWNRRSRSCGGVLRDPPGKIFNSDGPQKDCVWTIKVKPHFHVVIAIPPLNLSCGKEYVE
ASFP_BOVIN MDWLPRNTNCGGILKEESGVIATYYGPKTNCVWTIQMPPEYHVRVSIQYLQLNCNKESLE
h h a h h h h h h h h
60 70 80 90 100 110
--------------|
AQN1_PIG VQDGLPGAGNYGKLCSGIGLTYQSSSNALSIKYSRTAGHSASSFDIYYYGDS--------
AQN3_PIG VRDQRAGPDNFLKVCGGTGFVYQSSSNVATVKYSRDSHHPASSFNVYFYGIPQGAKA---
AWN_PIG LLDGPPGSEIIGKICGGISLVFRSSSNIATIKRLRTSGHRASPFHIYYYADPEGPLPFPY
PSPI_PIG VLEGAPGSKSLGKFCEGLSILNRGS-SGMTVKYKRDSGHPASPYEIIFLRDSQG------
PSPII_PIG VFDGLLSGPSYGKLCAGAAIVFLSTANTMTIKYNRISGNSSSPFLIFYGSSPGSEY----
HSP7_HORSE LLDGPPGSEIIGKICGGISLVFRSSSNIATIKYLRTSGQRASPFHIYYYADPEGPLPFPY
ASFP_BOVIN IIDGLPGSPVLGKICEGSLMDYRSSGSIMTVKYIREPEHPASFYEVLYFQDPQA------
h h h h a a h a
120 130
AQN1_PIG --------------
AQN3_PIG --------------
AWN_PIG FERQTIIATEKNIP
PSPI_PIG --------------
PSPII_PIG --------------
HSP7_HORSE FERQTIIATEKNIP
ASFP_BOVIN --------------

Fig. 17.1. Comparison of the amino acid sequences of boar (AQN-1, AQN-3, AWN, PSP-I, and PSP-II), stallion (HSP-7) and bull (aSFP) spermadhesin molecules. Absolutly conserved (*) and conservative amino acids (.) are labelled. The arrangement of the two disulphide bonds is indicated by S-S. Glycosylated residues of PSP-I (Asn^{50}) and PSP-II (Asn^{98}) are in boldface and doubly underlined. h and a below the sequence alignment represent amino acid positions occupied in all CUB domains by hydrophobic and aromatic residues, respectively, which display interior locations in the crystal structures of PSP-I/PSP-II and aSFP, and define the domain signature (Romero et al. 1997)

(1998) have provided evidence that neutral tri- and/or tetraantennary N-linked carbohydrate chains of sow egg endo-β-galactosidase-treated zona pellucida glycoprotein ZPB with terminal Galβ(1–4)-GlcNAc sequences show homologous sperm-binding activity. Hence, zona pellucida glycoproteins display both terminal and internal carbohydrate sequences that could act as spermadhesin molecule-binding epitopes.

Boar spermadhesins are peripheral membrane proteins. As the external surface of the sperm plasma membrane undergoes continuous changes from spermatogenesis to fertilisation, the complex milieu of the ampullary-isthmic junction of the Fallopian tube, the place where presumably fertilisation *in vivo* takes place, could induce remodelling of the sperm surface. Spermadhesin AWN has been demonstrated by immunoelectron microscopy on remnants of the plasmalemma of acrosome-reacted spermatozoa bound *in vivo* to the zona pellucida of oocytes recovered from natural mated sows (Rodríguez-Martínez H, Calvete JJ, submitted). This represents the first demonstration of a sperm-zona pellucida binding protein at the site of *in vivo* fertilisation.

3
HSP-7, a Stallion Seminal Plasma Molecule of the Spermadhesin Family

HSP-7 is a stallion seminal plasma protein, whose primary structure differs in just three amino acids from that of porcine AWN (Fig. 17.1) (Reinert et al. 1996). Purified stallion HSP-7, like its porcine homologue, binds isolated equine *zonae pellucidae*. This finding, together with the observation that stallion frozen/thawed epididymal spermatozoa possess fertilising capacity, suggested that spermadhesin HSP-7 may be one of the factors contributing to the reproductive capability of horse epididymal sperm. Moreover, since perissodactyls (e.g. horse) and artiodactyls (e.g. pig) had a common ancestor over 50 million years ago, the unusually low mutational rate might suggest that the whole AWN (HSP-7) structure is under strong selective pressure. This points to a highly conserved conformation and, most probably, a common biological function too for AWN and HSP-7 in both vertebrate species. Although the actual involvement of HSP-7 in horse fertilisation deserves further investigation, the homologous zona pellucida-binding activity displayed by the isolated protein supports its classification as a putative primary sperm-egg adhesion molecule.

The hypothesis that porcine AWN and equine HSP-7 were diverging under functional constraints, whereas the other polypeptides of the boar seminal plasma spermadhesin family were diverging more rapidly in their amino acid sequences, is in line with the proposal that, following gene duplication, one copy of the gene may divergently have evolved under pressure dictated by the ancestral function while the duplicate gene(s), unencumbered by a functional role, are free to search for new functional roles (Trabesinger-Ruef et al. 1996).

4
aSFP and PSP-I/PSP-II Heterodimer: Orphan Spermadhesin Molecules

Acidic seminal fluid protein (aSFP) is the only polypeptide of the spermadhesin family found in bovine seminal plasma. aSFP displays neither carbohydrate- nor zona pellucida glycoprotein-binding activity, nor does it bind to the bovine sperm surface (Dostálová et al. 1994), and may therefore not be involved in sperm-egg interactions. From the observations that (i) aSFP showed a concentration-dependent protection of bovine sperm from lipid peroxidation in *in vitro* experiments, and (ii) one of the two disulphide bridges of aSFP displayed re-dox equilibrium, which could account for the protective ability, it has been proposed that the major function of the bovine spermadhesin would be to act as a buffer dampening sperm damage caused by active molecules i.e. produced by sperm metabolism and liberated by occasional sperm death and membrane leakage (Einspanier et al. 1994). Although this attractive hypothesis deserves further study, it indicates that spermadhesin function has not been conserved during evolution.

Porcine seminal plasma (PSP) proteins I and II are major secretory components of the boar seminal vesicle epithelium. Both, PSP-I and PSP-II are glycoproteins and each display site heterogeneity in their single glycosylated residues, Asn^{50} and Asn^{98}, respectively (Solís et al. 1997). In porcine seminal plasma, PSP-II forms non-covalent heterodimers with specific glycoforms of PSP-I (Calvete et al. 1995b). The PSP-I/PSP-II heterodimer displays carbohydrate- and zona pellucida-binding activity linked to the PSP-II subunit (Calvete et al. 1995b). However, like bovine aSFP, the PSP-I/PSP-II complex does not bind to the sperm surface excluding its role in gamete interaction. Nonetheless, glycobiological interactions in the sow's uterus mediated by non-sperm-binding boar seminal plasma lectins might play a role in other aspects of porcine reproductive physiology. Noteworthy, although the biological functions of PSP-I/PSP-II remain to be elucidated, Yang et al. (1998) have provided evidence showing that PSP-I could be involved in the regulation of uterine immune activity. Thus, PSP-I bound to a small (3–5 %) subpopulation of porcine lymphocytes and displayed stimulatory effects on peripheral lymphocyte activities initiated by pokeweed mitogen (PWM) in a dose-responsive manner, e.g. 600 % increase of $[^3H]$thymidine uptake at 250 ng/well. Our own unpublished results (Assreuy AMS, Calvete JJ, et al.) demonstrate that intraperitoneal administration of spermadhesin PSP-I/PSP-II acts as a dose-dependent inflammatory stimuli inducing neutrophil migration to the site of inflammation in rats. These findings are parallel to previous observations that

(i) a low molecular mass fraction of porcine seminal plasma enhanced the growth of lymphocytes induced by PWM, and
(ii) after mating, porcine seminal plasma mediates an inflammatory response in the female reproductive tract.

Recently, Solís and coworkers (1998) have reported that, in addition to its zona pellucida glycoprotein- and β-galactoside-binding site, isolated PSP-II displays a specific recognition pocket for mannose-6-phosphate and glucose-6-phosphate, which is distinct but overlaps a binding site for sulphated polysaccharides. Both sites are cryptic in the heterodimer. Docking of Man-6-P and of a tetrasaccharide fragment of heparin onto the crystal structure of the PSP-II subunit suggested that PSP-II Arg^{43}, a residue involved in heterodimer formation with PSP-I, may play a pivotal role in both heparin- and (mannose/glucose)-6-phosphate complexation. The possible biological significance of these cryptic ligand-binding sites remains to be explored. Hence, the possibility that the cryptic recognition pockets in the PSP-I/PSP-II heterodimer could be exposed in response to a specific physiological environment cannot be ruled out.

5 The Crystal Structure of Spermadhesin PSP-I/PSP-II Reveals the CUB Domain Fold

The crystal structure of porcine seminal plasma spermadhesin PSP-I/PSP-II heterodimer has been determined in two crystal forms by multiple isomorphous replacement and has been refined at 2.4 Å (Varela et al. 1997; Romero et al. 1997). Although spermadhesin PSP-I/PSP-II heterodimer itself may not participate in sperm-egg recognition, the large amino acid sequence identity shared with any other spermadhesin molecule (Fig. 17.1) indicates that the conformation of PSP-I (and PSP-II) may be highly conserved in other spermadhesins. In addition, the carbohydrate- and zona pellucida glycoprotein-binding capabilities of PSP-II makes this spermadhesin a paradigm molecule for the whole protein family. Interestingly, besides its potential importance in mammalian fertilisation, spermadhesins are interesting targets for structure elucidation because this group of proteins is built by a single CUB domain architecture (Bork, Beckmann 1993). The CUB domain is a novel 100–110-residue module first reported in complement subcomponents C1r/C1s, embryonic sea urchin protein Uegf, and bone morphogenetic protein 1 (Bmp1). CUB domains have been subsequently identified in 16 functionally diverse proteins, many of which are known to be involved in developmental processes (Bork, Beckmann 1993). Analysis of conserved structural features in a multiple sequence alignment of CUB domains revealed the presence of several rather conserved blocks interrupted by variable regions of flexible length. Conserved characteristics include the four cysteine residues, which in spermadhesins have been shown to form disulphide bridges between neighbour cysteine residues, and various conserved hydrophobic and aromatic positions.

Both subunits of the PSP-I/PSP-II heterodimer are built by a single CUB domain architecture. The CUB domain is built by 10 β-strands arranged into a sandwich of two 5-stranded β-sheets (Fig. 17.2). The two β-sheets of the sandwich are comprised of strands 1, 3, 5, 8, 10, and 2, 4, 6, 7, 9, respectively. Each

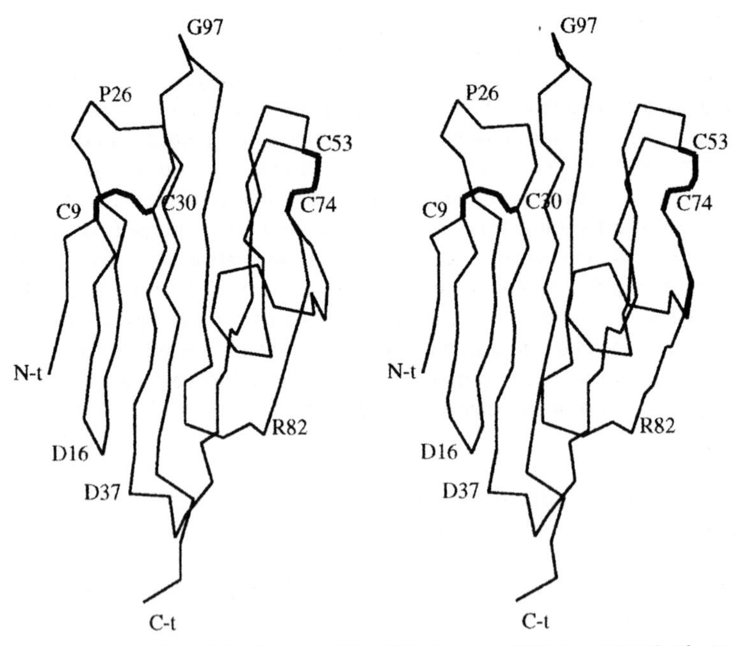

Fig. 17.2. Stereo view of the Ca trace of the CUB domain of PSP-I (or PSP-II). The N- and C-termini are labelled N-t and C-t, respectively. Disulphide bonds are depicted with thick lines

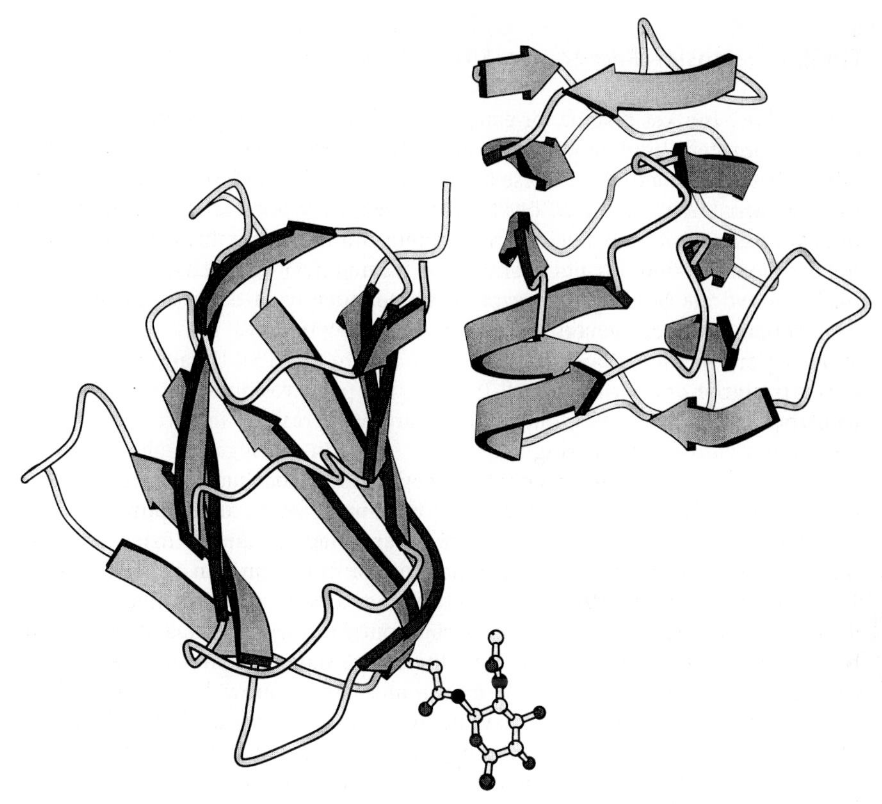

Fig. 17.3. MOLSCRIPT representation of the PSP-I/PSP-II heterodimer displaying the N-acetylglucosamine residue attached to asparagine 50 of PSP-I

sheet of the CUB β-sandwich contains two parallel (1 and 3, and 2 and 4) and four antiparallel (3, 10, 5, and 8, and 4, 9, 6, and 7) β-strands. Disulphide bridges between cysteine residues 9 and 30 and 53 and 74, which are conserved in all known spermadhesin molecules (Calvete et al. 1995a), crosslink loop LA and strand β_4 and loops LE and LG, respectively, at opposite edges of the same face of the CUB β-sandwich (Fig. 2). Glycosylated residues (PSP-I Asn^{50} and PSP-II Asn^{98}) are located at the end of strand β5 and loop LI, respectively. However, only the innermost N-acetylglucosamine of PSP-I is defined in the crystal structure (Fig. 17.3).

PSP-I/PSP-II represents the first crystal structure of a mammalian zona pellucida-binding protein and of a polypeptide built by a CUB domain architecture. The four highly conserved aromatic residues and 15 out of 17 invariant hydrophobic residues, which define the CUB domain signature (Fig. 17.1), display an interior location, suggesting that this hydrophobic core may be essential for maintaining the overall folding of the domain.

6
The β-Galactoside Recognition Site of Spermadhesin PSP-II?

Asparagine^{50} is a key residue for expression of the carbohydrate- and zona pellucida glycoprotein-binding capability of boar spermadhesins (Calvete et al.1995; Töpfer-Petersen et al. 1996). These activities are blocked in PSP-I due to glycosylation of Asn^{50}. Asparagine 50 of PSP-II is not glycosylated and the glycoprotein indeed possesses zona pellucida glycoprotein-binding capability, suggesting that the carbohydrate-binding pocket might be located near this residue. Mechanisms for carbohydrate binding have evolved independently in diverse protein structural frameworks, but nevertheless share some key features. Although the actual arrangement of the carbohydrate-binding residues of PSP-II can not be established from the present PSP-I/PSP-II structure, polar residues hydrogen-bonded to carbohydrate hydroxyl groups and an aromatic residue involved in stacking interaction with the sugar ring are usually found in the sugar combining site of leguminous and animal lectins of the galectin family. Such arrangement is potentially found in the peripheral domain of the dimer interface composed of PSP-I Gln^{112}, and Tyr^{48} and Asn^{50}y of PSP-II. Similar to legume and animal lectins, the putative carbohydrate recognition domain of PSP-II is located at a shallow groove on the protein surface. Residues PSP-II Tyr^{48} and Asn^{50} co-ordinate a structural water molecule, which may play a dual role contributing to the stabilisation of the heterodimer interface and maintaining the conformation of the side chains of the carbohydrate recognition domain in the absence of sugar ligands. Structural studies on a variety of lectins have revealed that the binding sites of lectins appear to be preformed with ordered water molecules forming hydrogen bonds with the unligated proteins in a pattern that closely mimics the hydrogen bonding by sugar hydroxyl groups. Binding of sugars often displaces those ordered water molecules. Mutagenesis experiments can now be carried out to investigate more directly the carbohydrate binding role of PSP-I Gln^{112}, and Tyr^{48} and Asn^{50} of PSP-II, and additional residues from the surroundings. In addition, we are engaged in generating PSP-I/PSP-II-oligosaccharide complexes for studying protein-carbohydrate interactions, which may provide clues on the molecular mechanism underlying sperm-egg binding mediated by spermadhesin molecules.

7
Structural Characterisation of the Glycan Chains of PSP-I and PSP-II

The structures of the N-linked oligosaccharides of PSP-I and PSP-II were elucidated by combination of methylation analysis and collision-induced dissociation by tandem electrospray ionisation mass spectrometry of reduced and permethylated oligosaccharides, released by hydrazine treatment of the isolated proteins, and purified by anion-exchange and amino-bonded phase chromatography (Nimtz M, Calvete JJ, in preparation). Selected proposed structures were further confirmed by ^{1}H-NMR spectroscopy. Both glycoproteins bear the same neutral, mono- and disialylated glycans although in different molar ratios. 22 neutral oligosaccharides, 11 monosialylated glycans, and 3 disialylated carbohydrate chains were characterised. Except for a few monoantennary structures, all the oligosaccharides are di- or triantennary glycans sharing a $(GlcNAc)_2(Man)_3GlcNAc$ (Fuc)GlcNAc core. Each antenna of diantennary structures displays combinations of $[NeuAc\alpha2-6]_{0-1}[Gal\alpha1-3]0-1[(Gal/GalNAC)\beta1-4]_{0-1}[GlcNAc(Fuc\alpha1-3)_{0-1}\beta1-(2/4)]_{0-1}$ Man sequence. A monosialylated oligosaccharide structure bearing terminal NeuGly instead of NeuAc was also characterised. The third branch of triantennary structures consist of $(Gal\alpha1-3)_{0-1}Gal\beta1-4$-GlcNAc sequences $\beta1-4$-linked to the mannose residue or $\beta1$-(3/6)-linked to the outer galactose residue of the 1–3 antenna. Disialylated glycans display NeuAcα2–6Galβ1–4GlcNAc or NeuAcα2–6GalNAcβ1–4GlcNAc sequences.

Despite being PSP-I/PSP-II heterodimer the major protein of boar seminal plasma, its function remains elusive. An interesting possibility is that the oligosaccharides participate in PSP-I/PSP-II function. Structural analysis of the oligosaccharides derived from glycodelin, a human glycoprotein with potent immunosuppressive and contraceptive activities, showed that the major non-reducing epitopes in the complex-type glycans are Galβ1–4GlcNAc (lacNAc), GalNAcβ1–4 GlcNAc (lacdiNAc), NeuAcα2–6Galβ1–4-GlcNAc (sialylated lacNAc), NeuAc α2–6GalNAcβ1–4-GlcNAc (sialylated lacdiNAc), Galβ1–4(Fucα1–3)GlcNAc (Lewis^x), and GalNAcβ1–4(Fucα1–3)GlcNAc (lacdiNAc analogue of Lewis^x) (Dell et al. 1995). The authors have put forward the hypothesis that the oligosaccharides bearing sialylated lacNAc or lacdiNAc antennae may manifest immunosuppressive effect by blocking adhesive and activation-related events mediated by CD22, the human B cell associated receptor, and that oligosaccharides with fucosylated lacdiNAc antennae might block selectin-mediated adhesion [40]. Boar seminal plasma contains immunosuppressive activity associated with 14 kDa protein(s) (Veselsky el al. 1996). Both, the molecular mass and the type of glycosylation of the PSP-I/PSP-II subunits would support immunosuppressive function for PSP-I and PSP-II glycoforms. The structural characterisation of the oligosaccharides attached to these boar seminal plasma glycoproteins provides the necessary foundation for investigating the potential biological role of PSP-I and PSP-II glycans.

References

Benoff S (1997) Carbohydrates and fertilization: an overview. Mol Hum Reprod 3: 599–637

Bork P, Beckmann G (1993) The CUB domain. A widespread module in developmentally regulated proteins. J Mol Biol 231: 539–545

Calvete JJ, Sanz L, Dostálová Z, Töpfer-Petersen E (1995a) Spermadhesins: sperm-coating proteins involved in capacitation and zona pellucida binding. Fertilität 11: 35–40

Calvete JJ, Mann K, Schäfer W, Raida M, Sanz L, Töpfer-Petersen E (1995b) Boar spermadhesin PSP-II: location of posttranslational modifications, heterodimer formation with PSP-I glycoforms, and effect of dimerization on the ligand-binding capabilities of the subunits. FEBS Lett 365: 179–182

Calvete JJ, Sanz L, Ensslin M, Töpfer-Petersen E (1996a) Sperm surface proteins. Repr Dom Anim 31: 101–106

Calvete JJ, Carrera E, Sanz L, Töpfer-Petersen E (1996b) Boar spermadhesins AQN-1 and AQN-3: oligosaccharide and zona-pellucida binding characteristics. Biol Chem 377: 521–527

Dell A, Morris HR, Easton RL, Panico M, Patankar M, Oehninger S, Koistinen R, Koistinen H, Seppala M, Clark GF (1995) Structural analysis of the oligosaccharides derived from glycodelin, a human glycoprotein with potent immunosuppressive and contraceptive activities. J Biol Chem 270: 24116–24126

Dostálovà Z, Calvete JJ, Sanz L, Hettel C, Riedel D, Einspanier R, Schönek C., Töpfer-Petersen E (1994) Immunolocalization and quantitation of spermadhesin aSFP in ejaculated, swim-up, and capacitated bull spermatozoa. Biol Chem Hoppe-Seyler 375: 457–461

Dostálová Z, Calvete JJ, Sanz L, Töpfer-Petersen E (1995) Boar spermadhesin AWN-1: oligosaccharide and zona-pellucida binding characteristics. Eur J Biochem 230: 329–336

Einspanier R, Krause I, Calvete JJ, Töpfer-Petersen E, Klostermeyer H, Karg H (1994) Bovine seminal plasma aSFP: localization of disulphide bridges and detection of three different isoelectric forms. FEBS Lett. 344: 61–64

Ensslin M, Vogel T, Calvete JJ, Thole HH, Schmidtke J, Matsuda T, Töpfer-Petersen E (1998) Molecular cloning and characterization of P47, a novel boar sperm-associated zona pellucida-binding protein homologous to a family of mammalian secretory proteins. Biol Reprod 58: 1057–1064

Green DPL (1997) Three-dimensional structure of the zona pellucida. Rev Reprod 2: 147–156

Kudo K, Yonezawa N, Katsumata T, Aoki H, Nakano M (1998) Localization of carbohydrate chains of pig sperm ligand in the glycoprotein ZPB of egg zona pellucida. Eur J Biochem 252: 492–499

Lillie FR (1913) The mechanism of fertilization. Science 38: 524–528

Reinert M, Calvete JJ, Sanz L, Mann K, Töpfer-Petersen E (1996) Primary structure of stallion seminal plasma protein HSP-7, a zona pellucida binding protein of the spermadhesin family. Eur J Biochem 242: 636–640

Romero A, Romão MJ, Varela PF, Kölln I, Dias JM, Carvalho AL, Sanz L, Töpfer-Petersen E, Calvete JJ (1997) The crystal structures of two spermadhesins reveal the CUB domain fold. Nature Struc Biol 4: 783–788

Sinowatz F, Töpfer-Petersen E, Calvete JJ (1997) Glycobiology of fertilization. In: Gabius H-J, Gabius S (eds) Glycosciences: status and perspectives. Chapman and Hall, Weinheim, pp 595–610

Solís D, Calvete JJ, Sanz L, Hettel C, Raida M, Díaz-Mauriño T & Töpfer-Petersen E (1997) Fractionation and characterization of boar seminal plasma spermadhesin PSP-II glycoforms reveal the presence of uncommon N-acetylgalactosamine-containing N-linked oligosaccharides. Glycoconjugate J 14: 275–280

Solís D, Romero A, Jiménez M, Díaz-Mauriño T, Calvete JJ (1998) Binding of mannose-6-phosphate and heparin by boar seminal plasma PSP-II, a membr of the spermadhesin protein family. FEBS Lett 431: 273–278

Spijker NM, Keuning CA, Hooglut M, Veenernan GH, van Boekel CAA (1996) Synthesis of a hexasaccharide corresponding to a zona pellucida fragment that inhibits porcine sperm-oocyte interaction in vitro. Tetrahedron 52: 5945–5069

Töpfer-Petersen E, Calvete JJ (1996) Sperm-associated protein candidates for primary zona pellucida-binding molecules: structure-function correlations of boar spermadhesins. J Reprod Fertil Suppl 50: 55–61

Töpfer-Petersen E, Romero A, Varela PF, Ekhlasi-Hundrieser M, Dostálová Z, Sanz L, Calvete JJ (1998) Spermadhesins: a new protein family. Facts, hypotheses and perspectives. Andrologia 30: 217–224

Trabesinger-Ruef N, Jermann T, Zankel T, Durrant B, Frank G, Benner SA (1996) Pseudogenes in ribonuclease evolution; a source of new biomolecular function? FEBS Lett 382: 319–322

Tulsiani DRP, Yoshida-Komiya H, Araki Y (1997) Mammalian fertilization: a carbohydrate-mediated event. Biol Reprod 57: 487–494

Varela PF, Romero A, Sanz L, Romão MJ, Töpfer-Petersen E, Calvete JJ (1997) The 2.4 Å resolution crystal structure of boar seminal plasma PSP-I/PSP-II: a zona pellucida-binding glycoprotein heterodimer of the spermadhesin family built by a CUB domain architecture. J Mol Biol 274: 635–649

Veselsky J, Dostál J, Holán V, Soucek J, Zelezná B (1996) Effect of boar seminal immunosuppressive fraction on B-lymphocyte and on primary antibody response. Biol Reprod 55: 194–199

Yang WC, Kwok SCM, Leshin S, Bollo E, Li WI (1998) Purified porcine seminal plasma protein enhances in vitro immune activities of porcine peripheral lymphocytes. Biol Reprod 59: 202–207

Yurewicz EC, Pack BA, Sacco AG (1991) Isolation, composition, and biological activity of sugar chains of porcine oocyte zona pellucida 55K glycoproteins. Mol Reprod Dev 30: 126–134

CHAPTER 18

Expression and Characterization of Saposin-Like Proteins

S. Zaltash[1] and J. Johansson[1]

1 Abstract

The 79 residue mature surfactant protein B is formed by proteolytic cleavage from a larger precursor. SP-B belongs to the family of saposin-like proteins and has unique functional roles in pulmonary surfactant. The 381-residue human proSP-B fused to an N-terminal poly-His tag was expressed in *E. coli,* and purified from inclusion bodies by resolubilisation with 2.5 % (w/v) SDS followed by metal affinity chromatography after removal of SDS. Recombinant proSP-B solubilised in sodium phosphate buffer exhibits about 35 % α-helical structure and is preferentially proteolytically cleaved between the three tandem saposin-like domains that have been proposed from amino acid sequence comparisons. These results give experimental support to the possibility that proSP-B contains, in addition to SP-B, two further saposin-like domains.

2 Introduction

Lung surfactant is a complex mixture of phospholipids and proteins. The main function of this system is to reduce the surface tension at the alveolar air/liquid interface. Four different surfactant-associated proteins have been purified. The larger surfactant proteins A (SP-A) and D (SP-D), are hydrophilic, while SP-B and SP-C are small and insoluble in water, see (1). SP-B and SP-C probably have unique functional roles in the formation of the surface-active monolayer. These two hydrophobic proteins are unrelated in structure, but both of the mature proteins are formed by proteolytic cleavage from larger precursors. The mature SP-B polypeptide chain consists of 79 residues, forms disulphide-dependent homodimers, and displays about 45 % α-helical secondary structure in phospholipid bilayers (2, 3). SP-B belongs to the family of saposin-like proteins (Table. 18.1). SP-B and the other members of the saposin-like family exhibit 17–24 % pairwise residue identities (4). From sequence alignments with saposins, the 42 kDa precursor of surfactant protein B (proSP-B) has been proposed to contain three tandem repeats of about 90 residues (5), where mature SP-B corresponds to the sec-

[1] Department of Medical Biochemistry and Biophysics, Karolinska Institutet, S-171 77 Stockholm, Sweden.

Table 18.1. The saposin family. SP-B, amoebapores (which are pore-forming polypeptides from *Entamoeba hisytolytica*), parts of acid sphingomyelinase, plant aspartic protease and acyloxyacylhydrolase, the saposins (which promote enzymatic degradation of sphingolipids in lysosomes) and NK-lysin (which is an antibacterial and tumourolytic polypeptide from nature killer cells), exhibit 17 to 24 % pairwise residue identities

- * NK-lysin
- * SP-B
- * Amoebapores
- * Saposins
- * Parts of acyloxyacylhydrolase
- * Parts of acid sphingomyelinase
- * Parts of plant aspartic protease

ond of these repeats (residues 201–279 in proSP-B). Moreover, the intramolecular disulphide patterns in SP-B, NK-lysin (which is an antibacterial and tumourolytic polypeptide from Natural Killer cells), and saposins B and C are identical (2, 6, 7). We have expressed human proSP-B in *E. coli*, and characterized the recombinant protein in terms of overall secondary structure and susceptibility to limited proteolysis with trypsin. This shows that proSP-B exhibits about 35 % α-helical structure and is preferentially cleaved between the three proposed saposin-like domains. Currently we aim to express single saposin domains, especially NK-lysin, for studies of structural and functional properties.

3 Analysis of Recombinant proSP-B

We have analyzed the domain organization of recombinant proSP-B. The cDNA coding for the precusor of human SP-B has been cloned and sequenced and codes for a protein of 381 amino acid residues (8). The 381-residue human proSP-B fused to an N-terminal poly-His tag was expressed in *E. coli*. The recombinant protein was purified from inclusion bodies by resolubilisation with 2.5 % (w/v) SDS and subsequent metal affinity chromatography after removal of SDS by dialysis (Fig. 18.1). Recombinant proSP-B solubilised in sodium phosphate buffer

Purification of Recombinant Poly-His-proSP-B

1. IPTG-induced *E. coli* BL21cells harvested.
2. Bacteria resupended in 20 mM Tris-HC1, 100 mM NaC1, pH 8.0 (buffer A).
3. Sonication 10 times, 60 Hz, 30 secs with an interval of 30 secs.
4. Centrifugation at 15000 rpm, 15 mins.
5. Pellet resuspended in buffer A containing 2.5 % SDS.
6. Sonication and incubation at 37 °C, 30 mins.
7. Centrifugation 15000 rpm, 15 mins. Supernatant divided into two aliquots.

↓ ↓

+SDS −SDS

↓ ↓

8. Dialysis against 20 mM Tris-HCl, pH 8.0 in the presence or absence of 2.5 % SDS, respectively.
9. Metal affinity chromatography. Elution with 100 mM imidazole.

Fig. 18.1. Purification strategy for rproSP B. The stategy for purification of rproSP-B is outlined

exhibits about 35 % a-helical structure, which is similar to the approximatly 45 % helix of SP-B in dodecylphosphocholine micelles estimated by CD spectroscopy. Limited proteolysis of rproSP-B occur predominantly between the three tandem saposin-like domains that were proposed from amino acid sequence comparisons. These results give experimental support to the possibility that proSP-B contains, in addition to SP-B, two further saposin-like domains (9).

4 Expression of Recombinant NK-Lysin

We aim to express single saposin domains, especially NK-lysin. For this purpose fragments corresponding to NK-lysin were PCR-amplified from porcine bone marrow cDNA. Probes which correspond to the 5' and 3' ends of NK-lysin (nucleotides 360–380 and 573–593 of porcine proNK-lysin) were used. The resulting product was inserted into the pET15b vector, which gives an N-terminal poly-His tag fused to the recombinant proteins, and sequenced by the dideoxy chain-termination method. The ligation product was transferred into *E.coli* JM 109 cells for selection of plasmid-carrying clones. Finally, pET15b vector with inserted NK-lysin DNA was transferred to *E.coli* BL21 DE3 cells, which contain an integrated copy of the T7 DNA polymerase gene under control of the lac UV5 promoter. The recombinant protein was purified from inclusion bodies by resolubilisation with 8 M urea and subsequent metal affinity chromatography, and reduced/reoxidised by treatment with DTT followed by dialysis against DTT-free buffer. The recombinant protein was subjected to SDS/PAGE which gives detectable bands at about 11 kDa, which is in agreement with the expected mass of 10.954 kDa of His-tag/NK-lysin fusion protein.

5 Discussion

ProSP-B fused to an N-terminal poly-His tag was expressed in *E. coli* and purified from inclusion bodies by metal affinity chromatography after resolubilisation with 2.5 % (w/v) SDS (9). This yields approximately 1.5 mg rproSP-B/L cell culture and the recombinant protein is concluded to be folded into a native-like conformation and to form disulphide-dependent oligomers. The present CD spectrum and limited proteolysis data lend experimental support to the suggestion that proSP-B is composed of three tandem saposin-like domains. Of the three saposin-like domains present in proSP-B, only SP-B has been isolated. It is therefore an open question whether the remaining two saposin-like domains in proSP-B are processed to yield unique entities, or if the sole function of proSP-B is to give rise to SP-B. Sequence alignments of canine, rabbit, rat and human proSP-B show that the first and second (i.e. SP-B) domains exhibit a high degree of conservation, while the third domain is little conserved. Notably, the proSP-B N-terminal part, but not the C-terminal region, is required for its processing and intracellular targeting (10,11). The conservation of the first saposin-like domain may indicate that it is particularly important for proSP-B functions, or that it plays a role also after processing of proSP-B to SP-B.

Porcine NK-lysin was expressed in *E. coli.* Western blotting of the recombinantNK-lysin after reduction and reoxidation shows a band corresponding to the monomeric protein. The recombinant protein is currently studied in terms of structural and functional properties.

References

Johansson, J. and Curstedt, T. (1997) Eur. J. Biochem. 244, 675–693.
Johansson, J., Curstedt, T. and Jörnvall, H. (1991) Biochemistry. 30, 6917–6921.
Vandenbussche, G., Clercx, A., Clercx, M., Curstedt, T., Johansson, J., Jörnvall, H. and Ruysschaert, J. -M. (1992) Biochemistry. 31, 9169–9176.
Andersson, M., Curstedt, T., Jörnvall, H. and Johansson, J. (1995) FEBS Lett. 362, 328–332
Patthy, L. (1991) J. Biol. Chem. 266, 6035–6037
Andersson, M., Gunne, H., Agerberth, B., Boman, A., Bergman, T., Sillard, R., Jörnvall, H., Mutt, V., Olsson. B., Wigzell, H., Dagrlind, Å., Boman, H.G. and Gudmundsson, G.H. (1995) EMBO J. 14, 1615–1625
Vaccaro, A. M., Salvioli, R., Barca, A., Tatti, M., Ciaffoni, F., Maras, B., Siciliano, R., Zappacosta, F., Amoresano, A. and Pucci, P. (1995) J. Biol. Chem. 270, 9953–9960
Jacobs, K. A., Phelps, D. S., Steinbrink, R., Fisch, J., Kriz, R., Mitsock, L., Dougherty, J. P, Taeusch, H. W. and Floros, J. (1987) J. Biol. Chem. 262, 9808–9811
Zaltash, S. and Johansson, J. (1998) FEBS Lett. 423, 1–4
Lin, S., Akinbi, H. T., Breslin, J. S. and Weaver, T. E. (1996) J. Biol. Chem. 271, 19689–19695
Lin, S., Phillips, K. S., Wilder, M. R. and Weaver, T. E. (1996) Biochim. Biophys. Acta 1312, 177–185

Part III
Protein-Protein and Protein-DNA Interaction

Part III
Protein-Protein and Protein-DNA Interactions

CHAPTER 19

A Structural Model for the P53 Complex with DNA Response Elements: Implications for P53 Function and Future Research Directions

R. E. HARRINGTON[1], V. B. ZHURKIN[2] and S. R. DURELL[2], R. L. Jernigan[2], A. K. NAGAICH[1,3] and E. APPELLA[3]

1 Introduction

Wild type p53 is a nuclear phosphoprotein that occurs in a wide variety of organisms and plays a central role in the regulation of cellular growth and in tumor suppression. It was first described as a cellular protein that co-precipitated with the large T antigen of SV40 (Lane and Crawford, 1979; Linzer and Levine, 1979) whose synthesis was enhanced in chemically induced tumors (DeLeo *et al.*, 1979). It is a potent, pleiotropic transcription factor that is activated in response to a variety of DNA damaging agents. Such activation can lead to cell cycle arrest at the G1/S phase checkpoint (Kuerbitz *et al.*, 1992; Hartwell and Kastan, 1994) or to induction of apoptosis (Lin *et al.*, 1992; Lowe *et al.*, 1993). It has long been known that p53 inactivation by mutation or deletion (along with loss of the wild type allele) or by interaction with cellular or viral proteins is highly correlated with a wide variety of human cancers (Hollstein *et al.*, 1991; Levine *et al.*, 1991; Lane, 1992; Vogelstein and Kinzler, 1992; Meltzer, 1994) through the development of dominant negative tumorogenic phenotypes (Harris, 1993; Levine, 1993, 1997).

Virtually all of the presently known biological functions of p53 depend critically upon its DNA binding properties. Studies of tumorogenic p53 mutants have shown that most of these are defective in DNA binding and consequently cannot activate transcription (Pavletich *et al.*, 1993; Arrowsmith and Morin, 1996; Jayaraman *et al.*, 1997). Additional evidence is based upon the role of p53 as a transcription factor or transcriptional enhancer for genes that mediate DNA damage repair and growth arrest through their gene products (El-Deiry *et al.*, 1993; Milner, 1994; Prives, 1994). The latter include Gadd45, which is implicated in the stimulation of DNA repair (Marx, 1994; Smith *et al.*, 1994) and *Waf1/Cip1/Sdi1* which codes for p21, a protein that inhibits several cyclin-dependent protein kinases necessary for cell cycle progression from G1 into S phase (Hartwell and Kastan, 1994; Pines, 1994). p53 can also induce apoptosis by responding to the induction of oncogenes such as *c-Myc* (Hermeking and Eick, 1994; Symonds *et al.*, 1994; Wu and Levine, 1994). Thus, it is becoming clear that sequence specific DNA binding and transactivation are the key activities that control most of the biological functions of p53 (Prives *et al.*, 1994; Jayaraman *et al.*, 1997).

[1] Department of Microbiology, Arizona State University, Tempe, AZ 85287-2701.
[2] Laboratory of Experimental and Computational Biology.
[3] Laboratory of Cell Biology, NCI, National Institutes of Health, Bethesda, MD 20892 USA.

Wild type p53 contains at least four functional domains. A schematic of the domain structure and known sites of post-translational modifications as currently envisioned is shown in Fig. 19.1. From the N-terminus on the left are:

(1) an acidic N-terminal region extending approximately from amino acids 1–73 containing a transactivation domain (amino acids 1–43), which mediates the binding of proteins such as MDM2, E1B and TBP, followed by a proline-rich flexible linker segment;
(2) a minimal core DNA binding domain (p53DBD) from amino acids 102–286 contained within the DNA binding region (amino acids 96–308) followed by another flexible linker region (amino acids ~300–318);
(3) a tetramerization domain from amino acids 319–360; and
(4) a 33 amino acid, lysine-rich basic domain at the C-terminus whose function has been ascribed to non-specific DNA binding and/or negative regulation of specific DNA binding. A variety of studies have shown that this C-terminus region strongly influences the DNA binding properties of p53 (Foord *et al.*, 1991; Anderson *et al.*, 1997; Wolkowicz and Rotter, 1997) and that this is further moderated by its charge state as determined by its acetylation level (Gu and Roeder, 1997; Sakaguchi *et al.*, 1998). This is of special interest since the C-termini undergo post-translational modifications including acetylation, phosphorylation and dephosphorylation, which appear to be initiated by DNA damage (Sakaguchi *et al.*, 1998; Kapoor and Lozano, 1998; Lu *et al.*, 1998; Waterman *et al.* 1998). The flexible hinge regions on either side of the core DNA binding domain are relatively unstructured (Unger *et al.*, 1992; Picksley *et al.*, 1994; Wang *et al.*, 1994; Appella and Anderson, 1994). At least four serine residues in human p53 C-terminus are phosphorylated, and the kinases that phosphorylate them have been identified (Bishoff *et al.*, 1990; Wang and Prives, 1995; Baudier *et al.*, 1992; Hupp *et al.*, 1994; Hermann *et al.*, 1991). The role of phosphorylation is not yet entirely clear (Meek, 1998; Prives, 1998), and although no phosphorylation occurs in the p53DBD, the fact that phosphorylation sites are located in other highly conserved regions of p53 suggests a functional role.

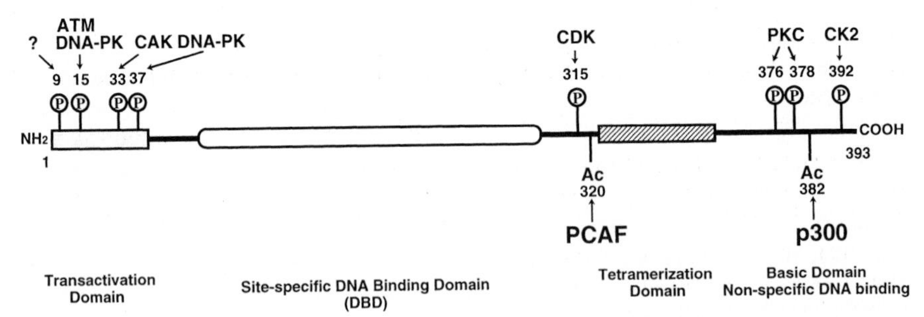

Fig 19.1. Schematic structure of the human p53 protein. The amino terminus is an acidic transcriptional activation domain. The central region encompasses a sequence-specific DNA binding domain and has five highly conserved regions. The carboxyl terminal region contains a tetramerization domain and has a non-specific DNA binding activity. The sites of phosphorylation and acetylation are shown and the kinases and acetylases that have been identified are indicated

2
Structure of the p53-DNA Nucleoprotein Complex

2.1
Tetrameric Nature of the p53 Complex

Wild type p53 binds as a tetramer (El-Deiry *et al.*, 1992; Friedman *et al.*, 1993; Halazonetis and Kandil, 1993; Wang *et al.*, 1994) to over 100 different naturally occurring response elements of which approximately 60 exhibit functionality. It has been estimated that the human genome contains approximately 200 to 300 such sites (Tokino *et al.*, 1994). Response elements differ in details of specific base sequence, but all contain two tandem decameric elements, each a pentameric inverted repeat. Most decamers follow the consensus sequence pattern (El-Deiry *et al.*, 1992) PuPuPuC(A/t)|(T/a)GPyPyPy, where Pu and Py are purines and pyrimidines, respectively, and the vertical bar denotes the center of pseudodyad symmetry. These relationships are illustrated in Fig.s 19.2a and 19.2b for the *p21/Waf1/Cip1* response element and for a symmetric consensus binding sequence (Tokino *et al.*, 1994).

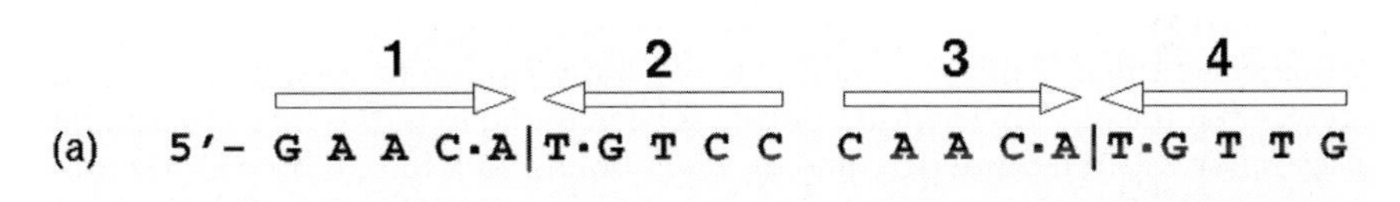

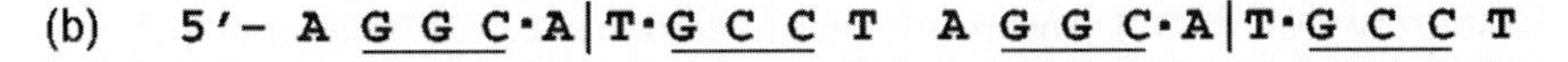

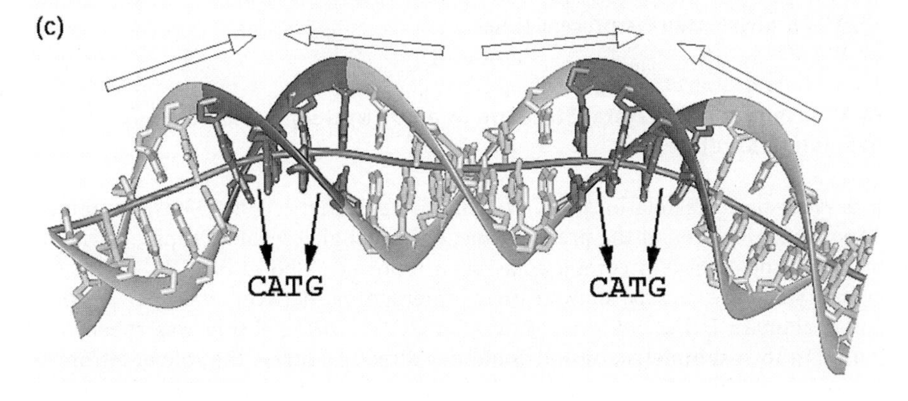

Fig. 19.2. Structure of the p53 response elements. (**a**) Response element in the *p21/Waf1/Cip1* promoter studied by Nagaich *et al.*, (1997). (**b**) Consensus symmetric element used in phasing experiments (Nagaich *et al.*, 1999) and model building (Durell *et al.*, 1998). The arrows indicate orientations of the pentamers. The vertical bars are for the "pentameric junctions". The underlined trimers are the same as in the pentamer specifically bound to p53DBD in the cocrystal (Cho *et al.*, 1994). (**c**) DNA bent in the tetrameric p53DBD complex according to the model. Small arrows show the major groove bending in the CA:TG dimers and are emphasized in black (c) and marked by bullets (a, b). The overall DNA bend is ~40° (as measured between the ends of the duplex axis going through the centers of the base pairs)

The decameric elements may be separated by as much as 21 bp without complete loss of p53 binding affinity (Waterman *et al.*, 1995), but functional sites, defined as those able to activate transcriptionally a nearby reporter gene, evidently follow very closely the consensus decamer pattern with no or only very short intervening spacers (Tokino *et al.*, 1994). Furthermore, the sequences of the tetrameric elements that span the pseudodyad in each half site assume unusual importance in determining the binding properties of p53 (Durell *et al.*, 1998; Nagaich *et al.*, 1998). These elements are most commonly CATG, but may also include CAAG (Tokino *et al.*, 1994). It is well known that these sequences, especially the former, exhibit unusual flexibility for bending or kinking into the major groove (McNamara *et al.*, 1990; Zhurkin *et al.*, 1991; Nagaich *et al.*, 1994; El Hassan and Calladine, 1998; Olson *et al.*, 1998). Since many architectural proteins utilize PyPu sequence elements (Steitz, 1990; Werner *et al.*, 1996; Bewley *et al.*, 1998), it is natural to suspect that these elements play an important role in specific DNA recognition by transcription factors, especially those with high functional multiplicities such as p53. Furthermore, the many regulatory roles for p53 and its large variety of binding sites suggest that its specificity of binding to individual DNA binding sites provides a clue about p53 function in its interactions with other regulatory proteins.

The tetrameric nature of the p53 nucleoprotein complex is one of the aspects of this system that is unique among known transcription factors. It is of special interest that the p53DBD peptide alone has been found to self-assemble with high cooperativity as a tetrameric complex when bound to a full 20 bp p53 response element, although it exists as a monomer in the absence of cognate DNA (Balagurumoorthy *et al.*, 1995). Thus, although the tetramerization domain may in part mediate tetramerization of the wild type protein, this cannot be its sole function, and the full role of the tetramerization domain in p53 function remains somewhat of a mystery at the present time.

2.2
Relationship of a Co-Crystal Structure of a p53 Nucleoprotein Complex to Solution Studies

It is clear that the relationships of structure to function in p53 are extremely important. However, at the present time, only a single crystallographic structure of a p53 nucleoprotein complex has been reported (Cho *et al.*, 1994). This co-crystal structure clearly showed direct interactions between p53DBD peptides and a cognate DNA, but since only a single p53DBD peptide was specifically bound in the asymmetric unit, it could not directly address the role of tetramerization in specific DNA recognition. In addition, it did not predict bending of the response element DNA in the complex, which was subsequently demonstrated using cyclization methods (Balagurumoorthy *et al.*, 1995) and later confirmed by both cyclic permutation (Nagaich *et al.*, 1997a) and by phase sensitive detection (Nagaich *et al.*, 1999). These limitations suggested that an advanced model for the complex, which would encompass these additional features, was required for a more complete understanding of structure-function relationships in the complex.

A relatively detailed model for the tetrameric p53DBD complex with the *p21/Waf1/Cip1* DNA response element has been proposed recently (Durell *et al.*, 1998; Nagaich *et al.*, 1998). To develop this model, chemical probe analysis (Nagaich *et al.*, 1997b, 1998; Appella *et al.*, 1998), cyclic permutation studies (Nagaich *et al.*, 1997a), and advanced phasing studies (Nagaich *et al.*, 1999) were combined with sophisticated molecular modeling that preserved the dynamical character of the DNA response element (Nagaich *et al.*, 1997b, 1999; Durell *et al.*, 1998). The chemical probe methods can assign protein-DNA contacts at single base resolution as well as identify certain bases involved in the specific protein-DNA recognition. Both cyclic permutation (Wu and Crothers, 1984) and A-tract-based phase sensitive detection (Zinkel and Crothers, 1987; Niederweis and Hillen, 1993) are robust methods for identifying and locating bent DNA, and the latter is especially powerful for the determination and quantitation of DNA twist changes associated with complex formation. The tetrameric structural model based upon these biochemical methods (Durell *et al.*, 1998; Nagaich *et al.*, 1999) is significantly more detailed than that obtained directly from the co-crystal study (Cho *et al.*, 1994), and it offers further insights into a number of additional structural and functional properties of the wild type p53 complex.

2.3 A Structural Model for the Tetrameric p53-DNA Complex

The structural model for the p53DBD complex developed as discussed above is shown in Fig. 19.3a. All nomenclature associated with structural features of the p53DBD used subsequently in this review is taken from the co-crystal structure (Cho *et al.*, 1994). The response element used for the modeling was a symmetric site in which both half sites were obtained from the p53 consensus response element (El-Deiry *et al.*, 1992). This site symmetry was necessary in order to keep the modeling within tractable bounds but does not affect seriously the generality of the model structure itself. In the development of the model, protein-DNA contacts were determined from high resolution chemical probes studies on the *p21/Waf1/Cip1* sequence. These contacts were used along with the crystallographic coordinates of the DNA-bound p53DBD (Cho *et al.*, 1994) to assemble a molecular model for the tetrameric complex that was energetically reasonable and preserved known energy constraints upon the DNA. The DNA is assumed to be in a B-conformation (Olson *et al.*, 1998). The studies illustrate how limited co-crystal structural data can be combined with relatively low resolution solution results to develop a simple mechanistic model that can rationalize a wide variety of experimental findings.

Four p53DBD peptides are arranged in a staggered array along the 20 base pair response element with each peptide bound to a single pentameric element using its H2 recognition helix and other protein-DNA contacts as shown in the co-crystal structure (Cho *et al.*, 1994). Peptides immediately adjacent along the response element are bound in opposite (antiparallel) orientations whereas non-adjacent peptides bind in the same (parallel) orientation. Energetic considerations dictate that the binding of four p53DBD peptides to each of the four response element pentamers requires an overall bending of the DNA through

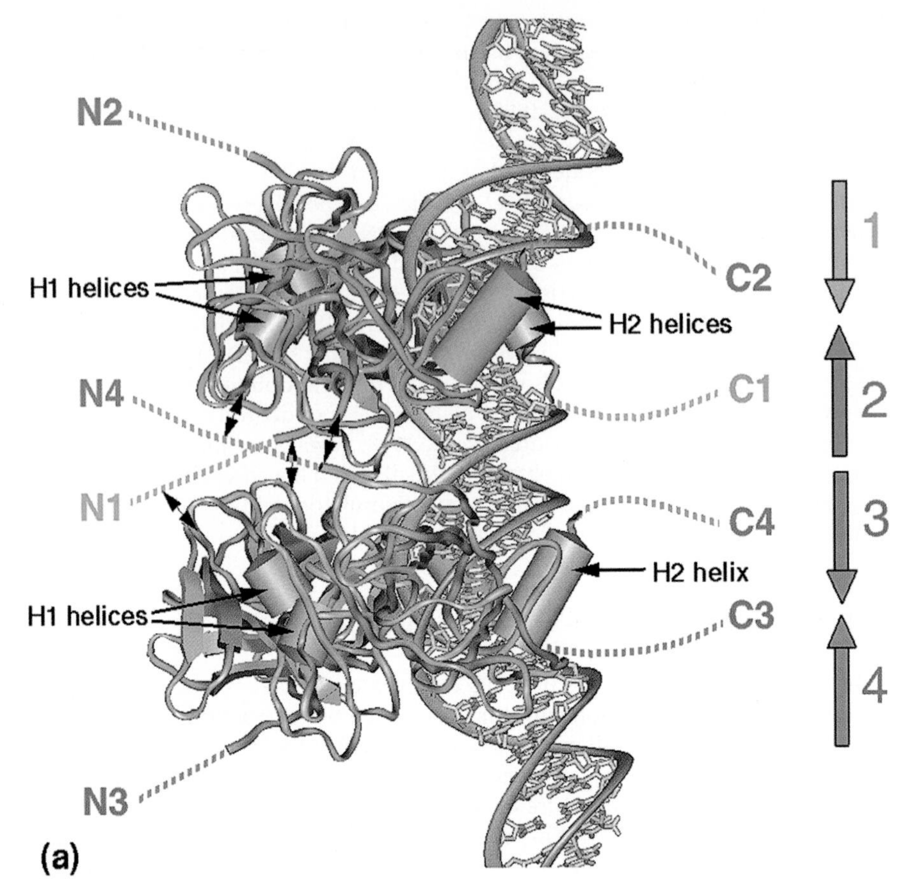

Fig. 19.3. Tetrameric p53DBD binding to bent B-DNA (a) and to "straight" B-DNA (b). (**a**) Four p53DBD subunits bound to bent DNA according to the model (Durell *et al.*, 1998). The recognition H2 helices interact with the major groove of DNA (Cho *et al.*, 1994). The H1-H1 interactions are operative in causing DNA bending and twisting (Fig. 19.4). The broken lines indicate that in the wild type p53 tetramer bound to DNA, the N-termini are located on the external side of the DNA loop, thus being accessible for interacting with transactivating factors. Whereas, the C-terminal domains are likely to be on the inside of the loop, facilitating tetramerization of the wild-type p53 protein and bringing the positively charged basic regions of p53 close to DNA (the tetramerization and basic domains are not shown). Small arrows denote putative interactions between the proline-rich N-fragments and the p53 core domains (Nagaich *et al.*, 1999). Large arrows indicate the orientations of the p53 subunits; they are directed in the same way as the arrows in Fig. 19.2

approximately a 30° to 40° angle. It is also required that the bending occurs into the major groove of DNA at a tetrad of bases at the pentameric junctions in each half site, specifically at flexible sequence elements at these positions, *i.e.*, at the CATG tetrads shown in Fig. 19.2c. This specific DNA bending is required to avoid two major types of stereochemical clash between the bound p53DBD moieties (Nagaich *et al.*, 1997b; Durell *et al.*, 1998). These are illustrated in Fig. 19.3b: clashes occur between the H1 helices of p53DBD peptides bound antiparallel and

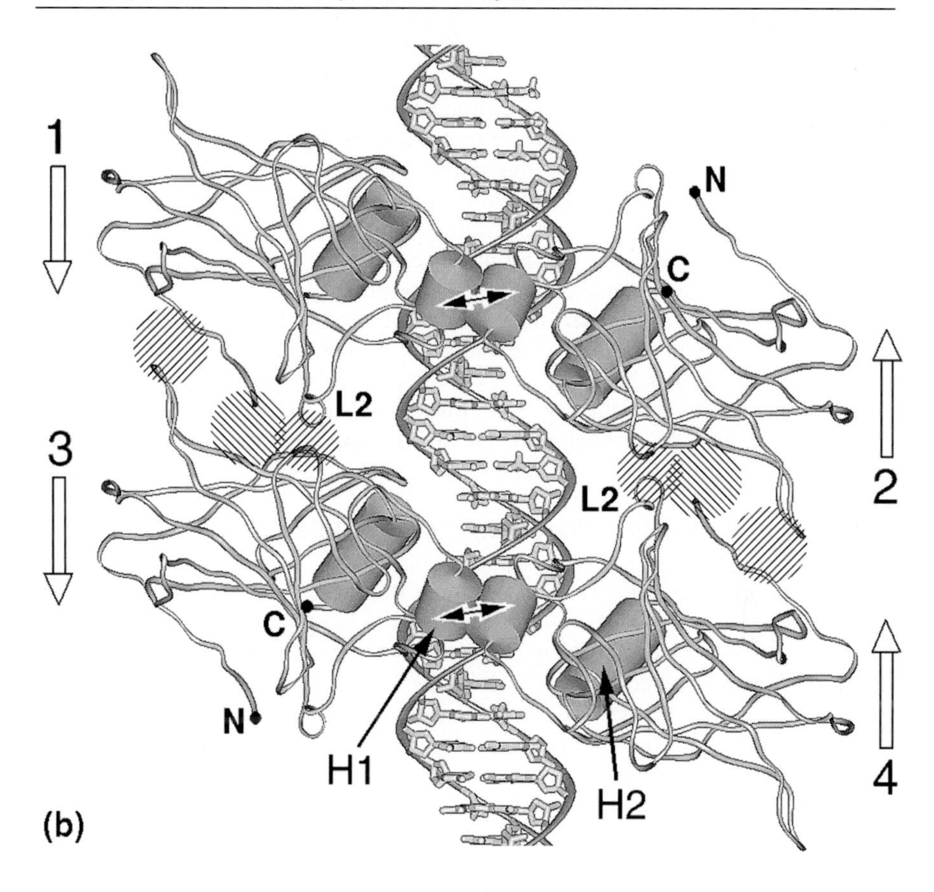

Fig. 19.3 b. Binding of the p53DBD tetramer to "straight" DNA results in steric clashes between the parallel core domains (subunits 1 and 3, 2 and 4; shown by hatched areas) and antiparallel domains (1 and 2, 3 and 4). The clashes between the parallel p53 domains involve the L2 loops, and between the antiparallel domains, the H1 helices. The idealized DNA structure corresponds to the average B-DNA conformation in fibers: Twist = 36° and Rise = 3.38 A°

involve the L2 loop regions of parallel bound moieties. When the peptides are bound to optimally bent DNA, a "hydrophobic pocket" (see Fig. 19.3a) between adjacent peptides stabilizes the complex and presumably accounts for the extremely high binding cooperativity observed in this system.

Comparisons of cyclic permutation studies with experimentally determined binding affinities for several p53 response elements demonstrated that the binding constant of the tetrameric p53DBD complex correlates with DNA bending, and that response elements with greatest flexibility at the pentameric junctions, *i.e.*, with one or more CATG elements at these positions, bind with the highest affinity (Nagaich *et al.*, 1997a). In addition, advanced applications of A-tract phase-sensitive detection methods (Niederweis and Hillen, 1993) have confirmed the DNA bending directionality predicted by the molecular modeling (Nagaich *et al.*, 1999).

The structural model for the complex as formulated (Nagaich *et al.*, 1997b, 1999; Durell *et al.*, 1998) is fully consistent with solution experiments as noted. This is summarized in Table 1 where the various experimental determinations are compared with predictions of the molecular model. All experimental methods, sensitive to DNA bending, show that the response element DNA is bent in the complex, although various methods provide somewhat different specific bending angles. We believe that the bending angle of 32–36°, based upon phasing studies, (Nagaich *et al.*, 1999) is the most accurate value presently available for the p53DBD-DNA tetrameric complex. The bending directionality for the DNA, originally proposed and based upon chemical probes results (Nagaich *et al.*, 1997b), is confirmed by the phasing studies. Finally, the importance of the CATG tetrads as bending loci in the complex is supported by studies of bending as a function of response element sequence (Nagaich *et al.*, 1997a) as well as by recent cyclization studies (P. Balagurumoorthy, manuscript submitted).

2.4
DNA Twisting and Bending Directionality in the p53-DNA Complex

Bending and twisting in the response element DNA upon complexation with both p53DBD and wild type p53 was examined using A-tract-based phase sensitive detection methods. A basic "three-segment" phasing construct (Zinkel and Crothers, 1987) was used to determine both bending directionality and the bend angle. Overtwisting in the DNA was identified and quantitated using a "four-segment" construct (Niederweis and Hillen, 1993). Together, the two constructs allow for a relatively precise determination of these important quantities (Nagaich *et al.*, 1999). The predictions of the original model are confirmed by these experiments: the bending is into the major groove in both the p53DBD and wild type complexes, but both bending and twisting in the DNA is appreciably greater in the wild type complex. A substantial increase in twist upon protein binding accompanied by bending into the major groove has been reported previously only for the nucleosome (Luger *et al.*, 1997) and does not appear to be a common feature of transcription complexes in which DNA bending is generally accompanied by undertwisting. This has important implications for possible p53 interaction with chromatin structures since it suggests that the p53 tetramer might bind to nucleosomal DNA sites without completely unraveling the nucleosome structure.

The fact that response element DNA bending in the nucleoprotein complex is correlated with overtwisting in both types of complex is of special importance, since molecular modeling suggested that DNA overtwisting occurs in a very specific manner with complex formation, and that local supercoiling in genomic DNA (Liu and Wang, 1987) might therefore play a role in the regulation of specific sequence recognition (Durell *et al.*, 1998). The coupling of bending and overtwisting may also have important implications in the interaction of p53 with chromatin. That these effects are significantly larger in the wild type p53 complex supports the notion that DNA binding in the wild type complex may be affected by regions of the N- and C-termini. This appears consistent with the proposal, based on other types of evidence, that p53 binding may be regulated by more than one allosterically related conformational state of the entire protein molecule (Vojtesek *et al.*, 1995; Waterman *et al.*, 1995). It may also rationalize the apparent discrepancy between the observations that

(i) wild type p53 can bind to "split" response elements containing various amounts of spacer DNA (Cook *et al.*, 1995), which in some cases may be functional (Tokino *et al.*, 1994), and that
(ii) p53DBD binds to split sites only if the spacers contain an integral number of helical turns of DNA and if both half sites incorporate the flexible tetrad CATG at the pentameric junctions (P. Balagurumoorthy *et al.*, manuscript submitted).

The reason that the tetrameric model requires a coupling of bending and twisting is illustrated in Fig. 19.4. When the p53DBD subunits are bound to the "straight" B-DNA, the antiparallel H1-helices produce unacceptable steric clashes (Fig. 19.4a). One of the ways to relieve these clashes is to bend the DNA into the major groove (Fig. 19.4b). In such a case, however, the H1-H1 interactions can be lost completely, which is also unfavorable. Energy calculations (Durell *et al.*, 1998) indicate that the optimum is achieved when the DNA is both bent and twisted, and the two H1 helices are closely juxtaposed, forming a two helix bundle as shown in Fig. 19.4c. Each antiparallel dimer in the tetrameric complex is stabilized by van der Waals interactions and hydrogen bonds involving His178, Glu180 and Arg181 (Durell *et al.*, 1998). Notably, the asymmetric interactions of Arg248 with the DNA backbone predicted by this model (Fig. 19.4d), are consistent with the OH radical protection results (Nagaich *et al.*, 1997b). This is not apparent from the co-crystal structure in which only a single p53DBD is bound to the DNA and the Arg248 residue is symmetrically bound in the minor groove (Cho *et al.*,

Table 19.1. Structural features of the p53DBD-DNA and wild type p53-DNA complexes

p53DBD-DNA Complex	WT p53-DNA complex
A. Bending angle[$]	
1. Cyclization assay (~60 °)[1]	
2. Cyclic permutation assay (~50 °)[2]	Phasing analysis (~55 °)[3]
3. Phasing analysis (~35 °)[3]	
4. Molecular modeling (~40 °)[4]	
B. Directionality of Bending*	
1. Phasing analysis (Major groove)[3]	Phasing analysis (Major groove)[3]
2. Molecular modeling (Major groove)[4]	
C. Overtwisting	
1. Phasing analysis (~35 °)[3]	Phasing analysis (~70 °)[3]
2. Modeling (~25 °)[3,5]	

[1] Balagurumoorthy *et al.* (1995). [2]Nagaich *et al.* (1997a). [3]Nagaich *et al.* (1999). [4]Nagaich *et al.* (1997b). [5]Durell *et al.* (1998).
[$] The estimated bend angles for p53DBD-DNA complex decrease in the order: (Cyclization assay) > (Cyclic permutation assay) > (Phasing analysis), in accord with earlier observation for the other nucleoprotein complexes (Van der Vliet and Verrijzer, 1993; Kerppola and Curran, 1993).
* The directionality of the global bending is consistent with the major groove bending in the CATG tetramers in the two half sites (Figure 2c).

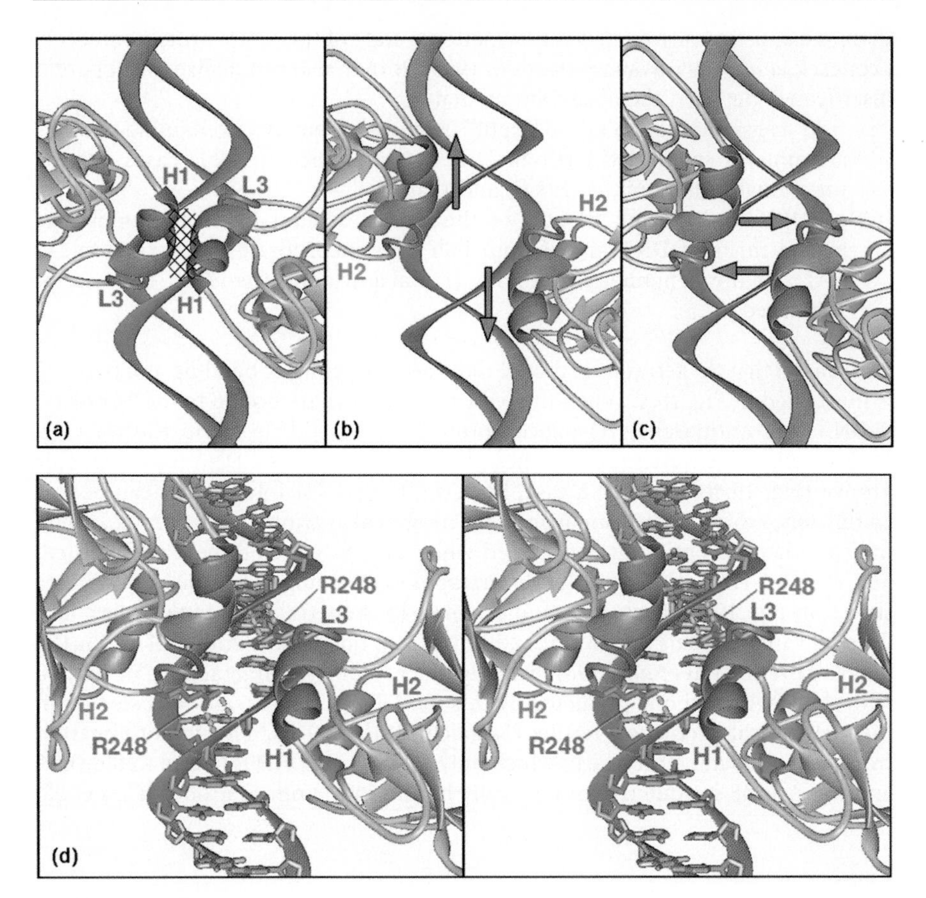

Fig. 19.4. Schematic representation of the Twist-Bend correlation in the p53-DNA complex. The H1 helices and L3 loops are shown with the minor groove in the background. The H2 helices shown in blue and green interact specifically with the DNA pentamers shown in the same color. (**a**) Binding of two p53DBD domains to "straight" B-DNA causes the sterical clash between the antiparallel H1 helices as indicated by the cross-hatched area. DNA is bent into the major groove (from the viewer) as shown by the arrows. Minor groove is enlarged and H1 helices are separated. (**c**) DNA is overtwisted and the minor groove size is thus restored. The contact between the H1 helices becomes favorable (Durell *et al.*, 1998). (**d**) Structural model accounting for the hydroxyl cleavage data for the *p21/Waf1/Cip1* response element (Nagaich *et al.*, 1997b). Each of the Arg248 residues interacts predominantly with one strand of DNA, protecting the GT dimers (shown in magenta) from cleavage (see Fig. 19.2a). Due to the correlated changes in DNA bending and twisting, the minor groove remains relatively narrow, in agreement with cleavage data

1994). It is likely that these H1-H1 contacts make an important contribution to the observed p53DBD-DNA binding cooperativity. Furthermore, these interactions are delicately balanced in terms of both bend and twist as shown in Figures 19.4a-c, and uncorrelated changes in either variable are costly to the complex stability. Thus, the response element DNA must be bent and twisted by precise amounts in order to relieve the steric clashes between the antiparallel p53DBD subunits, and at the same time, to maintain the stabilizing H1-H1 interactions.

Finally, the molecular model requires that bend and twist angles be positively correlated to optimize the antiparallel H1-H1 interactions, and that bending into the major groove in the CATG tetrads, therefore, be accompanied by an increase in twisting as observed in the gel studies (Table 19.1). However, it is not known precisely how the bend-twist correlation depends upon the specific sequence of the bending tetrads. DNA bending in the complex is maximal for CATG and is reduced in CAAG (Nagaich *et al.*, 1997a), which implies somewhat less twist in the latter tetrad also because of the observed positive correlation. It is quite possible that differential twist-bend correlation effects among sequence tetrads found in different p53 response elements may govern binding affinities, sequence specificities, and the responses of these elements to local supercoiling effects in promoter regions. Clearly, additional studies are required to more fully understand these complicated questions.

3
Implications of the Molecular Model to p53 Function

The basic structural model described here for a sequence specific nucleoprotein complex between p53DBD and a response element is not a fully determined structure but nevertheless is consistent with several types of independent biochemical experiments. These include ligase-mediated cyclization (Balagurumoorthy *et al.*, 1995), cyclic permutation (Nagaich *et al.*, 1997a), a variety of chemical probes studies (Nagaich *et al.*, 1997b, 1998; Appella *et al.*, 1998), and the use of A-tract phase sensitive detection analysis (Nagaich *et al.*, 1999). The latter has allowed determination for the first time of both the magnitude and directionality of the DNA bend in the complex. The bending directionality confirms our earlier prediction based on molecular modeling only (Nagaich *et al.*, 1997b, 1998; Durell *et al.*, 1998) and follows the pattern of major groove bending at conserved CATG sequence elements that has been observed in other nucleoprotein systems (El Hassan and Calladine, 1998; Dickerson, 1998; Olson *et al.*, 1998) including nucleosomes (Satchwell *et al.*, 1986).

An especially important observation from the phasing studies is that the DNA bending and twisting angles are significantly larger in the wild type p53 complex than in the complex with the p53 core domains only. This suggests that the p53 domains flanking the p53DBD are also involved in DNA binding, a reasonable concept since p53 function depends upon its DNA binding properties, and these domains retain a high level of evolutionary conservation. Such a view is also consistent with allosteric control in the binding of wild type p53 (Vojtesek *et al.*, 1995; Waterman *et al.*, 1995) and offers another possible facet of indirect control in the binding specificity of this protein (Lefstin and Yamamoto, 1998).

The model may also offer insights into the role of p53 in promoting DNA looping (Prives, 1994; Stenger *et al.*, 1994; Jackson *et al.*, 1998). This may occur through N-terminal interactions, since these are exposed outside the DNA loops. The model also suggests that direct binding of p53 to nucleosomes might occur, and that p53 binding to unfolded chromatin might offer a possible mechanism for the detection of DNA damage. The staggered external location of the p53DBD moieties on the DNA, and the critical importance of the major groove bending

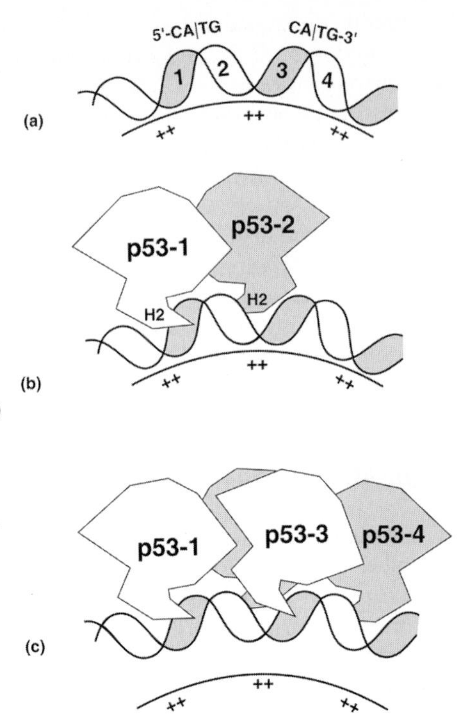

Fig. 19.5. Putative model for the p53 binding to nucleosomal DNA. (**a**) The nucleosomal DNA wrapped around the histone core is shown schematically. Pluses are for the positively charged histone tails interacting with the minor groove facing inside (Luger *et al.*, 1997). Based on the sequence-dependent anisotropic bendability of DNA, the p53 response element is expected to be packaged in a nucleosome so that the consensus CATG tetramers would have their minor grooves outside (Olson *et al.*, 1998; McNamara *et al.*, 1990). (**b**) Two p53 core domains are shown interacting with the two consensus pentamers (1 and 2). The H2 helices (depicted as protruding triangles) penetrate into the major groove on the "top" and "bottom" sides of the duplex in accord with the co-crystal p53DBD-DNA structure (Cho *et al.*, 1994), (see Fig. 19.3a) In this model, the "external" location of the CATG tetramers assures the easy access of the recognition H2 helices to the donor and acceptor groups in the major groove. (**c**) Four p53 core domains interacting with the response element. DNA is partially "peeled away" from the histone core and the bend angle is 35–55° per 20 bp (Nagaich *et al.*, 1999), which is approximately twice as small as in the nucleosome

dimers CA:TG at helically phased sites in the response elements, are suggestive of this since it is now well known that these dimers lie on the outside of nucleosomal DNA (Satchwell *et al.*, 1986; Luger *et al.*, 1997) and hence would be accessible to the protein. It is also likely that p53 could bind to response elements located on nucleosomes without serious nucleosomal disruption. A possible mechanism for the latter is illustrated in Fig.s 19.5. In this scheme, nucleosomal DNA (Fig. 19.5a) is first bound by an antiparallel p53DBD dimer (Fig. 19.5b) since energetic considerations predict an exceptionally high level of conformational flexibility for this system (Durell *et al.*, 1998). If this binding occurs in the flanking region of the nucleosome, subsequent binding of a second antiparallel dimer to form the tetrameric complex (Fig. 19.5c) could catalyze a release of this DNA from the histone octamer, since the DNA in this region of the nucleosome is known to be less tightly bound than in regions closer to the nucleosomal dyad (Weischet *et al.*, 1978; Simpson, 1979). Such a release would allow the response element DNA to assume the optimal curvature suggested by the model, which is less than in a normal nucleosome (Fig. 19.5c). Alternatively, binding might occur to internucleosomal linker regions of partially unfolded chromatin fibers.

Another possible function of p53 may be related to the formation of the antiparallel H1-H1 interface, which can serve as a recognition site for other binding proteins. It is well known that p53 interacts with a wide variety of upstream regulatory proteins including the TFIID complex (Chen *et al.*, 1993; Farmer *et al.*, 1996) and with the acetyltransferase p300 (Scolnick *et al.*, 1997; Sakaguchi *et al.*,

1998; Grossman *et al.*, 1998). A clue as to the importance of specific DNA binding to p53 function may therefore lie in the fact that this interface is formed only upon the association of multiple p53DBD moieties with DNA response elements. Furthermore, it is exposed outside the DNA loop and is therefore easily accessible in the complex. Similarly, the interactions between the p53DBD subunits involving the N-terminal fragments adjacent to the core domain (e.g., the proline rich fragments) might be operative in creating a "signal", indicating that p53 docking has occurred. On the other hand, it may be important in certain functions that the interface be hidden in order to prevent certain proteins from binding. For example, a protein involved in the p53 degradation pathway binding to the H1-H1 interface surface or to the proline-rich fragment of the N-terminus would be effectively prevented from binding, thus increasing the lifetime of p53. This might represent an alternative pathway for transcriptional regulation.

The binding motif for p53 appears to be unique among transcription factors so far investigated. The closest similarity to currently reported structures is with the lac repressor complex (Lewis *et al.*, 1996). The implications for allosteric control, which has been suggested for p53 (Halazonetis *et al.*, 1993; Vojtesek *et al.*, 1995; Waterman *et al.*, 1995), may also share similarities with this system (Horton *et al.*, 1997). The possibility for allosteric control of p53 binding is implicit in the location of the N- and C-termini of the p53DBD in Fig. 19.3a (and see also Fig. 19.8 of Nagaich *et al.*, 1997b) where the N-termini lie on the outside of the DNA loop and the C-termini on the inside. The phasing experimental results show clear differences in both DNA bending and twisting between the p53 core domain and wild type p53 complexes, suggesting that DNA binding is moderated by other regions of the p53 molecule. These termini can, in principle, interact either with the DNA binding domain (DBD) or with the DNA, either within or flanking the specific binding site.

Based on the structural model, it seems likely that the N-terminus interacts with the p53DBD, thereby affecting the p53DBD-p53DBD interactions, whereas the C-terminus is likely to interact directly with the DNA. Each N-terminus may interact either with its own p53DBD or with an adjacent one, a reasonable postulate since the N-terminal region, although variable among species, is proline-rich in the same region. In the model, this region is positioned in the vicinity of the adjacent p53DBD, and such interactions could lead to an increased bend in the bound DNA. Such a postulate might also help clarify the role of spacer DNA between half-sites (Tokino *et al.*, 1994), since the N-terminus potentially can form a bridge between p53 dimers separated by spacers up to 21 bp in length (Cook *et al.*, 1995; Miner and Kulesz-Martin, 1997). The latter could therefore be an aspect of indirect recognition for the binding specificity of p53. Finally, since the N-termini are involved in downstream signal transduction pathways, structural information on them is important. In the present model, they are relatively exposed and accessible to other downstream proteins after p53 is bound to DNA.

Functional roles of the C-termini are equally important and include moderating p53 DNA binding properties. This view is supported by a variety of studies showing that the basic, 30-residue C-terminal region strongly influences the DNA binding properties of p53 (Foord *et al.*, 1991; Anderson *et al.*, 1997), and that this binding is further moderated by the charge state as determined by its acetylation level and phosphorylation state. In the present model, the C-termini interacting electrostatically with the DNA can increase the bend simply by pulling the DNA toward the center of the bend. This is of special interest since the C-termini undergo post-translational modifications including acetylation, phosphorylation and dephosphorylation, apparently initiated by DNA damage (Sakaguchi *et al.*, 1998; Kapoor *et al.*, 1998; Lu *et al.*, 1998; Waterman *et al.*, 1998).

The structural model also offers insights into the remarkable binding specificity and selectivity of p53. The range of p53 function is so broad, including that of pleiotropic transcription factor and enhancer, that the demands for binding specificity and selectivity are necessarily extraordinary. This is likely accomplished through its tetrameric association with a repetitive binding site in which precise steric fit is extremely important. Steric fit is accommodated through both DNA bending and twisting which are evidently tightly coupled as discussed above. Therefore, it becomes possible for binding specificity in the p53 system to be regulated by local supercoiling in promoter regions (Liu and Wang, 1987) as well as by specific bending in larger transcriptional complexes or by nearby architectural elements (Jayaraman *et al.*, 1998).

4 Acknowledgements

The authors are grateful to Dr. Carl W. Anderson and Dr. Wilma K. Olson for valuable discussions. The work was supported by research grants CA70274 and GM53517 from the National Institutes of Health (REH).

References

Anderson, M.E., Woelker, B., Reed, M., Wang, P. and Tegtmeyer, P. (1997) *Mol. Cell. Biol.*, 17, 6255–6264.
Appella, E. and Anderson, C.W. (1994) *Biochim. Italia*, 1, 19–28.
Appella, E., Nagaich, A.K., Zhurkin, V.B. and Harrington, R.E. (1998) J. *Prot. Chem.*, 17, 527–528.
Arrowsmith, C.H. and Morin, P. (1996) *Oncogene*, 12, 1379–1385.
Balagurumoorthy, P., Sakamoto, H., Lewis, M.S., Zambrano, N., Clore, G.M., Gronenborn, A.M., Appella, E. and Harrington, R.E. (1995). *Proc. Natl. Acad Sci.* USA, 92, 8591–8595.
Baudier, J., Delphin, C., Grunwald, D., Khochbin, S. and Lawrence, J.J., *Proc. Natl. Acad. Sci. USA*, 89, 11627–11631
Bewley, C.A., Gronenborn, A.M. and Clore, G.M. (1998) *Annu. Rev. Biophys. Biomol. Struct.*, 27, 105–131.
Bishoff, J.R., Friedman, P.N., Marshak, D.R., Prives, C. and Beach, D. (1990) Proc. Natl. Acad. Sci. *USA*, 87, 4766–4770
Chen, X.B., Farmer, G., Zhu, H., Prywes, R. and Prives, C. (1993) *Genes Develop.*, 7, 1837–1849.
Cho, Y.J., Gorina, S., Jeffrey, P.D. and Pavletich, N.P. (1994) *Science*, 265, 346–355.
Cook, J.L., Re, R., Giardina, J.F., Fontenot, F.E., Cheng, D.Y. and Alam, J. (1995) Oncogene, 11, 723–733.
DeLeo, A.B., Jay, G., Appella, E., DuBois, G.C., Law, L.W. and Old, L.J. (1979) *Proc. Natl. Acad. Sci. USA*, 76, 2420–2424.
Dickerson, R.E. (1998) *Nucleic Acids Res.*, 26, 1906–1926.
Durell, S.R., Appella, E., Nagaich, A.K., Harrington, R.E., Jernigan, R.L. and Zhurkin, V.B. (1998) In Sarma, R.H. and Sarma, M.H. (eds.), *Structure, Motion, Interaction and Expression of Biological Macromolecules. Proceedings of the Tenth Conversation.* Adenine Press, Schenectady, NY, 2, 277–295.
El-Deiry, W.S., Kern, S.E., Pietenpol, J.A., Kinzler, K.W. and Vogelstein, B. (1992) *Nature Genetics*, 1, 45–49.

El-Deiry, W.S., Tokino, T., Velculescu, V.E., Levy, D.B., Parsons, R., Trent, J.M., Lin, D., Mercer, W.E., Kinzler, K.W. and Vogelstein,B. (1993) l*Cel*, 75, 817–825.
El Hassan, M.A. and Calladine, C.R. (1998) *J. Mol. Biol.*, 282, 331–343.
Farmer, G., Colgan, J., Nakatani, Y., Manley, J.L. and Prives, C. (1996) *Mol. Cellular Biol.*, 16, 4295–4304.
Foord, O.S., Bhattacharya, P., Reich, Z. and Rotter, V. (1991) *Nucleic Acids Res.*, 19, 5191–5198.
Friedman, P.N., Chen, X.B., Bargonetti, J. and Prives, C. (1993) *Proc. Natl. Acad. Sci.* USA, 90, 3319–3323.
Grossman, S.R., Perez, M., Kung, A.L., Joseph, M., Mansur, C., Xiao, Z.X., Kumar, S., Howley, P.M. and Livingston, D.M. (1998) *Mol. Cell*, 2, 405–415.
Gu, W. and Roeder, R.G. (1997) *Cell*, 90, 595–606.
Halazonetis, T.D. and Kandil, A.N. (1993) *EMBO J.*, 12, 5057–5064.
Halazonetis, T.D., Davis, L.J. and Kandil, A.N. (1993) *EMBO J.*, 12 (3), 1021–1028.
Harris, C.C. (1993) *Science*, 262, 1980–1981.
Hartwell, L.H. and Kastan, M.B. (1994) *Science*, 266, 1821–1828.
Hermann, C.P., Kraiss, S. and Montenarth, M. (1991) *Oncogene*, 6, 877–884
Hermeking, H. and Eick, D. (1994) *Science*, 265, 2091–2093.
Hollstein, M., Sidransky, D., Vogelstein, B. and Harris, C.C. (1991*) Science*, 253, 49–53.
Horton, N., Lewis, M. and Lu, P. (1997) *J. Mol. Biol.*, 265, 1–7.
Hupp, T.R. and Lane, D.P. (1994) *Cold Spring Harbor Symp. Quant. Biol.*, 59, 195–206.
Jackson, P., Mastrangelo, I., Reed, M., Tegtmeyer, P., Yardley, G. and Barrett, J. (1998) *Oncogene*, 16, 283–292.
Jayaraman, L., Freulich, E. and Prives, C. (1997) *Methods in Enzymology: Cell Cycle Control*, 283, 245–256.
Jayaraman, L., Moorthy, N.C., Murthy, K.G.K., Manley, J.L., Bustin, M. and Prives, C. (1998) *Genes Develop.*, 12, 462–472.
Kapoor, M. and Lozano, G. (1998) *Proc. Natl. Acad. Sci. U S A*, 95, 2834–2837.
Kuerbitz, S.J., Plunkett, B.S., Walsh, W.V. and Kastan, M.B. (1992) *Proc. Natl. Acad. Sci. U S A* 89, 7491–7495.
Kerppola, T.K. and Curran, T. (1983) *Mol. Cell. Biol.,* 13, 5479–5489.
Knippschild, U., Milne, D.M., Campbell, L.E., DeMaggio, A.J., Christenson, E., Hoekstra, M.F. and Meek, D.W. (1997), *Oncogene*15, 1727–1736.
Ko, L.J., Shieh, S.Y., Chen, X.B., Jayaraman, L., Tamai, K., Taya, Y., Prives, C. and Pan, Z.Q. (1997) *Mol. Cellular Biol.*, 17, 7220–7229.
Lane, D.P. (1992) Nature (London), 358, 15–16.
Lane, D.P. and Crawford, L.V. (1979) *Nature (London)*, 278, 261–263.
Lefstin, J.A. and Yamamoto, K.R. (1998) *Nature (London*), 392, 885–888.
Levine, A.J. (1993) *Annu. Rev. Biochem.*, 62, 623–651.
Levine, A.J. (1997) *Cell*, 88, 323–331.
Levine, A.J., Momand, J. and Finlay, C.A. (1991) *Nature (London*), 351, 453–456.
Lewis, M., Chang, G., Horton, N.C., Kercher, M.A., Pace, H.C., Schumacher, M.A., Brennan, R.G. and Lu, P.Z. (1996) *Science*, 271, 1247–1254.
Lin, D., Shields, M.T., Ullrich, S., Appella, E. and Mercer, W.E. (1992) *Proc. Natl. Acad. Sci. USA*, 89, 9210–9214
Linzer, D.I. and Levine, A.J. (1979) *Cell* 17, 43–52
Liu, L.F. and Wang, J.C. (1987) *Proc. Natl. Acad. Sci. USA*, 84, 7024–7027.
Lowe, S.W., Schmitt, E.M., Smith, S.W., Osborne, B.A. and Jacks, T. (1993) *Nature (London*), 362, 847–849.
Luger, K., Mader, A.W., Richmond, R.K., Sargent, D.F. and Richmond, T.J. (1997) *Nature (London*), 389, 251–260.
Lu, H., Taya, Y., Ikeda, M. and Levine A.J. (1998) *Proc. Natl. Acad. Sci. USA*, 95, 6399–6404
Marx, J. (1994) *Science*, 266, 1321–1322.
McNamara, P.T., Bolshoy, A., Trifonov, E.N. and Harrington, R.E. (1990) J. Biomol. Struct. Dynam., 8, 529–538.
Meek, D.W. (1997) *Patholog. Biol.*, 45, 804–814.
Meek, D.W. (1998) *Int. J. Radiat. Biol.* 74,729–737.
Meltzer, P.S. (1994) *J. Natl. Cancer Inst.*, 86, 1265–1266.
Milner, J. (1994) *Seminars Can. Biol.*, 5, 211–219.
Miner, Z. and Kulesz-Martin, M. (1997) *Nucleic Acids Res.*, 25, 1319–1326.
Nagaich, A.K., Bhattacharyya, D., Brahmachari, S.K. and Bansal, M. (1994) *J. Biol. Chem.*, 269, 7824–7833.
Nagaich, A.K., Appella, E. and Harrington, R.E. (1997a*) J.* Biol. Chem., 272, 14842–14849.

Nagaich, A.K., Zhurkin, V.B., Sakamoto, H., Gorin, A.A., Clore, G.M., Gronenborn, A.M., Appella, E. and Harrington, R.E. (1997b) *J. Biol. Chem.*, 272, 14830–14841.
Nagaich, A.K., Balagurumoorthy, P., Miller, W.M., Appella, E., Zhurkin, V.B. and Harrington, R.E. (1998) In Sarma, R.H. and Sarma, M.H. (eds.), *Structure, Motion, Interaction and Expression of Biological Macromolecules. Proceedings of the Tenth Conversation.* Adenine Press, Schenectady, NY, 2, 249–273.
Nagaich, A.K., Zhurkin, V.B., Durell, S.R., Jernigan, R.L., Appella, E. and Harrington, R.E. (1999) *Proc. Natl. Acad. Sci. USA*, 96, 1875–1880.
Niederweis, M. and Hillen, W. (1993) *Electrophoresis,* 14, 693–698.
Olson, W.K., Gorin, A.A., Lu, X.J., Hock, L.M. and Zhurkin, V.B. (1998) *Proc. Natl. Acad. Sci. USA*, 95, 11163–11168.
Pavletich, N.P., Chambers, K.A. and Pabo, C.O. (1993) *Genes Develop.*, 7, 2556–2564.
Picksley, S.M., Vojtesek, B., Sparks, A. and Lane, D.P. (1994) Oncogene, 9, 2523–2529.
Pines, J. (1994) *TIBS*, 19, 143–145.
Prives, C. (1994*) Cell*, 78, 543–546.
Prives, C. (1998) *Cell*, 95, 5–8.
Prives, C., Bargonetti, J., Farmer, G., Ferrari, E., Friedlander, P., Wang, Y., Jayaraman, L., Pavletich, N. and Hubscher, U. (1994) In *Cold Spring Harbor Symposia on Quantitative Biology* Cold Spring Harbor Lab Press, Plainview, 59, 207–213.
Sakaguchi, K., Herrera, J.E., Saito, S., Miki, T., Bustin, M., Vassilev, A., Anderson, C.W. and Appella, E. (1998) *Genes Develop.*, 12, 2831–2841.
Satchwell, S., Drew, H.R. and Travers, A.A. (1986) *J. Mol. Biol.*, 191, 659–675.
Scolnick, D.M., Chehab, N.H., Stavridi, E.S., Lien, M.C., Caruso, L., Moran, E., Berger, S.L. and Halazonetis, T.D. (1997) *Cancer Res.*, 57, 3693–3696
Simpson, R.T. (1979) *J. Biol. Chem.*, 254, 10123–10127.
Smith, M.L., Chen, I.T., Zhan, Q.M., Bae, I.S., Chen, C.Y., Gilmer, T.M., Kastan, M.B., Oconnor, P.M. and Fornace, A.J. (1994) Science, 266, 1376–1380.
Steitz, T.A. (1990) *Quart. Rev. Biophys.*, 23, 205–280.
Stenger, J.E., Tegtmeyer, P., Mayr, G.A., Reed, M., Wang, Y., Wang, P., Hough, P.V.C. and Mastrangelo, I.A. (1994) *EMBO J.*, 13, 6011–6020.
Symonds, H., Krail, L., Remington, L., Saenz-Robles, M., Lowe, S., Jacks, T. and Van Dyke, T. (1994) Cell, 78, 703–711.
Tokino, T., Thiagalingam, S., El-Deiry, W.S., Waldman, T., Kinzler, K.W. and Vogelstein, B. (1994) *Human Mol. Genet.*, 3, 1537–1542.
Unger, T., Nau, M.M., Segal, S. and Minna, J.D. (1992) *EMBO J.*, 11, 1383–1390.
Van der Vliet, P.C. and Verrijzer, C.P. (1993) *BioAssays*, 15, 25–32.
Vogelstein, B. and Kinzler, K.W. (1992) *Cell,* 70, 523–526.
Vojtesek, B., Dolezalova, H., Lauerova, L., Svitakova, M., Havlis, P., Kovarik, J., Midgley, C.A. and Lane, D.P. (1995) *Oncogene,* 10, 389–393.
Wang, P., Reed, M., Wang, Y., Mayr, G., Stenger, J.E., Anderson, M.E., Schwedes, J.F. and Tegtmeyer, P. (1994) *Mol. Cell. Biol.*, 14, 5182–5191.
Wang, Y. and Prives, C. (1995*) Nature*, 376, 88–91.
Waterman, J.L.F., Shenk, J.L. and Halazonetis, T.D. (1995) *EMBO J.*, 14, 512–519.
Waterman, M.J., Stavridi, E.S., Waterman, J.L. and Halazonetis, T.D. (1998) Nat. Genet., 2,175–178
Weischet, W.O., Tatchell, K., Van Holde, K.E. and Klump, H. (1978) *Nucleic Acids Res.*, 5, 139–160.
Werner, M.H., Gronenborn, A.M. and Clore, G.M. (1996) *Science*, 271, 778–784.
Wolkowicz, R. and Rotter, V. (1997) *Patholog. Biol.*, 45, 785–796.
Wu, H.M.. and Crothers, D.M. (1984) *Nature (London)*, 308, 509–513.
Wu, X. and Levine, A.J. (1994) *Proc. Natl. Acad. Sci. USA*, 91, 3602–3606.
Zhurkin, V.B., Ulyanov, N.B., Gorin, A.A. and Jernigan, R.L. (1991) *Proc. Natl. Acad. Sci. USA*, 88, 7046–7050.
Zinkel, S. and Crothers, D.M. (1987) *Nature (London)*, 328, 178–181.

CHAPTER 20

Overproduction, Purification and Structural Studies on the Zn Containing S14 Ribosomal Protein from *Thermus thermophilus*

P. Tsiboli[1], D. Triantafillidou[1], F. Leontiadou[1], M. Simitsopoulou[1], K. Anagnostopoulos[2], F. Franceschi[2] and T. Choli-Papadopoulou[1]

Introduction

Ribosomes are complexes of ribosomal RNAs and proteins. As more primary structures of ribosomal proteins have been compiled amino acid motifs associated with nucleic acid binding have been identified in ribosomal proteins (Wool *et al.* 1995). Furthermore, the cloning and overexpression of ribosomal proteins has resulted in the elucidation of the motifs or domains implicated in rRNA binding (Ramakrishnan *et al.* 1995). For example the *Bacillus stearothermophilus* L11 was recently proven to bind rRNA through a helix-turn-helix domain identical to that of the "homeodomain" regulatory proteins (Xing *et al.* 1997).

One motif of special interest is the zinc-finger, which is formed by a central zinc ion co-ordinated tetrahedrally by four amino acid residues, which are either cysteines or histidines (Kohn *et al.* 1997). The zinc-finger motifs in ribosomal proteins are primarily of the C4 variety, such as those of the rat ribosomal proteins S27 and S29 (Wool *et al.* 1995). Rat ribosomes have been proven to contain zinc but there has been no evidence so far that the zinc-finger motifs of the rat ribosomal proteins bind zinc (Chan *et al.* 1993). Rat S29 is related to the family of procaryotic S14 ribosomal proteins.

Not all members of the S14 family share the zinc-finger motif (Wool *et al.* 1995; Herfurth *et al.* 1994; Tsiboli and Choli 1994). Cross-linking studies have indicated an interaction of the amino-terminal domain of the S14 protein of *Bacillus stearothermophilus* (Urlaub *et al.* 1995) primarily with the 990–1045 domain of the 16S rRNA, which is highly conserved in all organisms. Although it has not been proven, it is plausible that a zinc-finger could contribute to the interaction of the S14 family members with the rRNA.

In this article we present our results on the cloning, overexpression and purification of the *Thermus thermophilus* S14 protein and we discuss the problems encountered at every purification step as well as the rationale for the structural studies on the overexpressed protein.

[1] Laboratory of Biochemistry, School of Chemistry, Aristotle University of Thessaloniki, Thessaloniki 54006, Greece.
[2] Max-Planck Institut for Molecular Genetics, AG Ribosomen, Berlin 14195, Germany.

2
Cloning

The TthS14 gene was amplified from Thermus thermophilus genomic DNA using primers based on the known sequence of the TthS14 gene (Tsiboli and Choli 1995). The resulting 200 bp PCR product was cloned into the pET11d vector and the recombinant plasmid (pET11d/TthS14) was used for the transformation of *E. coli* BL21(DE3)pLysS cells. Transformants were selected from Luria-Bertani plates – supplemented with 40 μg/ml ampicillin and 50 μg/ml chloramphenicol – analysed by restriction analysis and the positive clones were subjected to sequencing. For more details see: Tsiboli *et al.* (1998).

3
Overproduction

During the overproduction of a protein several problems may occur. Firstly, the protein might form inclusion bodies. From a purification standpoint, the accumulation of protein in an aggregated form can be advantageous. After lysing the cells and centrifuging the resulting lysate the agreggated protein can be recovered in the pellet fraction about 50 % pure, although mostly in an inactive form. The protein in the inclusion bodies can be a mixture of monomeric and multimeric forms, both reduced and oxidized. The major problem then becomes one of recovering biologically active protein in high yield. In order to accomplish this, the protein in the inclusion bodies must be solubilized, refolded and purified in a specific order. The common stages in processes designed to recover biologically active, soluble protein from such aggregates include
(1) cell lysis,
(2) isolation of inclusion bodies,
(3) solubilization of protein in inclusion bodies and
(4) refolding of solubilized protein.

Sometimes the overproduction of a protein at lower temperatures alleviates the problem and for this reason it is recommended to grow the cells at different temperatures, to control – after centrifugation of the lysate – the supernatant by SDS electrophoresis and finally to chose the right temperature for cell growth.

After all these controls we did not detect the S14 protein in the centrifugation pellet and therefore decided to grow the transformed *E. coli* cells at 37 °C, which is the optimal temperature for *E. coli*.

For vectors carrying the lac-z promoter, namely for IPTG inducible vectors, two factors that have to be taken into account are the IPTG concentration and the absorption of the culture where the induction takes place. In Fig. 20.1 we show that at concentrations of IPTG over 1.0 mM there were no significant differences in the amount of the overproduced cloned protein. The chosen absorption for the induction was between 0.7–0.9 because at lower values the protein was in very low yield (Fig. 20.2).

In the case of small proteins, such as TthS14, proteases might partially or fully digest the overproduced protein. Extended protein structures are even more suc-

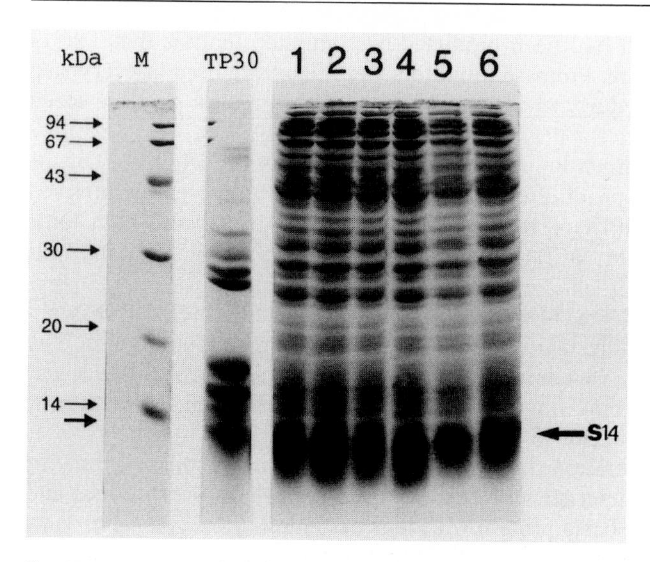

Fig. 20.1. SDS-PAGE (15 %-discontinuous system) of crude extracts from E. coli pEt/TthS14 cultures, which had been grown for 3 h at 37 °C after induction with different concentrations of IPTG. Lane 1: 0.5 mM IPTG, Lane 2: 1.0 mM IPTG, Lane 3: 1.5 mM IPTG, Lane 4: 2.0 mM IPTG, Lane 5: 2.5 mM IPTG, Lane 6: 3.0 mM IPTG

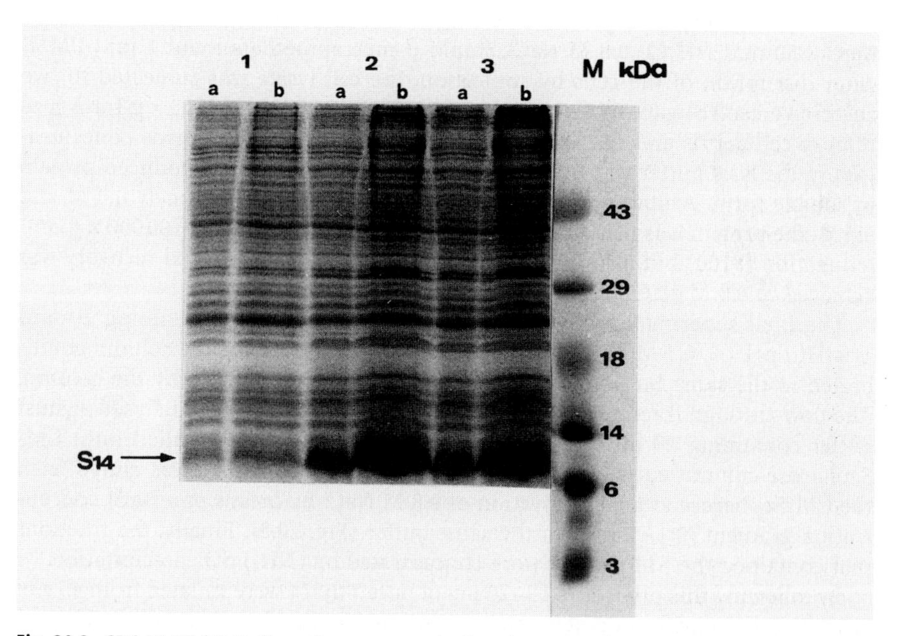

Fig. 20.2. SDS-PAGE (15 %-discontinuous system) of crude extracts from E. coli pEt11d/TthS14 cultures, which had been grown for 3 h at 37 °C after induction with IPTG at different A_{600}.. Lane 1: A_{600} 0.60, lane 2: A_{600} 0.70, lane 3: A_{600} 0.95, M: molecular weight markers. In all cases (a) corresponds to 50 μg and (b) to 100 μg of total protein. Staining of the gel was performed with Coomassie blue

ceptible to proteolysis. It had been predicted by computer analysis that TthS14 has an extended structure. Protease inhibitors were therefore used at every step of the purification procedure, which took place with the greatest possible speed in the coldroom.

Additional attention must be paid to the conditions used during lysis of the cells containing the overproduced protein to ensure that the protein is retrieved in a soluble form. TthS14 is a very basic protein (pI = 11,94) and high ionic strength conditions (0.8 M NaCl) were needed to avoid its precipitation after lysis of the cells.

A problem peculiar to TthS14 was the presence of four cysteines. These might form inter and/or intramolecular bonds leading to the aggregation of the protein. During the purification procedure no aggregation took place probably because the cysteines form the zinc finger domain with the zinc atom. Nevertheless, β-mercaptoethanol was added to protect the protein from oxidation.

By taking into consideration the problems common to overproduced proteins and the individual characteristics of TthS14 it was possible to overproduce and purify this small, highly basic protein with four cysteines.

4 Purification

The recombinant TthS14 protein was purified according to the following procedure: the BL21 *E. coli* cells carrying the TthS14 gene were harvested 3 h after induction and suspended in the following buffer: 20 mM Tris-HCl pH 7.5, 10 mM $MgCl_2$, 50 mM NH_4Cl, 0.8 M NaCl, 7 mM β-mercaptoethanol and 1 mM PMSF. After disruption of the cells by sonication, the cell lysate was subjected to two successive centrifugations at 10000 × g for 20 mins and at 100000 × g for 3 h to remove cell debris and ribosomes, respectively. The sodium chloride concentration in the lysis buffer was 0.8 M in order to maintain the overproduced protein in soluble form. At lower concentrations of sodium chloride, namely 0.4 M and 0.6 M, the protein was detected not only in the supernatant of the 100000 × g centrifugation (S100) but also in the pellet (data not shown). Optimal recovery was achieved at 0.8 M NaCl.

The final supernatant (S100) was dialysed against buffer containing 20 mM Tris-HCl pH 7.5, 0.1 M NaCl, and applied onto a DEAE-Sepharose column equilibrated in the same buffer. The TthS14 protein was not retained by the column. The flow-through fractions of the DEAE-Sepharose column were dialysed against buffer containing 20 mM Tris-HCl pH 6.5, 0.1 M NaCl and applied on a CM-Sepharose column equilibrated in the same buffer. The TthS14 was eluted from the CM-Sepharose at a concentration of 0.8 M NaCl by means of a NaCl concentration gradient (0.1–1.5 M) in the same buffer (Fig. 20.3). Finally, the fractions that contained the S14 protein were concentrated by $(NH_4)_2SO_4$ precipitation.

By following this procedure 15–20 mg of pure TthS14 were isolated from 45 g of cells. At every purification step the protein was identified by Edman degradation (after electrotransfer onto PVDF membranes (Choli *et al.*, 1989)) and by immunoblotting using the antiserum raised against the overproduced protein. The purified protein was homogeneous as shown by SDS-PAGE analysis (Fig. 20.4).

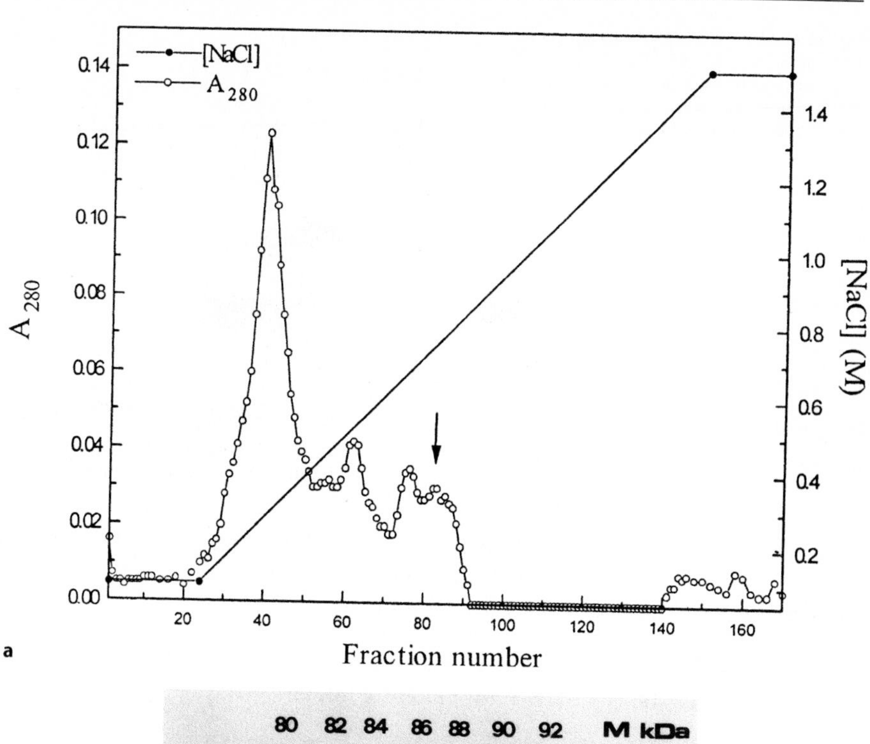

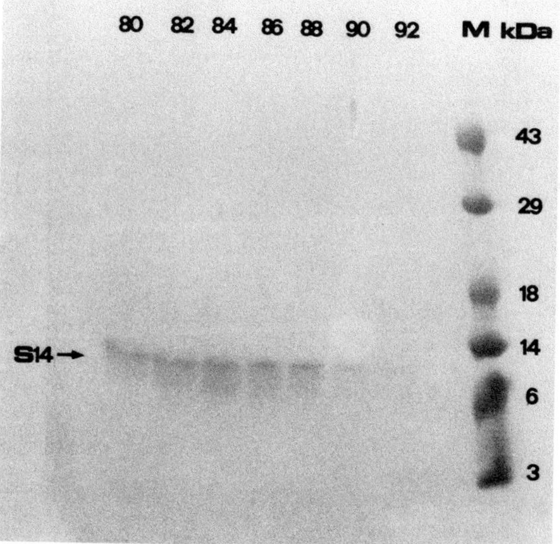

Fig. 20.3. (a) Chromatographic profile of TthS14 purification on a CM-Sepharose column (4 cm × 10 cm). Flow rate 70 ml/h. Fraction volume 4 ml. The total protein amount applied was 800 mg. (-○-) absorbance at 280nm, (-●-) concentration of NaCl in the gradient. The arrow indicates the elution peak of TthS14. (b) SDS-PAGE (15 %-discontinuous system) of CM-Sepharose fractions 80–92 (see Fig. 20.2. containing the TthS14 protein. 50 μl of each fraction were loaded on the gel. M: molecular weight markers

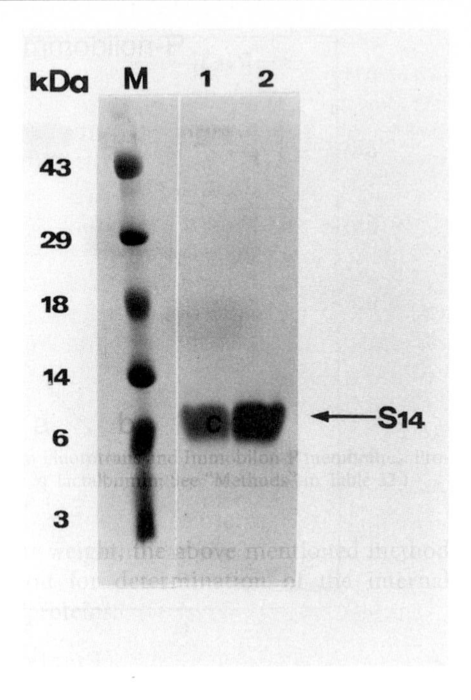

Fig. 20.4. SDS-PAGE (15 %-discontinuous system) of the purified TthS14. 5 μg and 10 μg of protein were loaded on the gel. M: molecular weight markers. Staining of the gel was performed with Coomassie blue

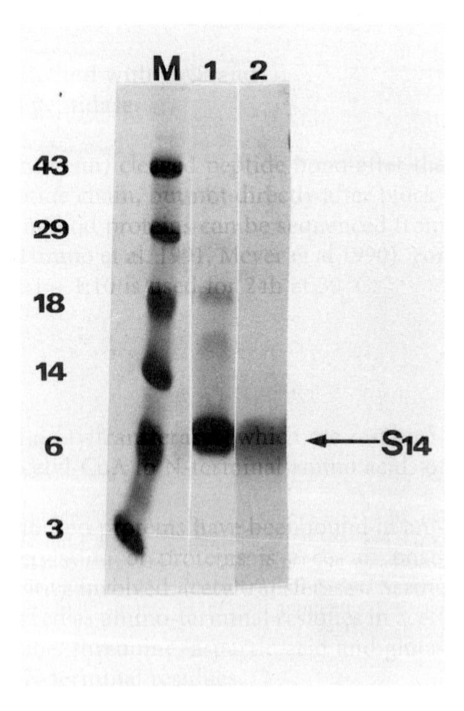

Fig. 20.5. SDS-PAGE (15 %-discontinuous system) of pure TthS14. Lane 1: 5 μg of pure TthS14 without heating and in the absence of β-mercaptoethanol. Lane 2: 5 μg of pure TthS14 after heating at 100 °C for 5 mins and in the presence of 5 mM β-mercaptoethanol

Concerning the diffusion apparent in the purified protein (Fig. 20.4), which could be attributed to degradation, the results of the amino-terminal sequencing showed that the protein is intact, at least at the amino-terminal region. Therefore, if degradation exists, it should be limited very close to the carboxy-terminal region. An alternate explanation for the diffusion might be the anomalous behaviour of highly charged polypeptides.

An interesting remark is that pure S14 protein, analysed on SDS-PAGE without previous heating of the samples, exhibited several bands (Fig. 20.5, Lane 1). One possible explanation is the formation of intramolecular disulphide bridges after the removal of Zn from some TthS14 molecules during purification. Indeed, after addition of β-mercaptoethanol and heating (at 90 °C-100 °C for 3–5 mins), only one band appears at the expected molecular mass of 7008 Da (Fig. 20.5, Lane 2).

5 Structural Studies

5.1 Limited Proteolysis Experiments

The structure of proteins can be deduced by limited proteolysis experiments. A certain amount of purified "structured" protein is treated with different enzymes at 0 °C and in a ratio of (enzyme: protein) 1:50.

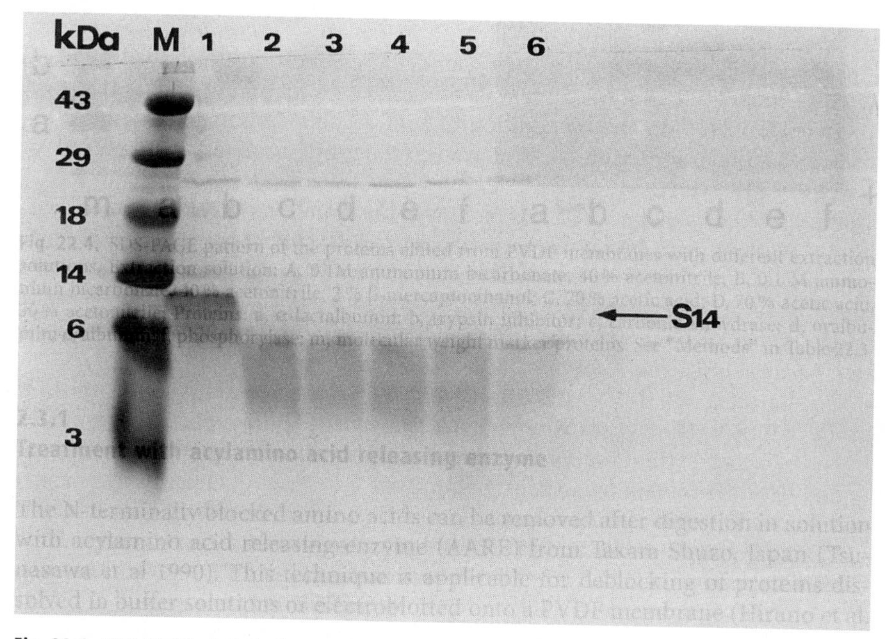

Fig. 20.6. SDS-PAGE of limited proteolysis products of TthS14 with trypsin for various digestion periods (0–60 min). 180 μg of TthS14 were digested with 3.6 ìg of trypsin at 0 °C. Lane 1: 0 min, lane 2: 5 mins, lane 3: 10 mins, lane 4: 20 mins, lane 5: 30 mins, lane 6: 60 mins digestion time

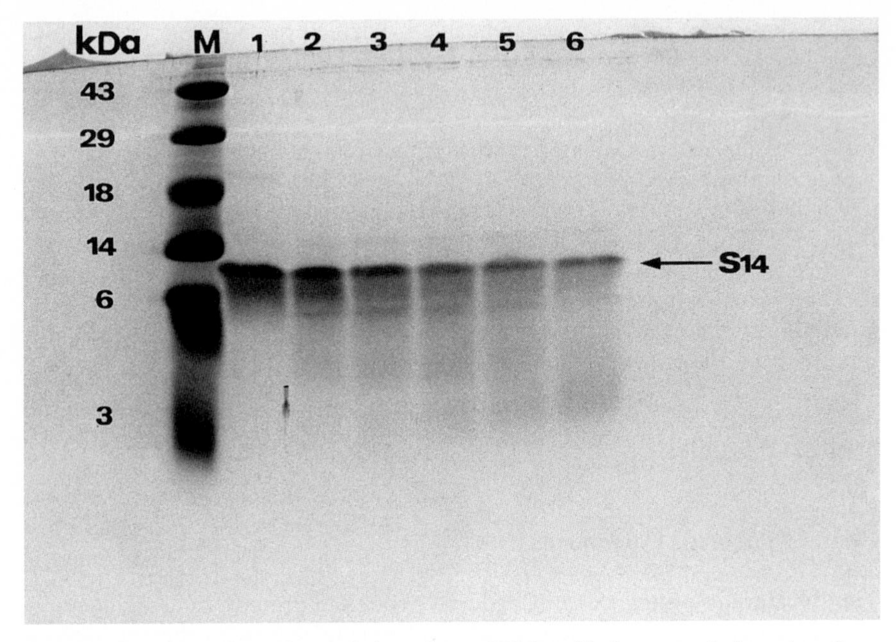

Fig. 20.7. SDS-PAGE of limited proteolysis products of TthS14 with chymotrypsin for various digestion periods (0–60 min). 180 µg of TthS14 were digested with 3.6 ìg of chymotrypsin at 0 °C. Lane 1: 0 min, lane 2: 5 mins, lane 3: 10 mins, lane 4: 20 mins, lane 5: 30 mins, lane 6: 60 mins digestion time

180 µg of TthS14 were incubated – under the conditions described by Choli (1989) – and at different time intervals the same amount of protein was removed, the digestion stopped by the addition of SDS gel sample buffer, containing 1 % SDS, and boiling for 5 minutes. The produced peptides were analysed onto SDS-PAGE as is shown in Fig. 20.6 and 20.7.

Fig. 20.6 shows that a 5 kDa fragment remains even after digestion for 60 mins at 0 °C with trypsin. In the primary structure of TthS14 there are a lot of basic amino acids i.e. lysines and arginines, but the most probable target for tryptic digestion is Arg-19 resulting in the release of a 2 kDa peptide. The remaining 5 kDa domain contains the zinc-finger which is less accessible to further digestion because of its more compact nature.

On the other hand, Fig. 20.7 shows that the 5 kDa domain which is initially produced by chymotrypsin is not stable as in the case of tryptic digestion. Our possible interpretation would be that chymotryptic hydrolysis at position 21 destabilizes the remaining structure and starts a cascade of chymotryptic enzymic reactions within the zinc-finger.

Finally, we hypothesize that TthS14 protein consists of two parts, a structured C-terminal domain and an extended N-terminal part.

5.2 The Presence of Zinc in TthS14 Protein

As has already been mentioned above, the primary structure of this protein leads to the suggestion that the four cysteine residues are involved in a C4 zinc-finger motif. The atomic absorption results support this hypothesis, because analysis of the S14 ribosomal protein from *Thermus thermophilus* indicated the presence of Zn ions in the stoicheiometry 1:1 (moles of Zn per mole of TthS14 protein).

Table 20.1 shows the Zn content of TthS14 determined by atomic absorption under various experimental conditions (in all cases the concentration of the protein was 0.4 mg/ml).

Table 20.1. Atomic absorption measurements for the presence of Zn in ribosomal protein TthS14. The amount of protein used in all cases was 80 μg

1. TthS14: TthS14 dialysed against deionized H_2O.
2. TthS14: TthS14 dialysed against H_2O, $ZnCl_2$ (10 mg/ml) and H_2O successively.
3. TthS14: TthS14 dialysed against H_2O, EDTA (50 mM) and H_2 successively.
4. BSA: Bovine serum albumin dialysed against H_2.
5. BSA: Bovine serum albumin dialysed against H_2O, $ZnCl_2$ (10 mg/ml) and H_2O, successively.
6. TthS14: TthS14 dialysed against 6 M guanidine-HCl in 20 mM Tris-HCl pH 7.5 and 50 mM EDTA, 50 mM and H_2O, successively.
7.: Zn content of the guanidine-HCl 6 M solution in 20 mM Tris-HCl pH 7.5 and 50 mM EDTA after dialysis of TthS14.
8.: Zn content of the EDTA 50 mM solution in 20 mM Tris-HCl pH 7.5 after dialysis of TthS14.
9. TthS14: TthS14, denatured as in sample 6, dialysed against 0.57 mM $ZnCl_2$ in Tris-HCl pH 7.5, 0.1 NaCl, twice against 50 mM EDTA and H_2O, successively.

Zn standard solution: 3 ppm → 0.392 $A_{214.5}$. In all cases where 6 M guanidine-HCl was used a background value of 0.12 ppm Zn was subtracted. In all cases where 50 mM EDTA was used a background value of 0.08 ppm Zn was subtracted.

Sample	Amount of protein (moles)	Absorbance (214.5 nm)	Zn		Molar ratio Zn/protein
			ppm	moles	
1. TthS14	11×10^{-9}	0.104	0.796	12×10^{-9}	1
2. TthS14	11×10^{-9}	22.40	176	2.7×10^{-6}	245
3. TthS14	11×10^{-9}	0.095	0.730	12×10^{-9}	1
4. BSA	1.1×10^{-9}	0	0	0	0
5. BSA	1.1×10^{-9}	37.50	294	4.5×10^{-6}	4090
6. TthS14 denatured	11×10^{-9}	0.008	0.06	0.9×10^{-9}	0.08
7. Guanidine-HCl 6 M (after dialysis of denatured TthS14)	–	0.061	0.45	6.9×10^{-9}	–
8. EDTA 50 mM (after dialysis of denatured TthS14)	–	0.047	0.35	4.7×10^{-9}	–
9. TthS14 renatured	11×10^{-9}	0.097	0.59	7.4×10^{-9}	0.81

5.2.1
Atomic Absorption of the Overproduced Protein

As we mentioned above, atomic absorption experiments indicated the presence of Zn ions in the stoicheiometry 1:1 (moles of Zn per mole of TthS14 protein). In order to study the behaviour of the protein against Zn, the protein was dialysed against $ZnCl_2$ in a concentration of 10 mg/ml for 3–4 hrs, followed by another dialysis against water (with two changes,the first lasting 2 hrs and the second 16 hrs) in order to remove traces of $ZnCl_2$. About 245 moles of Zn per mole of S14 were absorbed onto the surface of the protein (Table 20.1, sample 2). A similar behaviour is shown by albumin (Table 20.1, sample 5), which was subjected to the same treatment.

5.2.2
Attempts for Zn Detachment Without Denaturation

To study the "affinity" of Zn to the protein we used EDTA – as chelating agent – or acidic conditions as described below.

In an attempt to remove the Zn bound to TthS14 the protein was dialysed extensively against water (16 h), against 20 mM Tris-HCl pH 7.5 buffer containing 50 mM EDTA – as chelating agent – (8 h) and once more against water (16 h), successively. In this case the measured zinc content is the same as that of the protein dialysed only against water, i.e. a Zn-protein molar ratio of one was calculated, indicating that EDTA alone is not enough to remove the Zn ion bound to TthS14 (Table 20.1, Sample 3).

It was also attempted to remove Zn under acidic conditions because exposure to acid pH has been shown to remove Zn from zinc-finger proteins (Woods *et al.* 1995). However, dialysis of TthS14 against pH 5.0 failed to remove the Zn ion from the protein. The Zn-protein stoicheiometry remained 1:1. This indicates a stronger than usual affinity of the TthS14 protein for the zinc ion. Fifty mM EDTA could not be used at this pH (5.0) because it is barely soluble at room temperature at 25 °C and precipitation might occur at 4 °C–where all the Zn detachment procedures took place.

5.2.3
Zn Detachment Under Denaturating Conditions

Now the question arises "How does Zn bind to the protein? Is it covalently bound or absorbed onto the surface of the protein?".

In an attempt to remove the Zn bound to TthS14, the protein was denatured by dialysis against 6 M guanidine hydrochloride in 20 mM Tris-HCl pH 7.5, 50 mM EDTA (the buffer volume was 500 ml and the duration of the dialysis 4 h), followed by dialysis against 500 ml 50 mM EDTA in 20 mM Tris-HCl pH 7.5 for 4 h and then against water (16 h). In this case the drastic disturbance of the protein conformation with 6 M guanidine hydrochloride, followed by dialysis against 50 mM EDTA and finally against water, leads to an almost complete removal of Zn (0.06 ppm) as shown in Table 20.1, Sample 6. On the other hand, the guanidine hydrochloride and the EDTA solutions, used for dialysis of TthS14, contained 0.45 ppm (Table 20.1, Sample 7) and 0.35 ppm (Table 20.1, Sample 8) Zn, respectively. The addition of the two values gives a result (0.8 ppm Zn) that is extremely close to the value obtained for the non-denatured protein (0.796 ppm, Table 20.1, Sample 1). This allows us to infer that the zinc ion has been quantitatively removed from the protein through the concerted action of a denaturing and a chelating agent.

5.2.4 Zn Attachment – Protein Renaturation

The denatured protein was dialysed against a buffer of 20 mM Tris-HCl pH 7.5, 0.1 M NaCl, which contained $ZnCl_2$ in various concentrations. In order to remove absorbed zinc ions, the protein was dialysed twice against the same buffer as above containing 50 mM EDTA and finally against water. In all cases zinc remained bound on the protein and could not be removed even by the action of EDTA, which is in accordance with our results for the native protein (Table 20.1, Sample 3). The Zn/protein molar ratio was lower than one when the zinc concentrations that were used were not in large excess of the protein concentration. The ratio approached unity, when a 10:1 excess of $ZnCl_2$ concentration over TthS14 concentration was used (Table 20.1, Sample 9). Similar values were obtained in the case of the zinc-finger peptides of the estrogen-receptors (Archer *et al.* 1990). Based on the failure of EDTA to remove zinc we conclude that TthS14 was renatured. The percentage of protein that was renatured was limiting for the amount of Zn that was bound. The molar ratio exceeded unity only when an excess of $ZnCl_2$ over 1:20 was used and this could be attributed to non-specific absorption of Zn ions on the protein.

Thus, the existence of a zinc ion bound to the TthS14 zinc-finger like motif is supported by experimental evidence and we conclude that the protein bears tetrahedrally coordinated zinc in a molar ratio of one, which is typical for zinc-finger motifs. The zinc ion can only be removed by denaturation and the protein rebinds Zn upon renaturation.

5.3 The Possible Role of the Zinc-Finger Domain in TthS14

The existence of zinc-finger motifs in several ribosomal proteins (Wool *et al.* 1995) and the detection of zinc in rat ribosomes (Chan *et al.* 1993) have led to the question: do ribosomal proteins and, more specifically, their zinc-finger motifs bind Zn?

In this work it is proven, as explained above, that *T. thermophilus* S14, a protein with a zinc-finger like motif, contains Zn. The data are in agreement with the participation of the zinc ion in the formation of a zinc-finger structure. The 1:1 stoicheiometry and the absolute requirement for denaturation in order to detach the Zn from the protein are very strong evidence in support of the participation of the motif and the Zn in the formation of the zinc-finger domain.

Zinc-fingers have been implicated in nucleic acid binding (Kohn *et al.* 1997). What is the possible role of the zinc-finger in the S14 ribosomal protein? One answer might be the participation of the structure in the binding of rRNA. The C2H2 domain of the transcription factor IIIA (TFIIIA) has been shown to bind 5S rRNA (Travers 1993) and the C3H domain of the human immunodeficiency virus gag protein is involved in the packaging of the RNA genome of the virus (Demene *et al.* 1994). On the other hand, zinc-finger motifs of the C4 variety, such as those of the hormone receptor and transcription factor GATA (Chan *et al.* 1993), have been primarily implicated in DNA binding. In the above mentioned proteins, zinc-fingers are in the form of tandem repeats whereas ribosomal proteins only contain a single motif. However, it has been recently shown by Pedone *et al.* (Pedone e*et al.* 1996) that the combination of an amino-terminal basic region with a single C2H2 zinc-finger in the GAGA protein is sufficient for high affinity DNA binding domain.

An interaction between the 16S rRNA and the S14 has been proven. In *Bacillus stearothermophilus* S14 (BstS14) Lys16 has been cross-linked to 16S rRNA proving it to be in direct contact to the RNA (Urlaub *et al.* 1995). Lys16 of BstS14 is homologous to Lys17 in TthS14 (Fig. 20.8) and is in close proximity to the zinc-finger motif. S14 from *E. coli* protects several base pairs in the 990–1050 nt region of 16S rRNA from chemical modification (Stern *et al.* 1989). Still, the zinc-finger might not be absolutely required for the rRNA-protein interaction. The *E. coli* S14 lacks the zinc-finger motif. Only one of the four cysteines has been conserved. On the other hand, both the sequence which corresponds to the linker region of the zinc-finger motif (Fig. 20.8) and the sequence of the rRNA domain known to interact with S14 are highly conserved (Noller *et al.* 1995, and Stern *et al.* 1988). Therefore, the interaction might take place between the linker region and the rRNA and consequently the zinc-finger would not be absolutely required for the assembly and function of the 30S subunit. It might be that it is only a "fossil" of the function of the ancestral molecule, which was recruited into the ribosome and evolved into S14. The existence of the zinc-finger in the S14 from archaea and in the related S29 from eucaryotes (Fig. 20.8) points to an ancient origin (Wool *et al.* 1995).

The question remains as to the reason why several organisms have retained this structure in S14. One possible reason is that certain organisms, such as the thermophilic *Thermus thermophilus* might need the extra stabilization of the rRNA structure by the interaction with the zinc-finger. A second possibility is that the zinc-finger has an alternate functionality. The S27 ribosomal protein from human, which contains a similar C4 zinc-finger, has been shown to interact specifically with cap elements and to act in the stimulation of several genes in

→

Fig. 20.8. Alignment of the ribosomal protein TthS14 (Tsiboli and Choli 1995), using the PC-Gene program, with its counterparts from the following organisms, as well as with the ribosomal proteins S29 and S27: B. stearothermophilus (Herfurth 1992), B. subtilis (Henkin *et al.* 1989), T. aquaticus (Jahn *et al.* 1991), M. vanniellii (Auer *et al.* 1989), M. capricolum (Ohkubo *et al.* 1987), E. coli (Yagushi *et al.* 1983), Vicia faba (Wahleithner and Wolstenholme 1988), Oryzia sativa (Coté and Wu 1988), S. cerevisiae (Larkin *et al.* 1987), H. marismortui (Scholzen and Arndt 1991), Human S29 (Frigerio *et al.* 1995), Entamoeba histolytica S27 (Stanley and Li 1992)

```
TTHS14  - A R K A L I E K A K R T P - - - - - - - - - - - K F K
BSTS14  - A K K S M I I K O K R T P - - - - - - - - - - - K F K
BSUS14  M A K K S M I A K O O R T P - - - - - - - - - - - K F K
TAQS14  M A R K A L I E K A K R T P - - - - - - - - - - - K F K
MCAS14  M A K K S L K V K O A K H P - - - - - - - - - - - K F N
ECOS14  - A K O S M K A R E V - - - - - - K R V A L A D K Y F
VFAS14  M E S K R N I - R D H - - - - - - K R R L L A A K Y E
OSAS14  M A K K S L I O R E R - - - - - - K R O K L E O K Y H
SCES14  M G N F R F P I K T K L P P G F I N A R I L R D N F K
MVAS14  M - - - - T K E - - - - - - - - - - - - - - - - - - -
HMAS14  M S E S E T T D E - - - - - - - - - - - - - - - - - -

TTHS14  V R A Y T - - - - - - - - - - - - - - - - - -
BSTS14  V R A Y T - - - - - - - - - - - - - - - - - -
BSUS14  V O E Y T - - - - - - - - - - - - - - - - - -
TAQS14  V R A Y T - - - - - - - - - - - - - - - - - -
MCAS14  V R N Y T - - - - - - - - - - - - - - - - - -
ECOS14  A K R - A E L K A I I S D - - - - - V N A S -
VFAS14  L R R - K L Y K A F C K D - - - - - S D L P L
OSAS14  I R - R S S K K K I R S - - - - - K V Y P L L
SCES14  I Q Q F K E N E I L V K S L K F I A R N M N L
MVAS14  - - - - - - - - - - - - - - - - - - - - - P F
HMAS14  - - - - - - - - - - - - - - - - - - - - - P D
HUMS29  - - -   - - - - - - - - - - - - - - - - - G H

TTHS14  - - - - - - - - - - - - - - - - - - - - - - -
BSTS14  - - - - - - - - - - - - - - - - - - - - - - -
BSUS14  - - - - - - - - - - - - - - - - - - - - - - -
TAQS14  - - - - - - - - - - - - - - - - - - - - - - -
MCAS14  - - - - - - - - - - - - - - - - - - - - - - -
ECOS14  D E D R W N A V R - K L O T L P R D S S P S R
VFAS14  P S D M W D K L R Y K L S K L P R N S S F A R
OSAS14  S L S E K T K M R E K L O S L P R N S A P T R
SCES14  P T K L R L E A Q L K L N A L P N Y M R S T Q
MVAS14  K T K Y G O G S K V - - - - - - - - - - - - -
HMAS14  S E T A S S E R T G Q L E S - - - - - - - - -
HUMS29  O O L Y W S H P R K F G O G S R S - - - - - -

TTHS14  - - - R C V R C G R - A R S V Y R F F G - - L C R I C
BSTS14  - - - R C E R C G R - P H S V Y R K F K - - L C R I C
BSUS14  - - - R C E R C G R - P H S V I K K F K - - L C R I C
TAQS14  - - - R C V R C G R G A R S V Y R Y F G - - L C R I C
MCAS14  - - - R C N H C G R - P H A V L K K F G - - I C R K C
ECOS14  O R N R C R O T G R - P H G F L K R F G - - L S R I K
VFAS14  V R M R C I S T G R - P R S V Y E L F R - - A S R A V
OSAS14  L H R R C F L T G R - P R A N Y R D F G - - L S G H I
SCES14  I K N R C V D S G H I A R F V L S D F R - - L C R Y Q
MVAS14  - - - - C K R C G R K G P G I I R K Y G L D L C R O C
HMAS14  - - - - C Q R C G R E - Q G L V G K Y D I W L C R Q C
HUMS29  - - - - C R V C S N R H - G L I R K Y G L N M C R O C
EHIS29  - - - - C R K C G A R K - G L I R K Y G L D L C R R C
EHIS27  - - - - C P K C G A T T T T F S H A H R O I L C O K C

TTHS14  L R E L A H K G O L P G V R K A S W
BSTS14  F R E L A Y K G O L P G I K K A S W
BSUS14  F R E L A Y K G O I P G V K K A S W
TAQS14  L R E L A H K G O L P G V K K A S W
MCAS14  F L K F A Y E G O I P G I K A A S -
ECOS14  V R E A A M R G O I P G L K K G - -
VFAS14  F R S L A S R G P L M G I K K S S W
OSAS14  L R E M V Y A C L L P G A T R S S W
SCES14  F R E N A L K G N L P G V K K G I W
MVAS14  F R E L A P K - - - L G F K K Y D -
HMAS14  F R E I S R G - - - M G F R K Y S -
HUMS29  F R O Y A   K D I - - G F I K L D -
```

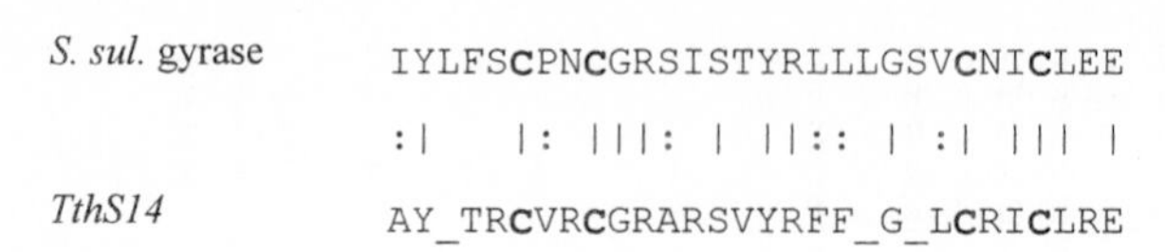

Fig. 20.9. Alignment of the zinc-finger domain of TthS14, using the BLAST program, with the C4 zinc-finger domain of the S. sulfataricus gyrase-topoisomerase

human carcinomas (Fernandez-Pol 1993). Additionally, S27 from *Entamoeba histolytica* has been shown to bind Zn (Zhang 1993). The DNA-binding ability of the S27 from eucaryotes raises the possibility that the members of the S14/S29 family that contain a zinc-finger participate in a second function through interaction with DNA sequences.

Interestingly, a search of the protein data banks for sequences that show high homology to the zinc-finger motif of *T. thermophilus* S14 using the BLAST program (Altschul *et al.* 1990) indicated great similarities between the TthS14 zinc-finger domain and a C4 zinc-finger like motif in the amino-terminal (helicase) domain of the reverse gyrases-topoisomerases (Confalonieri *et al.* 1993). The similarities are obvious in the alignment of the *Sulfolobus sulfataricus* gyrase-topoisomerase and the T. thermophilus S14 C4 zinc-finger domains (Fig. 20.9). There is an identity of 14 out of 30 amino acids and a similarity of an additional 6 amino acids.

In conclusion, the S14 ribosomal protein from T. thermophilus has been proven to bind Zn in a molar ratio of one. This and the absolute requirement of the protein structure for the binding of Zn indicate that the C4 motif of TthS14 participates in the formation of a zinc-finger domain. Additional experiments are needed to elucidate the role the domain might play in the interaction of the TthS14 protein with rRNA and/or DNA.

7 Acknowledgements

This work was supported by a grant from the General Secretariat of Reserach and Technology. The authors wish to thank the Laboratory of Analytical Chemistry of the Chemistry Department for the atomic absorption measurements and Prof. Ada Yonath for interest and support in this work.

References

Altschul SF, Gish W, Miller W, Myers EW & Lipman DJ (1990) Basic local alignment search tool. J Mol Biol 215: 403–410

Archer TK, Hager GL & Omichinski JG (1990) Sequence-specific DNA binding by glucocorticoid receptor "zinc-finger peptides". Proc Natl Acad Sci USA 87: 7560–7564

Auer J, Spicker G & Boeck A (1989) Organization and structure of the Methanococcus transcriptional unit homologous to the E. coli "spectinomycin operon". Implications for the evolutionary relationship of 70S and 80S ribosomes. J Mol Biol 209: 21–28

Chan YL, Suzuki K, Olvera J & Wool IG (1993) Zinc-finger-like motifs in rat ribosomal proteins S27 and S29. Nucl Acids Res 21(3): 649–655

Choli, T. (1989) Structural properties of ribosomal protein L11 from Escerichia coli. Biochem. Int. 19, 1323–1338

Choli T, Kapp U & Wittmann-Liebold B (1989) Blotting of proteins onto Immobilon membranes – In situ characterization and comparison with high performance liquid chromatography. J Chromat 476: 59–72

Confalonieri F, Elie C, Nadal M, de La Tour C, Forterre P & Duguet M (1993) Reverse gyrase: a helicase-like domain and a type I topoisomerase in the same polypeptide. Proc Natl Acad Sci USA 90: 4753–4757

Coté J C & Wu R (1988) Sequence of the chloroplast psbF gene encoding the photosystem II 10kDa phosphoprotein from Oryza sativa. Nucl Acids Res 16: 10384

Demene H, Dong CZ, Ottmann M, Rouyez MC, Jullian N, Morellet N, Mely Y, Darlix JL, Fournie-Zaluski MC, Saragosti S & Roques BP (1994) 1H NMR structure and biological studies of the His23→Cys mutant nucleocapsid protein of HIV-1 indicate that the conformation of the first zinc-finger is critical for virus infectivity. Biochemistry 33: 11707–11716

Fernandez-Pol JA (1993) A growth factor-inducible gene encodes a novel nuclear protein with zinc-finger structure. J Biol Chem 268(28): 21198–21204

Frigerio JM, Dagorn JC & Iovanna JL (1995) Cloning, sequencing and expression of the L5, L21, L27a, L28, S5, S9, S10, S29 human ribosomal protein mRNAs. Bioc Biophys Acta 1262: 64–68

Henkin TM, Moon SH, Mattheakis LC & Nomura M (1989) Cloning and analysis of the spc ribosomal protein operon of Bacillus subtilis: comparison with the spc operon of Escherichia coli. Nucl Acids Res 17: 7469–7486

Herfurth E (1992) Untersuchungen zur feinstruktur und topographie von ribosomen. PhD Thesis, Freie Universität, Berlin

Herfurth E, Briesemeister U & Wittmann-Liebold B (1994) Complete amino acid sequence of ribosomal protein S14 from B. stearothermophilus and homology studies to other ribosomal proteins. FEBS Lett 351: 114–118

Jahn O, Hartmann RK & Erdmann VA (1991) Analysis of the spc ribosomal protein operon of Thermus aquaticus. Eur J Biochem 197: 733–40

Kohn WD, Mant CT & Hodges RS (1997) Alpha-helical protein assembly motifs. J Biol Chem 272 (5): 2583–2586

Larkin JC & Wu R (1987) Structure and expression of the S. cerevisiae CRY1 gene: a highly conserved ribosomal protein gene. Mol Cell Biol 7: 1764–1775

Noller HF, Green R, Heilek G, Hoffarth V, Huettenhoffer A, Joseph S, Lee I, Lieberman K, Mankin A, Merryman C, Powers T, Puglisi EV, Samaha RR & Weiser B (1995) Structure and function of ribosomal RNA. In: Matheson AT, Davies JE, Dennis PP and Hill PE (eds) Frontiers in Translation. An International Conference on the Structure and Function of the Ribosome. National Research Council Canada, Ottawa, pp 997–1009

Ohkubo S, Muto A, Kawauchi Y, Yamao F & Osawa S (1987) The ribosomal protein gene cluster of Mycoplasma capricolum. Mol Gen Genet 210: 314–322

Pedone PV, Ghirlando R, Clore MG, Gronenborn AM, Felsenfeld G & Omichinski JG (1996) The single Cys2-His2 zinc-finger domain of the GAGA protein flanked by basic residues is sufficient for high-affinity specific DNA binding. Proc Natl Acad Sci USA 93: 2822–2826

Ramakrishnan V, Davies C, Gerchman G, Golden BL, Hoffmann DW, Jaishree TN, Kycia JH, Porter S & White SW (1995) Structures of prokaryotic ribosomal proteins: implications for RNA binding and evolution. In: Matheson AT, Davies JE, Dennis PP and Hill PE (eds) Frontiers in Translation. An International Conference on the Structure and Function of the Ribosome. National Research Council Canada, Ottawa, pp 979–986

Scholzen T & Arndt E (1991) Organization and nucleotide sequence of ten ribosomal protein genes from the region equivalent to the spectinomycin operon in the archaebacterium Halobacterium marismortui. Mol Gen Genet 228: 70–80

Stanley SLJr & Li E (1992) Isolation of an Entamoeba histolytica cDNA clone encoding a protein with a putative zinc-finger domain. Mol Biochem Parasitol 50: 185–187

Stern S, Changchien LM, Craven GR & Noller HF (1988) Interactions of proteins S16, S17 and S20 with 16S ribosomal RNA. J Mol Biol 200: 291–299

Stern S, Powers T, Changchien LM & Noller HF (1989) RNA-protein interactions in 30S ribosomal subunits: folding and function of 16S rRNA. Science 244: 783–790

Travers A (1993) Zinc-containing binding domains. In: DNA-Protein Interactions, Chapman & Hall (eds), London, pp 64–73

Tsiboli P & Choli T (1995) Studies on S14 protein from Thermus thermophilus possessing zinc-finger like motifs. Biol Chem Hoppe-Seyler 376: 127–130

Tsiboli P, Triantafillidou D, Franceschi F & Choli-Papadopoulou T (1998) Studies on the Zn-containing S14 ribosomal protein from Thermus thermophilus. Eur J Biochem 256:136–41

Urlaub H, Kruft V, Bischof O, Mueller EV & Wittmann-Liebold B (1995) Protein-rRNA binding features and their structural and functional implications in ribosomes as determined by cross-linking studies. EMBO J 14(18): 4578–4588
Wahleithner JA & Wolstenholme DR (1988) Ribosomal protein S14 genes in broad bean mitochondrial DNA. Nucl Acids Res 16: 6897–6913
Woods AS, Buchsbaum JC, Worrall TA, Berg JM & Cotter RJ (1995) Matrix-Assisted Laser Desorption/Ionization of Noncovalently Bound Compounds. Anal Chem 67: 4462–4465
Wool I, Chan Y-L & Glueck A (1995) Structure and evolution of mammalian ribosomal proteins. In: Matheson AT, Davies JE, Dennis PP and Hill PE (eds) Frontiers in Translation. An International Conference on the Structure and Function of the Ribosome. National Research Council Canada, Ottawa, pp 933–947
Xing Y, Guha Thakurta D & Draper DE (1997) The RNA binding domain of ribosomal protein L11 is structurally similar to homeodomains. Nat Struct Biol 4: 24–27
Yagushi M, Roy C, Reithmeier RAF, Wittmann-Liebold B & Wittmann HG (1983) The primary structure of protein S14 from the small ribosomal subunit of E. coli. FEBS Lett 154: 21–30
Zhang Y (1993) Entamoeba histolytica: the EHZc3 cDNA clone encodes a zinc-binding protein. Exp Parasitol 77(1): 118–120

CHAPTER 21

Biomolecular Interaction of Matrix Metalloproteinases and Their Inhibitors TIMPs

H. Tschesche[1] and M. Farr[1]

1 Abstract

Matrix metalloproteinases are a family of enzymes that altogether degrade almost all constituents of the extracellular matrix including collagens, laminin, elastin and fibronectin. Therefore, they are highly involved in the remodelling of the extracelluar matrix during embryogenesis, growth and development, and repair of tissues. Otherwise dysregulated proteolytic activity of matrix metalloproteinases is a key feature of a variety of diseases encompassing rheumatoid arthritis, multiple sclerosis, liver fibrosis, parodontosis, tumor formation and metastasis. In physiological processes the matrix metalloproteinases are precisely regulated by gene expression, secretion and activation of proenzymes and inhibition by their endogenous inhibitors, the tissue inhibitors of metalloproteinases (TIMPs). These proteins control the activation of the zymogens, especially of progelatinases, and the proteolytic activity of the mature enzymes. Besides ternary and quaternary complexes of gelatinases the typical complex between a matrix metalloproteinase and a TIMP is bimolecular and leads to a high affinity, quasi irreversible inhibition of the enzyme. Within the last few years several three-dimensional structures of MMPs and MMP-inhibitor complexes have been determined showing the polypeptide fold, domain organization, the architecture of the active site and binding modes of TIMPs and synthetic inhibitors. Based on these structural data enzyme kinetics and biomolecular analysis have clarified structure-funcion correlations of activation, proteolytic activity and inhibition by TIMPs.

2 Structure and Function of Matrix Metalloproteinases

The matrix metalloproteinases (MMPs, matrixins) form a family of about twenty zinc- and calcium-dependent endopeptidases that degrade most of the constituents of the extracellular matrix, for example interstitial and basement membrane collagens, gelatin, proteoglycans, aggrecan, elastin, fibronectin, entactin, vitronectin and laminin. Therefore they are highly involved in physiological processes

[1] University of Bielefeld, Biochemistry I, D-33615 Bielefeld, Germany.

like pregnancy, embryonic development, angiogenesis, formation of bones and organs, the remodelling of tissues, leukodiapedesis, wound healing and pathological processes like rheumatoid arthritis, osteoarthritis, parodontosis, multiple sclerosis, tumor formation, metastasis, liver fibrosis, cystic fibrosis and aneurysm (for reviews see Birkedal-Hansen et al. 1993). Because all MMPs have a conserved zinc binding motif (HExGHxxGxxH) and an adjacent 1,4-tight "Met-turn" they belong to the "metzincins" as a subgroup of metalloproteinases. They are

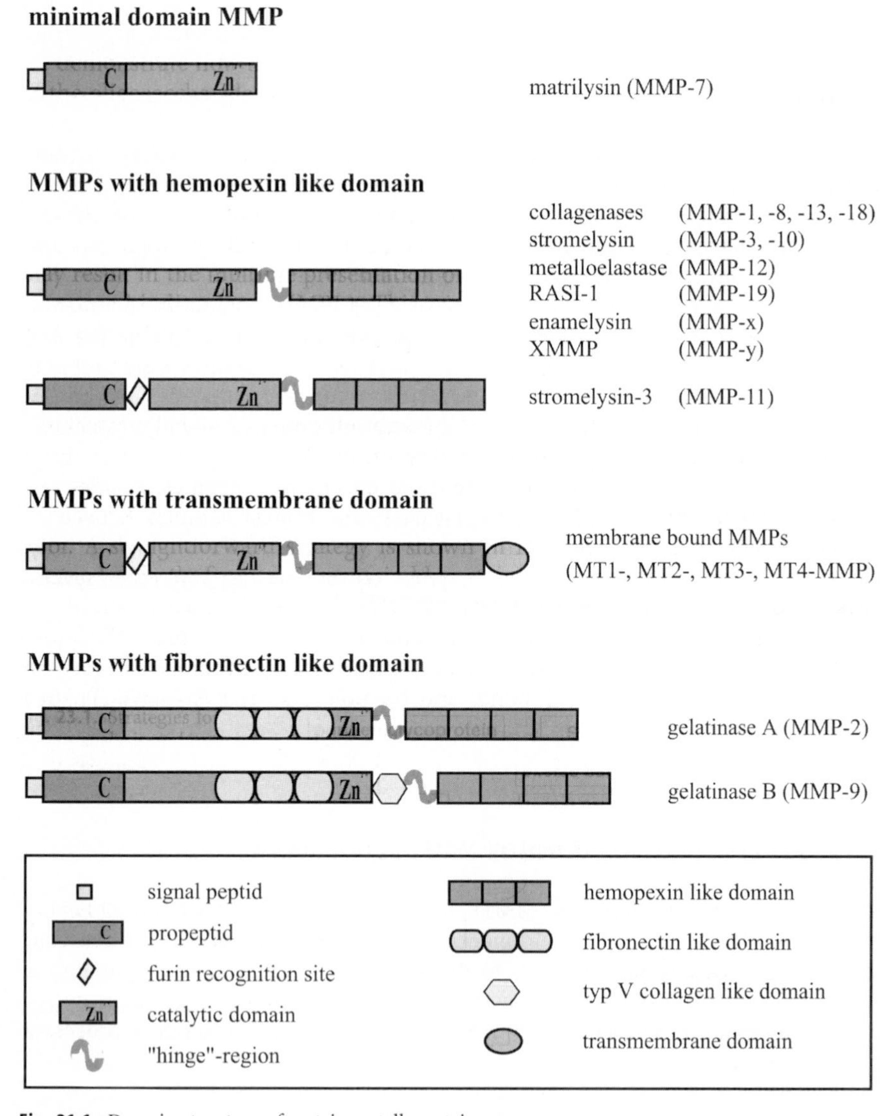

Fig. 21.1. Domain structure of matrix metalloproteinases

related to the serralysins (bacterial proteases from *Serratia, Pseudomonas* and *Erwinia*), the adamalysins (also called reprolysins or ADAMs) and the astacins (meprins, tolloids, "hatching enzymes") (Bode et al. 1993).

Because of their individual domain structure matrix metalloproteinases can be divided into matrilysin, MMPs with a hemopexin like domain, MMPs with a transmembrane domain and MMPs with a fibronectin like domain.

The hydrophobic signal peptide mediates intracellular sorting and is removed during secretion of the proenzyme. Its latency is maintained by the propeptide with the highly conserved sequence PRCGVPD that complexes the active site zinc ion with the thiol group of the cysteine residue and is removed in a stepwise process upon activation (Knäuper et al. 1990, Tschesche et al. 1992, Nagase et al. 1990) including the disruption of the thiol-zinc interaction, the so-called "cysteine-switch" (Springman et al. 1990). For example, *in vitro* activation of the neutrophil procollagenase (MMP-8) initiated by proteinases, mercurials or oxidants leads to active species having either Phe79, Met80 or Leu81 as the N-terminal residue and differing significantly in activity (Bläser et al. 1991, Grant et al. 1987, Knäuper et al. 1990, Knäuper et al. 1993, Mallya et al. 1990, Suzuki et al. 1990). The species upon stromelysin-1 activation is the Phe79 form and because it is approximately 3.5-fold more active than the two other species, this phenomenon has been called "superactivation" (Knäuper et al. 1993, Reinemer et al. 1994). Likewise fibroblast collagenase (MMP-1) can be "superactivated" by stromelysin-1 with an increase of proteolytic activity up to 12-fold (He et al. 1989, Murphy et al. 1987, Suzuki et al. 1990). Both the catalytic domain of MMP-8 with the N-terminal Phe79 and the N-terminal Met80 have been expressed in E. coli (Reinemer et al. 1994, Schnierer et al. 1993) and crystallized (Bode et al. 1994, Reinemer et al. 1994).

Determination of their X-ray structures has revealed that the N-terminal Phe79 ammonium group makes a salt link with the side chain carboxylate group of Asp232. Therefore, the N-terminal peptide is tightly packed against a hydrophobic surface groove made by a C-terminal helix and a descending segment centering around the third His ligand of the catalytic zinc ion. The attachment of this N-terminal heptameric segment probably results in stabilization of the catalytic site via strong hydrogen bonds mediated by the adjacent Asp233 with the "Met-turn", that forms the base of the active site residues (Reinemer et al. 1994). Correspondingly, the N-terminal peptide of the Met80 variant is too short to form a salt bridge with Asp 232, so that the N-terminal segment is less ordered and not localized in the X-ray structure analysis (Fig. 21.2, arrow). The lack of active site stabilization or interference of this segment with substrate binding might be an explanation for the lower enzymatic activity.

In vivo the activation of matrix metalloproteinases is a relatively complex process, because several soluble and membrane-bound proteases, receptors and inhibitors can be involved in the processing of the propeptide. Activation cascades by cathepsins and serine proteases, (pro)plasminogen activator and receptor, plasminogen and plasminogen receptor and finally active plasminogen have been described as well as induction of the "cysteine switch" by bacterial proteases and activated MMPs (for a review see Nagase 1997). Moreover furin seems to activate prostromelysin-3 (MMP-11) already intracellularly, so that it is secreted

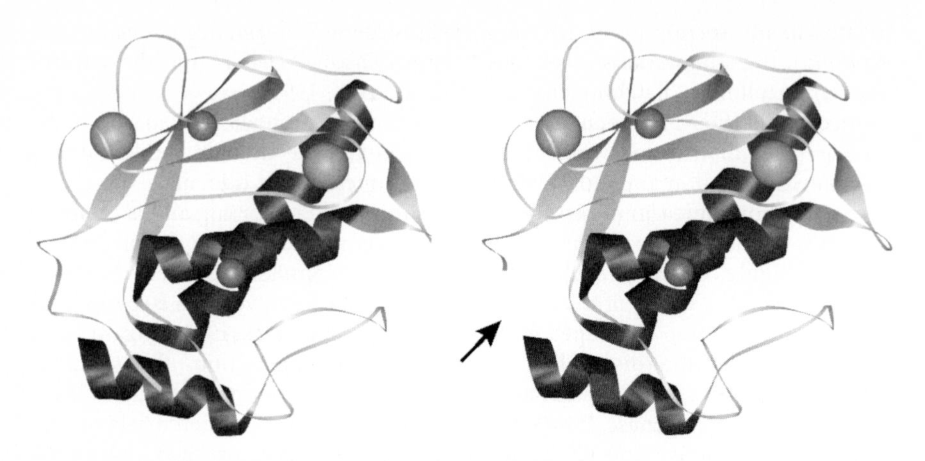

Fig. 21.2. Structure of the catalytic domain of MMP-8 (Bode et al. 1994, Reinemer et al. 1994). Left side: cdMMP-8^{Phe79}; right side: cdMMP-8^{Met80} (The arrow indicates that the N-terminal less ordered amino acids are not localized in the X-ray structure analysis.)

as an active enzyme (Pei and Weiss 1995). Because the membrane-type MMPs possess a potential furin recognition site too (Fig. 21.1), they might also be activated within cells (Sato et al. 1996b). On the other hand activation of progelatinase A (MMP-2) on cell surfaces is catalyzed by the membrane-type MMPs, which are anchored in the cell membrane by a hydrophobic, C-terminal transmembrane domain (Fig. 21.1). This activation process is mediated by TIMP-2, that binds N-terminally to one molecule MT-MMP and captures C-terminally the progelatinase and presents it to an adjacent active molecule of MT-MMP for activation (Sato et al. 1996a). Other domains of MMPs like the hemopexin like domain and the fibronectin like domain of gelatinases determine individual substrate specificities. For example, the hemopexin like domain of the collagenases is essential for their tripelhelicase activity towards native collagen type I, II and III (Schnierer et al. 1993).

3 Biomolecular Interaction with TIMPs

The physiological antagonists of the matrix metalloproteinases are the tissue inhibitors of metalloproteinases (TIMPs) comprising at least four proteins. The primary structure of their 184 to 195 amino acids polypeptide chain is dominated by twelve highly conserved cysteine residues which form six disulfide bonded loops (Williamson et al. 1990).

The MMP inhibitory activity of TIMPs is located to the three N-terminal loops (DeClerck et al. 1993, Murphy et al. 1991, O'Shea et al. 1992, Williamson et al. 1996), while the C-terminal loops include additional binding sites for the gelatinases outside their catalytic domain (Huang et al. 1996, Murphy et al. 1991, Willenbrock et al. 1993). Because of these two major domains there are several binding modes comprising binary complexes like (pro)gelatinase B/TIMP-1 (Wilhelm

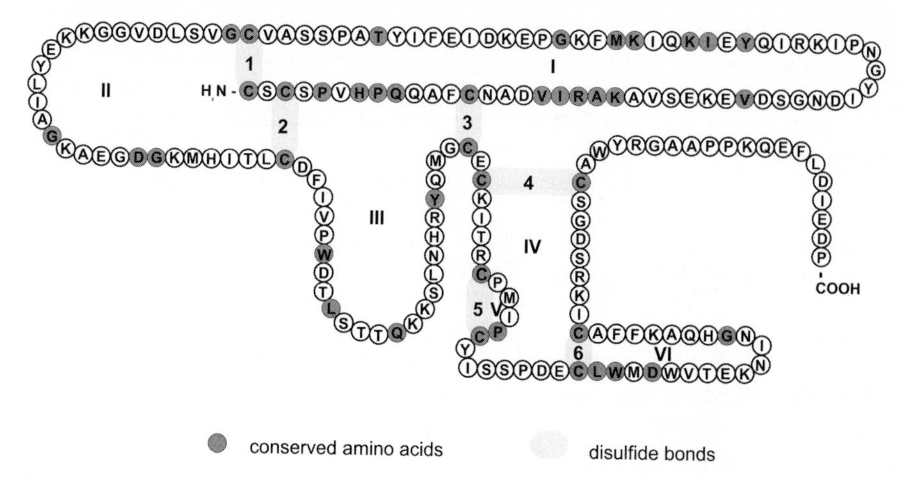

Fig. 21.3. Primary structure of TIMP-2. 1–6: disulfide bonds; I-VI: disulfide bonded loops

et al. 1989) or (pro)gelatinase A/TIMP-2 (Goldberg et al. 1989), ternary complexes like progelatinase A/TIMP-2/MMP-8 (Kolkenbrock et al. 1991) or progelatinase A/TIMP-2/MT1-MMP (Sato et al. 1996b) and quaternary complexes like progelatinase B/lipocalin/TIMP-1/gelatinase B (Kolkenbrock et al. 1996). Besides their inhibitory activity towards MMPs TIMPs exhibit growth-promoting activity for a wide range of cells, promote erythropoesis, modulate cell morphology and adhesion, might be involved in cell cyclus regulation and apoptosis and regulate angiogenesis (for a review see Gomez et al. 1997).

Recently two three dimensional structures of TIMPs complexed with MMPs have been determined by X-ray crystallography (Fernandez-Catalan et al. 1998, Gomis-Rüth et al. 1997), demonstrating that TIMP-1 and TIMP-2 have the shape of an elongated continuous wedge. According to functional experiments the N-terminal inhibitory domain and the C-terminal domain form two opposing subdomains. A characteristic of the N-termial part is a closed β-barrel of elliptical cross-section that is formed by a five-stranded β-pleated sheet. In complexes with stromelysin-1 and MT1-MMP the TIMPs bind with their edge into the whole active site cleft of the matrix metalloproteinases forming most of the intermolecular contacts by their N-terminal segment Cys1-Pro5 in a substrate like manner. In particular the conserved Cys1 of TIMP is involved in inhibition of the active site, because its "α-amino nitrogen and carbonyl oxygen atoms are localized directly above the catalytic zinc and coordinate this ion together with the three conserved His of the MMP. Moreover the water molecule that is necessary for the proteolytic attack is excluded in the complex, because the α-amino group of Cys-1 occupies the binding site by formation of a hydrogen bond to one carboxylate oxygen atom of the conserved catalytic Glu. Moreover the Thr-2 or Ser-2 side chain of TIMP extends partially into the specificity pocket of the MMP and is also hydrogen bonded to the catalytic Glu.

Although TIMPs have about 40% overall sequence identity and the same domain structure, their topologies differ considerably. Overlay plots of the TIMP-1

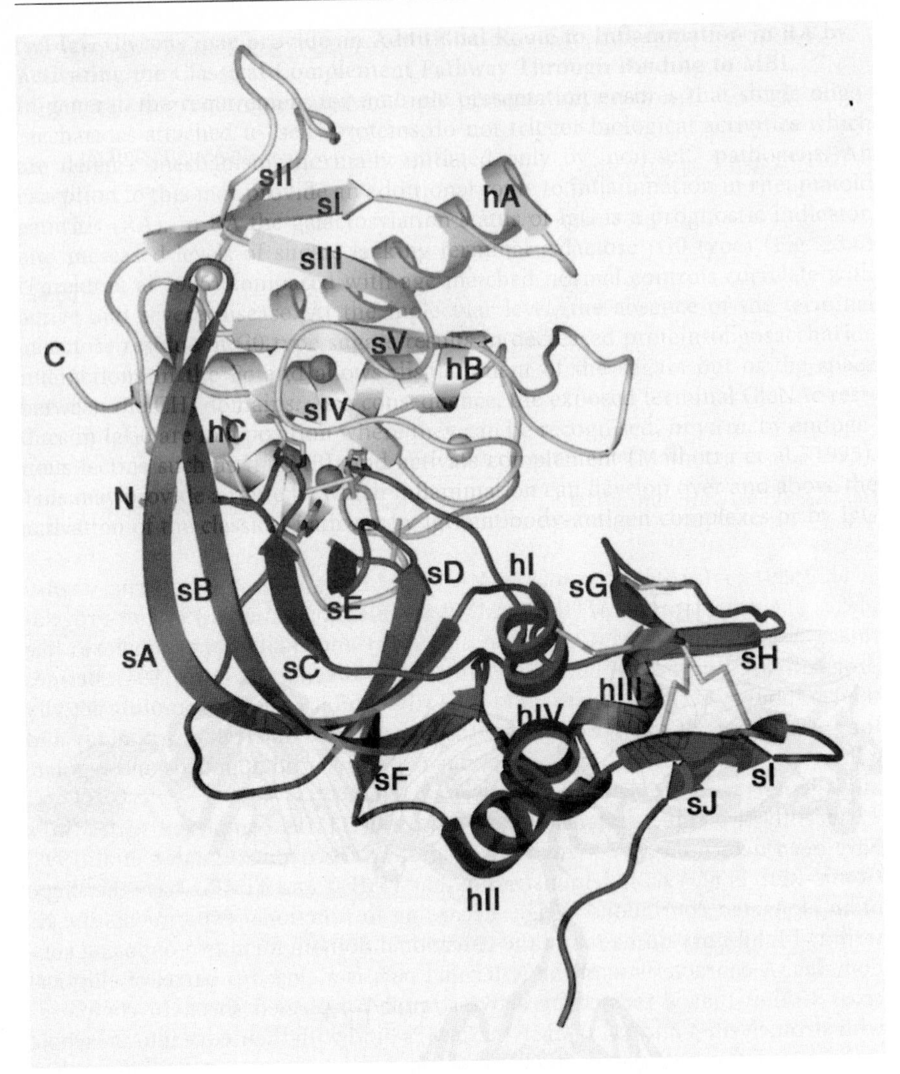

Fig. 21.4. Structure of the TIMP-2/MT1-MMP complex (Fernandez-Catalan et al. 1998). Bright structure: catalytic domain of MT1-MMP; dark structure: bovine TIMP-2; dark balls: zinc ions; bright balls: calcium ions; h: helix; s: β-pleated sheet

and TIMP-2 complexes show that both inhibitors are tilted relatively to each other, that TIMP-2 has a quite elongated sA-sB β-hairpin loop that rises above the β-barrel (Fig. 21.4), and that TIMP-2 possesses a much longer negatively charged flexible C-terminal tail (Fernandez-Catalan et al. 1998).

Before these structural details were resolved a variety of experiments was performed to analyze the thermodynamical and kinetic aspects of TIMP/MMP interaction using labelled synthetic or physiological substrates. One of the recent methods to analyze these biomolecular interactions are the BIA-systems

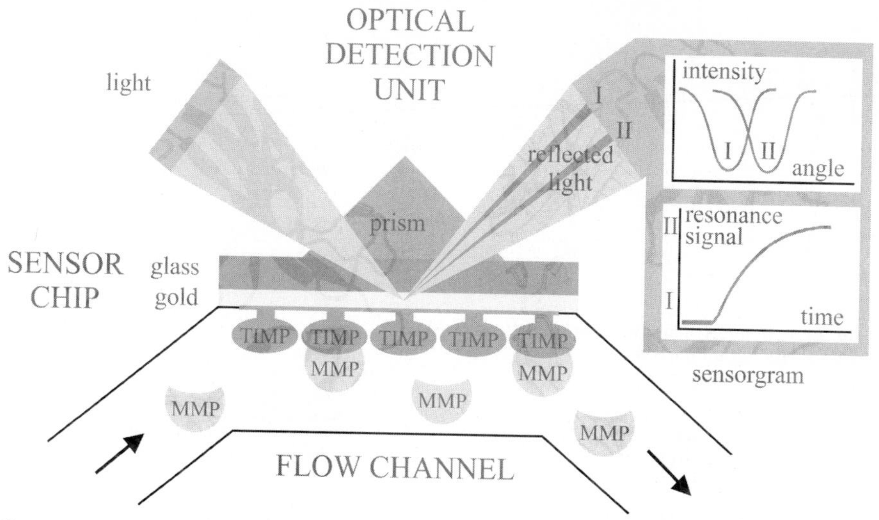

Fig. 21.5. BIAcore system (Jäger et al. 1997, modified)

(Fig. 21.5) combining surface plasmon resonance detection, a suitable sensor chip chemistry and an integrated flow system (Fängerstam and O'Shannessy 1993, Jönsson und Malmquist 1992). The biomolecular interaction between an analyte (e.g. MMP) and a ligand (e.g. TIMP) coupled to the sensor chip surface can be detected, because surface plasmon resonance measures changes in refractive index close to the sensor surface proportional to the surface concentration (g/m^2) of the analyte.

Besides kinetic studies the BIA measurement can be analyzed by a thermodynamic approach based on the saturation effect of the TIMP coated sensor chip surface by the MMP. Because the recombinant catalytic domain of MMP-8 used in this study does not comprise the C-terminal binding site for TIMPs, the mathematical term of the increase of response (Fig.21.7) is a relatively simple function derived from the equilibrium term of the K_i value. K_iY values are determined in a three-step procedure:

1. Sensorgram RU = f(t):
   Plot of the signal (RU) as a function of time for each MMP concentration (Fig. 21.6).
2. Saturation curve ΔRU = f([MMP]):
   Plot of the RU increase ([ordf]RU) versus the MMP concentration [MMP] (Fig. 21.6).
3. Computing of the Ki value:
   Simultaneous fitting of K_i, ΔRUmax (maximal possible RU increase), and [TIMP] (concentration of TIMP) to the saturation curve (2.) using the mathematical term for ΔRU (Fig. 21.7).

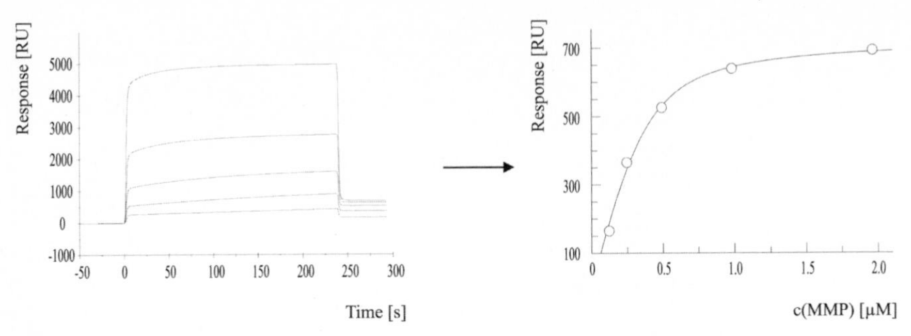

Fig. 21.6. Biomolecular interaction analysis of MMP-8^{Phe79} and idTIMP-2. Left side: sensorgram; right side: saturation curve

$$\Delta RU = \frac{\Delta RU_{max}}{[TIMP]} \cdot \left(\frac{[K_i] + [TIMP] + [MMP]}{2} - \sqrt{-[TIMP] \cdot [MMP] + \left(\frac{[K_i] + [TIMP] + [MMP]}{2} \right)^2} \right)$$

Fig. 21.7. Computation of K_i values. Mathematical term that describes the saturation curve (see text)

Table 21.1. Inhibition constants determined by BIA. Ki values are ± S.D.

	cdMMP-8^{Phe79} ("superactivated")	cdMMP-8^{Met80} (" activated")
idTIMP-2	81.4 nM ± 4.8 nM	18.8 nM ± 3.0 nM
bTIMP-2	37.3 nM ± 2.4 nM	9.7 nM ± 0.8 nM

With this approach the complex formation of the recombinant inhibitory domain of TIMP-2 (idTIMP-2) and bovine TIMP-2 (bTIMP-2) from semial plasma (Calvete et al. 1996) with the "activated" Met80 variant and the "superactivated" Phe79 variant of the recombinant catalytic domain of neutrophil collagenase was analyzed. The K_i values obtained from the biomolecular interaction analysis (Tab. 21.1) correlate with the results from fluorescence kinetics (data not shown) and reveal differences in the inhibition of the activated and superactivated collagenase.

It is obvious that the full-length TIMP (bTIMP-2) is the better inhibitor and this correlates with previous studies demonstrating that C-terminal interaction of TIMPs and MMPs can contribute to a decrease in the association constants of only N-terminal interacting molecules by about one up to two magnitudes (Baragi et al. 1994, Huang et al. 1996, Knäuper et al. 1997, Nguyen et al. 1994, Willenbrock et al. 1993). A remarkable feature is that both TIMPs inhibit the "activated" collagenase with N-terminal Met80 better than the "superactivated" variant with the additional N-terminal Phe79. The ammonium group of this phenylalanine makes a salt link with the side chain carboxylate group of Asp232, so that the N-terminal segment is bound to the globular protein structure as shown by X-ray analysis (Fig. 21.2). This putative more rigid structure of the Phe79-stabilized catalytic domain of human neutrophil collagenase seems to be not only

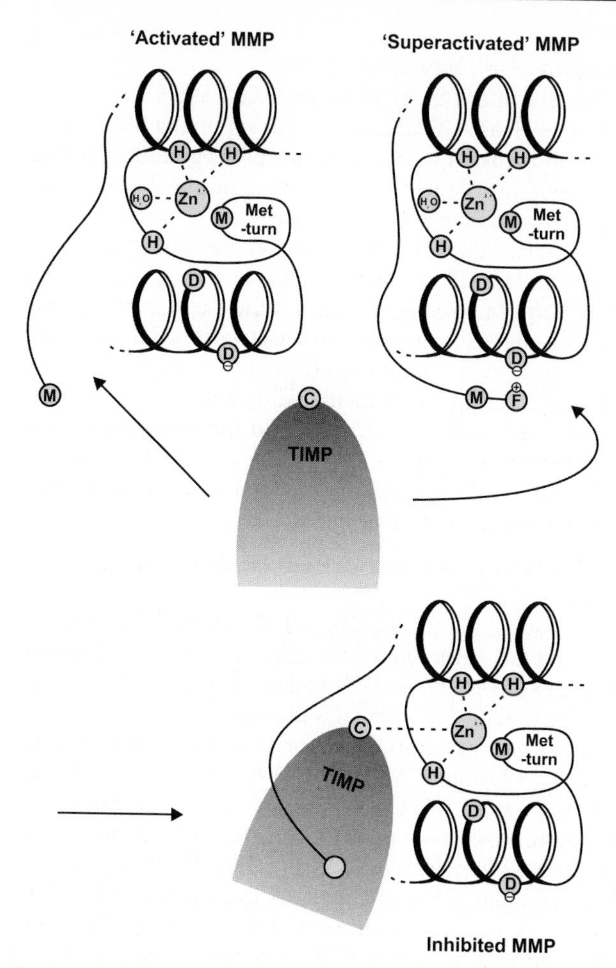

Fig. 21.8. Inhibition scheme of the "activated" and "superactivated" collagenase. Left side: inhibition of the "activated" MMP-8; right side: inhibition of the "superactivated" MMP-8 with disruption of the salt bridge between Phe79 (F) and Asp232 (D)

the reason for the higher activity of this variant, but also for a more difficult induced fit in the complex formation of enzyme and inhibitor. As the catalytic domain of stromelysin-1 comprises the homologous residues Phe83 and Asp237 which form no salt bridge due to the binding of TIMP-1 demonstrated by X-ray analysis (Gomis-Rüth et al. 1997), we assume that the salt link and the N-terminal segment binding have to be disrupted when TIMP-2 binds to the "superactivated" collagenase variant (Fig. 21.8). The other way round, binding to the "activated" collagenase is facilitated by the absence of this salt bridge. Therefore we propose that there is a functional correlation of the three-dimensional enzyme structure as determined by X-ray crystallography, the enzymatic property of superactivity, and the weaker inhibition by TIMPs.

4 Acknowledgements

We thank Dr. W. Jäger and Dr. B. Haase (BIACORE AB) for supplying the BIAcore instrument. The financial support by the SFBs 549 and 223 is greatly acknowledged.

References

Baragi VM, Fliszar CJ, Conroy MC, Ye QZ, Shipley JM, Welgus HG (1994) Contribution of the C-terminal domain of metalloproteinases to binding by tissue inhibitor of metalloproteinases. J. Biol. Chem. 269: 12692–12697

Birkedal-Hansen H, Moore WGI, Bodden MK, Windsor LJ, Birkedal-Hansen, B, DeCarlo A, Engler JA (1993) Matrix metalloproteinases: A review. Crit. Rev. Oral Biol. Med. 4: 197–250

Bläser J, Knäuper V, Osthues A, Reinke H, Tschesche H (1991) Mercurial activation of human polymorphonuclear leucocyte procollagenase. Eur. J. Biochem. 202: 1223–1230

Bode W, Gomis-Rüth FX, Stöckler W (1993) Astacins, serralysins, snake venom and matrix metalloproteinases exhibit identical zinc-binding environments (HEXXHXXGXXH and Met-turn) and topologies and should be grouped into a common family, the 'metzincins'. FEBS Letters 331(1,2): 134–140

Bode W, Reinemer P, Huber R, Kleine T, Schnierer S, Tschesche H (1994)The X-ray crystal structure of the catalytic domain of human neutrophil collagenase inhibited by a substrate analogue reveals the essentials for catalysis and specificity. EMBO J. 13: 1263–1269

Calvete JJ, Varela PF, Sanz L, Romero A, Mann K, Töpfer-Petersen E (1996) A procedure for the large-scale isolation of major bovine seminal plasma proteins. Prot. Exp. Purif. 8: 48–56

DeClerck YA, Yean TD, Lee Y, Tomich JM, Langley KE (1993) Characterization of the functional domain of tissue inhibitor of metalloproteinases-2 (TIMP-2). Biochem. J. 289: 65–69

Fängerstam LG, O'Shannessy (1993) Handbook of affinity chromatography 63: 229–252, Marcel Dekker Inc.

Fernandez-Catalan C, Bode W, Huber R, Turk D, Calvete JJ, Lichte A, Tschesche H, Maskos K (1988) Crystal structure of the complex formed by the membrane type 1-matrix metalloproteinase with the tissue inhibitor of metalloproteinases-2, the soluble progelatinase A receptor. The EMBO J. 17: 5238–5248

Goldberg GI, Barry LM, Marmer BL, Grant GA, Eisen AZ, Wilhelm S, He C (1989) Human 72-kDa type IV collagenase forms a complex with a tissue inhibitor of metalloproteinases designated TIMP-2. Proc. Natl. Acad. Sci. USA 86: 8207–8211

Gomez DE, Alonso DF, Yoshiji H, Thorgeirsson UP (1997) Tissue inhibitors of metalloproteinases: structure, regulation and biological functions. Eur. J. Cell Biol. 74: 111–122

Gomis-Rüth FX, Maskos K, Betz M, Bergner A, Huber R, Suzuki K, Yoshida N, Nagase H, Brew K, Bourenkov GP, Bartunik H, Bode W (1997) Mechanism of inhibition of the human matrix metalloproteinase stromelysin-1 by TIMP-1. Nature 389: 77–81

Grant GA, Eisen AZ, Marmer BL, Roswit WT, Goldberg GI (1987) The activation of human skin fibroblast procollagenase. Sequence identification of the major conversion products. J. Biol. Chem. 262: 5886–5889

Hayakawa T (1994) Tissue inhibitors of metalloproteinases and their cell growth-promoting activity. Cell Struct. Funct.19: 109–114

He C, Wilhelm SM, Pentland AP, Marmer BL, Grant GA, Eisen AZ, Goldberg GI (1989) Tissue cooperation in a proteolytic cascade activating human interstitial collagenase. Proc. Natl. Acad. Sci. USA 86: 2632–2636

Huang W, Suzuki K, Nagase H, Arumugam S, Van Doren SR, Brew K (1996) Folding and characterization of the amino-terminal domain of human tissue inhibitor of metalloproteinases-1 (TIMP-1) expressed at high yield in *E. coli*. FEBS Letters 384: 155–161

Jäger W, Haase B, Lutz V, Herberg FW, Zimmermann B (1997) Biomolekulare Interaktionsanalyse (BIA) mit SPR-Biosensortechnologie. Biospektrum 4/1997: 82–87

Jönsson U, Malmqvist M (1992) Real time biospecific interaction analysis. The integration of surface plasmon resonance detection, general biospecific interface chemistry and microfluidics into one analytical system. Advances in Biosensors 2: 291–336

Knäuper V, Krämer S, Reinecke H, Tschesche H (1990) Characterization and activation of procollagenase from polymorphonuclear leucocytes. N-terminal sequence determination of the proenzyme and various proteolytically active forms. Eur. J. Biochem. 189: 295–300

Knäuper V, Wilhelm SM, Seperak PK, DeClerk YA, Langley KE, Osthues A, Tschesche H (1993) Direct activation of human neutrophil procollagenase by recombinant stromelysin. Biochem. J. 295: 581–586

Knäuper V, Cowell S, Smith B, López-Otin C, O'Shea M, Morris H, Zardi L, Murphy G (1997) The role of the C-terminal domain of human collagenase-3 (MMP-13) in the activation of procollagease-3, substrate specificity, and tissue inhibitor of metalloproteinase interaction. J. Biol. Chem. 272: 7608–7616

Kolkenbrock H, Orgel D, Hecker-Kia A, Noack W, Ulbrich N (1991) The complex between a tissue inhibitor of metalloproteinases (TIMP-2) and 72-kDa progelatinase is a metalloproteinase inhibitor. Eur. J. Biochem. 198: 775–781

Kolkenbrock H, Hecker-Kia A, Orgel D, Kinawi A, Ulbrich N (1996) Progelatinase B forms from human neutrophils. Complex formation of monomer/lipocalin with TIMP-1. Biol. Chem. 377: 529–533

Mallya SK, Mookhtiar KA, Gao Y, Brew K, Dioszegi M, Birkedal-Hansen H, Van Wart HE (1990) Characterization of 58-kilodalton human neutrophil collagenase: Comparison with human fibroblast collagenase. Biochemistry 29: 10628–10634

Murphy G, Cockett MI, Stephens PE, Smith BJ, Docherty AJP (1987) Stromelysin is an activator of procollagenase. Biochem. J. 248: 265–268

Murphy G, Houbrechts A, Cockett MI, Williamson RA, O'Shea M, Docherty AJ P (1991) The N-terminal domain of tissue inhibitor of metalloproteinases retains metalloproteinase inhibitory activity. Biochemistry 30: 8102–8108

Nagase, H , Enghild J J , Suzuki K , Salvesen G (1990) Stepwise activation mechanism of the precursor of matrix metalloproteinase 3 (stromelysin-1) by proteinases and (4-aminophenyl)mercuric acetate. Biochemistry 29: 5783–5789

Nagase H (1997) Activation mechanism of matrix metalloproteinases. Biol. Chem. 378: 151–160

Nguyen Q, Willenbrock F, Cocket MI, O'Shea M, Docherty AJP, Murphy G (1994) Different domain interactions are involved in the binding of tissue inhibitor of metalloproteinases to stromelysin-1 and gelatinase A. Biochemistry 33: 2089–2095

O'Shea M, Willenbrock F, Williamson RA, Crokett MI, Freedman RB, Reynolds JJ, Docherty AJ, Murphy G (1992) Site-directed mutations that alter the inhibitory activity of the tissue inhibitor of metalloproteinases-1: importance of the N-terminal region between cysteine 3 and cysteine 13. Biochemistry 31: 10146–10152

Pei D, Weiss SJ (1995) Furin-dependent intracellular activation of the human stromelysin-3 zymogen. Nature 375: 244–247

Reinemer P, Grams F, Huber R, Kleine T, Schnierer S, Pieper M, Tschesche H, Bode W (1994) Structural implications for the role of the N terminus in the "superactivation" of collagenases. A crystallographic study. FEBS Letters 338: 227–233

Sato H, Takino T, Kinoshita T, Imai K, Okada Y, Stevenson WG, Seiki M (1996a) Cell surface binding and activation of gelatinase A induced by expression of membrane-type-1-matrix metalloproteinase (MT1-MMP). FEBS Letters 385: 238–240

Sato H, Kinoshita T, Takino T, Nakayama K, Seiki M (1996b) Activation of a recombinant membrane type 1-matrix metalloproteinase (MT1-MMP) by furin and ist interaction with tissue inhibitor of metalloproteinases (TIMP)-2. FEBS Letters 393: 101–104

Schnierer S, Kleine T, Gote T, Hillemann A, Knäuper V, Tschesche H (1993) The recombinant catalytic domain of human neutrophil collagenase lacks type I collagenase activity. Biochem. Biophys. Res. Commun. 191: 319–326

Springman EB, Angleton EL, Birkedal-Hansen H, Van Wart HE (1990) Multiple modes of activation of latent human fibroblast collagenase: Evidence for the role of Cys^{73} active-site zinc complex in latency and a "cysteine switch" mechanism for activation. Proc. Natl. Acad. Sci. U.S.A. 87: 364–368

Suzuki K, Enghild JJ, Morodomi T, Salvesen G, Nagase H (1990) Mechanisms of activation of tissue procollagenase by matrix metalloproteinase 3 (stromelysin). Biochemistry 29: 10261–10270

Tschesche H, Knäuper V., Krämer S, Michaelis J, Oberhoff R, Reinke H (1992) Latent collagenase and gelatinase from human neutrophils and their activation. In: Birkedal-Hansen H, Werb Z, Welgus H, Van Wart H (eds) Matrix metalloproteinases and inhibitors. Gustav-Fischer Verlag, Stuttgart, New York, Matrix Supplement 1: 245–255

Wilhelm SM, Collier JE, Marmer BL, Eisen AZ, Grant GA, Goldberg GJ (1989) SV-40 transformed human lung fibroblasts secrete a 92-kDa type IV collagenase which is identical to that secreted by normal human macrophages. J. Biol. Chem. 264: 17213–17221

Willenbrock F, Crabbe T, Slocombe PM, Sutton CW, Docherty AJP, Cockett MI, O'Shea M, Brocklehurst K, Phillips IR, Murphy G (1993) The activity of the tissue inhibitors of metalloproteinases is regulated by C-terminal domain interactions: A kinetic analysis of the inhibition of gelatinase A. Biochemistry 32: 4330–4337

Williamson RA, Marston FAO, Angal S, Koklitis P, Panico M, Morris HR, Carne AF, Smith BJ, Harris TJR, Freedman RB (1990) Disulfide bond assignment in human tissue inhibitor of metalloproteinases (TIMP). Biochem. J. 268: 267–274

Williamson RA, Natalia D, Gee CK, Murphy G, Carr MD, Freedman RB (1996) Chemically and conformationally authentic active domain of human tissue inhibitor of metalloproteinases-2 refolded from bacterial inclusion bodies. Eur. J. Biochem. 241: 476–484

Part IV
Posttranslational Modifications

CHAPTER 22

Enzymatic and Chemical Deblocking of N-Terminally Modified Proteins

R. M. KAMP[1] and H. HIRANO[2]

1 Introduction

Proteins are complicated structures, forming three dimensional folds of the linear polypeptide chain, which is specific and unique for each of the various proteins. The specifity is fully imprinted in the amino acid sequence and firstly the primary structure has to be established. Additionally, the knowledge of amino acid sequence is the basis for identification of purified proteins, for construction of oligonucleotides and further isolation of the corresponded genes, for preparation of antibodies and for correct interpretation of structural analysis data as X-ray data and folding of the linear structure. Sequence information can be obtained by direct amino sequence analysis using pmol amounts of protein. Unfortunately 80 % of all intracellular soluble proteins from eukaryotic cells and a few proteins in prokaryotes are N-terminally blocked (Kraft 1997) and cannot be sequeced by direct Edman degradation (Aitken 1990, Tsunasawa and Sakiyama 1992). The most common modification is the blocking with acyl- and alkyl-groups. Tsunasawa and Hirano (1993) described commonly blocked amino acids in polypeptide chain (see Table 22.1). The sequence data can be obtained from the N-terminally blocked proteins only after removing these blocking groups. Thus, there is a need for a simple and rapid technique for obtaining sequence information of blocked proteins. Most techniques proposed for N-terminally blocked polypeptides involve internal cleavage of the respective proteins and subsequent sequencing of the single peptide chains. The yields of these methods are often very low and inefficient. Additionally *in vitro* modifications of proteins were observed. It is important to prevent artificial blocking generated during protein sample preparation, extraction, purification, electrophoresis or blotting (Hirano 1997). Pure reagents, preelectrophoresis and use of thioglycolic acid as free-radicals scavenger during extraction should be used to prevent *in vitro* blocking. However, if proteins are posttranslationally blocked, the special procedure using chemical and enzymatic cleavage is required to determine the N-terminal amino acid sequence (Hirano 1997). Depending on the blocked amino acid, several chemical or enzymatic methods can be used for removing of block-

[1] TFH-University of Applied Sciences (Dep. of Biotechnology), Seestrasse 64, 13347 Berlin, Germany.
[2] Yokohama City University, Kihara Institute for Biological Research, Maioka 641-12, Totsuka, Yokohama, 244 Japan.

Table 22.1. N-blocking groups in proteins (from Tsunasawa and Hirano 1993)

Blocking group	Modified amino acid
Acyl-	
Formyl	Gly, Met
Acetyl	Ser, Ala, Met, Asp, Glu, Thr, Val, Pro
Mirystoyl	Gly
α-Ketoacyl	Pyruvoyl(Ser), α-Ketobutyl (Thr)
Glucoronyl	Gly
Pyroglutamyl	Glu, Gln
Murein	Lys
Alkyl-	
Methyl(mono)	Met, Ala, Phe
Methyl(di)	Pro
Methyl(di)	Pro
Methyl(tri)	Ala
Glucosyl	Val:[1-deoxy, 1-(N^{α}-Val)-fructose]

ing groups. The reaction can be performed in solution or for proteins separated by 2D-electrophoresis on inert membranes for proteins after transfer by tank or semi-dry blotting procedures.

The characterization and deblocking of in situ N-terminally modified protein can be performed

- direct on the membrane or
- after elution from the membranes.

With very sensitive Edman sequencing or mass spectrometry, it is possible to sequence proteins after chemical or enzymatic deblocking reaction. In this review we describe several methods for deblocking of N-teminally modified proteins.

2 Enzymatic Deblocking of Proteins

2.1 A Novel Deblocking Enzymatic Method Using Deblocking Aminopeptidase for N-Acylated Proteins

Several methods have been established to remove blocked amino-terminal groups, but never was a universal procedure developed to release modified N-terminal residues.

A novel method using deblocking aminopeptidase from the archaeon *Pyrococcus furiosus* (DAP, Takara Shuzo, Japan) is a widely applicable procedure, which allows removing of different acyl groups (Tsunasawa 1998). The method is suitable for deblocking in solution or for proteins eluted from gels or PVDF membranes (Kamp et al. 1998).

DAP was applied for deblocking of several polypeptides in the molecular weight range 1700–42 000 to study the optimal conditions for specific removal of acetylated amino acid, including buffer, pH, temperature, and time of cleavage.

To determine the efficiency of deblocking and optimize the cleavage conditions, fluorescence labelled amino acids such as Ac-Met-MCA {MCA-(4-methylcoumaryl-7-amide}, Met-MCA, Pyr-MCA and peptides such as Ac-TVG, Ac-GVDEK, Ac-DEVD-MCA were used. It is very important to find the optimal balance between cleavage time and the sequence initial yields. If longer time and higher temperature (50–80 °C) were used, the deacetylation was more efficient, but high background level significantly decreased the sensitivity and made identification of amino acid impossible.

Since acetylation is one of the most common modifications in proteins, a variety of samples with N-terminally acylated amino acids were investigated, such as cytochrome c, lentinus protease inhibitor, ovalbumin, superoxide dismutase, neurotensin and α-MSH. The following acylated amino acids occur in these proteins and peptides: Ac-Ala, Ac-Gly, Ac-Ser Pyr-Glu and Ac-Met. Additionally, formyl- and myristoyl groups can be removed using the above procedure (personal communication Dr. S. Tsunasawa). We carried through similar tests with nonblocked proteins to determine the yields of deblocking reaction. Depending on blocking group and length of the polypeptide chain, the time of cleavage differed from hours to several days. The deblocking aminopeptidase removed acylated amino acids very slowly, and after cleavage of the first blocked amino acid, the speed of protein degradation rapidly increased. At present, it is impossible to remove only the N-terminal blocking group with DAP. DAP can cleave all peptide bonds sequentially from N-terminus, except the peptide bond at the N-terminal side of X-Pro sequence. The optimized conditions for digestion with DAP are:

50mM ethylmorpholine buffer pH 8.0 containing 0.1 % CoCl2, temperature 50 °C. The time of the cleavage varies from hours for small proteins to 2 days for large proteins. The optimized enzyme:substrate molar ratio increased from 1:100 for small peptides to 1:1 for large proteins. The treatment with DAP under optimized conditions results in sufficient deblocking with initial yields up to 50 %. However, the initial yields in the sequencer for some large proteins were below 5 %. Fig. 1 shows the amino acid sequence analysis after deblocking of 100 pmol of neurotensin. The conditions were: 50mM N-ethylmorpholine pH 8.0, time 24h, temperature 50 °C and substrate: enzyme ratio 1:1. Because the amount of the used enzyme is relatively high, not only the sequence of neurotensin after cleavage one amino acid before proline (Lys-Pro-Arg-Arg-Pro-Tyr..), but also the sequence of DAP was analyzed in the run (Met-Tyr-Asp-Tyr-Glu-Leu...). The first four amino acids PyrLeu-Tyr-Glu-Asn cannot be analysed, because the cleavage with DAP is very fast and only appearance of proline in the sequence can stop this rapid digestion. More advantageous is to apply blocked, acetylated-DAP, whose sequence does not appear in the chromatogram, but unfortunately such enzyme is commercially not available.

Although the cleavage can be performed in the solution or after elution of proteins from the PVDF membrane, unfortunately the direct cleavage on PVDF membrane was not successful.

We found that proteins are efficiently eluted from the PVDF membranes with 70 % acetic acid and sonication for 1/2 hour (Table 22.2, Fig. 4). The extraction efficiency is higher than that of 0.1 M ammonium bicarbonate in 40 % acetonitrile, 0.1 M ammonium bicarbonate in 40 % acetonitrile and 2 % β-mercapto-

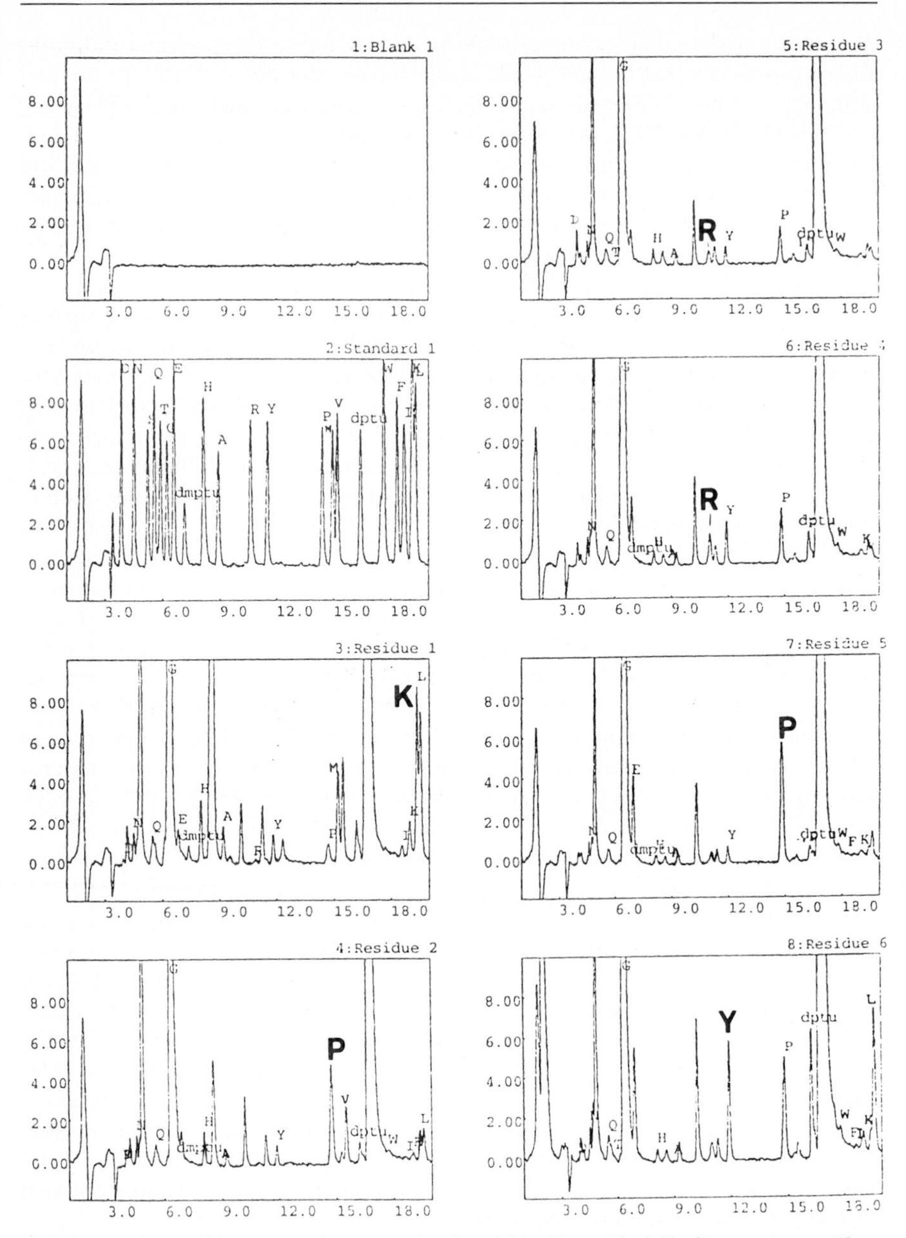

Fig. 22.1. Amino acid sequence of neurotensin after deblocking with deblocking aminopeptidase (DAP). 50 μl of 50mM ethylmorpholine buffer pH 8.0 were added to 100 pmol neurotensin in 50 μl 30 % pyridin. Enzyme: substrate ratio 1:1, cleavage time 72h, temperature 50 °C. Deblocked neurotensin was sequenced using Procise Sequencer (Applied Biosystems, USA). The sequence KPRRPY corresponds to neurotensin after deblocking with DAP and the sequence MVDYLL corresponds to intact DAP, which was used in high substrate: enzyme ratio 1 : 1. It is the reason for simultaniously identification of deblocking enzyme DAP and neurotensin during amino acid sequence analysis

Table 22.2. Extraction of α-lactalbumin from PVDF membranes with different volatile solutions

	Immobilon-P	Fluorotrans
0.1 % TFA, 40 % acetonitrile	–	–
0.1 M ammonium bicarbonate, 40 % acetonitrile	+	+
0.1 M ammonium bicarbonate, 40 % acetonitrile, 2 % β-ME	++	++
30 % acetic acid	+	+
40 % acetic acid	++	++
70 % acetic acid	+++	+++
90 % acetic acid	+++	+++
70 % acetic acid, 30 % acetonitrile	++	+++

Methods: A vial of low molecular marker proteins (Pharmacia Biotec) was dissolved in 200 μl of SDS sample buffer. 5 μl of the protein solution was applied to SDS-PAGE and the separated proteins were electroblotted onto a PVDF membrane, staining with Coomassie Blue. Three blots for α-lactalbumin were used to elute the proteins out of the wet membrane with different extraction solutions at room temperature for 1 h with sonication. The relative recovery of proteins was measured: +++, >50 %, ++, >30–50 %, +, >10–30 %, -,none.

Table 22.3. Extraction yields of proteins from different type of membranes

	Carbonic anhydrase	α-lactalbumin
α-Immobilon-P	+++	+++
Hyperbond	++	+
Immobilon-CD	+	+
Fluorotrans	+	–
ProBlott	++	+++
Transblot	++	+++
Polypropylene	+	–
Teflon tape	+++	+++
Teflon (GoreTex)	++	+++

Methods: A vial of low molecular marker proteins (Pharmacia Biotech) was dissolved in 200 μl of SDS sample buffer. 5 μl of the protein solution was applied to SDS-PAGE and the separated α-lactalbumin and carbonic anhydrase were electroblotted onto different PVDF membranes. The proteins were extracted with 70 % acetic acid at room temperature for 1 h with sonication. The relative recovery of proteins was measured: +++. >50 %, ++, >30–50 %, +, >10–30 %, -none.

ethanol or 70 % acetic acid in 30 % acetonitrile. It is recommended to use the Immobilon-P membrane or teflon membrane (GoreTex) for more efficient elution than Immobilon-PSQ, -CD, Hyperbond, Fluorotrans, Problott, Transblott and polypropylene membranes (Table 22.3, Fig. 3). Fig. 2 shows amino acid sequence of deblocked super oxide dismutase (SOD) eluted from Immobilon-P membrane using 70 % acetic acid by sonication for 1h. The condition for deblocking are the same as for neurotensin, except that for deblocking acetylated-DAP was used. For this reason only sequence of SOD after cleavage with DAP (GPVQGTI..) was analysed and not additionally the sequence of DAP, as described for deblocking of neurotensin. The interpretation of data obtained in this way is much easier.

Summarizing the results, DAP can be used for the cleavage of acetylated peptides and proteins. Although yields may still vary from 1–50 %, depending on the

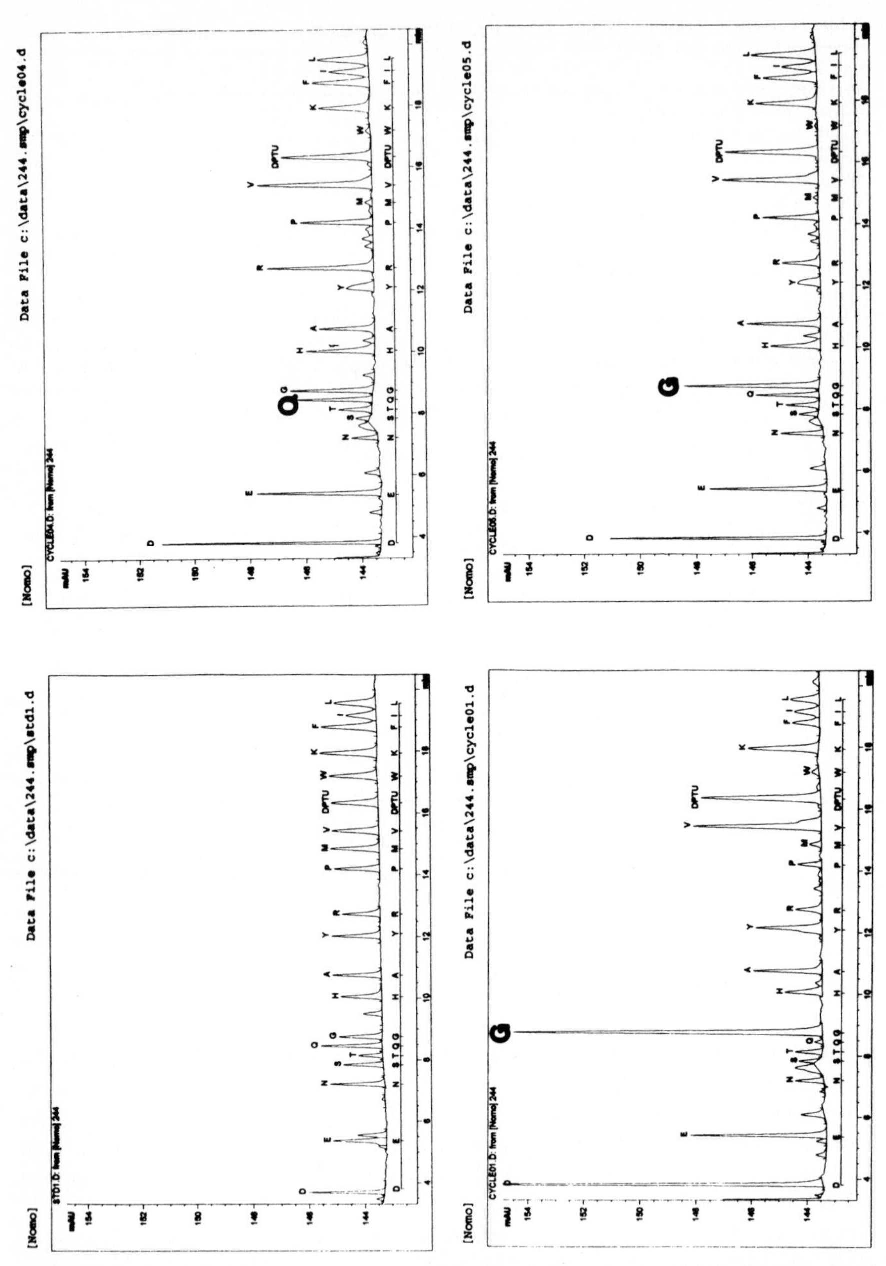

Fig. 22.2. Amino acid sequence of super oxide dismutase (SOD) after elution from the SDS gel and deblocking using acetylated deblocking aminopeptidase (DAP). 500 pmol supeoxide dismutase were separated in SDS-PAGE gel, blotted to Immobilon-P membrane and eluted using 70 % acetic acid and sonication for 1h. The cleavage was performed in 50mM ethymorpholine buffer pH 8.0 for 48h at 50 °C using acetylated-DAP and enzyme: substrate ratio 1:1. For sequencing of deblocked SOD Hewlett-Packard sequencer was used. Amino acid sequence GPVQGTT corresponds to SOD after deblocking by DAP. Because acetylated DAP was used, no DAP-sequence appears in the chromatograms

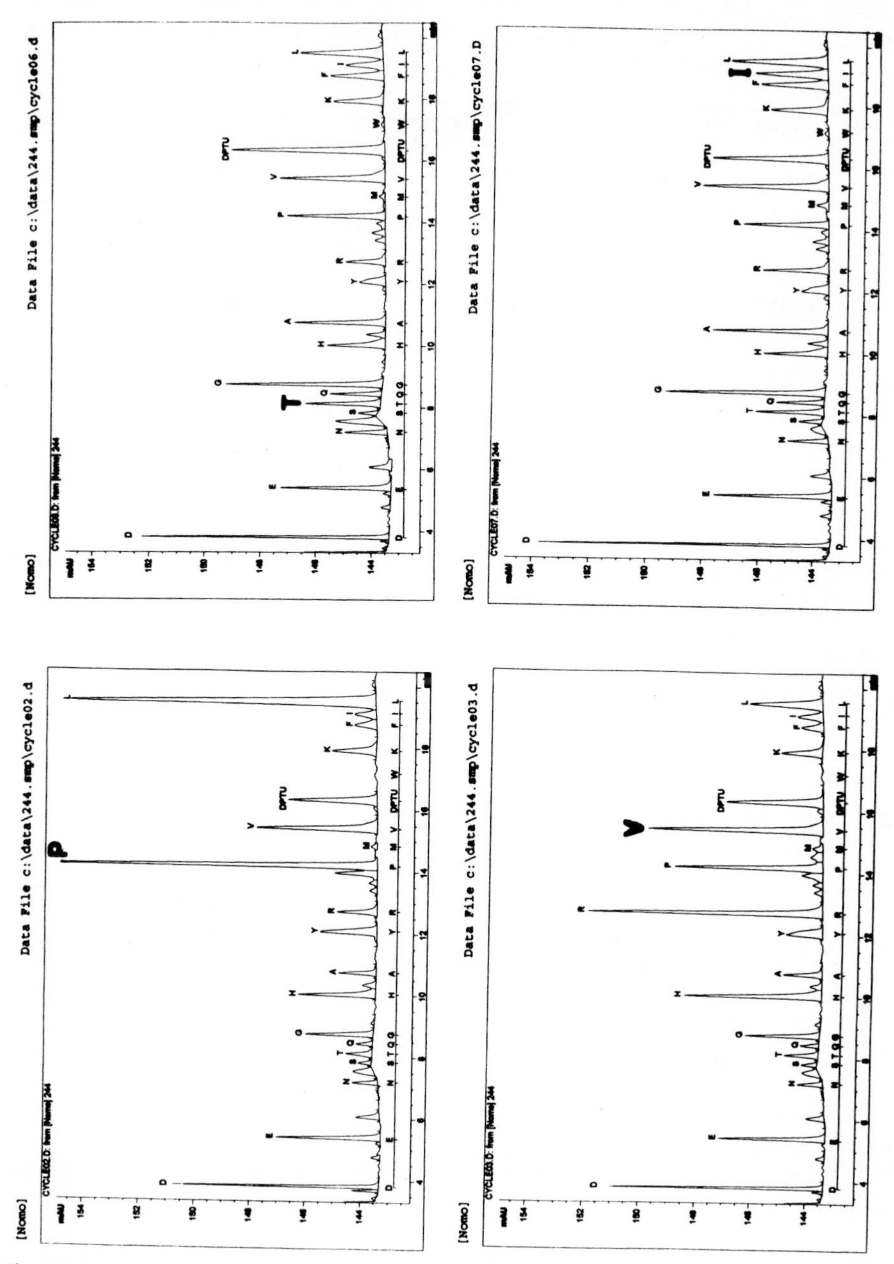
Data File c:\data\244.smp\cycle06.d
Data File c:\data\244.smp\cycle07.D
Data File c:\data\244.smp\cycle02.d
Data File c:\data\244.smp\cycle03.d
[Nomo]
[Nomo]
[Nomo]
[Nomo]

Fig. 22.2 b.

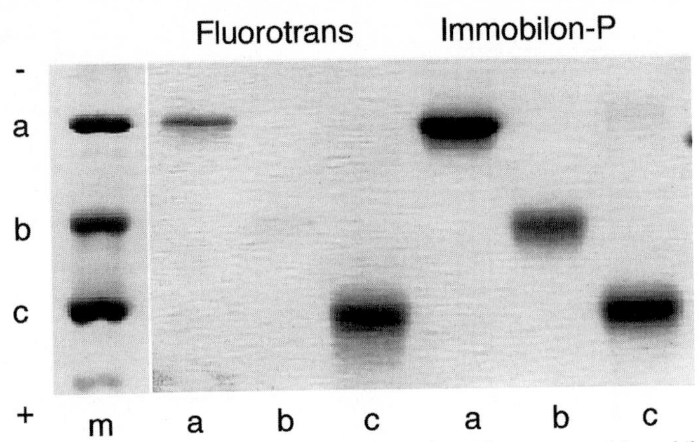

Fig. 22.3. SDS-PAGE pattern of proteins eluted from Fluorotrans and Immobilon-P membranes. Proteins: a, carbonic anhydrase; b, trypsin inhibitor; c, α-lactalbumin; See "Methods" in Table 22.1

sequence, blocking groups and molecular weight, the above mentioned method can be applied as a convenient method for determination of the internal sequence of posttranslationally acylated proteins.

2.2 Deblocking of Proteins with N-Terminal Pyroglutamic Acid

For removing of pyroglutamic acid from N-terminal end of proteins two different methods can be applied:
- the above discussed novel enzymatic method with DAP and
- enzymatic cleavage with pyroglutamyl peptidase.

Pyroglutamyl peptidase (Boehringer Mannheim) cleaved peptide bond after the N-terminal pyroglutamic acid in polypeptide chain, but not directly after blocking group. For this reason by using this method proteins can be sequenced from the second amino acid after deblocking (Hirano et al. 1991, Meyer et al 1990). For the deblocking, enzyme/substrate ratio 1:1 – 1:10 is used for 24h at 30 °C.

2.3 Deblocking of N-Acetylated Proteins

N-Terminal acetylation is catalyzed by N-acetyltransferases, which are responsible for transport of acetyl groups from acetyl-CoA to N-terminal amino acids of proteins.

A high number of amino terminally blocked proteins have been found in animals, plants, bacteria and humans. Acetylation of proteins is a co- or posttranslational process, depending on specifity involved acetyltransferases. Serine and alanine are the most frequently observed as amino-terminal residues in acetylated proteins. Futher, methionine, glycine, threonine, aspartic acid and glutamine acid were also found as acetylated N-terminal residues.

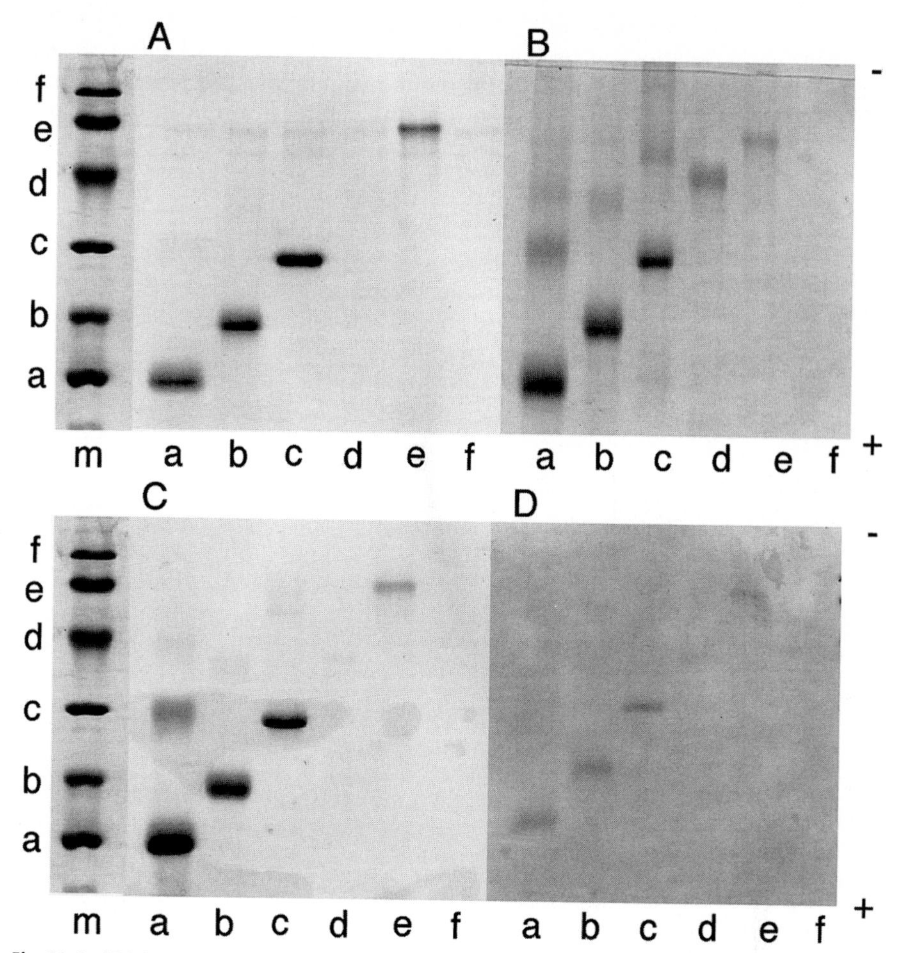

Fig. 22.4. SDS-PAGE pattern of the proteins eluted from PVDF membranes with different extraction solutions. Extraction solution: **A**, 0.1M ammonium bicarbonate, 40 % acetonitrile; **B**, 0.1 M ammonium bicarbonate, 40 % acetonitrile, 2 % β-mercaptoethanol; **C**, 70 % acetic acid; **D**, 70 % acetic acid, 30 % acetonitrile; Proteins: a, α-lactalbumin; b, trypsin inhibitor; c, carbonic anhydrase; d, ovalbumin; e, albumin; f, phosphorylase; m, molecular weight marker proteins. See "Methods" in Table 22.3

2.3.1 Treatment with acylamino acid releasing enzyme

The N-terminally blocked amino acids can be removed after digestion in solution with acylamino acid releasing-enzyme (AARE) from Takara Shuzo, Japan (Tsunasawa et al 1990). This technique is applicable for deblocking of proteins dissolved in buffer solutions or electroblotted onto a PVDF membrane (Hirano et al. 1991).The mechanism of such reaction is shown in Fig. 22.5.

The treatment with AARE is applicable without any restrictions for proteins with low molecular weights. AARE does not directly digest proteins with high molecular mass (Tsunasawa et al. 1990). In this case, prior to AARE digestion, proteins in the solutions or on the membrane should be digested with proteases such as trypsin. When the treatment is performed, non blocked peptides and N-acetylated peptides are detected. The resulting peptides can be easily removed

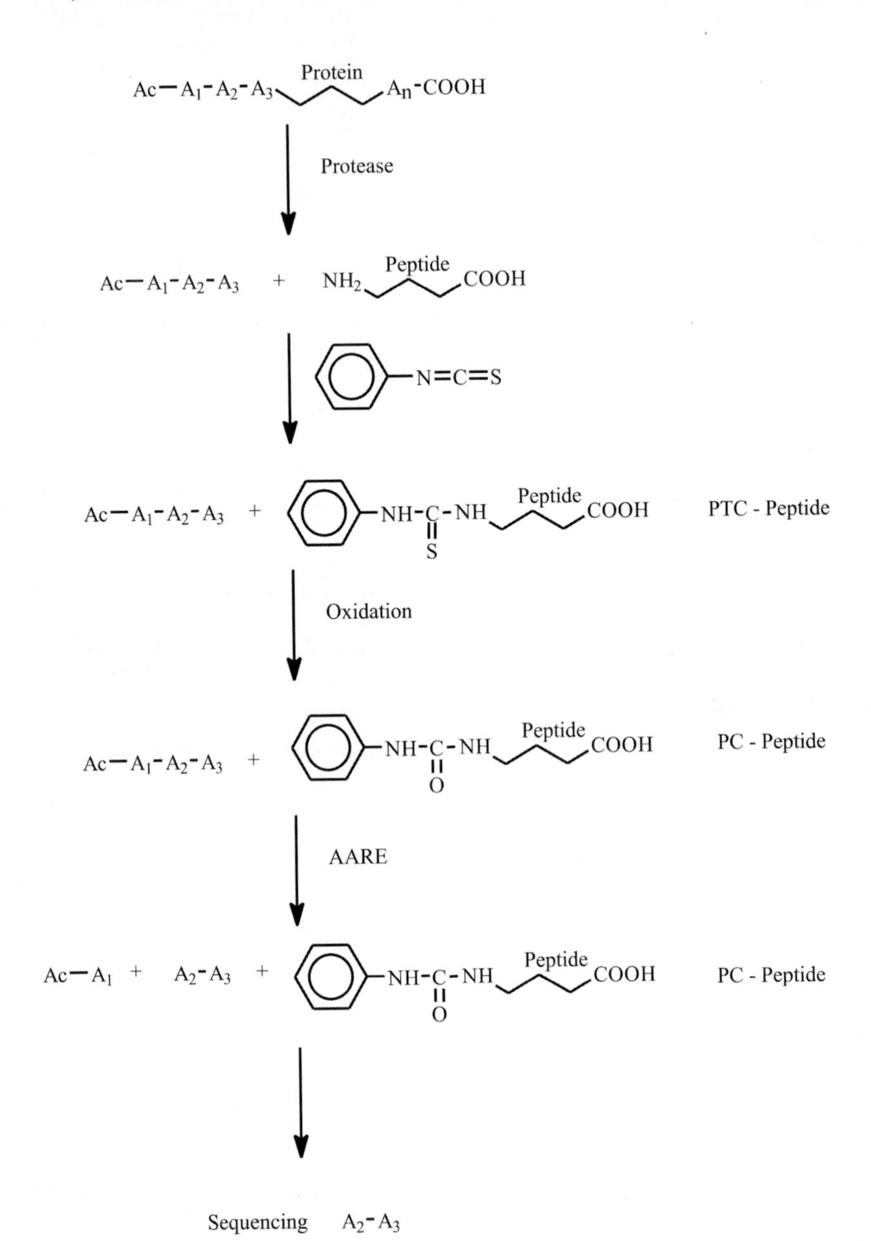

Fig. 22.5. Deblocking of N-acetylated proteins with acylamino acid releasing enzyme (AARE). (From Tsunasawa et al. 1990)

from the membrane. For better elution of peptides and to reduce hydrophobic interaction between peptide and PVDF membrane, 10 % acetonitrile is added in the digestion buffer. Non blocked peptides are reacted with PITC in the next step and form phenylthiocarbamyl peptides, which are oxidized to phenylcarbamyl peptides and are not detected during subsequent Edman degradation. The acetylated amino acid of the N-terminal peptide is selectively removed by digestion with AARE.

The above described method works very well for peptides smaller than 10 residues. The extraction from the PVDF membrane for larger peptides is difficult. To overcome this problem, second digestion with another protease may be performed using the same membrane.

Instead of reaction with PITC, the non blocked peptides after protease digestion can be succinylated and the N-terminal blocked peptide is then deblocked with AARE (Krishna et al. 1991).

After releasing of N-terminaly blocked amino acid by AARE, the blocked residue can be identified by mass spectrometry.

The efficiency of deblocking and sequencing depends primarily on tryptic digestion and elution yields from the PVDF membrane.

2.4 Deblocking of Myristoyl Group

Mirystoyl group of the blocked proteins can be removed when peptide N-fatty acylase (WAKO Pure Chemicals) is used. The procedure can be use as described for AARE.

Also treatment with deblocking aminopeptidase (DAP) can be applied for deblocking of mirystoyl groups. Optimized conditions see under DAP (see chapter 2.1).

3 Chemical Deblocking Methods

3.1 Deblocking of Proteins with N-terminal Acetylserine and Acetylthreonine

Proteins with acetylserine and acetylthreonine can be deblocked by trifluorocetic acid (TFA) treatment (Wellner et al. 1990). Comparison with all N-blocked proteins shows, that 36 % of known acetylated amino acids are acetylserine and 4 % acetylthreonine. The yields of chemical deblocking reaction with TFA vary depending on proteins by up to 40 %. The protein sample is first applied to the glass fiber filter, dried and saturated with anhydrous TFA. The deblocking reaction is performed for 2–3 days. The possible mechanism proposed by Wellner 1990 of the deblocking reaction is shown in Fig. 22.6.

The advantage of this method is that deblocking is simple and rapid, only the sequencing initial yields of proteins treated with TFA are rather low, about 7 % (Wellner et al. 1990).

The deblocking reaction can be performed in solution or on the membrane (Hirano 1997). When the membrane e.g PVDF membrane is exposed to TFA

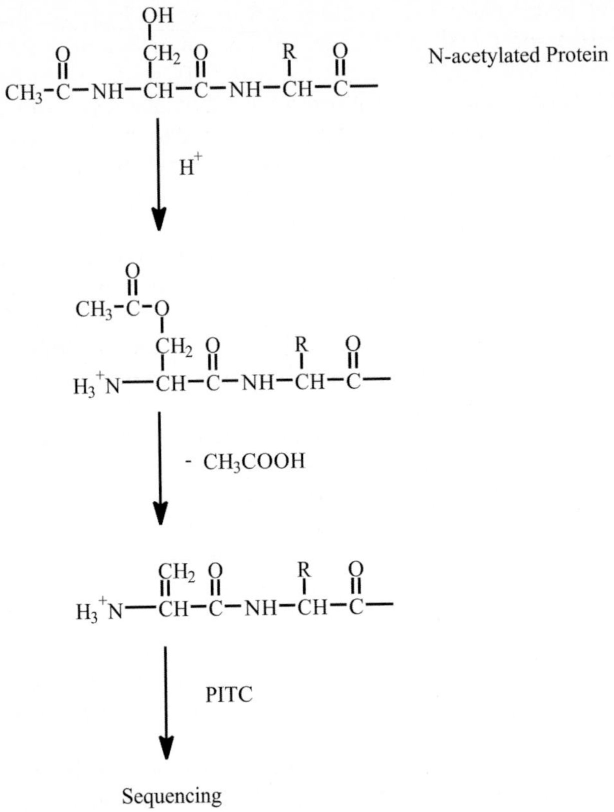

OH
O CH2 O R O
CH_3–C–NH–CH–C–NH–CH–C– N-acetylated Protein

H^+

O
CH_3–C–O
CH_2 O R O
$H_3{}^+N$–CH–C–NH–CH–C–

- CH_3COOH

CH_2 O R O
$H_3{}^+N$–CH–C–NH–CH–C–

PITC

Sequencing

Fig. 22.6. Mechanism of deblocking reaction in N-acetylated proteins by trifluoroacetic acid (Wellner et al. 1990)

vapor, the time of cleavage is decisive for sequencing yields. The yields of PTH increase, when the reaction is longer than 2h, but too long reaction causes higher formation of reaction by-products which hinder identification of PTH amino acids. Reaction by-products may be generated primarily through unspecific cleavage of polypeptide chain (Hulmes and Pan 1991).

Although the mechanism of the deblocking reaction is not known, Wellner suggested that N-O acyl shift may be involved (see Fig. 22.6).

Takakura et. al. 1992 described other treatment with TFA for deblocking of acetylated proteins. Proteins separated by 2D-electrophoresis and blotted onto the PVDF membrane were exposed to vapour TFA for 16h at 65 °C. After deblocking reaction the PVDF membrane can be used directly for automated sequence analysis (Takakura et al. 1992).

3.2 Alcoholytic Deacetylation

N-terminally blocking of native proteins and peptides through acetylation can be removed using incubation with trifluoroacetic acid in methanol (v/v 1:1) for 2–3 days at 47 °C (Bergman el al. 1996, Gheorghe MT et al. 1997). This alcoholytic deacetylation allows direct application of deblocked proteins and peptides to Edman degradation with initial yields of up to 50 %.

The lower temperature gives a clear background, but also much reduced yields. The cleavage time should be prolonged to several days or a week. The higher temperature results in a better recovery, but the unspecific cleavage of internal peptide bonds increases and hinders the sequence data interpretation. It is difficult to balance between deacetylation and general alcoholysis. The deacetylation for 2 days results in sufficient cleavage and in low level of internal peptide bond cleavage. It is possible to use ethanol and isopropanol instead of methanol for deacetylation, but the initial yields are about 20–30 % lower in comparison to 60 % with methanol. For this reason short chain alcohols are optimal for more rapid reactions. The optimal ratio TFA to methanol is 1:1. Higher concentrations of TFA causes non-specific cleavage. The yields of deacetylation are sequence dependent, they are highest with acetyl-Ser, and very low with acetyl-Ala. Acetyl-Ser is the most common modification in native proteins and therefore most important for deacetylation.

3.3 Deblocking of N-Formylated Proteins

The N-formyl group of proteins can be removed , when the protein is incubated in a dilute 0.6 M HCl at 25 °C for 24 hours. The deblocking reaction can be performed in solution or on PVDF membrane after blotting (Ikeuchi and Inoue 1988).

Milder treatment with anhydrous hydrazine vapor at –5 °C for 8h may be useful to deblock the N-formylated amino acids (Miyatake et.al 1992).

3.4 Deblocking of Proteins with N-Terminal Pyroglutamic Acid

For proteins blotted onto PVDF membranes, 8h treatment with anhydrous hydrazine at 20 °C is required to remove the N-terminal pyroglutamic acid (Miyatake et al. 1997). During the hydrozynolysis, the pyroglutamic acid converts to Glu-γ-hydrazine, which can be detected using mass spectrometry or Edman-degradation. As undesired by-products, ornithin, appears after conversion of arginine, asparagine and glutamine are detected as their hydrazide.

4
Sequential Deblocking of Proteins

The above mentioned deblocking enzymatic and chemical methods may be used in combination to allow step-by-step deblocking for sequencing of proteins with unknown blocking groups, which are immobilized on a PVDF membrane (Hirano 1997). The protein sample after separation on polyacrylamide gel is transferred onto a PVDF membrane by electroblotting and subjected to gas-phase sequencing.

If sequencing does not show results, the membrane is removed from the sequencer and treated with TFA vapor to remove the acetyl groups of acetylserine and acetylthreonine. After this deblocking reaction, the PVDF membrane is applied to the sequencer. If sequencing fails, the membrane is incubated in diluted HCl to remove formyl groups and then subjected to protein sequencer. If the protein is N-blocked with formyl-group, the proteins can be sequenced. When not, the sample is applied for pyroglutamyl peptidase digestion to remove N-terminal pyroglutamic acid. If sequencing after above treatment was impossi-

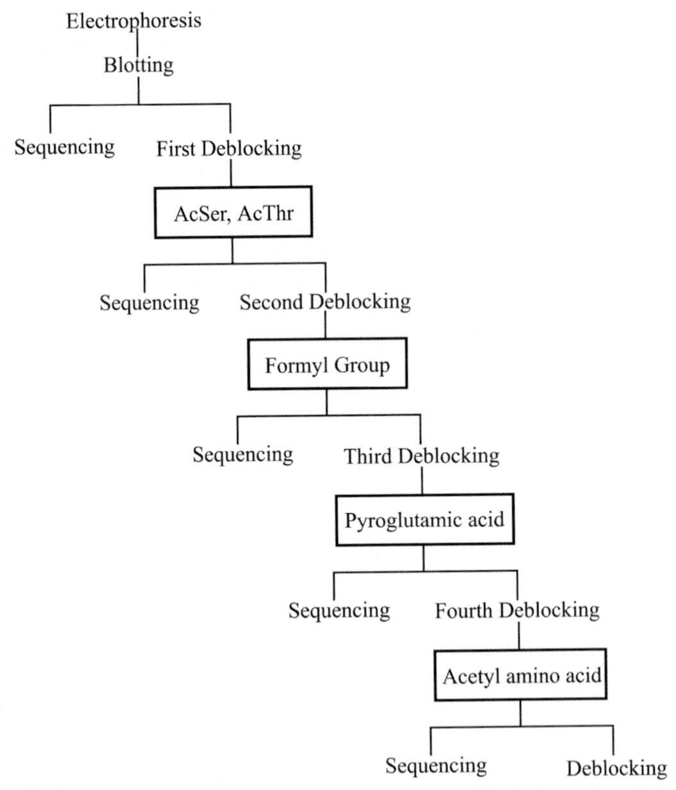

Fig. 22.7. Step-by-step deblocking for N-terminally blocked proteins electroblotted onto a PVDF membrane by combined chemical and enzymatic methods

ble, deblocking with AARE is performed to remove acetylamino acids. If sequencing still fails, other methods should be used or developed. Fig. 22.7 shows the strategy of sequencial deblocking for N-terminally blocked proteins electroblotted onto a PVDF-membrane.

5 Sample Preparation of Minute Amounts of Proteins

If the N-terminally blocked protein is available only in picomole amounts, the biotin-avidin method may be used to remove selectively the blocked N-terminal peptide (Mitsunaga et al. 1993).

After digestion of N-terminally blocked proteins with lysylendopeptidase, the resulted peptides were digested in the second step with carboxypeptidase B to remove the C-terminal lysine from each peptide. In the next reaction, the peptides were bound to phenylendiisothiocyanate (DITC) glass to eliminate the peptides and lysines except for the N-terminal blocked peptides.

The unabsorbed peptides by DITC glass treatment were reacted with the N-hydrosuccinimide-biotin to biotinylate the α- and ε-amino groups of remained peptides and lysines. The biotinylated peptides and lysine residues were immobilized on an avidin-agarose column and the N-terminally blocked peptides were obtained as an unabsorbed fraction, which can be futher applied to Edman degradation or mass spectrometry. Fig. 22.8 shows the schemé for above described isolation of the N-terminal peptide from N-terminally blocked proteins.

6 Conclusion

Summarizing the methods for deblocking of N-terminally modified proteins, several enzymes or chemicals can be used for the cleavage of different blocking groups from N-terminal end. The above mentioned methods can be used for deblocking in solution or on the inert membrane, which may be used directly for subsequent amino acid sequence analysis. Although yields may still vary, depending on the sequence, blocking groups, and molecular weight, the above methods can be applied as a convenient way of deblocking posttranslationally modified N-terminal amino acids. It is important to choose a suitable method for the deblocking dependent of the blocking group, because no generally applicable method for deblocking has yet been developed.

7 Acknowledgment

We would like to thank Dr. Susumi Tsunasawa Takara Shuzo, Japan for enthusiastic meetings, valuable discussion and for sequencing of SOD.

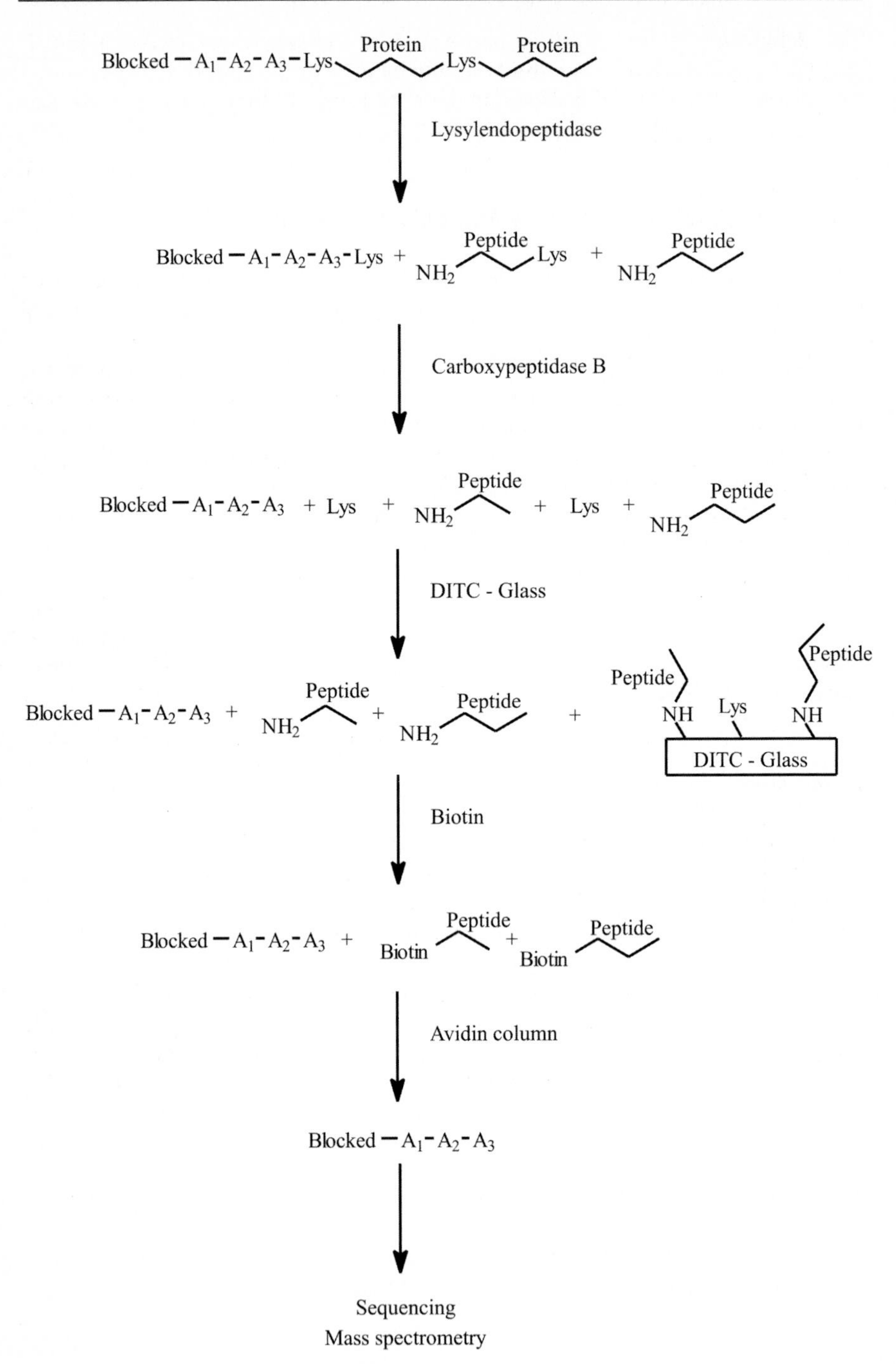

Fig. 22.8. Isolation of the N-terminal peptide from N-terminally blocked proteins (from Mitsunaga et al. 1993)

References

Aitkin A (1990) Identification of protein consensus sequences, Ellis Horwood, Chichester, West Sussex

Gheorghe MT, Jörnvall H, Bergman T (1997) Optimized alcoholytic deacetylation of N-acetyl-blocked polypeptides for subsequent Edman degradation. Anal Biochem 254: 119–125

Hirano H, Komatsu S, Nakamura A, Kikuchi F, Kajiwara H, Tsunasawa S, Sakiyama F (1991) Structural homology between semidwarfism-related proteins and glutelin seed protein in rice (Oryza sativa L.,). Theor Appl Genet 83: 153–158

Hirano H (1997) Sequence analysis of the NH2-terminally blocked proteins immobilized on PVDF membranes from polyacrylamide gels In: Kamp RM, Papadopoulou T, Wittmann-Liebold B (eds) Protein Structure Analysis. Springer Verlag Berlin Heidelberg, pp 167–179

Hulmes JD, Pan Y-C (1991) Selective cleavage of polypeptides with trifluoroacetic acid: Applications for microsequencing. Anal Biochem 197: 368–376

Ikeuchi M, Inoue Y (1988) A new photosystem II reaction center component (4,8 kDa protein) encoded by chloroplast genome. FEBS Lett 241: 99–104

Kamp RM, Tsunasawa S, Hirano H (1998) Application of new deblocking aminopeptidase from *Pyrococcus furiosus* for microsequence analysis of blocked proteins. J.Prot. Chem. 17: 512–513

Kraft R (1997) Enzymatic and chemical cleavages of proteins. In: Kamp RM, Papadopoulou T, Wittmann-Liebold B (eds) Protein Structure Analysis. Springer Verlag Berlin Heidelberg, pp63–71

Krishna RG, Christopher CQ, Wold F (1991) N-terminal sequence analysis of N-acetylated proteins after unblocking with N-acylaminacyl peptide hydrolase. Anal Biochem 199: 45–50

Mitsunaga K, Miyagi T, Ishimizu T, Kato I, Tsunasawa S (1993) A novel method for specific isolation of the N-terminally blocked peptide from proteins. Proc. Jap. Prot. Symp.

Miyatake N, Kamo M, Satake K, Uchiyama Y, Tsugita A (1992) Removal of N-terminal formyl groups and deblocking of pyrrolidone carboxylic acid of proteins with anhydrous hydrazine vapor. Eur J Biochem 212: 785–789

Takakura H, Tsunasawa S, Miyagi M, Warner J (1992) NH2-terminal acetylation of ribosomal protein of *Saccharomyces cerevisiae.* J Biol Chem 267: 5442–5445

Tsunasawa S (1998) Purification and application of a novel N-terminal deblocking aminopeptidase (DAP) from *Pyrococcus Furiosus.* J Protein Chem 17: 521–5 22

Tsunasawa S, Hirano H (1993) Deblocking and subsequent microsequence analysis of N-terminally blocked proteins immoblized on PVDF membrane, In: Imahori K, Sakiyama F (eds) Methods in protein sequence analysis. Plenum, New York, pp 43–51

Tsunasawa S, Sakiyama F (1992) Amino-terminal acetylation of proteins. In: Tuboi S, Taniguchi N, Katsumara N (eds) The posttranslational modification of proteins. Japan Scientific Societies, Tokyo, pp 113–121

Tsunasawa S, Takakura H, Sakiyama F (1990) Microsequence analysis of N-acetylated proteins. J Protein Chem : 265–266

Wellner D, Panneerselvam C, Horecker BL (1990) Sequencing of peptides and proteins with blocked N-terminal amino acids: N-acetylserine or N-acetylthreonine. Proc Natl Acad Sci USA 87: 1947–1949

CHAPTER 23

Exploring Functions for Glycosylation in Host Defence using Novel Oligosaccharide Sequencing Technology

P. M. Rudd[1], M. R. Wormald[1] and R. A. Dwek[1]

1 Abstract

A full understanding of the implications of glycosylation for the structure and function of any glycoprotein can only be reached when the molecule is viewed in its entirety. NMR solution and X-ray crystallography studies of glycoproteins do not normally yield detailed information about the sugars. Here protein structural data have been complemented by oligosaccharide analysis of the sugars released from 5–10μg of protein. The data discussed in this paper were obtained using rapid glycan sequencing technology and the linkage structure data base (both developed in the Institute), which provides the dimensions of the sugars. In this way it has been possible to obtain a more complete view of some glycoproteins in the immune system and the roles which the oligosaccharides play in their functions. Roles for the sugars include stabilising the protein structure, modifying the activity of effector functions, orienting the protein on the cell surface, shielding the protein from proteases and providing specific epitopes for recognition events. The glycoproteins which will be discussed include the immunoglobulins IgG and IgA1, the inhibitors of the complement pathway CD59 and DAF (CD55), and the cell adhesion molecules CD2 and CD48. CD2 and CD48 mediate the alignment of the cell surfaces of cytolytic T-lymphocytes carrying the TCR complex with those of target cells carrying loaded HLA class 1 molecules.

2 Introduction

Mammals have developed a variety of strategies with which to respond to the challenges posed by pathogenic viruses or bacteria. These organisms can replicate in the blood or mucosa or have evolved mechanisms for invading and replicating in host cells. A key event in every immune response to infection involves functional recognition of foreign proteins or lipids. Many proteins involved in the immune response are glycosylated, and the awareness of the importance of the attached sugars in the function of these molecules is increasing. By combining

[1] Glycobiology Institute, Department of Biochemistry, University of Oxford, South Parks Road, Oxford, OX1 3QU, UK.

glycan sequencing data and oligosaccharide structural information from the Glycobiology Institute data base with protein structural data many glycoproteins can now be modelled in their entirety. This means that it is now possible to visualise the sugars in the context of the glycoproteins to which they are attached.

Pivotal to this approach is rapid, sensitive state-of-the-art oligosaccharide sequencing technology. Strategies have been developed which enable the N-glycosylation of a few micrograms of glycoproteins of major biological interest purified by SDS PAGE gel (Kuster et al., 1997) to be analysed in a few days using MS or HPLC (Rudd et al., 1997a). The technology is now being extended to analyse O-glycans. These advances open up the possibility of ranking high through-put oligosaccharide sequencing alongside that of proteins and DNA. In this paper we demonstrate how these strategies have been used to determine the structures of the oligosaccharides attached to a range of glycoproteins which operate in the humoral or cellular immune systems. Some of the roles which have been proposed for these sugars include assisting in protein folding and loading of antigenic peptide, providing protease protection, stabilising protein structure and modulating the organisation and presentation of cell surface glycoproteins. In autoimmune disease, such as rheumatoid arthritis, altered glycosylation of IgG may result in the multiple presentation of terminal sugars which can bind to the mannose binding lectin (MBL). This may contribute to the inflammatory processes through the activation of complement by MBL.

3 Oligosaccharide Sequencing Technology

The steps involved in oligosaccharide analysis include releasing the sugars from the protein, and labelling, separating and analysing the components of the glycan pool. A straightforward strategy is shown in Fig. 23.1. N-linked sugars can be released directly from a Coomassie blue stained gel band and both N- and O-glycans by hydrazinolysis. The sugars may be analysed directly by MALDI MS or the non-reducing termini of the glycans may be labelled with 2-aminobenzamide,

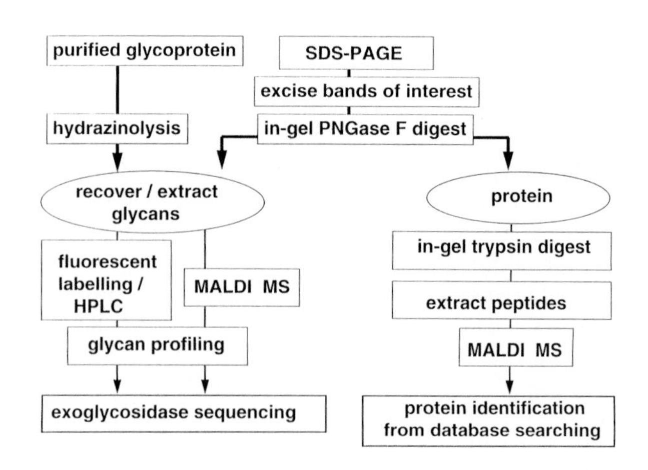

Fig. 23.1. Strategies for glycan analysis and protein identification

which allows direct quantitation from HPLC profiles. In the HPLC technology a column containing amide-silica resolves both charged and neutral N- and O-glycans in a single run. The glycans are separated on the basis of differences in hydrophilicity, which reflects arm specificity, linkage and monosaccharide composition in a single run. The system uses a volatile buffer system so that individual glycans can be recovered free from buffer salts for analysis by mass spectrometry. The elution positions of the glycans, determined in glucose units with refer-

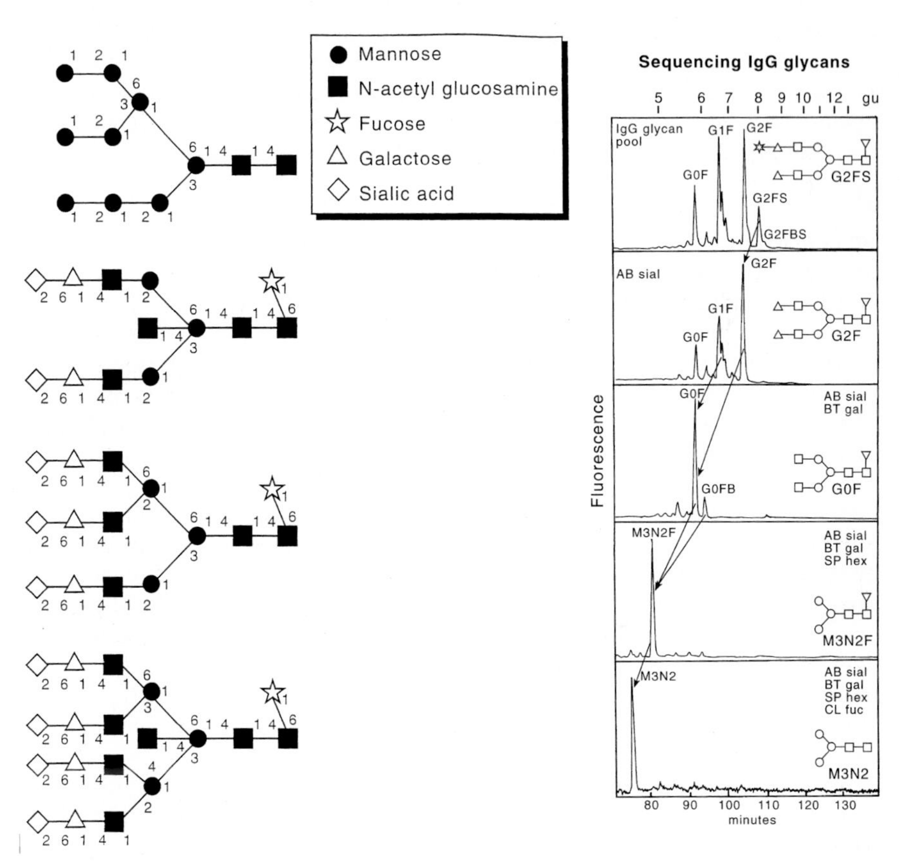

Fig. 23.2. a Schematic of a representative oligomannose sugar and of a bi-, tri and tetra-antennary complex oligosaccharide. In human serum IgG the most fully processed oligosaccharide is the disialylated, biantennary complex sugar shown here. G2 type sugars are those in which both arms contain galactose, in the G1 type structures either the α1,3 or the α1,6 arm terminates in galactose while those in which both arms lack galactose and terminate in N-acetylglucosamine are known as G0 type structures. **b** Normal phase HPLC analysis of glycans from normal human serum IgG. Peaks were assigned glucose unit values by comparison with a standard dextran ladder, the elution positions of which are shown at the top of the figure. The assignment of structures was made using a combination of mass spectrometry, co-injection with standard sugars and exoglycosidase digestions. Peaks were assigned gu values using the standard dextran ladder shown above the profiles. Aliquots of the total glycan pool were incubated with a series of enzyme arrays shown alongside the profiles. The arrays contained *Arthrobacter ureafaciens* sialidase (AB sial), bovine testes β-galactosidase (BT gal), *Streptococcus pneumonia* β-hexosaminidase (SPH), *Charonia lampas* a-fucosidase (CL fuc)

ence to a dextran ladder, are used to predict structure based on experimentally determined incremental values for the addition of all monosaccharide residues to basic core oligosaccharides. Thus preliminary structural assignments may be made from the profile of a single HPLC run (Fig. 23.2).

The preliminary assignments are confirmed by exoglycosidase digestions. Arrays of highly specific enzymes are used to digest aliquots of the whole pool of sugars simultaneously. The products of such digestions are identified using the same HPLC system, based on the incremental value for each type of monosaccharide residue and the specificity of the enzymes in the arrays. Typically, aliquots of the entire glycan pool are sequenced simultaneously using a standard panel of five enzyme arrays, after which the products are analysed using the same HPLC system (Fig. 23.2). Any less common or novel structures, such as sulphated sugars, are highlighted since these will not be digested to a common tri-mannosyl N-linked glycan core by any of the five standard enzyme arrays.

4 Glycoproteins exist as Populations of Glycoforms and the 3D Structure of the Individual Protein is a Factor which Influences its own Glycosylation Pattern

Glycoproteins generally exist as populations of glycosylated variants (glycoforms) of a single polypeptide (Review: Rudd and Dwek 1997). Although the same glycosylation machinery is available to all proteins which enter the secretory pathway in a given cell, most glycoproteins emerge with a characteristic glycosylation pattern and heterogeneous populations of glycans at each glycosylation site. The local protein structure plays a key role in glycan processing (Rudd et al., in press). In the immunoglobulin family for example, N-linked glycans are acquired by the transfer of the oligosaccharide precursor, $Glc_3Man_9GlcNAc_2$, to some Asn residues in the H and L chains. This event takes place co-translationally in the endoplasmic reticulum (ER) before the protein is fully folded. The glucose residues and some mannose residues are subsequently trimmed in the ER, where immunoglobulin monomers are assembled from H and L chains (Bole et al., 1986). In some cases, such as IgA and IgM, monomers assemble further in the ER forming dimers and pentamers, respectively.

The assembled immunoglobulins are transferred to the Golgi apparatus where the N-linked oligomannose sugars are processed to complex glycans and O-linked glycans are attached to some accessible Ser or Thr residues, beginning with the addition of GalNAc. The processing of N-linked sugars and the attachment and processing of O-glycans therefore depends both on the local 3D structure of the immunoglobulin fold around the glycosylation sites and on the structure of the fully assembled multimers.

5 The Structure and Function of the Fc Oligosaccharides in IgA1 and IgG is Influenced by the Location of the Glycans

(i) The sugars attached to IgA1 CH2 domains are larger than those on IgG and, in contrast to those on IgG, are not contained within the CH2 domains
Crystallographic studies of human IgG Fc (Deisenhofer, 1981) have shown that the two C_H2 domains do not form extensive lateral associations. The resulting interstitial region accommodates the oligosaccharides, which are attached to Asn297 on each heavy chain (Fig. 23.3a). In contrast, in IgA1 (Fig. 23.3b), the interstitial space is not large enough to accommodate the sugars. Moreover, molecular modelling (Mattu, et al., 1998) indicates that the Asn298 side chains point away from the protein surface such that the N-glycans are fully exposed on the outside of the molecule. As a consequence, IgA1 sugars can be larger than those on IgG (Fig 23.4) (Field et al., 1994, Mattu et al., 1998).

(ii) IgG Fc Oligosaccharides Maintain the Relative Geometry of the CH2 Domains.
In contrast to IgA1, in IgG Fc the oligosaccharides maintain the relative geometry of the CH2 domains. The oligosaccharides in the Fc region of IgG are of the complex bi-antennary type and can be classified according to the number of terminal galactose residues they contain. Each sugar has either two terminal galactose residues (G2), one galactose and one GlcNAc (G1) or two terminal GlcNAc residues (G0). The major determinant for the binding site for the Fc receptor is Leu235 and the receptor may interact directly with the region linking the CH2 domain to the hinge (Duncan et al., 1988; Burton and Woof, 1992). Both non-glycosylated and degalactosylated IgG bind less efficiently to the Fcγ receptors (Nose and Wigzell 1983, Leatherbarrow et al., 1984, Tsuchiya et al., 1989). This is consistent with the finding that protein-oligosaccharide interactions involving the conserved glycans at Asn297 (Padlan, 1994) play a role in maintaining the geometry of the hinge region and therefore of the CH2 domains in IgG Fc (Rudd et al., 1991). In contrast to IgG, in IgA1 no reduction in binding to its Fc receptor was detected in CHO-K1 mutants in which the N-linked sugar sites in the CH2 domains had been deleted (Mattu et al., 1998). This result is not unexpected since the sugars in the CH2 domains do not play a role in maintaining the hinge structure in IgA1.

(iii) Sugars Protect IgG and IgA1 Against Proteolysis
Although the N-glycans in the CH2 domains of IgA1 do not play a role in maintaining domain structure, just as the multiple O-glycans in the hinge have been shown to protect the hinge region of IgA1 against non-specific proteolysis (Mestecky and Kilian, 1985) so the N-glycans are expected to shield large areas of the Fc region from proteases.

In the case of IgG, which can be cleaved by papain into Fab (the antigen binding fragment) and Fc (the region responsible for effector functions of the molecule, such as Fcγ receptor binding and complement activation), non-glycosylated IgG, expressed in tunicamycin treated cells, is more readily cleaved by papain than normal IgG (Leatherbarrow and Dwek 1984). This indicates that the conserved glycans in the Fc provide protection against proteases.

(iv) IgG Glycans may provide an Additional Route to Inflammation in RA by Activating the Classical Complement Pathway Through Binding to MBL

In general, the requirement for multiple presentation ensures that single oligosaccharides attached to 'self' proteins do not trigger biological activities which are defence mechanisms normally initiated only by 'non-self' pathogens. An exception to this may provide an additional route to inflammation in rheumatoid arthritis (RA). In RA the galactosylation status of IgG is a prognostic indicator, and increased levels of sugars lacking terminal galactose (G0 type) (Fig. 23.6) (Parekh et al 1985) compared with age matched normal controls correlate with active and severe disease. At the molecular level, the absence of the terminal galactose residue in G0 type sugars results in decreased protein-oligosaccharide interactions in the Fc and allows displacement of the sugars out of the space between the CH_2 domains. As a consequence, the exposed terminal GlcNAc residues in IgG0 are in a position where they can be recognised, *in vitro*, by endogenous lectins such as the MBL and activate complement (Malhotra et al., 1995). This may provide a route by which inflammation can develop over and above the activation of the classical pathway by IgG antibody-antigen complexes or by IgG

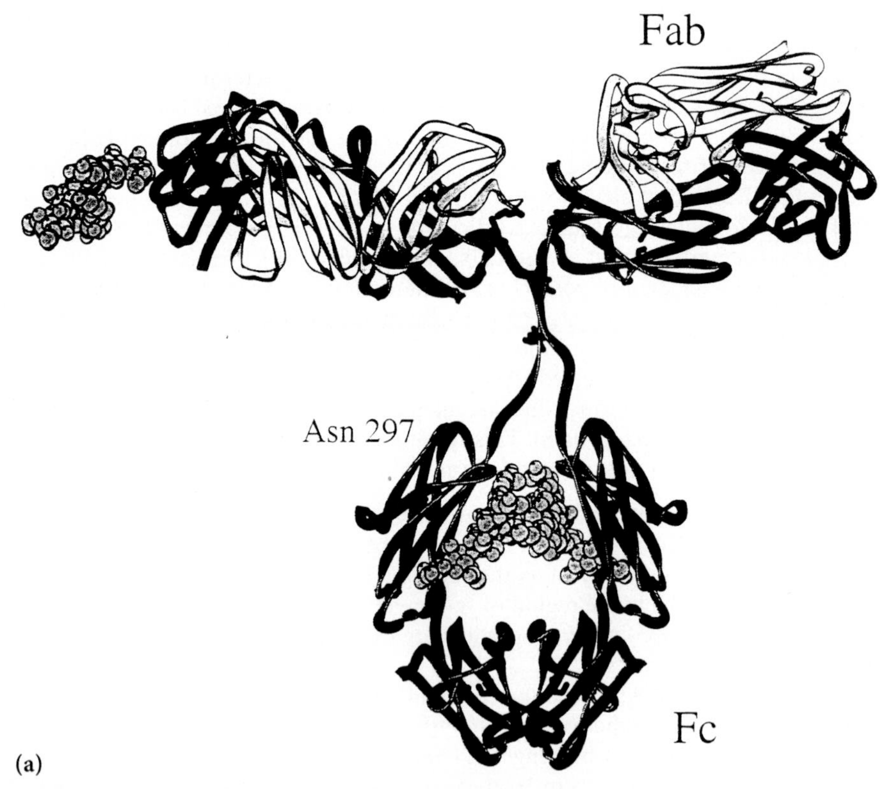

Fig. 23.3. Molecular models of (**a**) an IgG1 and (**b**) an IgA1 glycoform (based on Mattu *et al.*, 1998 and Deisenhofer 1981 respectively). In IgA1 the sugars are fully exposed on the surface of the Fc. In contrast, in IgG the sugars are contained within the CH2 domains

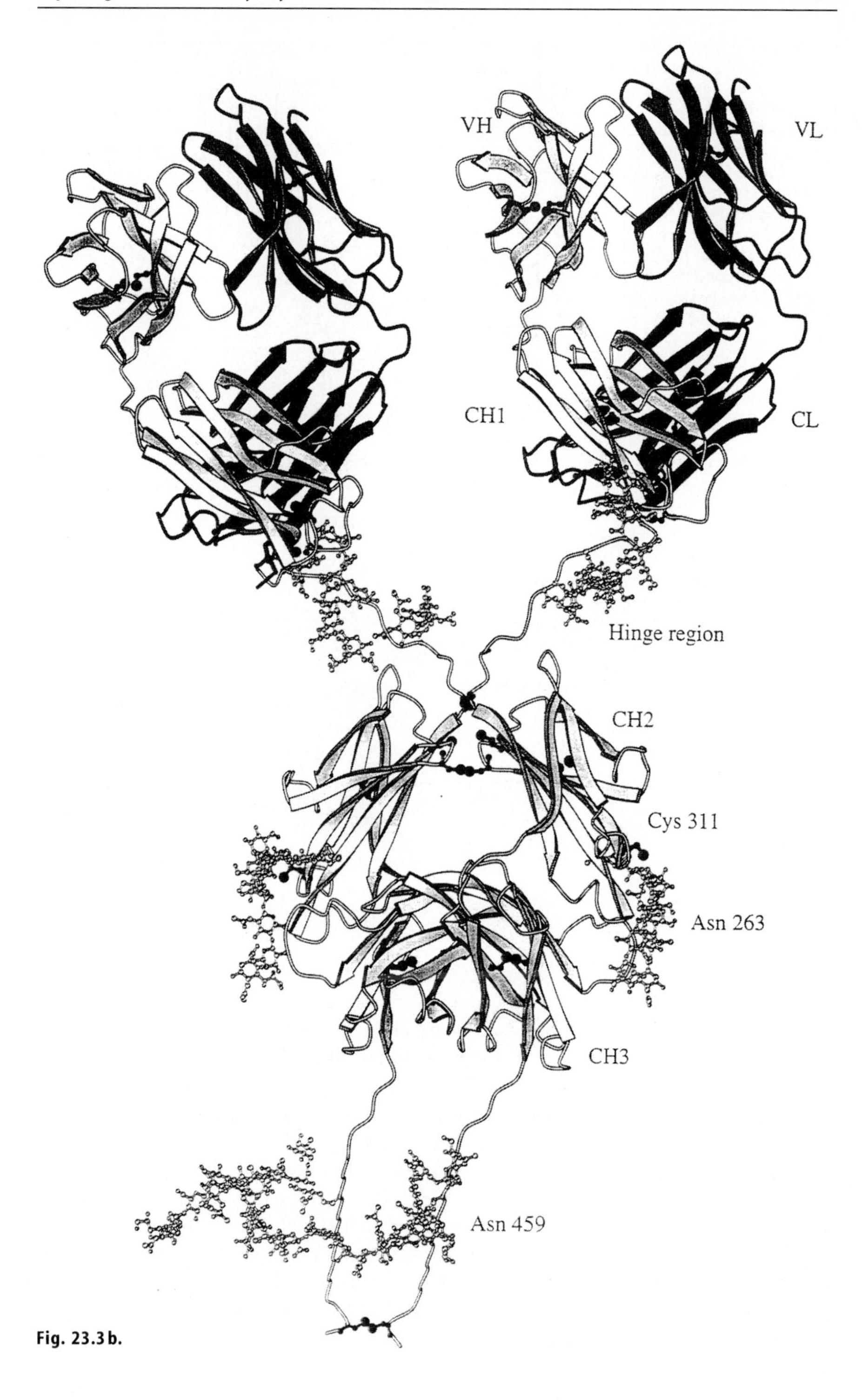

Fig. 23.3 b.

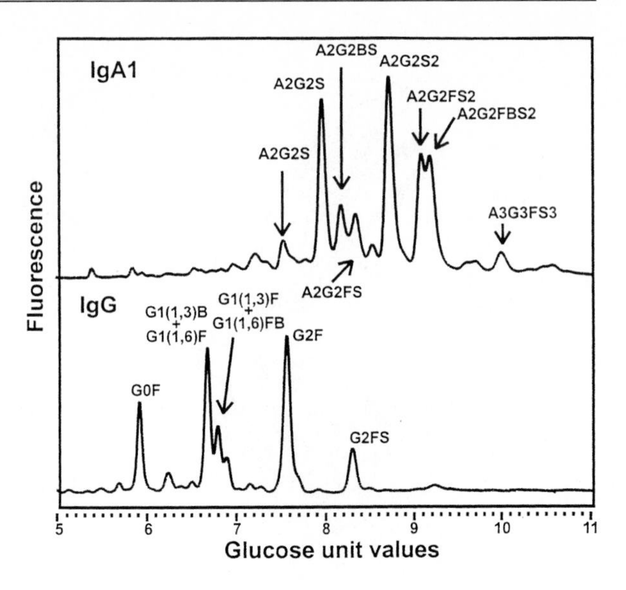

Fig. 23.4. IgA1 accommodates larger oligosaccharides than IgG. The normal phase HPLC profiles of 2AB labelled oligosaccharides released from IgA1 and IgG. 90 % of IgA1 N-linked sugars are of the larger sialylated type compared with only 15 % on IgG which are mainly located on the more accessible glycosylation sites in the Fab region (Mattu et al., 1998, Wormald *et al.*, 1997)

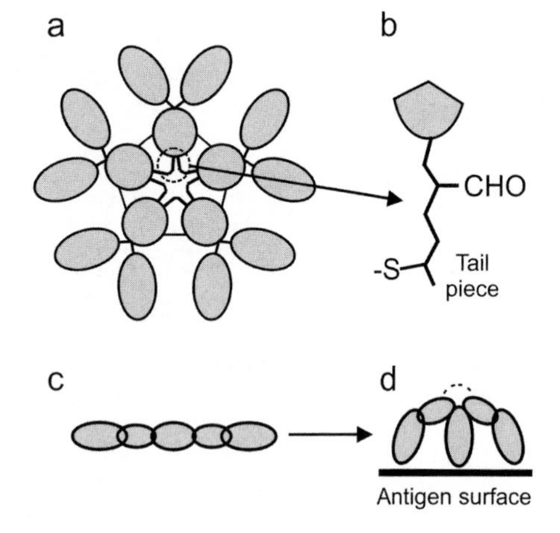

Fig. 23.5. Schematic representation of the IgM pentamer. (**a**) showing the location of the glycosylated tail (**b**) side on view of the "star" form of the pentamer found in solution is shown in (**c**) which converts to a "staple" form (**d**) on binding to an antigenic surface

rheumatoid factor (RF) complexes. The mechanism by which specific IgG sugars initiate complement activation in RA is expected to be important in the synovial cavity where IgG0 levels are elevated (Parekh et al., 1989), IgG0 is deposited on the synovium (Leader et al., 1996) so that the sugars are multiply presented, and MBL levels are increased (Malhotra et al., 1995).

Both the classical and alternative complement pathways terminate with the formation of the membrane attack complex (MAC) in the cell surfaces of bacteria and other pathogens and this leads to cell lysis. Host cells are normally protected from destruction through inhibitors of the complement pathway such as decay accelerating factor (DAF, CD55), which destabilises the C3 convertase components C3b/Bb and C4b2a, and CD59, which binds C8 and/or C9 preventing the formation of the fully assembled MAC complex.

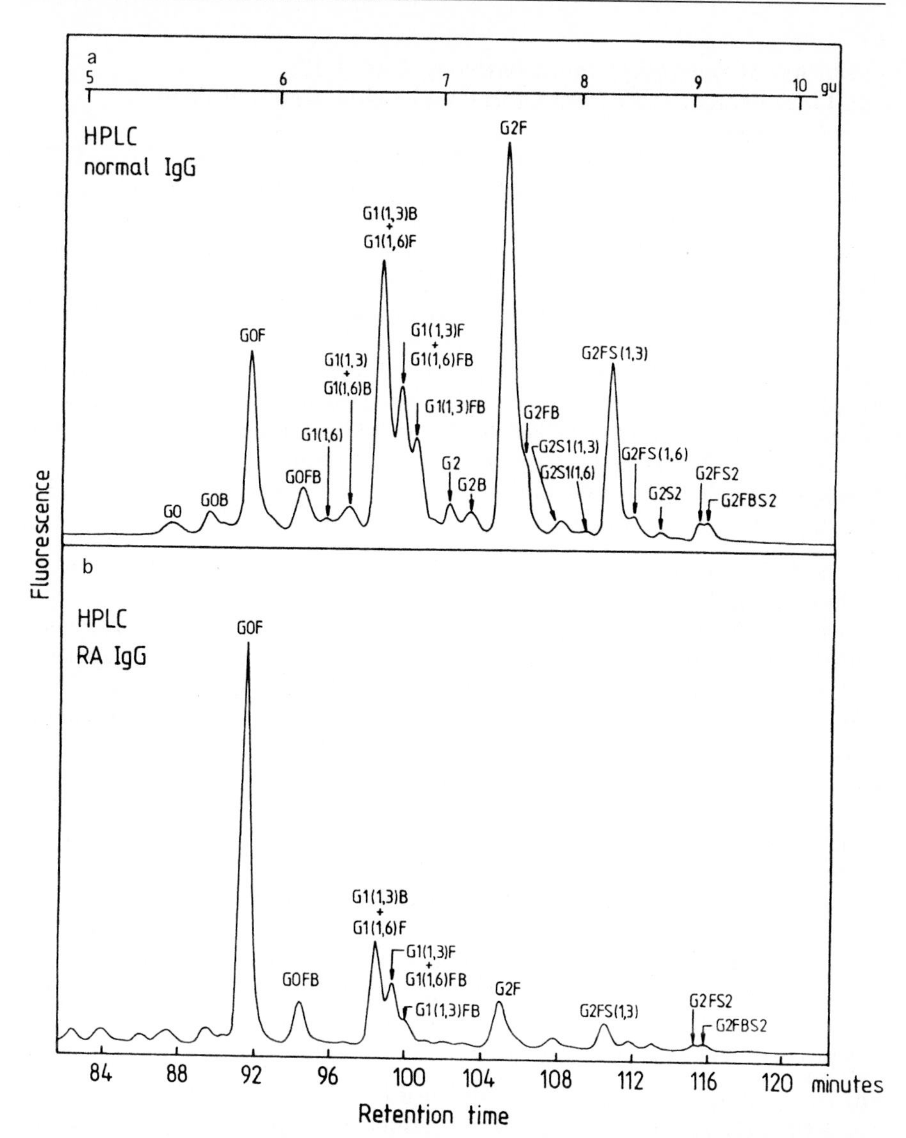

Fig. 23.6. Normal phase HPLC separation of total glycans from (**a**) pooled normal serum IgG and (**b**) rheumatoid serum IgG. Abbreviations used for describing oligosaccharide structures: G(0–2) indicates the number of terminal galactose residues in the structure; F: fucose; B: bisecting N-acetyl glucosamine (GlcNAc); S: sialic acid

6 Inhibitors of the Complement Pathway, DAF (CD55) and CD59 Protect Cells from Complement Mediated Cell Lysis

The complement system can be activated by either of two pathways. The classical pathway is triggered by antigen-antibody complexes, specifically those of IgM and IgG. The alternative pathway is activated by the surfaces of foreign bodies such as viruses, fungi and bacteria. Activation by either pathway leads to the production of the bimolecular complex, C3 convertase. The C3 convertase enzyme cleaves C3 into C3b and C3a. C3b can covalently attach to the particle or cell with the resultant opsonisation leading to clearance by phagocytic cells with C3b receptors. DAF and CD59 are glycosylphosphatidylinositol (GPI) anchored glycoproteins which protect cells from complement-mediated damage. The active site of CD59 is located close to the cell surface at 3.5 nm from the membrane while that of DAF is further away, at 16 nm (Fig. 23.7). While CD59 binds the C5b-8 component of the MAC, DAF both dissociates the classical and the alternative pathway C3/C5 convertases and inhibits their assembly.

(i) The O-Glycosylated Region of DAF Serves as a Protease Resistant Spacer to Position the Active Site

DAF is composed of four complement control protein (CCP) domains suspended above the membrane by a heavily O-glycosylated Ser/Thr rich region of approximately 70 amino acids which is linked to the GPI anchor (Fig. 23.7). The structures of the O-glycans have been determined by glycan sequencing of the sugars released by hydrazine under conditions optimised for O-glycan release (Rudd, P.M., Morgan, B.P., Harris, C., Wormald, M.R. and Dwek, R.A.- unpublished data). The regulatory activity against C4b2a is localised in CCP domains 2–4 while its activity against C3bBb also involves CCP domain 4. There is one N-glycosylation site which is located between CCP1 and CCP2. The N-glycan site has been deleted without any associated loss of activity (Coyne et al., 1992). However, deletion of the Ser/Thr rich region eliminated DAF function. This was restored in a fusion construct in which the four CCP domains were added to the HLA B44 molecule suggesting that the O-glycosylated region serves as an important protease resistant spacer which projects the DAF functional domains above the plasma membrane (Coyne et al., 1992).

(ii) Potential Roles for the Glycans Attached to CD59

CD59 (Fig. 23.7) belongs to the Ly-6 superfamily and is present on a wide variety of cell types, including leukocytes, platelets, epithelial and endothelial cells, placental cells and erythrocytes. CD59 is attached to the surfaces of these cells by means of a GPI anchor containing three lipid chains (Davies et al., 1989, Rudd et al., 1997). Human erythrocyte CD59 consists of a heterogeneous mixture of more than 120 glycoforms (Fig. 23.8) of which the major single sugar is a complex glycan containing both a bisecting GlcNAc residue and a core fucose (Rudd et al., 1997). A population of sialylated O-glycans was recovered from human erythrocyte CD59; the major species which were identified were two mono-sialylated forms of the disaccharide Galβ1,3GalNAc. 90 % of the GPI anchor glycans contain the tri-mannose sugar common to all mammalian anchors analysed to date.

The N-linked oligosaccharides (size range 3–6 nm in length) which are attached to the disc-like extra-cellular region of CD59 (diameter approximately 3 nm) project away from the protein domain in the plane of the active face and adjacent to the membrane surface. The glycans do not appear to restrict access to the proposed active site residues of human CD59 (Asp24, Trp 40, Arg 53 and

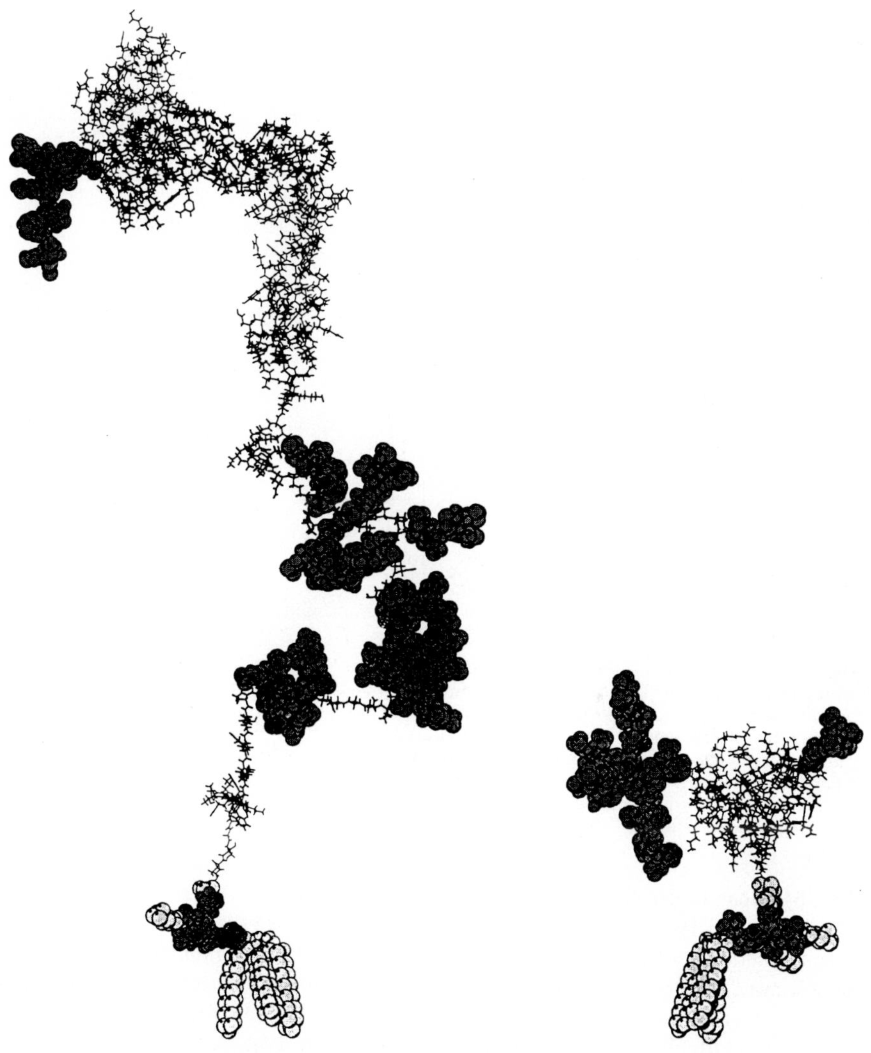

Fig. 23.7. Molecular models of DAF and CD59. DAF is based on the model of Kuttner-Kondo et al., 1996 with an A2G2FBS1 glycan on domain 1. The C-terminal tail, containing 12 O-linked glycans and a GPI anchor, is modelled in an extended conformation. CD59 is based on the protein co-ordinates from the solution structure (Fletcher et al., 1994). The glycan anchor is modelled with a tri-mannosyl core, an ethanolamine bridge at Man3 and additional ethanolamine groups at Man1 and Man2. Two lipids are attached to inositol via phosphate and the third is attached directly to the inositol ring through an ester linkage. A trisialylated, tetra-antennary complex N-glycan is shown attached to Asn 18 and the O-glycan, NeuNAc2,3Galβ1–3GalNAc, is attached to Thr51 (Rudd et al., 1997)

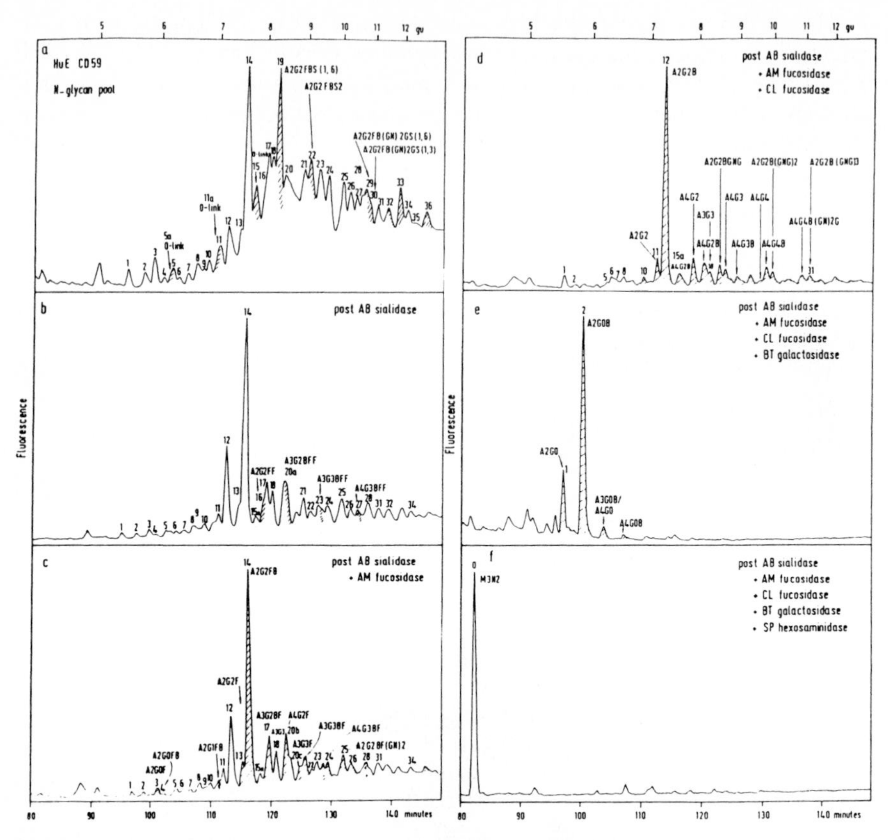

Fig. 23.8. HPLC profile of the N-and O-glycan populations of HuE CD59 simultaneously digested with a series of enzyme arrays. The figure shows the HPLC analysis of the total glycan pool released by hydrazinolysis. The gu value of each of the numbered peaks was calculated by comparison with the dextran hydrolysate ladder. Structures were first assigned from the gu values and previously determined incremental values for monosaccharide residues (Guile *et al.*, 1996) and confirmed by digests using five arrays of exoglycosidases (for details see Rudd et al 1997). The final enzyme array digested all the structures to the common tri-mannosyl core

Glu56) located on the membrane distal surface of the extracellular domain (Bodian et al., 1997). However, the glycans would be expected to restrict the rotational freedom of the extracellular domain around axes parallel to the membrane which may, in turn, stabilise an exposed location for the active face. Removal of the conserved N-linked glycan might therefore reduce the affinity of CD59 for the membrane attack complex without eliminating it completely. The effects of removing the N-linked glycan might therefore be expected to depend on the density of the glycoprotein at the cell surface and this may explain the observed variation in the activities of unglycosylated CD59 (Ninomyia et al.,1992, Bodian et al., 1997).

The heterogeneity of the sugars may influence the geometry of the packing and it is likely that they will also prevent CD59 molecules forming regular arrays on the cell surface. By limiting non-specific protein-protein interactions and controlling the spacing, the glycans may influence the distribution of CD59 molecules at the cell surface where GPI anchored proteins may associate in microdomains in dynamic equilibrium with isolated individual molecules (van den Berg et al., 1995). The large N-glycans may also be important in preventing proteolysis of the extracellular domain since N-glycosylation has been shown to increase the dynamic stability of a protein while different glycoforms variably increase its resistance to protease digestion (Rudd et al., 1994).

(iii) Function for GPI Anchors on DAF and CD59
GPI membrane anchors are used by a wide variety of cell surface glycoproteins (McConville et al., 1993). The tetrasaccharide backbone (Man_3 inositol) may be extended with other sugars as the anchored protein moves through the secretory pathway. In the anchored form of CD59 the bulky, hydrophilic glycans at the C-terminus would be expected to limit interactions with the lipid bilayer and in this way facilitate the diffusion of the glycosylated protein in the top leaflet of the cell membrane. DAF has been shown to reside in microdomains on the surface of MDCK cells. These domains contain at least 15 molecules (Friedrichson and Kurzchalia, 1998).

7
Glycosylation Provides a Mechanism for Fine Tuning the Affinity of CD2 and CD48 for their Receptors

One aspect of the cellular immune response involves the cell adhesion molecules CD2 and CD48 in rat (CD58 in human), which are present on cells expressing the TCR and on antigen presenting cells, respectively. CD2 and its ligand interact through homologous binding surfaces located in the amino-terminal domains distal to the membrane surface (Davis et al., 1999).

The N-glycans attached to the highly conserved, membrane proximal glycosylation site at the base of CD2 and CD48 limit the conformational space available to the protein to a cone of 52o described by CD2 (Dustin et al., 1996). This also promotes trans-interactions with rat CD48 on other cells and discourages cis-interactions with other human CD2 molecules on the same cell. The sugars attached to human soluble CD2 and rat soluble CD48 (Fig. 23.9) were analysed (Fig. 23.10). In neither case are any of the sugars close to the binding site, and they may therefore be expected to protect the protein surfaces against proteases without hindering the function of the protein.

Since the length of the CD2/CD48 cell adhesion pair is the same as that of the TCR/MHC pair it has been proposed that CD2 and CD48 mediate the alignment of the cell surfaces of cytolytic T-lymphocytes carrying the TCR complex with those of target cells carrying loaded HLA class 1 molecules. CD48 on rat (CD58 on human) molecules cluster when presented with CD2 receptors (Dustin et al., 1996). During clustering non-specific protein interactions may be inhibited by the heterogeneous arrays of sugars which variably shield the protein surface.

Human sCD2 and rat sCD48 have 3 and 5 N-glycosylation sites, respectively. Both molecules contain mainly complex type glycans (Fig. 23.10). Human sCD2

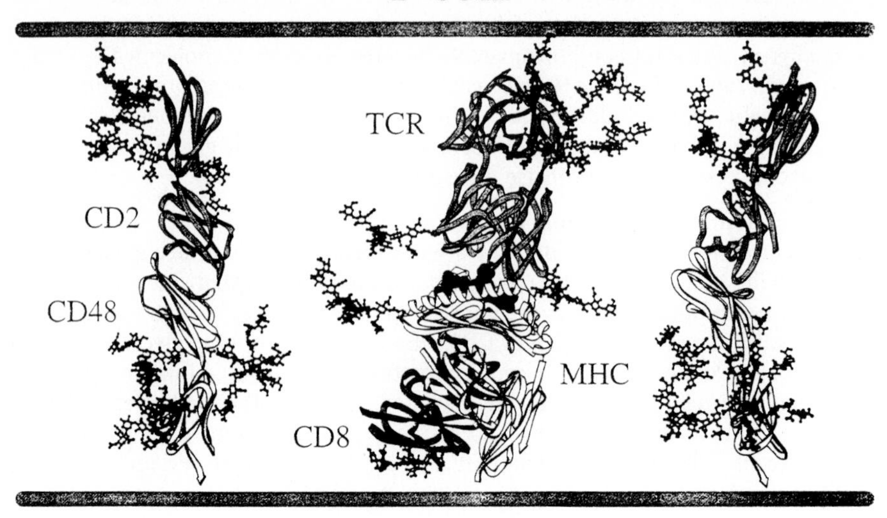

Fig. 23.9. The TCR/MHC/CD8 interaction showing the alignment at the junction between a natural killer T-cell with an antigen presenting cell which is mediated by CD2 and CD48. Four of the seven potential sites on the TCR are occupied with ordered sugars (Garcia et al, 1996), the MHC has one N-glycosylation site, CD8 has one N-glycan site and 8 potential sites for O-glycosylation

contains a series of oligomannose structures which have been located to Asn65 (Recny et al.,1992). This site is close to a cleft in the protein, where it may be relatively shielded from mannosidases and from GlcNAc transferase II, the key enzyme in the processing of complex type sugars. In addition the GlcNAc-1, which links the nitrogen in the side chain of Asn65 to the remainder of the oligosaccharide, is relatively inflexible with respect to the protein (Wyss et al., 1995a). This may provide a further basis for the finding that the site contains only oligomannose sugars. It has been suggested that a specific role for sugars at Asn65 on human CD2 may be to counterbalance the de-stabilising effect of the cluster of positive charges created by 5 lysine residues situated on the surface of domain 1 (Wyss et al., 1995b). This may be a specific mechanism for human CD2 since the location of the glycosylation site is not conserved in rat CD2.

8 The TCR/MHC Recognition Event Involves Heavily Glycosylated Glycoproteins

In cytotoxic CD8 +ve T-cells the function of the TCR is to recognise and bind class 1 MHC carrying specific peptide antigens. In the ER MHC heavy chains carrying monoglucosylated oligomannose sugars ($Glc_1Man_{7-9}GlcNAc_2$) bind calnexin through the terminal glucose residue. After the heavy chains associate with

Fig. 23.10. Glycan processing is influenced by the 3D structure of the protein: HPLC profiles of the glycan pools released from a range of leucocyte antigens expressed in CHO cells. CD2 contains an oligomannose series (Man5-Man8) not found in CD48. In CD2, Asn 65 has been shown to contain only oligomannose sugars (Recny et al., 1992)

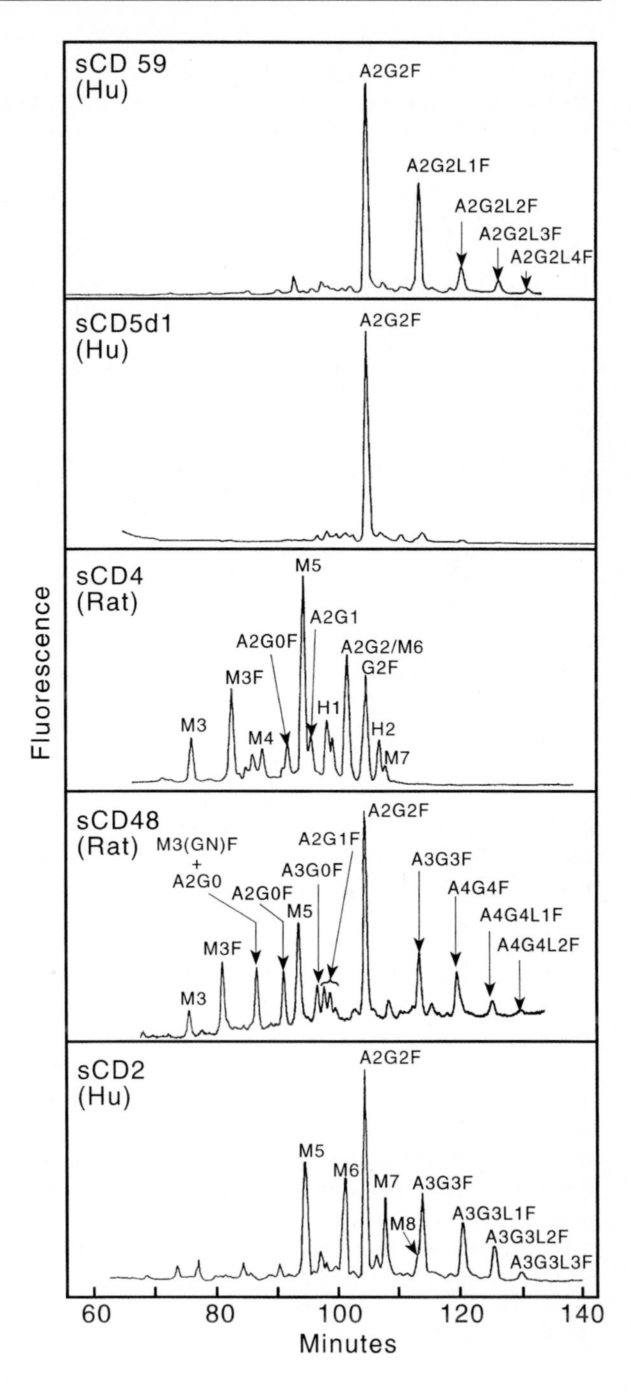

β-microglobulin the fully assembled MHC is released from calnexin and binds calreticulin, also through the terminal glucose residue. This event, which also involves tapasin, is necessary for the loading of antigenic peptide from the TAP transporter (Sadasivan et al., 1996).

Membrane bound CD8 is a glycoprotein which stabilises the interaction of the TCR with the human MHC molecule (HLA) class 1 carrying antigenic peptide. CD8 binds to peptide loaded HLA class 1 molecules, interfacing with the α2 and α3 domains of HLA-A2 and also contacting β_2-microglobulin (Gao et al., 1997). HLA B2705, prepared from the EBV transformed homozygous B-cell line LG-2, contains a very restricted set of glycans (Rudd, P.M., Pazmany, L., Strominger, J.L. and Dwek, R.A.- unpublished data) and this is consistent with the analysis of a set of B27 molecules analysed by (Barber et al., 1996). The predominant structure at Thr 86 was the fully galactosylated biantennary complex glycan containing bisecting GlcNAc and core fucose, (shown in the fully sialylated form in Fig. 23.2). The N-linked sugars occupying the seven potential glycosylation sites on the αβ TCR complex and the site on CD8 have not yet been analysed, but in order to demonstrate the extensive glycosylation of the interaction the proteins have been modelled with bi-antennary complex glycans at each potential glycosylation site (Fig 23.9).

9 Concluding Remarks

One of the essential elements of immunity is the recognition of foreign structures. Often, as in bacteria, the epitopes which interact with our immune system are mutiply presented oligosaccharides which activate the complement pathway both through the classical and the alternative pathways. Oligosaccharide recognition (e.g. by selectins) is also essential for the functioning of immune cells for this event initiates the recruitment of cells to inflamed endothelial cells prior to margination and extravasation. Here we have discussed the structures and functions of the oligosaccharides attached to effector or control molecules of the immune system. The roles of the sugars include maintaining the 3D structure and function of immunoglobulins and protection of proteins from proteolysis. Oligosaccharides are also associated with particular autoimmune diseases, and we have discussed the finding that specific glycoforms of IgG interact with MBL and activate the complement pathway, suggesting a role for glycosylation in the pathogenesis of RA. In complement control proteins such as CD59 and DAF, the attached oligosaccharides are important for the spatial orientation and function of the molecules. Spatial orientation of proteins by their attached sugars is also a factor in the intermolecular interactions between T-cells and antigen presenting cells, in particular those involving the cell adhesion molecules, CD2 and CD48. Finally the sugars attached to the MHC play a role in protein assembly and in the formation of the complex required for the transfer of antigenic peptide to the binding groove of the MHC.

Acknowledgements

The co-ordinates for the CCP domain of Factor H, on which the molecular model of DAF (Fig. 23.7) was based, were kindly sent to us by Professor E. Medof and Professor M. Shoham. DAF was a gift from Prof. Paul Morgan and Dr. Claire Harris. The full glycan analysis of DAF will be published elsewhere.

The authors would like to thank Professor Ghislain Opdenakker and Dr. Anthony Merry for their very helpful criticisms of the manuscript.

The GBI acknowledges support from the BBSRC and from the European Commission Grant BIO4-CT95-0138.

Abbreviations

C: constant; CCP: complement control protein; CRD: carbohydrate recognition domain; ER: endoplasmic reticulum; GPI: glycosylphosphatidylinositol; MBL: mannose binding lectin also known as mannose binding protein; RA: Rheumatoid arthritis; TCR: T-cell receptor; V: variable chain, H: heavy chain, L: light chain.

Abbreviations used for describing oligosaccharide structures: A: indicates the number of antennae (from 1–4); G(0–2) indicates the number of terminal galactose residues in the structure; F: fucose; B: bisecting N-acetyl glucosamine; GlcNAc: N-acetyl glucosamine; S: sialic acid; MAC: membrane attack complex; RF: Rheumatoid factor; DAF: decay accelerating factor; GPI: glycosylphosphatidyl inositol.

References

Barber LD, Patel TP, Percival L, Gumperz JE, Lanier LL, Phillips JH, Bigge JC, Wormald, MR, Parekh RB, Parham, P. (1996) Unusual uniformity of the N-linked oligosaccharides of HLA-A, -B, and -C glycoproteins. J Immunol 156: 3275–84

Bodian DL, Davis SJ, Rushmere NK, Morgan BP (1997) Mutational analysis of the active site and antibody epitopes of the complement-inhibitory glycoprotein, CD59. J Exp Med 185: 507–516.

Bole DG, Hendershot LM, Kearney JF (1986) Post-translational association of immunoglobulin heavy chain binding protein with nascent heavy chains in nonsecreting and secreting hybridomas. J Cell Biol 102: 1558–1566

Burton DR, Woof JM (1992) Human antibody effector function. Adv Immunol 51: 1–84

Coyne KE, Hall SE, Scott Thompson E, Arce MA, Kinoshita T, Fujita T, Anstee DJ, Rosse W, Lublin DM (1992) Mapping of epitopes, glycosylation sites, and complement regulatory domains in human decay accelerating factor. J Immunol 149: 2906–2913

Davies A, Simmons DL, Hale G, Harrison RA, Tighe H, Lachmann PJ, Waldman H. (1989) CD59, an LY-6-like protein expressed in human lymphoid cells, regulates the action of the complement membrane attack complex on homologous cells. J Exp Med 170: 637–654

The analyses referred to in this article were carried out in the Glycobiology Institute using an HPLC system consisting of a Douglas Scientific DG604 degasser, two Waters 510 pumps, a Waters 717 autoinjector, a Waters Temperature Control Module and a Jasco FP-920 fluorescence detector. The system was controlled via a Waters LAC/E box using Waters Expertease 3.1 software running on a DEC VAX 4000-200 computer. The Normal phase, and Weak Anion Exchange HPLC columns (GlycoSep N and GlycoSepC respectively) were obtained from Oxford GlycoSciences (OGS) UK Ltd). Fluorescent labelling with 2-aminobenzamide was carried out using a Signal Labelling Kit, glycans were released by hydrazine using the GlycoPrep 1000 or by PNGase F. Glycosidases were also obtained from OGS. Curve fitting software was Peak Time, developed by Dr. E. Hait, Glycobiology Institute, Oxford.

Deisenhofer J (1981) Crystallographic refinement and atomic models of a human Fc fragment and its complex with fragment B of protein A from Staphylococcus aureus at 2.9- and 2.8-A resolution. Biochemistry 20: 2361–2370
Duncan AR, Woof JM, Partridge LJ, Burton DR, Winter G (1988) Localization of the binding site for the human high-affinity Fc receptor on IgG. Nature 332: 563–564
Dustin ML, Ferguson LM, Chan PY, Springer TA, Golan DE (1996) Visualization of CD2 interaction with LFA-3 and determination of the two-dimensional dissociation constant for adhesion receptors in a contact area. J Cell Biol 132: 456–474
Field MC, Amatayakul-Chantler S, Rademacher TW, Rudd PM, Dwek, RA (1994) Structural analysis of the N-glycans from human immunoglobulin A1: comparison of normal human serum immunoglobulin A1 with that isolated from patients with rheumatoid arthritis. Biochem J 299: 261–275
Fletcher CM, Harrison RA, Lachmann PJ, Neuhaus D (1994) Structure of a soluble, glycosylated form of the human complement regulatory protein CD59. Current Biology Structure 2: 185–199
Friedrichson T, Kurzchalia TV (1998) Microdomains of GPI-anchored proteins in living cells revealed by crosslinking. Nature 394: 802–804
Gao GF, Tormo J, Gerth UC, Wyer JR, McMichael AJ, Stuart DI, Bell JI, Jones EY, Jakobsen BK (1997) Crystal structure of the complex between human CD8a (alpha) and HLA-A2. Nature 387: 630–634
Garcia KC, Degano M, Stanfield RL, Brunmark A, Jackson MR, Peterson PA, Teyton L, Wilson IA (1996) An alphabeta T cell receptor structure at 2.5 A and its orientation in the TCR-MHC complex. Science 274: 209–219
Kuster B, Wheeler SF, Hunter AP, Dwek RA, Harvey DJ (1997) Sequencing of N-linked oligosaccharides directly from protein gels: in-gel deglycosylation followed by matrix-assisted laser desorption/ionization mass spectrometry and normal-phase high-performance liquid chromatography. Anal Biochem 250: 82–101
Kuttner-Kondo L, Medof ME, Brodbeck W, Shoham M (1996) Molecular modeling and mechanism of action of human decay-accelerating factor. Protein Eng 9: 1143–1149
Leader KA, Lastra GC, Kirwan JR, Elson CJ (1996) Agalactosyl IgG in aggregates from the rheumatoid joint. Br J Rheumatol 35: 335–341
Leatherbarrow RJ, Dwek RA (1984) Binding of complement subcomponent C1q to mouse IgG1, IgG2a and IgG2b: a novel C1q binding assay. Molecular Immunology 21: 321–327
Malhotra R, Wormald MR, Rudd PM, Fischer PB, Dwek RA, Sim RB (1995) Glycosylation changes of IgG associated with rheumatoid arthritis can activate complement via the mannose-binding protein. Nature Medicine 1: 237–241
Mattu TS, Woof J M, Lellouch A, Rudd PM, Dwek, RA (1998) The glycosylation and structure of human serum IgA1, Fab, and Fc regions and the role of N-glycosylation on Fc alpha receptor interactions. J Biol Chem 273: 2260–2272
McConville MJ, Ferguson MAJ (1993) The structure, biosynthesis and function of glycosylated phosphatidylinositols in the parasitic protozoa and higher eukaryotes. Biochem J 294: 305–324
Mestecky J, Kilian M (1985) Immunoglobulin A (IgA). Methods Enzymol 116: 37–75
Ninomiya H, Stewart BH, Rollins SA, Zhao J, Bothwell ALM, Sims PJ (1992) Contribution of the N-linked carbohydrate of erythrocyte antigen CD59 to its complement-inhibitory activity. J Biol Chem 267: 8404–8410
Nose M, Wigzell H (1983) Biological significance of carbohydrate chains on monoclonal antibodies. Proc Nat Acad Sci USA 80: 6632–6636
Padlan EA (1994) Anatomy of the antibody molecule. Mol Immunol 31: 169–217
Parekh RB, Dwek RA, Sutton BJ, Fernandes DL, Leung A, Stanworth D, Rademacher TW, Mizuochi T, Taniguchi T, Matsuta K, Takeuchi F, Nagano Y, Miyamoto T, Kobata A (1985) Association of rheumatoid arthritis and primary osteoarthritis with changes in the glycosylation pattern of total serum IgG. Nature 316: 452–457
Parekh RB, Isenberg D, Rook G, Roitt I, Dwek RA, Rademacher TW (1989) A comparative analysis of disease-associated changes in the galactosylation of serum IgG. J Autoimmunity 2: 101–114
Recny MA, Luther MA, Knoppers MH, Neinhardt EA, Khandekar SS, Concino MF, Schimke, PA, Francis MA, Moebius U, Reinhold B, Reinhold VN, Reinherz EL (1992) N-glycosylation is required for human CD2 immunoadhesion functions. J Biol Chem 267: 22428–22434
Rudd PM, Guile GR, Küster B, Harvey DJ, Opdenakker G, Dwek RA (1997) Oligosaccharide Sequencing Technology. Nature 388: 205–208
Rudd PM, Leatherbarrow RJ, Rademacher TW and Dwek RA (1991) Diversification of the IgG molecule by oligosaccharides. Molecular Immunol 28: 1369–1378
Rudd PM, Joao HC, Coghill E, Fiten P, Saunders MR, Opdenakker G, Dwek RA (1994) Glycoforms modify the dynamic stability and functional activity of an enzyme. Biochemistry 33: 17–22

Rudd PM, Morgan BP, Wormald MR, Harvey DJ, van den Berg CW, Davis SJ, Ferguson MAJ, Dwek, RA (1997a) The glycosylation of the complement regulatory protein, human erythrocyte CD59. J Biol Chem 272: 7229–7244

Rudd PM, Wormald MR, Harvey DJ, Devashayem M, McAlister MSB, Barclay AN, Brown MH, Davis SJ and Dwek RA Oligosaccharide processing in the Ly-6, scavenger receptor and immunoglobulin superfamilies–implications for roles for glycosylation on cell surface molecules In press: Glycobiology

Rudd PM, Dwek RA (1997b) Glycosylation: heterogeneity and the 3D structure of proteins. Critical Reviews in Biochemistry and Molecular Biology 32: 1–100

Sadasivan B, Lehner PJ, Ortmann B, Spies T, Cresswell P (1996) Roles for calreticulin and a novel glycoprotein, tapasin, in the interaction of MHC class I molecules with TAP. Immunity 5: 103–114

Tsuchiya N, Endo T, Matsuta K, Yoshinoya S, Aikawa T, Kosuge E, Takeuchi F, Miyamoto T, Kobata, A (1989) Effects of galactose depletion from oligosaccharide chains on immunological activities of human IgG. J Rheumatol 16: 285–290

van den Berg CW, Cinek T, Hallett MB, Horejsi V, Morgan BP (1995) Exogenous glycosyl phosphatidylinositol-anchored CD59 associates with kinases in membrane clusters on U937 cells and becomes Ca(2+)-signaling competent. J Cell Biology 131: 669–677

Wormald MR, Rudd PM, Harvey DH, Chang S-C, Scragg IG, Dwek, RA (1997) Variations in oligosaccharide-protein interactions in immunoglobulin G determine the site-specific glycosylation profiles and modulate the dynamic motion of the Fc oligosaccharides. Biochemistry 36: 1370–1380

Wyss DF, Choi JS, Wagner G (1995a) Composition and sequence specific resonance assignments of the heterogeneous N-linked glycan in the 13.6 kDa adhesion domain of human CD2 as determined by NMR on the intact glycoprotein. Biochemistry 34: 1622–1634

Wyss DF, Choi JS, Li J, Knoppers MH, Willis KJ, Arulanandam AR, Smolyar A, Reinherz EL, Wagner G (1995b) Conformation and function of the N-linked glycan in the adhesion domain of human CD2. Science 269: 1273–1278

CHAPTER 24

Proteoglycans: Biological Roles and Strategies for Isolation and Determination of Their Glycan Constituents

N. K. Karamanos[1]

1 Summary

This chapter addresses issues on a major glycoconjugate family, the proteoglycans with particular interest in their types, the structural diversity of carbohydrates and the structure-function relationship. Procedures to isolate proteoglycans and strategies to characterize the fine chemical structure of their glycan moieties are also presented.

2 Glycoconjugates: General Aspects on Structural Diversity and Importance

Glycoconjugate is a generic term given to describe macromolecules in which carbohydrates having either the mono-, oligo- or polysaccharide form are covalently bound to polypeptides/proteins or lipids. They involve glycoproteins, e.g., enzymes, immunoglobulins, transport proteins and hormones, mucins or mucoproteins, peptidoglycans, which are mainly present in bacteria, glycolipids and proteoglycans.

Carbohydrate chemistry and analysis has to address a major problem arising from the great structural diversity of glycans. The heterogeneity in structure is mainly due to the following five parameters:

- variable number of monosaccharides present in oligo- and polysaccharides
- the anomericity of glycosidic bond, α- or β- form
- the position of the glycosidic linkage between the anomeric C-1 and any of the available -OH groups of the next monosaccharide
- the branching of the main chain with other sugars, and
- the substitution of -OH groups and amino groups with sulfate, phosphate or acetyl groups.

This structural diversity has no counterpart in proteins and nucleic acids. For example, four monosaccharides are able to form 36,000 tetrasaccharides, whereas four amino acids or nucleosides may form only 24 tetramers. This may well be

[1] Section of Organic Chemistry-Biochemistry and Natural Products, Department of Chemistry, University of Patras, 26 110 Patras, Greece.

explained by the absence of a template to determine the order of monomer units in the polymer and it could, therefore, be assumed that this structural diversity of carbohydrates is allowed during biosynthesis. As a result, the majority of glycoproteins consists of glycosylated variants, which are formed by the presence or absence of monosaccharides in the glycan chain of the same protein glycosylation site. RNAse B, for example, has one *N*-linked glycosylation site, but consists of a mixture of glycoforms due to the presence of numerous mannose residues and, therefore, it is characterized as RNAse Man 5–9. Human IgG is a mixture of a large number of glycoforms due to the presence or not of various monosaccharides. The importance of glycosylation can be deduced from the fact that the absence of Gal of the largest carbohydrate chain of human IgG is related with rheumatoid arthritis.

Glycoproteins and proteoglycans (PGs) constitute a family of glycoconjugates with great importance since they participate and regulate several cellular events

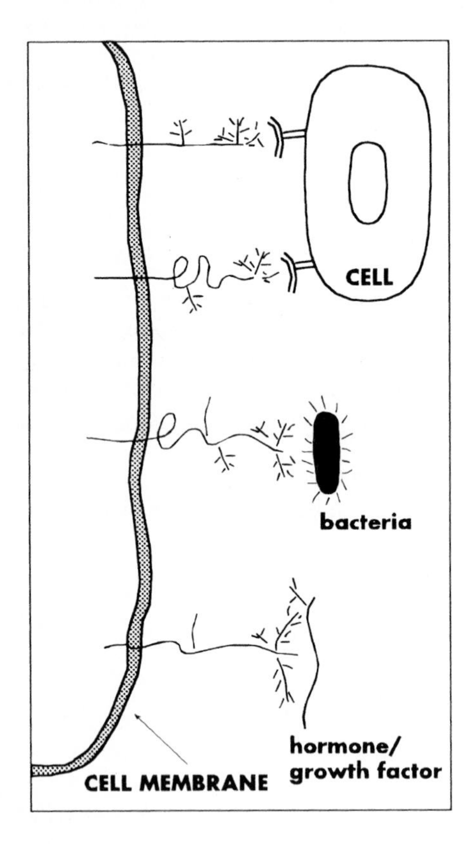

Glycoproteins / Proteoglycans

Determine:

- reproduction
- embryonic development and differentiation
- tissue organisation
- migration of malignant tumorous cells (metastasis)
- immunological reactions:
  - *serve as antigens triggering harmful immune reactions*
  - *lymphocyte activation, differentiation, migration to the site of infection and recognition of antigens*
- infection not only by bacteria but also by viruses and parasites (e.g. HIV infection)
- growth control, cell signalling

Fig. 24.1. Schematic diagram showing the participation of glycan-containing macromolecules (glycoproteins and proteoglycans) in several cellular events and processes

and processes (Karamanos 1997), which are summarized in Fig. 24.1. For further information on structure and function of PGs the excellent reviews of Khellen and Lindahl 1991, Hascall et al. 1994 and Iozzo 1997 are recommended.

PGs are named according to their respective protein core and they form a heterogeneous group of glycoconjugates that are either components of the extracellular matrix or associated with the cell membrane or even as storage macromolecules in cytoplasmic granules (Khellen and Lindahl 1991). Studies on their structures and interactions with other effective molecules revealed their *in vivo* importance and the results may trigger the development of new powerful pharmaceuticals and enhanced diagnostic techniques for various diseases. Apart from their protein part, complex-carbohydrate structures and particularly the glycosaminoglycans (GAGs) give unique properties to PGs. Via their GAG chains PGs interact with growth factors and growth factor receptors and participate in the regulation of cell proliferation and differentiation as well as in matrix synthesis. The effects of GAG constituents on these issues have been recently elucidated for human malignant mesothelioma cells (Tzanakakis et al. 1995 and 1997, Syrokou et al. 1999). To understand, therefore, in depth the mechanisms of PGs action as well as their involvement in life processes, the fine chemical structures of their carbohydrate constituents should be fully elucidated. The improvement in technology has helped scientists to develop strategies useful to determine PG composition and to elucidate their structure and interactions even when very low amount of PGs (a few μg) are available. In spite of the ever increasing knowledge of molecular biology which is helpful in characterizing the different protein cores of PGs, there is still a long way to go when elucidating the fine structure of GAGs/PGs and the biological functions driven from their complex-carbohydrate moieties (Karamanos and Hjerpe 1996).

3
Proteoglycans

PGs are considered to be one of the major family of structural glycoconjugates. They constitute a separate family from that of glycoproteins since PGs are highly charged macromolecules consisting of GAG and oligosaccharide chains covalently bound to a protein core. With the exception of hyaluronan (HA), all GAGs are synthesized as PGs. A representative model of a PG monomer is given in Fig. 24.2.

3.1
Glycosaminoglycans: Structure, Biosynthesis and Linkage to Protein

GAGs are linear polymers consisting of repeating disaccharide units that contain one hexosamine (glucosamine or galactosamine) and one uronic acid (glucuronic and/or iduronic acid) or galactose in the case of keratan sulfate. The definition of the various GAG types is based on the type of these monosaccharides and the glycosidic bonds between them (Figs. 24.3–5). There are four groups of GAGs:

(1) hyaluronic acid or hyaluronan with [→4GlcAβ1→3GlcNAcβ1→] as repeating disaccharide unit,

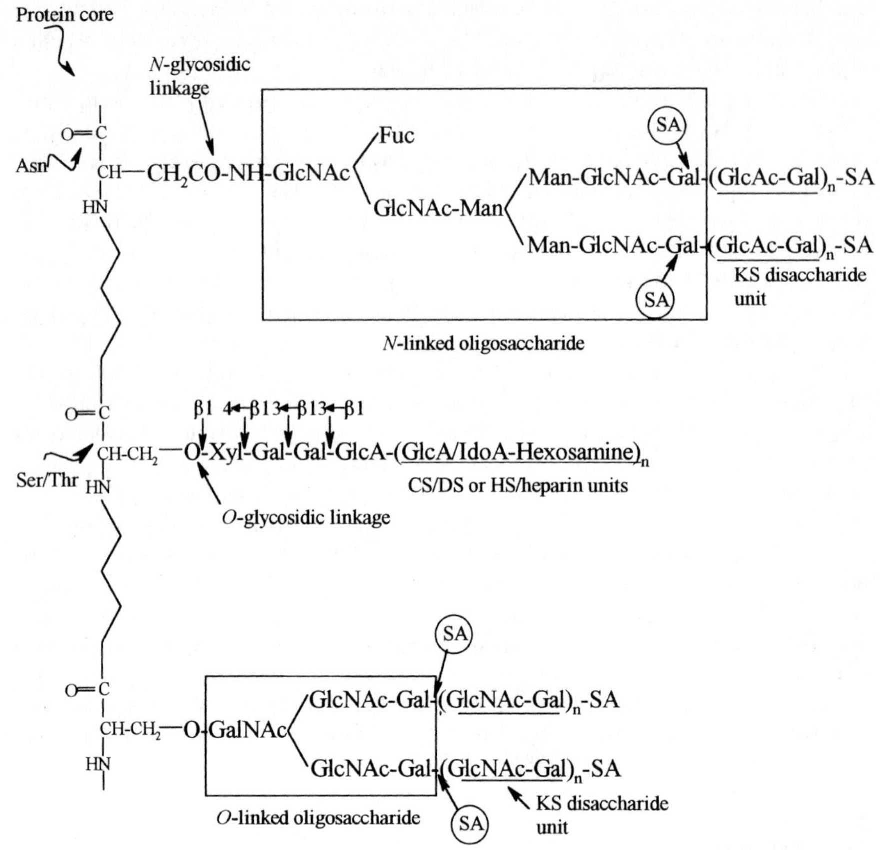

Fig. 24.2. Structure of a PG monomer. GAG and oligosaccharide chains are covalently bound in to a protein core via *O*- and *N*- glycosidic bonds to Ser/Thr and Asn, respectively

(2) chondroitin sulfate (CS) and dermatan sulfate (DS), which are known as galactosaminoglycans (GalAGs) since they contain galactosamine as the only hexosamine,
(3) heparan sulfate (HS) and heparin known as glucosaminoglycans (GlcAGs) since their structural units contain glucosamine as the only hexosamine, and
(4) keratan sulfate (KS) with a repeating unit [→4GlcNAcβ1→3Galβ1→].

HA is the simplest GAG since none of the hydroxyl groups is esterified with sulfates (Fig. 24.3A). Furthermore, this GAG is not synthesized covalently bound to a protein core (Laurent and Fraser 1992). Although HA is synthesized at sites near the plasma membrane, with a mechanism that differs from those used for the other protein bound GAGs (Prehm 1984), it is exclusively secreted to the extracellular matrix where it participates in the formation of large size aggregates with extracellular PGs, such as aggrecan and versican. HA is by far the largest GAG and can reach very large sizes (25,000 disaccharide units, Mr $\sim 10^6$).

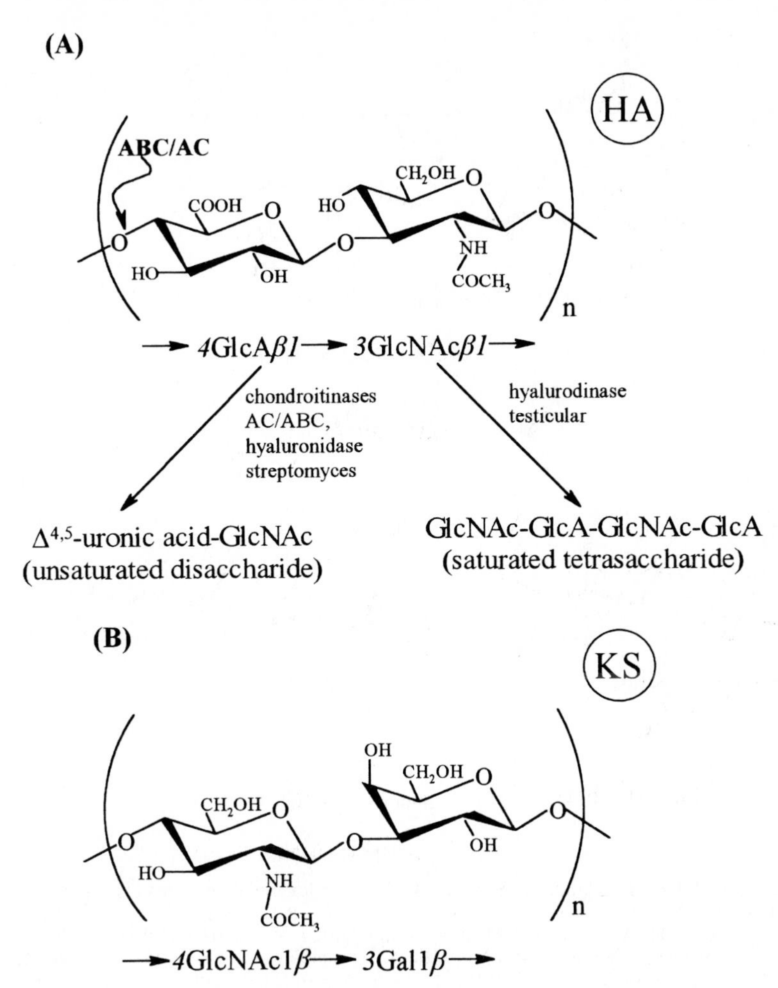

Fig. 24.3. Structure of hyaluronan, cleavage by chondroitinases and hyaluronidases and degradation products (**A**) and structure of the repeating disaccharide unit of keratan sulfate (**B**). GlcNAc in KS replaces the position of uronic acid and Gal that of hexosamine in HA and GalAGs. Alternating glycosidic bounds β1→3 and β1→4 are then the same with HA and GalAGs

Chondroitin sulfate differs from HA in two major structural features:

(1) GlcNAc of HA is replaced with GalNAc, i.e., the structure of the repeating unit is [→4GlcAβ1→3GalNAcβ1→] (Fig. 24.4A) and
(2) CS is normally sulfated at C-4 or C-6 positions of GalNAc forming a sulfated backbone. Furthermore, any one of the available hydroxyl groups may be sulfated. Thus, except nonsulfated disaccharide units, mono- and disulfated or even trisulfated may also be present, and the potential variations in terms of longer sequences are numerous (Karamanos et al. 1994). CS is synthesized on the characteristic linkage region oligosaccharide (-O-Xyl-Gal-Gal-GlcA) via an O-glycosidic linkage to serine or threonine residues (Fig. 24.2). Chain

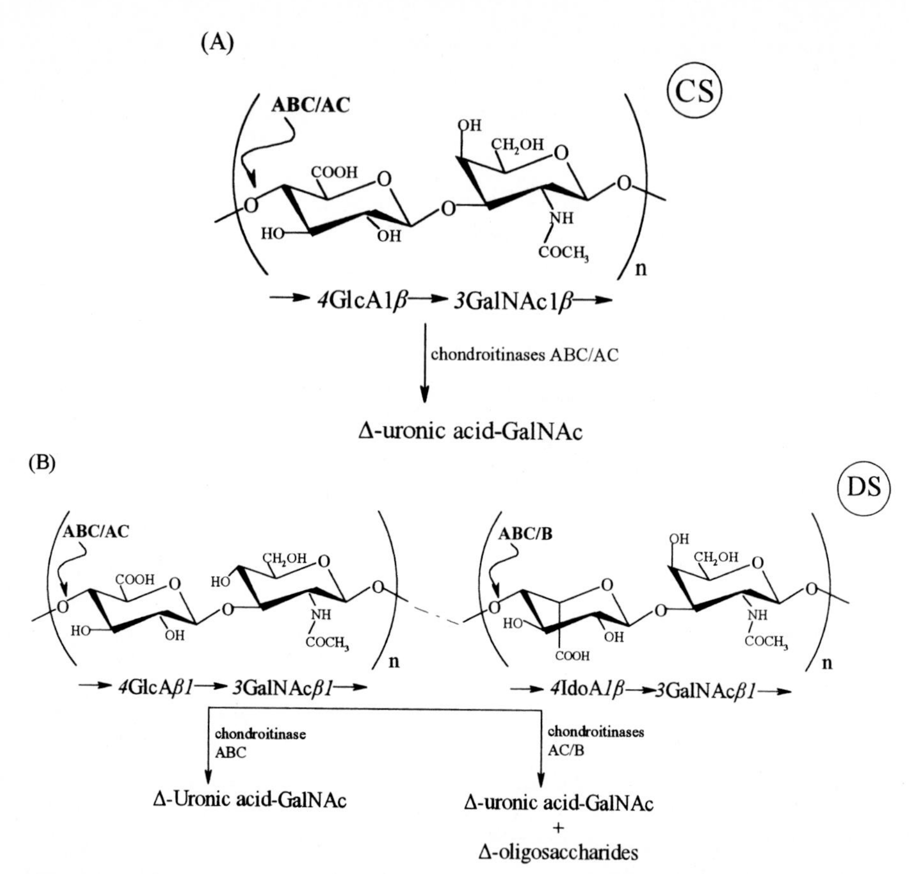

Fig. 24.4. Repeating disaccharide units of CS (**A**) and DS (**B**), their degradation sites and products by chondroitinases ABC, AC and B. The produced Δ-disaccharides of the repeating units may be non-sulfated or may contain one, two or three sulfate groups. Configuration of -COOH group at C-5 determines the presence of configuration of D-GlcA or L-IdoA. DS contains both uronic acid, whereas CS only GlcA. The CS- and DS-derived Δ-disaccharides have the same structure. Separate digestions with chondroitinases AC and B may be of great value in determining the presence of DS- or CS-derived Δ-disaccharides

elongation occurs by the transfer of the appropriate UDP-monosaccharides at the non-reducing terminal. This mechanism is quite different from that occurring in HA synthesis where elongation occurs at the reducing end. Addition of sulfate groups to the growing chondroitin chain takes place in the Golgi apparatus. The other GalAG member, i.e., dermatan sulfate is formed when the GlcA C-5 epimerase, which changes the configuration of the carboxyl group at the C-5 of D-GlcA, converts a proportion of D-GlcA residues to L-IdoA. Therefore, DS chains are constructed by the repeating units containing L-IdoA and D-GlcA (Fig. 24.4B). The IdoA-containing units are often sulfated at C-4 of galactosamine. A considerable proportion of IdoA is sulfated at C-2 mainly forming disulfated disaccharides. Galactosamine sulfated at C-6 is often associated with GlcA-containing disaccharides (Karamanos et al. 1995b). Twenty three non- and variously sulfated disaccharides have been so far identified (Karamanos et al. 1994). The molecular mass (M_r) of GalAGs is often between 20 and 30-kDa, although higher M_rs have been determined in GalAGs isolated from invertebrates (Karamanos et al. 1992).

The GlcAGs, heparin and HS, constitute another group of GAGs with a repeating disaccharide structure distinctly different from other GAGs. The basic unit is [→4 uronic acidβ1→4GlcNα1→]. Therefore, the glycosidic linkage between uronic acid and GlcN is β1→4 instead of β1→3 and that between GlcN and uronic acid is α1→4 instead of β1→4. The uronic acid of the basic unit being D-GlcA or L-IdoA (Fig. 25.5). In this group of GAGs, hexosamine is not only N-acetylated but also N-sulfonylated. $GlcNSO_3$- is generally less than 40–50 % of total GlcN in HS, whereas in heparin this structure accounts for more than 60–70 %. A minor portion of the amino group may also be underivatized. Any free hydroxyl group may carry a sulfate, producing twelve non- and variously sulfated disaccharides (Karamanos et al. 1996, 1997). The frequent oversulfated unit contains sulfate groups at C-2 of IdoA and at C-6 of glucosamine. Some other infrequent sulfation patterns at C-3 of glucosamine and C-2 of GlcA as a part of longer sequences have important biological functions, such as the binding of heparin to antithrombin-III (Lindahl 1989) and the regulation of heparin mitotic activity (Syrokou et al. 1999). The molecular sizes of these GlcAGs do not exceed 100-kDa and normally range from 15 to 30-kDa.

The repeating disaccharide structure of KS contains the same alternative β1→3 and β1→4 glycosidic bonds with those present in HA and GalAGs, but with galactose in the position of the hexosamine and GlcNAc in the position of uronic acid (Fig. 24.3B). KS is also a sulfated GAG. Sulfate esters are generally present at C-6 position of any one of the monosaccharides. This GAG is synthesized on two different linkage oligosaccharide precursor, the structure of which is dependent on the type of glycosidic linkage on the protein core. O-linked linkage-region oligosaccharide is characteristic for mucins and N-linked for cell-bound glycoproteins (Fig. 24.2). KS attached to the protein core via the O-linked branched hexasaccharide is found in the large cartilage PG, the aggrecan. O-linked oligosaccharides of this type are also present in aggrecan and versican. Sialic acids attached to this O-linked oligosaccharide regulate the synthesis of KS, i.e., KS elongation can occur by glycosyl-transferases and sulfotransferases only when sialic acids are not present in the non-reducing terminal. After chain elongation, sialic acids are also added to the non-reducing end of the KS chain (Fig. 24.2). The mannose rich oligosaccharide is the precursor for the synthesis at the biantennary N-linked oligosaccharide. The precursor is transferred to the asparagine residue which is a part of the peptide sequence Asn-xxx-Ser/Thr, in the rough endoplasmic reticulum (Wight et al. 1991). Here also, when one or both of the indicated sialic acids are not added (Fig. 24.2), the biantennary oligosaccharide can be elongated with KS chains. This type of N-linked KS is often in cartilage fibromodulin (KSPG) and cornea lumican (KSPG). The molecular mass of KS is often lower than that of the other GAGs and seldom exceeds 5-kDa.

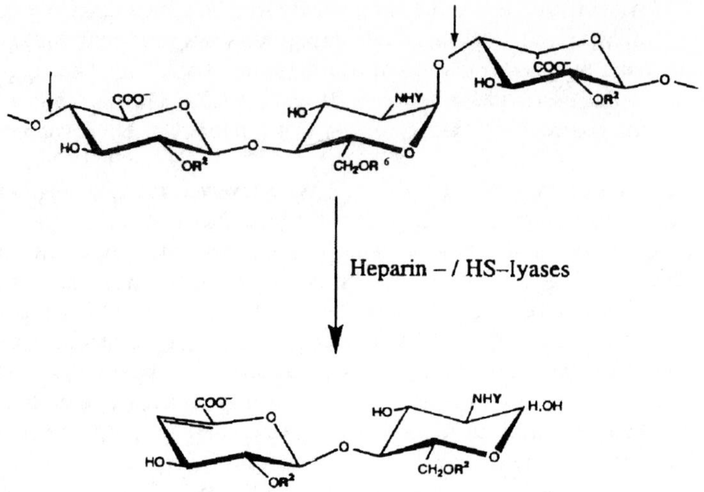

No	Formulas	Suggested terminology	R^2	R^6	Y
1	ΔUA-[1→4]-GlcN	Δdi-nonS	H	H	H
2	ΔUA-[1→4]-GlcNAc	aΔdi-nonS	H	H	Ac
3	ΔUA-2S-[1→4]-GlcN	Δdi-mono2S	SO_3^-	H	H
4	ΔUA-2S-[1→4]-GlcNAc	aΔdi-mono2S	SO_3^-	H	Ac
5	ΔUA-[1→4]-GlcN-6S	Δdi-mono6S	H	SO_3^-	H
6	ΔUA-[1→4]-GlcNAc-6S	aΔdi-mono6S	H	SO_3^-	Ac
7	ΔUA-[1→4]-GlcNS	Δdi-monoNS	H	H	SO_3^-
8	ΔUA-2S-[1→4]-GlcN-6S	Δdi-di(2,6)S	SO_3^-	SO_3^-	H
9	ΔUA-2S-[1→4]-GlcNAc-6S	aΔdi-di(2,6)S	SO_3^-	SO_3^-	Ac
10	ΔUA-2S-[1→4]-GlcNS	Δdi-di(2,N)S	SO_3^-	H	SO_3^-
11	ΔUA-[1→4]-GlcNS-6S	Δdi-di(6,N)S	H	SO_3^-	SO_3^-
12	ΔUA-2S-[1→4]-GlcNS-6S	Δdi-tri(2,6,N)S	SO_3^-	SO_3^-	SO_3^-

a = acetylated

Fig. 24.5. Primary structure of heparan sulfate and heparin and of all known Δ-disaccharides obtainable by digestion with heparin lyases I, II and III. Arrows indicate the sites degraded by heparin lyases. Reprinted from Karamanos et al. 1997

3.2 Protein Core of Proteoglycans

Distinct and non-genetically related proteins serve as protein cores of the various PGs. Their combination with certain GAG chains gives to each PG unique structural entity and biological functions. Some protein cores may contain one or more type of GAGs (syndecan-1 and aggrecan) and from a few (decorin and biglycan) to hundred chains (aggrecan and versican). Other protein cores may contain hydrophobic transmembrane domains (α-helix) and are found in the cell membrane PGs, such as the family of syndecans (David 1993). Molecular masses of protein cores are variable ranging from about 20-kDa for serglycin PG family to more than 200-kDa for aggrecan, versican and perlecan (Bourdon 1990).

3.3 Structure and Biological Significance of Proteoglycans and Hyaluronan

The importance of PGs in cell proliferation, differentiation and matrix synthesis is summarized in Fig. 24.6. Cell membrane PGs and/or PGs produced from the surrounding cells are responsible for growth factor binding to their receptors and via the intracellular signaling pathway the signal is transmitted to the nucleus and affects the cellular response.

The largest cartilage extracellular PG, aggrecan, contains 50–150 CS chains and some shorter KS chains. Aggrecan interacts with HA and link proteins to form PG aggregates with high molecular sizes. These aggregates contribute to hydration of the tissue. Aggrecan is the macromolecule responsible for the load bearing properties and the resilience of the tissue.

Syndecans constitute a family of cell membrane HSPGs (Syndecan-1, -2, -3 and -4). Syndecan-1 contains both CS and HS chains. Syndecans are responsible

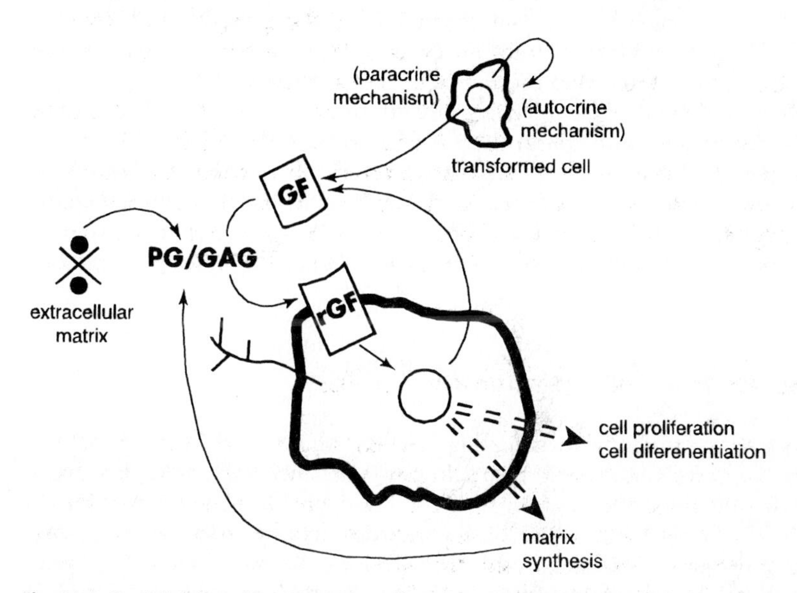

Fig. 24.6. Participation of PGs/GAGs in cell proliferation, differentiation and matrix synthesis via interactions with growth factors and growth factor receptors. Growth factors produced by cells binds to their receptors via direct interactions or mediated either by cell membrane or matrix PGs/GAGs. Cell membrane HSPGs bind to bFGF, protect the proteolytic degradation of growth factor and result in effective binding to bFGF receptor. Furthermore, PDGF stimulates the synthesis of versican, whereas the mitogenic activity of TGF is inactivated by decorin and biglycan

for migration of cells, act as growth factor receptors and their HS chains interact with various growth factors via specific domains (David 1993). Human immunodeficiency type 2 viruses, for example, bind selectively to HS chains of syndecans and in this way increase their effectiveness in infectivity. This PG family is considered as the bridge between the extracellular matrix (cell-cell and cell-matrix interactions) and the cell signal transduction pathways.

Perlecan, is a HSPG and constitutes a major structural component of the basement membrane. The negative charges of HS provide the basement membranes with a negative surface that acts either as antithrombogenic factor or as selective filter in the kidney.

Serglycin, the PG found in storage granules of mast cells contains a large number of heparin or oversulfated CS chains with high molecular size. The GAG chains are concentrated into a short protein core region that contains serine-glycine repeats. Serglycin is, therefore, a compact PG with very high negative charge density, able to bind and concentrate cationic histamines, proteases and glucosaminidases. The bound constituents are released when mast cells degranulate in host defense responses. The other small PGs decorin, biglycan, fibromodulin, as well as the KSPG found in cornea (lumican), are of importance in the network organization since they interact with collagen fibrils in most connective tissues (Hardingham and Fosang 1992).

Hyaluronan is a ubiquitous GAG with a wide variety of functions. This GAG forms the ligand structure for a family of aggregating proteins, the hyaladherins (Prehm 1984). Through such bindings, HA, for example, can form huge complexes with PGs of the ECM and it can participate in the aggregation of lymphocytes. HA also plays important roles in tissue morphogenesis, in angiogenesis and in tissue remodeling. Because of the universal presence of this molecule, attempts have also been made to analyze HA for diagnostic purposes (Nurminen et al. 1994, Karamanos and Hjerpe 1997, Lamari et al. 1998). High levels of this GAG in a pleural effusion seem to be pathognomonic for a malignant mesothelioma, but serum levels are also affected in rheumatoid conditions, in liver diseases and in malignancies (Cooper and Rathbone 1990). HA is now widely used pharmaceutically to relieve pain in osteoarthrosis and medico-technically in eye surgery.

4 Labelling, Isolation and Characterization of PGs

The steps we have to follow for studying PGs involve their labelling, extraction/solubilization, fractionation and characterization of their type. Although there are some definite procedures when studying PGs it should be noted that due to the structural heterogeneity of PGs these procedures may be modified for solving particular problems. Commonly used strategies are discussed below. For more details on the labelling and fractionation of PGs the excellent review of Hascall et al. 1994 is proposed.

4.1 Metabolic Labelling of Proteoglycans in Cell Cultures

When the system under consideration is focused on cell cultures, radiolabelling is the commonest and most convenient way for screening PGs through the various purification steps and measuring recoveries at every purification step. Selective labelling of PGs can be performed by [^{35}S]-sulfate. A high portion of radiosulfate (> 90 %) is incorporated to GAG chains. The carbohydrate precursor, [^{3}H]-glucosamine, is also frequently used together with radiosulfate to label covalently bound GAGs and oligosaccharides to the protein core. Glucosamine precursor is also of great value for evaluating HA synthesis by the cells and for subsequent steps, such as the determination of HA. Most cells that synthesize HA are able to synthesize CS/DSPG, and, therefore, combination of radiosulfate and [^{3}H]-glucosamine results in double labelling of CS/DS chains. [2-^{3}H]-mannose is another specific carbohydrate precursor useful in evaluating synthesis of N-linked oligosaccharides. Carbohydrate precursor, generally, can provide useful information for composition, structure and type of the carbohydrate moieties bound to the protein core.

Amino acid precursors can be used to label the protein core of PGs as well as other proteins/glycoproteins. [^{3}H]-serine and [^{3}H]-leucine are the most common labels for this purpose and can be used with [^{35}S]-sulfate. Labelling with amino acids and radiosulfate is useful for estimating the number of different protein cores in each PG population. Conversion of [^{3}H]-serine to alanine following alkaline borohydride treatment (Carlsson 1968) may also be used for determining the proportion of substituted serine residues of the core protein with GAG chains and *O*-linked oligosaccharides. Typical radioisotope concentration should be within the range from 50 to 250 μCi/ml of culture. To ensure that synthesized GAG chains are not under-sulfated [^{35}S]-sulfate concentration in the culture medium should be at least 0.1 mM. Incorporation of sulfate by cells is linear for 24–48 h following a short (often a few minutes) equilibration time between the medium and the intracellular precursor pool. Following this labelling procedure with sulfate, carbohydrate and amino acid precursor all PG and protein/glycoprotein populations present in the medium and the cell-matrix compartments are labelled.

4.2 Extraction/Solubilization of Proteoglycans

The most effective solutions contain both chaotropic reagents and detergents. The former will ensure dissociation of non-covalent interactions and denaturation of proteins, whereas the latter dissociation of hydrophobic interactions. Guanidine hydrochloride (GdnHCl), at a final concentration of 4 M, is the most effective chaotropic reagent for PGs extraction (Hascall and Kimura 1982). Compatible reagents with guanidine are Triton X-100 (1–2 % w/v) and CHAPS (2–4 % w/v). The latter detergent is more convenient since it can be removed by dialysis. When Triton X-100 is used, however, care should be taken to avoid formation of nondialyzable micelles. For extraction of cell bound PGs guanidine should be eliminated from the extraction solution since nucleic acids are also extracted and may interfere with subsequent steps. Protease activity is another crucial parameter that should be taken into consideration during extraction and further purification steps. A cocktail of protease inhibitors should be used for this purpose. An effective mixture which is widely used contains: 0.1 M 6-amino-hexanoic acid, 5 mM benzamidine hydrochloride, 0.2 to 1 mM phenylmethansulfonyl fluoride and 10 mM EDTA for cathepsin D-like activity, trypsin-like activity, serine-dependent proteases and metalloproteases, respectively. They are added just before use to the extraction buffer consisting of 4 M GdnHCl-0.1 M AcONa, pH 5.8, and the suitable detergent. It should be noted that 80–100 % of PGs is usually extracted and solubilized. When low yield is obtained, such as in some exceptional cases, for instance in invertebrate tissues, SDS (2 %) in the presence of other detergent may help to improve extractability (Vynios and Tsiganos 1990). Treatment of tissue with collagenase and/or elastase may also increase extractability disrupting the PG-collagen network.

Extraction should be performed at 4 °C overnight with gentle stirring and using 10 vol. per g of wet weight tissue or 1–2 ml per 35-mm dish for cell cultures. Following low-speed centrifugation the residue can be re-extracted using half the volume of the extractant for 3–4 h and the extracts can be combined. Extractability yield can be estimated by degrading the PGs present in the residue and the extract with papain and measuring the GAG content. This can be done by applying the digest to a microcolumn packed with an anion-exchange resin, such as DEAE-Sephacel (Pharmacia). Following washing with 0.1 M NaCl, GAGs are eluted with 1.2 M LiCl.

4.3 Fractionation and Purification of PGs

A schematic strategy followed to isolate and characterize PG is given in Fig. 24.7. Both the culture medium and the cell extract will contain labelled macromolecules and unincorporated precursors. To remove unincorporated precursors and quantitate the incorporated radioactivity in macromolecules as well as to exchange the extraction solvent with a solvent which will be suitable for subsequent anion-exchangers used in later steps, a molecular sieve step is necessary. Sephadex G-25 (PD-10 prepacked columns, BioRad) is equilibrated with chaotropic and non-ionic solvent, such as 8 M urea or 10 M formamide. Column is equilibrated with one column vol. of elution buffer: 10 M formamide-0.05 M sodium acetate, pH 6, containing 0.30 M NaCl, 0.5 % (w/v) CHAPS and protease inhibitors. Formamide is preferably used since it is more stable than urea, which decomposes with time. The sample is then applied to the column and following washing with 1.5 ml, the macromolecular fraction is recovered by applying a total of 3.5 ml of eluting solution (elution between 1.5 and 5.0 ml after sample application). This procedure ensures that macromolecules are free from unincorporated radioactivity). The total included substances are then eluted with 7.0 ml and this results in column re-equilibration. Instead of the last step, the column, which contains the majority of unincorporated radioactivity, can be discarded in radioactive waste, a step that effectively minimizes contamination of the laboratory

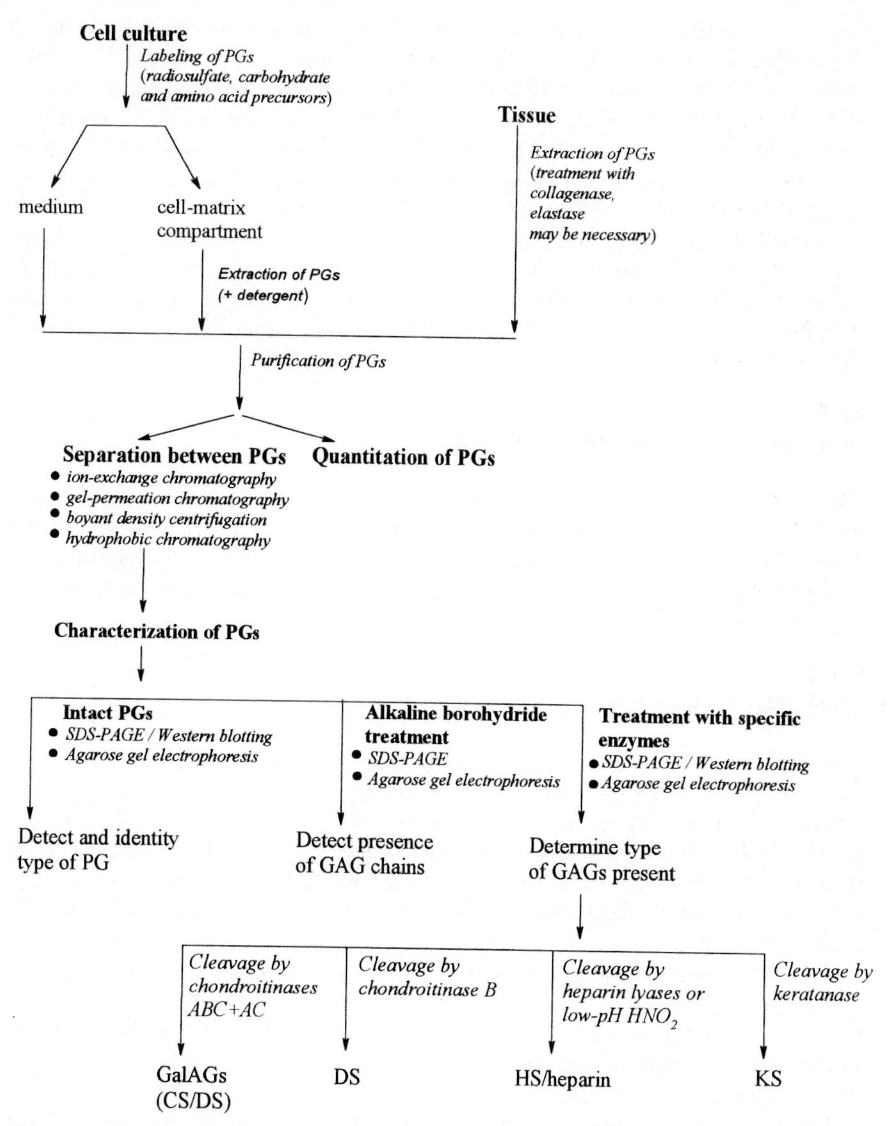

Fig. 24.7. Schematic diagram showing the strategy to isolate and characterize PGs from tissues and cell cultures expands

with radioactivity. Aliquots of the collected fractions are counted for radioactivity to determine total recovery of macromolecular radioactivity. This procedure is effective for concentrating PGs into smaller volumes.

Batch anion-exchange chromatography used after this molecular sieve step is effective for concentrating PGs and separating them from other macromolecules. Q-Sepharose or DEAE-Sephacel (Pharmacia) columns are first equilibrated with 10 vol. of solvent. The NaCl content in the eluant should be as high as possible to minimize nonspecific adsorption and simultaneously permit binding of PGs to the active groups of the resin. Small volume of resins (~1 ml), due to their high binding capacity, is often sufficient for ~3 mg of PGs. Proteoglycans are then eluted with a solvent that contains a sufficient salt concentration. 4 M GdnHCl or a 10 M formamide that contains ~1.0–1.2 M NaCl are commonly used. Utilization of detergents and salts in the eluants reduces nonspecific adsorption of contaminating macromolecules and results in high column recoveries.

In samples containing nucleic acids, e.g., cell extract obtained with guanidine extraction, they coelute with the PG-containing fraction in anion-exchange chromatography. In this case further purification steps, such as gel-permeation chromatography (GPC) and enzymic digestion with DNAse are often useful in removing nucleic acids.

4.4 Techniques to Separate PG Population

Due to the type of the chains present in various PG populations and the degree of substitution of PGs with GAG chains, PGs may differ in their charge density, average buoyant densities, average hydrodynamic sizes, and the hydrophobicities of their core proteins. Each of these properties can be used to separate and identify different PG classes.

4.4.1 Separation According to Charge Density

The batch ion-exchange step results in a reasonably high level of purification of PGs from other types of macromolecules. However, a second ion-exchange step with a continuous salt gradient elution often results in better purity and can separate different PG populations. A commonly used starting eluant is 10 M formamide-0.05 M sodium acetate, pH 6, containing 0.3 M NaCl and 0.5 % (w/v) CHAPS. The binary gradient system is composed of equal volumes of the starting buffer solution and of the same solution containing ~1.0–1.2 M NaCl.

PGs with different charge densities can be separated, at least partially, into separate peaks, as often occurs when both HSPGs and CS/DSPGs are present (Fig. 24.8A). Single peaks represent homogeneously sized PGs. Further purification of these PGs may be achieved by size exclusion chromatography and buoyant density gradient centrifugation.

Various types of ion-exchangers are available in high-performance liquid chromatography (HPLC) and membrane cartridge forms. They offer a variety of advantages, such as short analysis time, high capacity and recoveries and better separation profiles.

4.4.2 Separation According to Hydrodynamic Volume

GPC is widely used to separate and characterize macromolecular properties of PGs, such as average Mr and binding with HA. Furthermore, the presence of selectively degraded PGs can be assessed. The choice of support matrix and elu-

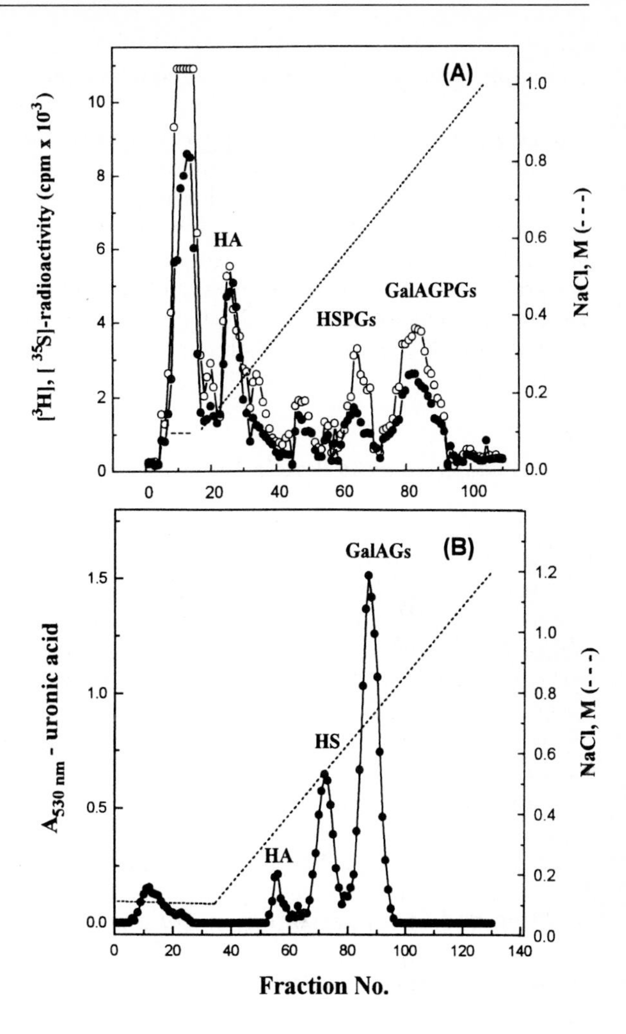

Fig. 24.8. Separation between [^{35}S]-sulfate and [^{3}H]-glucosamine labeled HSPGs and CS/DSPGs (A) and HA, HS and the GalAGs (B) using anion-exchange chromatography on DEAE-Sephacel column. Culture medium from a human malignant mesothelioma cell line was concentrated on a YM-10 membrane (Amicon) and chromatographed using a NaCl linear gradient in 8 M formamide-0.05 M sodium acetate (pH 6.0), containing 0.1 % Triton X-100. In tissue preparations, screening of GAGs is performed by analysis of uronic acid (Fig.s from personal data)

ants depends on the properties of the particular PG under consideration. Matrices, such as Superose 6 (Pharmacia) are designed for high pressure and flow rates, have distinct advantages in terms of speed and are compatible with chaotropic solvents. For PGs with large hydrodynamic volume, more porous matrices, such as Sepharose CL-2B and CL-4B or Sephacryl S-500 and S-1000 (Pharmacia), are used. GPC can separate PG populations in to more than one peak. For example, when PGs contain the same type of GAG, but different number of GAGs, e.g., versican and decorin, 10 M formamide-0.05 M sodium acetate (pH 6.0), containing 0.30 M NaCl and 0.5 % (w/v) CHAPS, is often used for column elution. For analytical purposes, a solution of 4 M GdnHCl containing 0.1 to 0.5 % Triton X-100 can also be used effectively.

4.4.3
Separation with Buoyant Density Gradient Centrifugation

GAGs with their highly anionic groups incorporate the cesium cations and therefore they present much higher buoyant densities in CsCl equilibrium density gradients than those of proteins. PGs with higher GAG-to-protein ratios have, therefore, higher buoyant densities. Gradients formed in the presence of chaotropic solvents (4 M GdnHCl, 8 urea or 8–10 M formamide) are referred to as dissociative gradients and those in the absence of chaotropic solvents are referred to as associative gradients (Hascall and Kimura 1982). When hydrophobic PGs, such as cell surface HSPGs, have to be analyzed addition of 0.5 % CHAPS results in better recoveries.

4.4.4
Separation According to Protein Core Hydrophobicity

Some PGs can be separated on the basis of differences of their protein cores hydrophobic properties. Following a detergent gradient elution of octyl-Sepharose column, decorin and the KSPG from cornea, as well as different types of cell surface HSPGs can be separated at different detergent concentrations (Takeuchi et al. 1992). Eluants, such as 10 M formamide-0.05 mM sodium acetate (pH 6.0), containing 0.3 M NaCl, or 4 M GdnHCl and 8 M urea can be used. After sample application, the binary gradient system is developed using equal volumes of the starting eluant and of the same solution containing 1.5 % (w/v) CHAPS. Concentrations of CHAPS in the collected fractions can be determined with a carbazole assay. Detergents such as Triton X-100 or other nonpolar reagents can also be used.

5
Isolation and Characterization of Glycans

The carbohydrate constituents of PGs – GAG chains and oligosaccharides – play key roles in regulating physical properties and biological functions of PGs. The structure of the most common carbohydrate constituents, the GAGs, is variable due to various factors, such as the sulfation degree, the position of sulfates in repeating disaccharide units, the presence or not of IdoA, the type of hexosamine present and the structure of the region that links GAGs to protein cores. These parameters are essential to determine the type of GAG and therefore the PG functional properties.

HA is the only GAG not bound to protein cores. This is secreted extracellularly and may be separated from PGs by ion-exchange chromatography on DEAE-Sephacel, where HA is eluted with 0.2 M NaCl in 10 M formamide buffer. Buoyant density centrifugation under dissociative conditions can also be used when HA-PG aggregates are present. All other GAGs should be liberated from their PGs using either digestion with papain and/or alkaline borohydride treatment. In case of papain-resistant PGs further treatment with alkali may be used to liberate protein-free GAG chains.

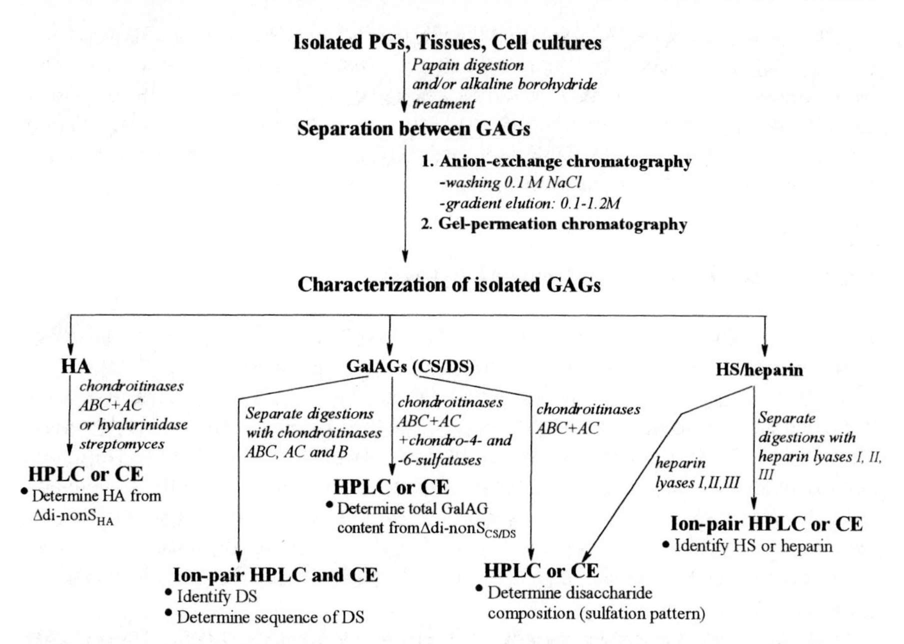

Fig. 24.9. Diagram showing the strategy to isolate, identify and characterize the fine chemical structure of GAGs present in isolated PGs from tissues and cell cultures

There are basic procedures to ensure the presence of a PG type. They are based on the same principle, but they can be performed using either SDS-PAGE, agarose gel electrophoresis or GPC. Treatment with specific lyases, such as chondroitinases AC, ABC, B or heparin lyases I, II, III will help to identify the protein core(s) of the PG, whereas alkaline borohydride treatment will liberate the GAG chains and will help to identify the type(s) of GAG(s) present in a PG. An overview of the strategies followed for characterization of the fine chemical structure of GAGs is presented in Fig. 24.9.

5.1 Procedure to Isolate Glycosaminoglycans

GAG chains liberated from PG populations, isolated from tissues or cell cultures following papain digestion and/or alkaline borohydride treatment are first separated according to their charge density by ion-exchange chromatography. DEAE-Sephacel is a commonly used resin. Non-sulfated oligosaccharides are removed by washing with 0.1 M NaCl and GAGs are often separated following a gradient elution ranging from 0.1 to 1.2 M NaCl. HA is eluted as a sharp peak with ~0.2 M NaCl and is completely resolved from the other sulfated GalAGs and GlcAGs. CS and DS are eluted together with a salt concentration depending on their sulfation degree. GalAGs bearing one sulfate residue per disaccharide unit are eluted with 0.5–0.6 M NaCl. An example of GAG separation is given in Fig. 24.8B.

GPC on Sepharose CL-6B, Sephacyl S-200 or S-500 of the GAG population isolated by ion-exchange chromatography may provide useful information on the homogeneity of the population according to its size. The average molecular mass can also be determined using standard GAG preparations with known M_rs. When low amount of GAGs is available, gel-permeation HPLC coupled with refractive index detection of eluting polymers could be used (Arvaniti et al. 1994).

5.2 Strategies to Characterize Glycosaminoglycans

Various strategies may be used to determine GAG structures. Analysis of intact polymers may give information on charge density, polydispersity and the molecular size of the GAG chain. Information on the chain composition can be obtained by depolymerization of GAGs to unit fragments. Modern separations allow most of this information to be deduced from the disaccharide fragments obtained by enzymic cleavage. Harsher hydrolytic procedures yielding monosaccharide components are rarely needed. When determining longer sequences, larger fragments must be analyzed. Such fragments can be obtained by employing enzymes, with specific cleavage sites distributed less frequently along the GAG chain.

The improved resolution provided by HPLC and CE permitted the accurate analysis of crucial parameters such as the pattern of sulfation, types of uronic acid and hexosamine (Karamanos et al. 1988, Hjerpe et al. 1980). Clinically, the simple HPLC or CE based determination of HA in effusion has proven to be of great value for diagnosis malignant mesotheliomas (Karamanos and Hjerpe 1997). Combining different HPLC and CE separations one could even identify a sequence characteristic of an entire DS chain (Karamanos et al. 1995a, 1995b). With chromatographic methods, flow is more or less laminar, while in CE flow is more homogeneous, giving narrower bands/peaks and, therefore, higher resolving power. When similar detector sensitivity in terms of concentration is needed, the narrower bands also provide improved sensitivity and consequently, better possibilities for further elucidation of GAG structures. Both HPLC and CE can be used to elucidate GAG structures, CE is advantageous when working with biological samples because of the trace amount of sample needed. Furthermore the low consumption of solvents seems desirable for environmental concerns. Applications of both HPLC and CE in combination to characterize the fine chemical structure of GAGs have been reviewed by Karamanos and Hjerpe, 1996. For further insight into principles of CE and applications to carbohydrate analysis the books written by R. Kuhn 1993 and Z. El Rassi 1996 are recommended. Some reviews (El Rassi and Mechref 1996, Grimshaw 1997, Karamanos and Hjerpe 1998 and Lamari and Karamanos 1999) on the structural characterization of GAGs/PGs and the relation of GAG/PGs to disease diagnosis (Hjerpe and Karamanos 1998), using CE could be recommended.

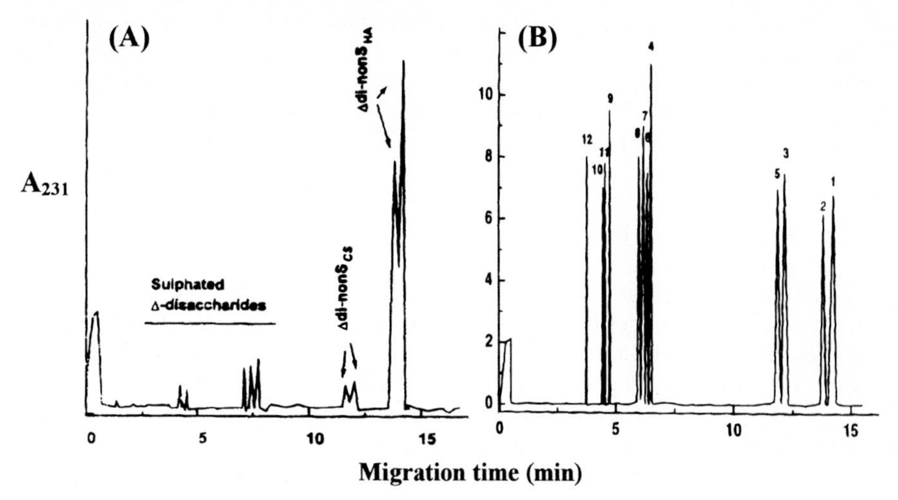

Fig. 24.10. Electropherograms showing the determination of HA and total CS/DS content by HPCE (A) and disaccharide composition in HS/heparin (B) using reversed polarity CE. When samples are digested with chondroitinases ABC/AC and chondro-4- and -6-sulfatases in combination the Δdi-nonSHA peak measures the amount of HA and that of Δdi-nonSCS/DS the amount of both GalAGs. For identity of peaks 1–12 see Fig. 24.5. Reprinted from Karamanos et al. 1995a and 1996, respectively

5.2.1
Analysis of Total GalAG Content

Enzymic treatment of crude tissue extracts, cell culture fractions or isolated GAGs with chondroitinases ABC/AC and chondro-4- and -6-sulfatases in combination will depolymerize HA, CS and DS (Figs. 24.3–4). In all cases, unsaturated-4,5-disaccharides (Δ-disaccharides) are produced. Using ion-suppression HPLC (Karamanos et al. 1994) or reserved polarity CE (Karamanos et al. 1995a) two peaks are obtained (Fig. 24.10A). Δdi-nonS_{HA} corresponds to Δ-disaccharides produced from HA and Δdi-non$S_{CS/DS}$ to those produced from both CS and DS. The combination of both chondro-4- and -6-sulfatases desulfates the Δ-disaccharides obtained by chondroitinases ABC and AC and therefore both GalAGs are concentrated in the peak of Δdi-non$S_{CS/DS}$ which measures the total content of CS and DS.

5.2.2
Disaccharide Composition of GalAGs

Chondroitinase ABC degrades both CS and DS to Δ-disaccharides (Fig. 24.4). The produced disaccharides may be non-sulfated or contain one to three sulfate groups. All possible combinations result in twenty-three differently sulfated Δ-disaccharides, twelve derived from DS and eleven from CS. All differently sulfated disaccharides can be analyzed with HPLC using three different eluting buffers for non-, mono- and oversulfated species (Karamanos et al. 1994). Using CE analysis, however, the most frequent non-, mono-, and oversulfated Δ-disaccharides can be determined by a single run (Karamanos et al. 1995a).

Chondroitinase B degrades DS chains only at the sites of IdoA, whereas chondroitinase AC only at the sites of GlcA (Fig. 24.4B). Using these two enzymes separately, CE analysis of the digests may provide useful information on the presence of IdoA-repeats and the distribution of GlcA in the DS chain (Karamanos et al. 1995a, Karamanos and Hjerpe 1996).

5.2.3
Sequence Analysis of DS

Chondroitinases AC and B degraded DS chain at GlcA and IdoA sites, respectively resulting in longer fragments that can be analyzed by ion-pair HPLC (Karamanos et al. 1995b). Complete separation of the variously sized non-sulfated Δ-disaccharides with sizes varying from Δ-(di-nonS)$_1$ to Δ-(di-nonS)$_8$ and variously sized sulfated Δ-saccharides with sizes ranging from Δ-(di-monoS)$_1$ to Δ-(di-monoS)$_4$ is obtained by one single HPLC run.

The profiles obtained following the separate action of these chondro- and dermato- lyases on the DS polymer may provide useful information on the size and degree of sulfation. For further information on the strategy to sequence uronic-acid containing GAGs see the survey by Karamanos and Hjerpe 1996. With the proposed strategy the structure of DS-18 fraction from pig skin has been elucidated (Karamanos et al. 1995b). Apart from HPLC analysis the electrophoretic method described by Cheng et al. 1994 may also be used to obtain information on the primary sequence of DS.

5.3.3
Determination of GAG Reducing Terminal Monosaccharides and Sulfated and/or Phosphorylated Linkage-Region Oligosaccharide

Following separate digestions with chondro- and dermato-lyases and HPLC analysis the separated Δ-saccharides can be further analyzed to determine the type of their disaccharide units. This can be easily performed by a further digestion of the isolated Δ-saccharides with chondroitinase ABC and analysis of the obtained Δ-disaccharides with HPLC and/or CE by the methods described above.

Linkage-region oligosaccharides of GAGs to protein core can be analyzed using the same ion-pair HPLC set-up. Due to the presence of only one linkage oligosaccharide per chain the available amount is limited and therefore it is necessary to increase sensitivity. Utilization of [^{3}H]-NaBH4 in a proportion of 25 % of the total NaBH4 used for alkaline borohydride treatment will convert the liberated reducing terminal monosaccharide to alditole, e.g., xylose to [^{3}H]-xylitole, making possible the detection and determination of the linkage-region fragment. Sulfated and/or phosphorylated linkage-region oligosaccharides are completely separated and determined by HPLC analysis (Gioldassi and Karamanos 1999). To determine the type of monosaccharide present in the reducing terminal, the reduced oligosaccharide fragment may further be hydrolyzed with 8 M trifluoroacetic acid at 100 °C for 6h and the obtained alditoles can be resolved by HPLC analysis as per-*O*-benzoylated derivatives (Karamanos 1992).

5.3.4 Disaccharide Composition and Identification of HS and Heparin

Specific degradation of heparin and HS to Δ-di- and Δ-oligosaccharide can be achieved by using the various heparin lyases. Heparin lyase I, commonly referred to as heparinase (EC 4.2.2.7), heparin lyase II (heparinase II, no EC number) and heparin lyase III, commonly referred to as heparitinase (EC 4.2.2.8), are all commercially available and each one of them has different specificity in respect to the uronic acid moiety and sulfation (Fig. 24.5). When these lyases are used in combination, the polymeric chain is cleaved to a large extent ($>$ 90 %) to Δ-disaccharides and the disaccharide composition can therefore be obtained. This approach has recently been used to separate and determine all commercially available heparin- and HS-derived Δ-disaccharides by a sensitive HPLC method (Karamanos et al. 1997).

Using CE in a reversed polarity mode and phosphate buffer (pH 3.5) all twelve known isomeric Δ-disaccharides are separated within 15 min (Fig. 24.10B) (Karamanos et al. 1996). This method is useful for determining structural differences between heparin and HS as well as for distinguishing them from each other in biologic samples by analyzing the Δ-disaccharides obtained by their separate depolymerization with heparin lyases.

References

Arvaniti A, Karamanos NK, Dimitracopoulos G, Anastassiou ED (1994) Isolation and characterization of a novel 20-kDa sulfated polysaccharide from the extracellular slime layer of *Staphylococcus epidermidis*. Arch Biochem Biophys 308: 432–438

Bourdon MA (1990) Extracellular matrix genes. In: Sandell L, Boyd C (eds) Academic press, San Diego, USA

Carlsson M (1968) Structure and immunological properties of oligosaccharides isolated from pig submaxillary mucins. J Biol Chem 243: 616–626

Cheng F, Heinegard D, Malmstrom A, Schmidtchen A, Yoshida K, Fransson LA (1994) Patterns of uronosyl epimerization and 4-/6-O-sulphation in chondroitin/dermatan sulphate from decorin and biglycan of various tissues. Glycobiology 4: 685–696

Cooper EH, Rathobone BJ (1990) Clinical significance of immunometric measurement of hyaluronic acid. Ann Clin Biochem 27: 444–451

David G (1993) Integral membrane heparan sulfate proteoglycans. FASEB J 7: 1023–1030

El Rassi, Z (1996) High-performance capillary electrophoresis of carbohydrates. Beckman Instruments, Fullerton, CA, USA

El Rassi Z, Mechref Y (1996) Recent advances in capillary electrophoresis of carbohydrates. Electrophoresis 17: 275–301

Gioldassi X, Karamanos NK (1999) Determination of phosphorylated and sulfated linkage-region oligosaccharides in chondroitin/dermatan and heparan sulfate proteoglycans by high-performance liquid chromatography. J Liq Chrom & Rel Technol 23 "in press"

Grimshaw J (1997) Analysis of glycosaminoglycans and their oligosaccharide fragments by capillary electrophoresis. Electrophoresis 18: 2408–2414

Hascall VC, Kimura JH (1982) Proteoglycans: isolation and characterization. Methods Enzym 82: 769–800

Hascall VC, Calabro A, Midura RJ, Yanagishita M (1994) Isolation and characterization of proteoglycans. Methods Enzym 230: 390–417

Hardingham T, Fosang AJ (1992) Proteoglycans: many forms and many functions. FASEB J 6: 861–870

Hjerpe A, Karamanos NK (1998) Capillary electrophoresis in structural analysis of proteoglycans and disease diagnosis. In: Karamanos NK (ed) European training program in microseparation techniques (ECOSEP1). Typorama, Patras, pp 136–154

Hjerpe A, Antonopoulos CA, Classon B, Engfeldt B (1980) Separation and quantitative determination of galactosamine and glucosamine at the nanogram level by sulphonyl chloride reaction and high-performance liquid chromatography. J Chromatogr 202: 453–459

Iozzo RV (1997) The family of the small leucine-rich proteoglycans: key regulators of matrix assembly and cellular growth. Crit Rev Biochem Mol Biol 32: 141–174

Karamanos NK, Hjerpe A, Tsegenidis T, Engfeld B, Antonopoulos CA (1988) Determination of iduronic and glucuronic acid in glycosaminoglycans after stoichiometric reduction and depolymerization using high-performance liquid chromatography and ultraviolet detection. Anal Biochem 172: 410–419

Karamanos NK (1992) High-performance liquid chromatographic determination of xylitole and hexosaminitols present in the reduced terminal of glycosaminoglycans. J Liq Chromatogr 16: 2639–2652

Karamanos NK, Aletras AJ, Tsegenidis T, Tsiganos CP, Antonopoulos CA (1992) Isolation, characterization and properties of the oversulphated chondroitin sulphate proteoglycan from squid skin with peculiar glycosaminoglycan sulphation pattern. Eur J Biochem 204: 553–560

Karamanos NK, Syrokou A, Vanky P, Nurminen M, Hjerpe A (1994) Determination of 24 variously sulfated galactosaminoglycan- and hyaluronan-derived disaccharides by high-performance liquid chromatography. Anal Biochem 221: 189–199

Karamanos NK, Axelson S, Vanky P, Tzanakakis GN, Hjerpe A (1995a) Determination of hyaluronan and galactosaminoglycan disaccharides by high-performance capillary electrophoresis at the attomole level. Application to analyses of tissue and cell culture proteoglycans. J Chromatogr A 696: 295–305

Karamanos NK, Vanky P, Syrokou A, Hjerpe A (1995b) Identity of dermatan and chondroitin sequences in dermatan sulfate chains by using fragmentation with chondroitinases and ion-pair high-performance liquid chromatography. Anal Biochem 225: 220–230

Karamanos NK, Hjerpe A (1996) Strategy to characterize sequences of uronic acid-containing glycosaminoglycans by high-performance liquid chromatography and capillary electrophoresis. In: Balduini C, Cherubikno C, De Luca G (eds) Advances in Biomedical Studies: Biology and Physiopathology of the Extracellular Matrix. La Goliardica Pavese, Pavia, vol II, pp 71–82

Karamanos NK, Vanky P, Tzanakakis GN, Hjerpe A (1996) High-performance capillary electrophoresis to characterize heparin and heparan sulfate disaccharides. Electrophoresis 17: 391–395

Karamanos NK (1997) Structure-function relationship of glycoconjugates and their role in diagnosis and treatment. Farmakeftiki 10: 50–61

Karamanos NK, Hjerpe A (1997) High-performance capillary electrophoretic analysis of hyaluronan in effusions from human malignant mesothelioma. J Chromatogr B 697: 277–281

Karamanos NK, Vanky P, Tzanakakis GN, Tsegenidis T, Hjerpe A (1997) Ion-pair high-performance liquid chromatography for determining disaccharide composition of heparin and heparan sulfate. J Chromatogr A 765: 169–179.

Karamanos NK, Hjerpe A (1998) An survey of methodological challenges for glycosaminoglycan/proteoglycan analysis and structural characterization by capillary electrophoresis. Electrophoresis 19: 2561–2571

Karamanos NK, Lamari F, Katsimpris J, Gartaganis S (1999) Development of an HPLC method for determining the alpha2-adrenergic receptor agonist brimonidine in blood serum and aqueous humor of the eye. Biomed. Chromatogr. 13: 86–88

Khellen L, Lindahl U (1991) Proteoglycans: structures and interactions. Ann Rev Biochem 60: 867–869

Kuhn R (1993) Capillary electrophoresis: principles and practise. Springer-Verlag, London

Lamari F, Karamanos NK (1999) High-performance capillary electrophoresis as a powerful analytical tool of glycoconjugates. J Liq Chrom & Rel Technol 22 "in press"

Lamari F, Katsimpris J, Gartaganis S, Karamanos NK (1998) Profiling of the eye aqueous humor in exfoliation syndrome by high-performance liquid chromatographic analysis of hyaluronan and galactosaminopglycans. J Chromatogr B 709: 173–178

Laurent T, Fraser JR (1992) Hyaluronan FASEB J 6: 2397–2404

Lindahl U (1989) Heparin: chemical and biological properties, clinical applications. In: Lane D, Lindahl U (eds). CRC press, Boca Raton, FL, USA

Nurminen M, Dejmek A, Martensson G, Thylen A, Hjerpe A (1994) Clinical utility of liquid-chromatographic analysis of effusions for hyaluronate content. Clin Chem 40: 777–780

Prehm P (1984) Hyaluronate is synthesized at plasma membrane. Biochem J 220: 597–600

Syrokou A, Tzanakakis GN, Tsegenidis T, Hjerpe A, Karamanos NK (1999) Effects of glycosaminoglycans in proliferation of epithelial and fibroblast human malignant mesothelioma cells. A structure-function relationship. Cell Prolif 32 " in press"

Takeuchi Y, Yanagishita M, Hascall VC (1992) Recycling of transferrin receptors and heparan sulfate proteoglycans in a rat parathyroid cell line. J Biol Chem 25: 14685–14690

Tzanakakis GN, Karamanos NK, Hjerpe A (1995) Effects on glycosaminoglycan synthesis in cultured malignant mesothelioma cells of transforming, epidermal and fibroblast growth factors and their combinations with platelet-derived growth factor. Exp Cell Res 220: 130–137

Tzanakakis GN, Hjerpe A, Karamanos NK (1997) Proteoglycan synthesis induced by transforming and basic fibroblast growth factors in human malignant mesothelioma is mediated through specific receptors and the tyrosine kinase intracellular pathway. Biochimie 79: 323–332.
Vynios DH, Tsiganos CP (1990) Squid proteoglycans: isolation and characterization of three populations from cranial cartilage. Biochim Biophys Acta 1033: 139–147
Wight TN, Hascall VC, Heinegard D (1991) Cell biology of extracellular matrix. In: Hay E (ed) Plenum, New York, USA

Subject Index

Z